Big Data over Networks

Utilizing both key mathematical tools and state-of-the-art research results, this text explores the principles underpinning large-scale information processing over networks and examines the crucial interaction between big data and its associated communication, social, and biological networks.

Written by experts in the diverse fields of machine learning, optimization, statistics, signal processing, networking, communications, sociology, and biology, this book employs two complementary approaches: first, analyzing how the underlying network constrains the upper layer of collaborative big data processing, and second, examining how big data processing may boost performance in various networks. Unifying the broad scope of the book is the rigorous mathematical treatment of the subjects, which is enriched by in-depth discussion of future directions and numerous open-ended problems that conclude each chapter.

Readers will be able to master the fundamental principles for dealing with big data over large systems, making it essential reading for graduate students, scientific researchers, and industry practitioners alike.

Shuguang Cui is Professor at Texas A&M University. He is a Fellow of the IEEE and was selected as a Highly Cited Researcher by Thomson Reuters, 2014.

Alfred O. Hero III is R. Jamison and Betty Williams Professor of Engineering at the University of Michigan, Ann Arbor, with appointments in the departments of Electrical Engineering and Computer Science, Biomedical Engineering and Statistics. He is a Fellow of the IEEE.

Zhi-Quan Luo is Professor at the University of Minnesota. He has served as the Editor-in-Chief of *IEEE Transactions on Signal Processing* and is a Fellow of the IEEE, SIAM, and the Royal Society of Canada.

José M. F. Moura is Philip L. and Marsha Dowd University Professor at CMU with appointments in the Departments of Electrical and Computer Engineering and, by courtesy, of Biomedical Engineering. He is a Fellow of the IEEE and the AAAS, a corresponding member of the Academy of Sciences of Portugal, and a member of the US NAE.

Big Data over Networks

Edited by

SHUGUANG CUI
Texas A&M University

ALFRED O. HERO III
University of Michigan, Ann Arbor

ZHI-QUAN LUO
University of Minnesota

JOSÉ M. F. MOURA
Carnegie Mellon University

CAMBRIDGE
UNIVERSITY PRESS

CAMBRIDGE
UNIVERSITY PRESS

University Printing House, Cambridge CB2 8BS, United Kingdom

Cambridge University Press is part of the University of Cambridge.

It furthers the University's mission by disseminating knowledge in the pursuit of education, learning and research at the highest international levels of excellence.

www.cambridge.org
Information on this title: www.cambridge.org/9781107099005

First published 2016

Printed in the United Kingdom by TJ International Ltd. Padstow Cornwall

A catalog record for this publication is available from the British Library

ISBN 978-1-107-09900-5 Hardback

Contents

List of contributors *page* xiii
Preface xvii

Part I Mathematical foundations 1

1 Tensor models: solution methods and applications 3
Shiqian Ma, Bo Jiang, Xiuzhen Huang, and Shuzhong Zhang

1.1 Introduction 3
1.2 Tensor models 5
1.2.1 Sparse and low-rank tensor optimization models 5
1.2.2 Tensor principal component analysis 6
1.2.3 The tensor co-clustering problem 8
1.3 Reformulation of tensor models 11
1.3.1 Low-n-rank tensor optimization 11
1.3.2 Equivalent formulation of tensor PCA 13
1.4 Solution methods 16
1.4.1 Directly resorting to some existing solver 16
1.4.2 First-order methods 18
1.4.3 The block optimization technique 22
1.5 Applications 24
1.5.1 Computational results on gene expression data 25
1.6 Conclusions 30
References 31

2 Sparsity-aware distributed learning 37
Symeon Chouvardas, Yannis Kopsinis, and Sergios Theodoridis

2.1 Introduction 37
2.2 Batch distributed sparsity promoting algorithms 39
2.2.1 Problem formulation 39
2.2.2 LASSO and its distributed learning formulation 40
2.2.3 Sparsity-aware learning: the greedy point of view 42
2.2.4 Other distributed sparse recovery algorithms 45

2.3 Online sparsity-aware distributed learning 46
2.3.1 Problem description 46
2.3.2 LMS based sparsity-promoting algorithm 47
2.3.3 The GreeDi LMS algorithm 49
2.3.4 Set-theoretic sparsity-aware distributed learning 51
2.4 Simulation examples 56
2.4.1 Performance evaluation of batch methods 57
2.4.2 Performance evaluation of online methods 58
References 61

3 Optimization algorithms for big data with application in wireless networks 66
Mingyi Hong, Wei-Cheng Liao, Ruoyu Sun, and Zhi-Quan Luo

3.1 Introduction 66
3.1.1 Motivation 66
3.1.2 The organization of the chapter 67
3.2 First-order algorithms for big data 67
3.2.1 The block coordinate descent algorithm 67
3.2.2 The ADMM algorithm 69
3.2.3 The BSUM method 70
3.3 Application to network provisioning problem 72
3.3.1 The setting 72
3.3.2 Network with an uncapacitated backhaul 75
3.3.3 Network with a capacitated backhaul 82
3.4 Numerical results 88
3.4.1 Scenario 1: Performance comparison with heuristic algorithms 89
3.4.2 Scenario 2: The efficiency of N-MaxMin WMMSE algorithm 91
3.4.3 Scenario 3: Multi-commodity routing problem with parallel implementation 92
3.4.4 Scenario 4: Performance evaluation for Algorithm 1 with zones of nodes 94
3.5 Appendix 94
References 97

4 A unified distributed algorithm for non-cooperative games 101
Jong-Shi Pang and Meisam Razaviyayn

4.1 Introduction 101
4.2 The nonsmooth, nonconvex game 104
4.3 The unified algorithm 106
4.3.1 Special cases 108
4.4 Convergence analysis: contraction approach 110
4.4.1 Probabilistic player choices 115

4.5 Convergence analysis: potential approach 116
4.5.1 Generalized potential games 121
References 122
Appendix 125

Part II Big data over cyber networks 135

5 Big data analytics systems 137
Ganesh Ananthanarayanan and Ishai Menache

5.1 Introduction 137
5.2 Scheduling 139
5.2.1 Fairness 139
5.2.2 Placement constraints 142
5.2.3 Additional system-wide objectives 145
5.2.4 Stragglers 146
5.3 Storage 148
5.3.1 Distributed file system 148
5.3.2 In-memory storage 151
5.4 Concluding remarks 156
References 158

6 Distributed big data storage in optical wireless networks 161
Chen Gong, Zhengyuan Xu, and Xiaodong Wang

6.1 Introduction 161
6.2 Big data distributed storage in a wireless network 163
6.2.1 Wireless distributed storage network framework 163
6.2.2 Optical wireless framework 165
6.2.3 Rateless coded distributed data storage 167
6.2.4 Network coded system with full downloading 167
6.2.5 Network coded system with partial downloading 168
6.3 Reconstructability condition for partial downloading 169
6.3.1 μ-Reconstructability for MSR point 169
6.3.2 μ-Reconstructability for a practical MBR point coding scheme 170
6.4 Channel and power allocation for partial downloading 173
6.4.1 Wireless resource allocation framework 174
6.4.2 Optimal channel and power allocation for the relaxed problem 174
6.5 Open research topics 176
6.5.1 General research topics for wireless distributed storage networks 176
6.5.2 Research topics for data storage in optical wireless networks 177
6.5.3 Research topics for data storage in named data networks 177
References 178

7 **Big data aware wireless communication: challenges and opportunities** 180
Suzhi Bi, Rui Zhang, Zhi Ding, and Shuguang Cui

7.1 Introduction 180
7.2 Scalable wireless network architecture for big data 182
7.2.1 Hybrid processing structure 182
7.2.2 Web caching in wireless infrastructure 185
7.2.3 Data aware processing units 187
7.3 Wireless system design in big data era 188
7.3.1 Analog vs. digital backhaul 188
7.3.2 Joint base station and cloud processing with digital backhaul 191
7.3.3 Section summary 198
7.4 Big data aware wireless networking 198
7.4.1 Wireless big data analytics 200
7.4.2 Data-driven mobile cloud computing 205
7.4.3 Software-defined networking design 207
7.4.4 Section summary 209
7.5 Conclusions 210
Acknowledgement 211
References 211

8 **Big data processing for smart grid security** 217
Lanchao Liu, Zhu Han, H. Vincent Poor, and Shuguang Cui

8.1 Preliminaries and motivations 217
8.2 Sparse optimization for false data injection detection 219
8.2.1 State estimation and false data injection attacks 219
8.2.2 Nuclear norm minimization 223
8.2.3 Low-rank matrix factorization 225
8.2.4 Numerical results 227
8.3 Distributed approach for security-constrained optimal power flow 232
8.3.1 Security-constrained optimal power flow 232
8.3.2 ADMM method 235
8.3.3 Distributed and parallel approach for SCOPF 236
8.3.4 Numerical results 238
8.4 Concluding remarks 241
Acknowledgement 241
References 242

Part III **Big data over social networks** 245

9 **Big data: a new perspective on cities** 247
Riccardo Gallotti, Thomas Louail, Rémi Louf, and Marc Barthelemy

9.1 Big data and urban systems 247
9.2 Infrastructure networks 249

9.2.1 Road networks 249
9.2.2 Subway networks 255
9.3 Mobility networks 257
9.3.1 A renewed interest 257
9.3.2 Individual mobility networks 258
9.3.3 From big data to the spatial structure of cities 261
9.4 Scaling in cities 268
9.5 Discussion: towards a new science of cities 272
Acknowledgments 273
References 273

10 High-dimensional network analytics: mapping topic networks in Twitter data during the Arab Spring 278
Kathleen M. Carley, Wei Wei, and Kenneth Joseph

10.1 Introduction 278
10.2 Arab Spring 280
10.3 General background 280
10.4 Data 281
10.5 The social pulse: geo-temporal trends in Twitter topics and users 284
10.5.1 Methodology 284
10.5.2 Topic overview 285
10.5.3 Over time analysis 287
10.5.4 Characterization of user–topic similarity network 288
10.5.5 Social interaction overview: the reply network 290
10.5.6 Characterization of group structure 291
10.5.7 Key actors 294
10.6 Discussion 295
10.7 Conclusion 297
Acknowledgements 298
References 299

11 Social influence analysis in the big data era: a review 301
Jianping Cao, Dongliang Duan, Liuqing Yang, Qingpeng Zhang, Senzhang Wang, and Feiyue Wang

11.1 Introduction 301
11.2 Social influence measurement 304
11.2.1 Network-based measures 304
11.2.2 Behavior-based measures 309
11.2.3 Interaction-based measures 312
11.2.4 Topic-based measures 313
11.2.5 Other measures 316
11.3 Influence propagation and maximization 317
11.3.1 Opinion leader identification 317
11.3.2 Influence maximization 319

11.3.3 Diffusion network inference 324
11.3.4 Challenges of IP&M 327
11.4 Challenges in big data 327
11.5 Summary 328
Acknowledgement 329
References 329

Part IV Big data over biological networks 335

12 Inference of gene regulatory networks: validation and uncertainty 337
Xiaoning Qian, Byung-Jun Yoon, and Edward R. Dougherty

12.1 Introduction 337
12.2 Background 338
12.2.1 Markov chains 339
12.2.2 Logical regulatory networks 340
12.2.3 Control policy for maximal steady-state alteration 341
12.2.4 Inference algorithms 341
12.3 Network distance functions 343
12.3.1 Semi-metrics 343
12.3.2 Rule-based distance 343
12.3.3 Topology-based distance 344
12.3.4 Transition-probability-based distance 344
12.3.5 Steady-state distance 345
12.3.6 Control-based distance 345
12.4 Inference performance 346
12.4.1 Measuring inference performance using distance functions 346
12.4.2 Analytic example 347
12.4.3 Synthetic examples 348
12.5 Consistency 352
12.6 Approximation 352
12.7 Validation from experimental data 353
12.7.1 Metastatic melanoma network inference 354
12.8 Uncertainty quantification 354
12.8.1 Mean objective cost of uncertainty 356
12.8.2 Intervention in yeast cell cycle network with uncertainty 358
References 360

13 Inference of gene networks associated with the host response to infectious disease 365
Zhe Gan, Xin Yuan, Ricardo Henao, Ephraim L. Tsalik, and Lawrence Carin

13.1 Background 365
13.2 Factor models in gene expression analysis 366

13.3 Factor models 367
13.3.1 Shrinkage prior 368
13.3.2 Multiplicative gamma process 369
13.4 Discriminative models 370
13.4.1 Bayesian log-loss 370
13.4.2 Bayesian hinge-loss 372
13.5 Discriminative factor model 372
13.5.1 Multi-task learning 374
13.6 Inference 374
13.7 Experiments 376
13.7.1 Performance measures 377
13.7.2 Experimental setup 377
13.7.3 Classification results 378
13.7.4 Interpretation 382
13.8 Closing remarks 384
13.9 Inference details 385
Acknowledgements 387
References 388

14 Gene-set-based inference of biological network topologies from big molecular profiling data 391
Lipi Acharya and Dongxiao Zhu

14.1 Introduction 391
14.2 Big data to network components 393
14.3 Gene sets related to network components 394
14.4 Reconstructing biological network topologies using gene sets 395
14.4.1 A general setting 395
14.4.2 Gene set Gibbs sampling 397
14.4.3 Gene set simulated annealing 397
14.5 Discussion and future work 403
References 406

15 Large-scale correlation mining for biomolecular network discovery 409
Alfred Hero and Bala Rajaratnam

15.1 Introduction 409
15.2 Illustrative example 414
15.2.1 Pairwise correlation 416
15.2.2 From pairwise correlation to networks of correlations 417
15.3 Principles of correlation mining for big data 419
15.3.1 Correlation mining for correlation flips between two populations 424
15.3.2 Large-scale implementation of correlation mining 426

15.4 Perspectives and future challenges 427
15.4.1 State-of-the-art in correlation mining 427
15.4.2 Future challenges in correlation mining biomolecular networks 429
15.5 Conclusion 431
Acknowledgements 431
References 432

Index 437

Contributors

Shiqian Ma
The Chinese University of Hong Kong, China

Bo Jiang
Shanghai University of Finance and Economics, China

Xiuzhen Huang
Arkansas State University, USA

Shuzhong Zhang
University of Minnesota, USA

Symeon Chouvardas
University of Athens, Greece

Yannis Kopsinis
University of Athens, Greece

Sergios Theodoridis
University of Athens, Greece

Mingyi Hong
Iowa State University, USA

Wei-Cheng Liao
University of Minnesota, USA

Ruoyu Sun
University of Minnesota, USA

Zhi-Quan (Tom) Luo
University of Minnesota, USA

Jong-Shi Pang
University of Southern California, USA

Meisam Razaviyayn
Stanford University, USA

Ganesh Ananthanarayanan
Microsoft Research, USA

Ishai Menache
Microsoft Research, USA

Chen Gong
University of Science and Technology of China, China

Zhengyuan Xu
University of Science and Technology of China, China

Xiaodong Wang
Columbia University, USA

Suzhi Bi
National University of Singapore, Singapore

Rui Zhang
National University of Singapore, Singapore

Zhi Ding
University of California at Davis, USA

Shuguang Cui
Texas A&M University, USA

Lanchao Liu
University of Houston, USA

Zhu Han
University of Houston, USA

H. Vincent Poor
Princeton University, USA

Riccardo Gallotti
Institut de Physique Théorique, CEA, France

Thomas Louail
Institut de Physique Théorique, CEA, France

Rémi Louf
Institut de Physique Théorique, CEA, France

Marc Barthelemy
Institut de Physique Théorique, CEA, France
and
Centre d'Analyse et de Mathématiques Sociales, EHESS, France

Kathleen M. Carley
Carnegie Mellon University, USA

Wei Wei
Carnegie Mellon University, USA

Kenneth Joseph
Carnegie Mellon University, USA

Jianping Cao
National University of Defense Technology, China

Dongliang Duan
University of Wyoming, USA

Liuqing Yang
Colorado State University, USA

Qingpeng Zhang
City University of Hong Kong, China

Senzhang Wang
Beihang Univerisity, China

Feiyue Wang
National University of Defense Technology, China
and
Chinese Academy of Science, China

Xiaoning Qian
Texas A&M University, USA

Byung-Jun Yoon
Hamad bin Khalifa University, Qatar

Edward R Dougherty
Texas A&M University, USA

Zhe Gan
Duke University, USA

Xin Yuan
Duke University, USA

Ricardo Henao
Duke University, USA

Ephraim L. Tsalik
Durham Veterans Affairs Medical Center, USA
and
Duke University Medical Center, USA

Lawrence Carin
Duke University, USA

Lipi Acharya
Dow AgroSciences LLC, USA

Dongxiao Zhu
Wayne State University, USA

Alfred Hero
University of Michigan, USA

Bala Rajaratnam
Stanford University, USA

Preface

In each day of our modern world, quintillions of bytes of data are generated. This rate keeps increasing, far outpacing the rate at which we can upgrade the computing systems. In fact, a 2011 study by IDC estimates that the amount of data available in the world doubles every two years. This rate of growth closely follows Moore's Law. If we cannot fully understand the issues involved and invent new data processing methods, society will soon be flooded with data, *big data*. Big data sets can exist within a single entity; but most such sets are distributed and can only be aggregated through some type of network. A bigger challenge is that the big data dynamics and the underlying network dynamics are almost always highly correlated. Therefore, understanding the interplay between big data and the associated networks is a critical step in our effort to tackle big data. However, to date, we do not have a systematic theory with which to study the problem thoroughly. Even worse, in some cases, we do not even know how to formulate and approach those problems. We are in need of a comprehensive book to survey and cover both the critical mathematical tools and the state of the art in related research fields.

This book focuses on large-scale information processing over networks, where the meaning of the term information processing can refer to data processing, data storage, or information retrieval, and the term networks may refer to cyber networks, social networks, or biological networks. We take three complementary angles to study the interaction between the data and the underlying network connections. First, we address ways that the underlying network can constrain the upper-layer collaborative big data processing; second, we show how certain big data processing perspectives can help boost the performance in various networks; third, we address the fundamental limits that govern statistical and computational bottlenecks in the analysis of big data. The book consists of chapters contributed by experts from diverse fields spanning machine learning, optimization, statistics, signal processing, networking, communications, sociology, and biology. The core unifying theme of the book is the rigorous mathematical treatment of various subjects, enriched by in-depth discussions of future directions at the end of each chapter. It is expected that this book will not only help researchers in related fields learn the basic tools and understand the state of the art, but also help practitioners formulate approaches to study other large systems that generate and process big data.

The book starts by introducing the recent development of several important mathematical tools for large-scale computation and learning. This provides a useful mathematical framework for studying the big data problem over different types of networks. Afterwards, the book turns to big data problems in several specific application domains: big data over cyber networks, big data over social networks, and big data over biological networks. One of our main goals is to encourage readers to ponder the large number of interesting interdisciplinary research efforts across different fields that illustrate the problem of big data over networks.

Specifically, in Part I of the book, we focus on some mathematical tools for large-scale data modeling and processing. We start with Chapter 1 on tensor models, which provides a powerful modeling framework for data of high dimensions, with special coverage on the tensor principal component analysis, the tensor low-rank and sparse decomposition models, and the tensor co-clustering problems. Chapter 2 is on the sparsity-aware distributed learning method, where both batch and online algorithms will be discussed. In the batch learning context, the distributed LASSO algorithm and distributed greedy technique will be presented. Furthermore, an LMS-based sparsity promoting algorithm, revolving around the l_1 norm, as well as a greedy distributed LMS will be discussed. In addition, a set-theoretic sparsity promoting distributed technique will be examined. Chapter 3 is on the introduction of optimization algorithms for big data, with the focus on modern first-order large-scale optimization techniques. A few popular first-order methods for large-scale optimization will be surveyed first, including the Block Coordinate Descent (BCD) method, the Block Successive Upper-Bound Minimization (BSUM) method, and the Alternating Direction Method of Multipliers (ADMM). Then the optimal management of a cloud-based densely deployed next-generation wireless network will be studied as a design example. Chapter 4 is on the distributed algorithms for game theoretical approaches, presenting a unified framework for the design and analysis of distributed algorithms for computing first-order stationary solutions of noncooperative games with non-differentiable player objective functions. These games are closely associated with multi-agent optimization wherein a large number of selfish players compete non-cooperatively to optimize their individual objectives under various constraints.

In Part II of this book, we focus on big data over cyber networks, which include computer networks, communication networks, and the cyber part of a smart grid. In Chapters 5 and 6, we discuss the architecture-level issues for big data analytics and storage; in Chapters 7 and 8, we discuss the big data challenges and opportunities in communication networks and smart grid design. Particularly, in Chapter 5, two fundamental aspects are surveyed on the architecture design for a big data analytics system: scheduling and storage. Their key principles, and how these principles are realized in widely-deployed systems, will be described. In Chapter 6, a big data distributed storage system is discussed, employing existing regenerate codes where the storage nodes are scattered in an optical wireless network. A partial downloading scheme is proposed, which allows downloading a portion of the symbols from any storage nodes. A cross-layer wireless resource allocation framework is then formulated and discussed for data reconstruction in such a distributed storage systems employing partial downloading. Channel and power allocation schemes are also investigated for partial downloading in wireless distributed storage systems. In Chapter 7, the interaction between big data and communication networks is the focus. The challenges and opportunities in the design of scalable wireless systems to embrace the big data era are discussed. The state-of-the-art techniques in wireless big-data processing are reviewed and the potential implementations of key technologies in the future wireless systems are studied. It is argued that proper wireless system designs could harness, and in fact take advantages of the mobile big data traffic. In Chapter 8, the cyber layer of a smart grid is studied from the point of view of data analytics for security. A distributed and user centric system will be introduced, which incorporates end-consumers into its decision processes to provide a cost-effective and reliable energy supply. The applications of big data processing techniques for smart grid security are investigated from two perspectives: how to exploit the inherent structure of the data, and how to deal with the huge size of the data sets.

In Part III of this book, we shift to big data over social networks, where the social network could refer to either the physical interaction or the virtual interaction (e.g., via Facebook or Twitter) among people. In Chapter 9, a city environment is used to illustrate what could be learned from data about this social network with physical interactions. By analyzing data over small, intermediate, and large time scales, different aspects about the city could be learned. At this period of human history that experiences a rapid urban expansion, such a scientific approach appears more important than ever in order to understand the impact of current urban planning on the future evolution of cities. In Chapter 10, we study a social network with virtual interactions, where large sets of Twitter discourses are analyzed. A high-dimensional network approach is presented for assessing such discourses and identifying not just what is being discussed, but the locality, the change, the associated groups, and the structure in these discourses. This approach is applied to data captured with respect to the Arab Spring. The results provide insight into the co-evolution of topics and groups across the region during a period of dramatic social change. In Chapter 11, a study of the social influence from big data is presented. First, different ways of making measurements of social influence are discussed, followed by the descriptions of the algorithms and models that can be used to quantify the propagation of social influences. Then research on optimization of social influence propagation is summarized. Finally, the method of diffusion network inference is presented, and several challenges and open problems are discussed.

In Part IV of this book, we turn to the data analytics over biological networks. In Chapter 12, a general paradigm is discussed for inference validation based on defining a distance between networks and judging validity according to the distance between the original network and the inferred network. Rather than assuming that a single network is inferred, one can take the perspective that the inference procedure leads to an uncertainty class of networks, which contains the ground truth network. Accordingly, a measure of uncertainty is defined in terms of the cost that uncertainty imposes on the objective, for which the model network is to be employed. The example discussed in the chapter involves interventions in the yeast cell cycle network. In Chapter 13, the importance of Bayesian modeling for gene expression analysis is highlighted. Discriminative factor models, which are the particular theme of this chapter, are presented within a principled framework to jointly build factor models and multiple classifiers. As an alternative to Bayesian classifiers based on the traditional probit link, logistic regression and support vector classification are integrated into the modeling scheme, using novel variable augmentation techniques. The factor models are equipped with global-local shrinkage priors, recently proposed within the machine learning community, with which the number of factors are inferred automatically from the data. Extensions to multi-task learning are also presented. Inference is developed using both MCMC and variational Bayes algorithms, while online learning is further investigated to scale the model to large datasets. In Chapter 14, the focus is on the inference of biological network topologies from big molecular profiling data that is divided into three stages: identification of network components from molecular data, derivation of gene sets related to a network component, and gene-set-based inference of the underlying network topology in the given network component. In Chapter 15, some of the main advances and challenges are discussed in correlation mining in the context of large-scale biomolecular networks with a focus on medicine. The chapter emphasizes that there are fundamental statistical limits to reliable extraction of information from the sample correlation. These limits are associated with phase transitions in the false detection rate when looking for edges and hubs in a correlation or partial correlation network. A new regime of sample complexity is introduced that is ideally suited for big data problems in biology: the purely-high-dimensional regime, in which the number of samples n is fixed while the number p of biomarker variables is very large. A new correlation mining application also discusses discovery of correlation sign flips between edges in a pair of correlation or partial correlation networks. The pair of networks could respectively correspond to a disease (or treatment) group and a control group.

All the above chapters include comprehensive lists of references related to their contents. We hope that by reading this book, the readers could gain a foothold on the exciting activities in the many networking fields that are coupled with big data. We also hope that this book will serve to inspire the readers to develop mathematical approaches to other big data problems not covered here.

Finally, we thank all the chapter authors and the editors from Cambridge University Press, who together made this book possible.

Co-Editors:
Shuguang, Alfred, Zhi-Quan, and José
Spring, 2015

Part I

Mathematical foundations

1 Tensor models: solution methods and applications

Shiqian Ma, Bo Jiang, Xiuzhen Huang, and Shuzhong Zhang

This chapter introduces several models and associated computational tools for tensor data analysis. In particular, we discuss: tensor principal component analysis, tensor low-rank and sparse decomposition models, and tensor co-clustering problems. Such models have a great variety of applications; examples can be found in computer vision, machine learning, image processing, statistics, and bio-informatics. For computational purposes, we present several useful tools in the context of tensor data analysis, including the alternating direction method of multipliers (ADMM), and the block variables optimization techniques. We draw on applications from the gene expression data analysis in bio-informatics to demonstrate the performance of some of the aforementioned tools.

1.1 Introduction

One rich source of *big data* roots is the high dimensionality of the data formats known as *tensors*. Specifically, a complex-valued m-dimensional or mth-order *tensor* (a.k.a. m-way multiarray) can be denoted by $\mathcal{F} \in \mathbb{C}^{n_1 \times n_2 \times \cdots \times n_m}$, whose dimension in the ith direction is n_i, $i = 1, \ldots, m$. Vector and matrix are special cases of tensor when $m = 1$ and $m = 2$, respectively. In the era of big data analytics, huge-scale dense data in the form of tensors can be found in different domains such as computer vision [1], diffusion magnetic resonance imaging (MRI) [2–4], the quantum entanglement problem [5], spectral hypergraph theory [6], and higher-order Markov chains [7]. For instance, a color image can be considered as 3D data with row, column, color in each direction, while a color video sequence can be considered as 4D data, where time is the fourth dimension. Therefore, how to extract useful information from these tensor data becomes a very meaningful task.

On the other hand, the past few years have witnessed an emergence of sparse and low-rank matrix optimization models and their applications in data sciences, signal processing, machine learning, bioinformatics, and so on. There have been extensive investigations on low-rank matrix completion and recovery problems since the seminal

Big Data over Networks, ed. Shuguang Cui, Alfred O. Hero III, Zhi-Quan Luo, and José M. F. Moura. Published by Cambridge University Press.

works of [8–11]. Some important variants of sparse and low-rank matrix optimization problems such as robust principal component analysis (PCA) [12, 13] and sparse PCA [14] have also been studied. A natural extension of the matrix to higher-dimensional space is the tensor. Traditional matrix-based data analysis is inherently two-dimensional, which limits its ability in extracting information from a multi-dimensional perspective. Tensor-based multi-dimensional data analysis has shown that tensor models can take full advantage of the multi-dimensional structures of the data, and generate more useful information. For example, Wang and Ahuja [1] reported that the images obtained by tensor PCA technique have higher quality than those from matrix PCA.

Stimulated by the need of big data analytics, and motivated by the success of compressed sensing and low-rank matrix optimization, it is important and timely to study methods for analyzing massive tensor data.

Before proceeding let us introduce notations that will be used throughout this chapter. We use $\mathbf{R}^n$ to denote the n-dimensional Euclidean space. A tensor is usually denoted by a calligraphic letter, as $\mathcal{A} = (\mathcal{A}_{i_1 i_2 \cdots i_m})_{n_1 \times n_2 \times \cdots \times n_m}$. The space where $n_1 \times n_2 \times \cdots \times n_m$-dimensional real-valued tensor resides is denoted by $\mathbf{R}^{n_1 \times n_2 \times \cdots \times n_m}$. We call $\mathcal{A}$ super-symmetric if $n_1 = n_2 = \cdots = n_m$ and $\mathcal{A}_{i_1 i_2 \cdots i_m}$ is invariant under any permutation of $(i_1, i_2, \ldots, i_m)$, i.e., $\mathcal{A}_{i_1 i_2 \cdots i_m} = \mathcal{A}_{\pi(i_1, i_2, \ldots, i_m)}$, where $\pi(i_1, i_2, \ldots, i_m)$ is any permutation of indices $(i_1, i_2, \ldots, i_m)$. The space where $\underbrace{n \times n \times \cdots \times n}_{m}$ super-symmetric tensors reside is denoted by $\mathbf{S}^{n^m}$. Special cases of tensors are vector ($m = 1$) and matrix ($m = 2$), and tensors can also be seen as a long vector or a specially arranged matrix. For instance, the tensor space $\mathbf{R}^{n_1 \times n_2 \times \cdots \times n_m}$ can also be seen as a matrix space $\mathbf{R}^{(n_1 \times n_2 \times \cdots \times n_{m_1}) \times (n_{m_1+1} \times n_{m_1+2} \times \cdots \times n_m)}$, where the row is actually an m_1-way array tensor space and the column is another $(m - m_1)$-dimensional tensor space. Such connections between tensor and matrix re-arrangements will play an important role in this chapter. As a convention in this chapter, if there is no other specification we shall adhere to the Euclidean norm (i.e. the L_2-norm) for vectors and tensors; in the latter case, the Euclidean norm is also known as the Frobenius norm, and is sometimes denoted as $\|\mathcal{A}\|_F = \sqrt{\sum_{i_1, i_2, \ldots, i_m} \mathcal{A}^2_{i_1 i_2 \cdots i_m}}$. For a given matrix X, we use $\|X\|_*$ to denote the nuclear norm of X, which is the sum of all the singular values of X. Regarding the products, we use $\otimes$ to denote the outer product for tensors; that is, for $\mathcal{A}_1 \in \mathbf{R}^{n_1 \times n_2 \times \cdots \times n_m}$ and $\mathcal{A}_2 \in \mathbf{R}^{n_{m+1} \times n_{m+2} \times \cdots \times n_{m+\ell}}$, $\mathcal{A}_1 \otimes \mathcal{A}_2$ is in $\mathbf{R}^{n_1 \times n_2 \times \cdots \times n_{m+\ell}}$ with

$$(\mathcal{A}_1 \otimes \mathcal{A}_2)_{i_1 i_2 \cdots i_{m+\ell}} = (\mathcal{A}_1)_{i_1 i_2 \cdots i_m} (\mathcal{A}_2)_{i_{m+1} \cdots i_{m+\ell}}.$$

The inner product between two tensors $\mathcal{A}_1$ and $\mathcal{A}_2$ residing in the same space $\mathbf{R}^{n_1 \times n_2 \times \cdots \times n_m}$ is denoted

$$\mathcal{A}_1 \bullet \mathcal{A}_2 = \sum_{i_1, i_2, \ldots, i_m} (\mathcal{A}_1)_{i_1 i_2 \cdots i_m} (\mathcal{A}_2)_{i_1 i_2 \cdots i_m}.$$

Under this light, a multi-linear form $\mathcal{A}(x^1, x^2, \ldots, x^m)$ can also be written in inner/outer products of tensors as

$$\mathcal{A} \bullet (x^1 \otimes \cdots \otimes x^m) := \sum_{i_1,\ldots,i_m} \mathcal{A}_{i_1,\ldots,i_m} (x^1 \otimes \cdots \otimes x^m)_{i_1,\ldots,i_m} = \sum_{i_1,\ldots,i_m} \mathcal{A}_{i_1,\ldots,i_m} \prod_{k=1}^{m} x^k_{i_k}.$$

1.2 Tensor models

1.2.1 Sparse and low-rank tensor optimization models

We first consider the common-background and sparse-foreground decomposition for the tensor data. To this end, we propose two tensor models below. The first model is to write a given tensor $\mathcal{A} \in \mathbf{R}^{n_1 \times n_2 \times \cdots \times n_m}$ as the sum of three tensors: $\mathcal{X}$, $\mathcal{Y}$, and $\mathcal{Z}$. That is $\mathcal{A} = \mathcal{X} + \mathcal{Y} + \mathcal{Z}$, while $\mathcal{X}$ is in the form of $\mathcal{X} = \bar{\mathcal{X}} \otimes e$ where $\bar{\mathcal{X}} \in \mathbf{R}^{n_1 \times n_2 \times \cdots \times n_{m-1}}$ is a $(m-1)$-dimensional tensor and e is the all-one vector, and $\mathcal{Z}$ is the noise tensor. Specifically, the model in question is given by [15]

$$\begin{array}{ll} \min & \|\mathcal{Y}\|_1 \\ \text{s.t.} & \bar{\mathcal{X}} \otimes e + \mathcal{Y} + \mathcal{Z} = \mathcal{A} \\ & \|\mathcal{Z}\|_F \leq \delta. \end{array} \tag{1.1}$$

Note that $\mathcal{A}$ thus has a common-tensor structure in the sense that all the $\mathbf{R}^{n_1 \times n_2 \times \cdots \times n_{m-1}}$-dimensional subtensors of $\mathcal{X}$ are the same. We now give more details about the physical meaning of model (1.1). For ease of presentation, we assume $m = 3$ at this moment. In this case, $\mathcal{A}$ consists of n_3 matrices $\mathcal{A}_1, \ldots, \mathcal{A}_{n_3}$ with the same size $n_1 \times n_2$. The equality constraint in (1.1) indicates that each matrix $\mathcal{A}_i$ can be decomposed into three parts: the common part (matrix $\bar{\mathcal{X}}$), the sparse part (matrix $\mathcal{Y}_i$), and the noisy part (matrix $\mathcal{Z}_i$). In many real applications, the last dimension in tensor $\mathcal{A}$ denotes time. Model (1.1) implies that the subtensors of $\mathcal{A}$ along time are almost the same, but different from each other with certain sparse changes captured in $\mathcal{Y}$ and small noises captured in $\mathcal{Z}$. By solving model (1.1) one can identify the common part and detect the changing part that results in significant difference among the subtensors. It should be pointed out that, even in the matrix case, our common-tensor model (1.1) is theoretically different from the low-rank + sparse decomposition of the robust PCA model proposed by Candes *et al.* [12] and Chandrasekaran *et al.* [13]. The L_1 norm in the objective of (1.1) naturally promotes the sparsity in tensor $\mathcal{Y}$. Recently, a similar model was also considered independently by Li *et al.* [16] in the context of image processing.

A common observation for huge-scale data analysis is that the data exhibit a low-dimensional property, or the most-representative part lies in low-dimensional subspace. Along with this line, we can model the background fluctuation by a low-rank tensor and achieve another optimization model:

$$\begin{array}{ll} \min & \operatorname{rank}(\bar{\mathcal{X}}) + \rho\|\mathcal{Y}\|_1 \\ \text{s.t.} & \bar{\mathcal{X}} \otimes e + \mathcal{Y} + \mathcal{Z} = \mathcal{A} \\ & \|\mathcal{Z}\|_F \leq \delta, \end{array} \tag{1.2}$$

where $\text{rank}(\bar{\mathcal{X}})$ denotes the CP rank of $\bar{\mathcal{X}}$ and its precise definition can be described as follows.

Definition 1.1 Suppose $\mathcal{X} \in \mathbf{R}^{n_1 \times n_2 \times \cdots \times n_m}$, the CP rank of $\mathcal{X}$ denoted by $\text{rank}(\mathcal{X})$ is the smallest integer r such that

$$\mathcal{F} = \sum_{i=1}^{r} a^{1,i} \otimes \cdots \otimes a^{m,i}, \tag{1.3}$$

where $a^{k,i} \in \mathbf{R}^{n_k}$ for all $1 \leq i \leq r$ and $1 \leq k \leq m$.

The idea of decomposing a tensor into an (asymmetric) outer product of vectors was first introduced and studied by Hitchcock in 1927 [17, 18]. This concept of tensor rank became popular after its rediscovery in the 1970s in the form of CANDECOMP (canonical decomposition) by Carroll and Chang [19] and PARAFAC (parallel factors) by Harshman [20]. Consequently, CANDECOMP and PARAFAC are further abbreviated as “CP” in the context of “CP rank” by many authors in the literature. In the next subsection, we will introduce the CP rank for super-symmetric tensors.

1.2.2 Tensor principal component analysis

Principal component analysis (PCA) plays an important role in applications arising from areas such as data analysis, dimension reduction, and bioinformatics, among others. PCA finds a few linear combinations of the original variables. These linear combinations, which are called principal components (PCs), are orthogonal to each other and explain most of the variance of the data. PCs provide a powerful tool to compress data along the direction of maximum variance to reach the minimum information loss.

Although the PCA and eigenvalue problem for the matrices have been well studied in the literature, the research of PCA for tensors is still underdeveloped. The tensor PCA is of great importance in practice and has many applications in computer vision [1], diffusion magnetic resonance imaging (MRI) [2–4], quantum entanglement problem [5], spectral hypergraph theory [6] and higher-order Markov chains [7]. Similar to its matrix counterpart, the problem of finding the PC that explains the most of the variance of a tensor $\mathcal{T}$ (with degree m) can be formulated as:

$$\begin{array}{ll} \min & \|\mathcal{T} - \lambda x^1 \otimes x^2 \otimes \cdots \otimes x^m\| \\ \text{s.t.} & \lambda \in \mathbf{R},\ \|x^i\| = 1, i = 1, 2, \ldots, m, \end{array} \tag{1.4}$$

which is equivalent to

$$\begin{array}{ll} \max & \mathcal{T}(x^1, x^2, \ldots, x^m) \\ \text{s.t.} & \|x^i\| = 1, i = 1, 2, \ldots, m. \end{array} \tag{1.5}$$

Let us call the above solution the *leading* PC. Once the leading PC is found, the other PCs can be computed sequentially via the so-called “deflation” technique. For instance, the second PC is defined as the leading PC of the tensor subtracting the leading PC from the original tensor, and so forth. The theoretical basis of such a deflation procedure for tensors is not exactly sound, although its matrix counterpart is well established (see

[21] and the references therein for more details). However, the deflation process does provide a heuristic way to compute multiple principal components of a tensor, albeit approximately. Thus in the rest of this paper, we focus on finding the leading PC of a tensor.

Problem (1.5) is also known as the best rank-one approximation of tensor $\mathcal{T}$, which has been studied in [22]. By embedding $\mathcal{T}$ into a larger tensor (for instance, see Section 8.4 in [23]), problem (1.5) can be reformulated as

$$\begin{array}{ll} \max & \mathcal{F}(x, x, \ldots, x) \\ \text{s.t.} & \|x\| = 1, \end{array} \tag{1.6}$$

where $\mathcal{F}$ is a super-symmetric tensor. Problem (1.6) is NP-hard and is called the maximum Z-eigenvalue problem in [24] and the nonlinear eigenproblem in [25]. Although a systematic study of the eigenvalues and eigenvectors for a real symmetric tensor was first conducted by Lim [26] and Qi [24] independently in 2005, Kofidis and Regalia in 2001 already showed that blind deconvolution can be formulated as a nonlinear eigenproblem [25]. Note that various methods have been proposed to find the Z-eigenvalues [27–31], which, however, may correspond only to local optimums, although some efforts on heuristics for finding global optimal solution were made (see, e.g., [25, 28]). In this chapter, we shall focus on finding the global optimal solution of (1.6).

In the subsequent analysis, for convenience we assume m to be even, i.e. $m = 2d$ in (1.6), where d is a positive integer, as this assumption is essentially not restrictive (see [23]). Therefore, we will focus on the following problem of largest eigenvalue of an even-order super-symmetric tensor:

$$\begin{array}{ll} \max & \mathcal{F}(\underbrace{x, \ldots, x}_{2d}) \\ \text{s.t.} & \|x\| = 1, \end{array} \tag{1.7}$$

where $\mathcal{F}$ is a $2d$th-order super-symmetric tensor. In particular, problem (1.7) can be equivalently written as

$$\begin{array}{ll} \max & \mathcal{F} \bullet \underbrace{x \otimes \cdots \otimes x}_{2d} \\ \text{s.t.} & \|x\| = 1. \end{array} \tag{1.8}$$

Now we introduce the so-called CP rank for even-order super-symmetric tensors.

Definition 1.2 Suppose $\mathcal{F} \in \mathbf{S}^{n^{2d}}$, the CP rank of $\mathcal{F}$ denoted by $\text{rank}(\mathcal{F})$ is the smallest integer r such that

$$\mathcal{F} = \sum_{i=1}^{r} \lambda_i \underbrace{a^i \otimes \cdots \otimes a^i}_{2d}, \tag{1.9}$$

where $a_i \in \mathbf{R}^n$, $\lambda_i \in \{1, -1\}$.

Thus, given any $2d$th-order super-symmetric tensor form $\mathcal{F}$, we call it *rank one* if $\mathcal{F} = \lambda \underbrace{a \otimes \cdots \otimes a}_{2d}$ for some $a \in \mathbf{R}^n$ and $\lambda \in \{1, -1\}$.

In the following, to simplify the notation, we denote

$$\mathbb{K}(n,d) = \left\{ k = (k_1, \ldots, k_n) \in \mathbb{Z}_+^n \;\middle|\; \sum_{j=1}^{n} k_j = d \right\}$$

and

$$\mathcal{X}_{1^{2k_1} 2^{2k_2} \ldots n^{2k_n}} := \mathcal{X}_{\underbrace{1\ldots1}_{2k_1}\underbrace{2\ldots2}_{2k_2}\ldots\underbrace{n\ldots n}_{2k_n}}.$$

By letting $\mathcal{X} = \underbrace{x \otimes \cdots \otimes x}_{2d}$ we can further convert problem (1.8) into:

$$\begin{array}{ll} \max & \mathcal{F} \bullet \mathcal{X} \\ \text{s.t.} & \sum\limits_{k \in \mathbb{K}(n,d)} \frac{d!}{\prod_{j=1}^n k_j!} \mathcal{X}_{1^{2k_1} 2^{2k_2} \ldots n^{2k_n}} = 1, \\ & \mathcal{X} \in \mathbf{S}^{n^{2d}}, \ \operatorname{rank}(\mathcal{X}) = 1, \end{array} \tag{1.10}$$

where the first equality constraint is due to the fact that

$$\sum_{k \in \mathbb{K}(n,d)} \frac{d!}{\prod_{j=1}^n k_j!} \prod_{j=1}^{n} x_j^{2k_j} = \|x\|^{2d} = 1.$$

Thus, the tensor PCA problem can be viewed as a tensor optimization problem with rank-one constraint, which is the extreme case of low-rank tensor optimization.

1.2.3 The tensor co-clustering problem

While genome data are relatively static, gene expression, which reflects gene activity, is highly dynamic. Patterns of gene expression change dramatically based on cell type, developmental stage, disease state, and in response to a wide variety of biological or environmental factors. In addition, both the kinetics and amplitude of changes in gene expression can have biological and biomedical significance. Gene expression of the cell could be used to infer the cell type, state, stage, and cell environment, and may indicate a homeostasis response or a pathological condition and thus relate to development of new medicines, drug metabolism, and diagnosis of diseases [32–34]. High-throughput gene expression techniques (such as microarray, next-generation sequencing and third-generation sequencing technologies) are generating huge amounts of high-dimensional genome-wide gene expression data (e.g. 4D with genes vs. timepoints vs. conditions vs. tissues). While the availability of these data presents unprecedented opportunities, it also presents major challenges for extractions of biologically meaningful information from the mountain-like gene expression data. In particular, it calls for effective computational models, equipped with efficient solution methods, to categorize gene expression data into biologically relevant groups in order to facilitate further functional assessment of important biological and biomedical processes. Classical clustering and co-clustering analysis of gene expression data is a worthy approach in this endeavor [35, 36] (Figure 1.1).

Gene expression of 10 genes at 5 time points.
(Values in the table are not real gene expression values; They just indicate the difference in gene expression levels.)

Four co-clusters

Two clusters

Genes\ time points	T1	t2	t3	t4	t5
Gene a	1	1	2	2	2
Gene b	1	1	2	2	2
Gene c	1	1	5	5	5
Gene d	1	1	5	5	5
Gene e	1	1	5	5	5
Gene f	1	1	5	5	5
Gene g	0	0	2	2	2
Gene h	0	0	2	2	2
Gene i	0	0	2	2	2
Gene j	0	0	2	2	2

Figure 1.1 This figure illustrates the idea of clustering and co-clustering analysis. This is a table of 10 genes expression at five different time points. According to classical clustering, there are two clusters of genes $\{a, b, c, d, e, f\}$, $\{g, h, i, j\}$. For co-clustering, there could be four co-clusters, as shown.

Clustering as an effective approach, is usually applied to partition gene expression data into groups, where each group aggregates genes with similar expression levels. A lot of research has been conducted in clustering: cf. [37] for classical clustering in gene expression analysis, where the author discussed two classes of clustering (hierarchical clustering and partitioning), and three popular clustering methods (Eisen hierarchical clustering [38], k-means [39], and self-organizing map (SOM) method [40]). The classical clustering methods cluster genes into a number of groups based on their similar expression on all the considered conditions.

The concept of *co-clustering* was first introduced to 2D gene expression data analysis by Cheng and Church [41]. The co-clustering method can cluster genes and conditions simultaneously and thus can discover the similar expression of a certain group of genes on a certain group of conditions and vice versa. Readers may refer to [42] for a comprehensive comparison of the popular co-clustering approaches. Recently there are developed approaches for 3D gene expression data clustering analysis [43–46].

Essentially, the principle of current clustering and co-clustering models is to conduct *partitions* based on the assignment of a gene and/or a condition to a specific cluster or co-cluster. However, even a slightly less explicitly expressed function of the gene, which may be very important to know, can get lost under the principle of *sole assignment* of each gene to one co-cluster in the clustering analysis. In fact, it is widely known that one enzyme or a group of enzymes may get involved in more than one pathway, and one particular gene may be co-regulated with different groups of genes under different conditions and different development stages. The current clustering and co-clustering models are not designed to allocate more than one assignment per gene. Note that post-processing for merging identified clusters or co-clusters into overlapping groups [41, 47] could not address the issue. Motivated by this urgent need from the real-world gene expression data analysis, we develop a novel identification model based on tensor optimization that is capable of recognizing more than one assignment for one element, to better accommodate the reality of complex biological systems.

To illustrate the ideas, let us start by considering the conventional co-clustering formulation. Suppose that $\mathcal{A} \in \mathbf{R}^{n_1 \times n_2 \times \cdots \times n_d}$ is a d-dimensional tensor. Let $I_j = \{1, 2, \ldots, n_j\}$ be the set of indices on the jth dimension, $j = 1, 2, \ldots, d$. We wish to find a p_j-partition of the index set I_j, say $I_j = I_1^j \cup I_2^j \cup \cdots \cup I_{p_j}^j$, where $j = 1, 2, \ldots, d$, in such a way that each of the *subtensor* $A_{I_{i_1}^1 \times I_{i_2}^2 \times \cdots \times I_{i_d}^d}$ is as tightly packed up as possible, where $1 \le i_j \le n_j$ and $j = 1, 2, \ldots, d$. The notion that plays an important role in our model is the so-called *mode product* between a tensor $\mathcal{X}$ and a matrix P. Suppose that $\mathcal{X} \in \mathbf{R}^{p_1 \times p_2 \times \cdots \times p_d}$ and $P \in \mathbf{R}^{p_i \times m}$. Then, $\mathcal{X} \times_i P$ is a tensor in $\mathbf{R}^{p_1 \times p_2 \times \cdots \times p_{i-1} \times m \times p_{i+1} \times \cdots \times p_d}$, whose $(j_1, j_2, \ldots, j_{i-1}, j_i, j_{i+1}, \ldots, j_d)$th component is defined by

$$(X \times_i P)_{j_1, j_2, \ldots, j_{i-1}, j_i, j_{i+1}, \ldots, j_d} = \sum_{\ell=1}^{p_i} X_{j_1, j_2, \ldots, j_{i-1}, \ell, j_{i+1}, \ldots, j_d} P_{\ell, j_i}.$$

Let $\mathcal{X}_{j_1, j_2, \ldots, j_{i-1}, j_i, j_{i+1}, \ldots, j_d}$ be the value of the co-cluster

$$(j_1, j_2, \ldots, j_{i-1}, j_i, j_{i+1}, \ldots, j_d) \text{ with } 1 \le j_i \le p_i, i = 1, 2, \ldots, d.$$

Let an assignment matrix $Y^j \in \mathbf{R}^{n_j \times p_j}$ for the indices for jth array of tensor $\mathcal{A}$ be:

$$Y_{ik}^j = \begin{cases} 1, & \text{if } i \text{ is assigned to the } k\text{th partition } I_k^j; \\ 0, & \text{otherwise.} \end{cases}$$

Then, we introduce a *proximity* measure $f(s) : \mathbf{R} \to \mathbf{R}_+$, with the property that $f(s) \ge 0$ for all $s \in \mathbf{R}$ and $f(s) = 0$ if and only if $s = 0$. The co-clustering problem can be formulated as

$$\begin{array}{ll} \min & \displaystyle\sum_{j_1=1}^{n_1} \sum_{j_2=1}^{n_2} \cdots \sum_{j_d=1}^{n_d} f\left(\mathcal{A}_{j_1, j_2, \ldots, j_d} - (\mathcal{X} \times_1 Y^1 \times_2 Y^2 \times_3 \cdots \times_d Y^d)_{j_1, j_2, \ldots, j_d}\right) \\ \text{s.t.} & \mathcal{X} \in \mathbf{R}^{p_1 \times p_2 \times \cdots \times p_d}, \\ & Y^j \in \mathbf{R}^{n_j \times p_j} \text{ is a row assignment matrix}, \ j = 1, 2, \ldots, d. \end{array} \tag{1.11}$$

We may consider a variety of proximity measures. For instance, if $f(s) = |s|^2$ then (1.11) can be written as

$$\begin{array}{ll} \min & \|\mathcal{A} - \mathcal{X} \times_1 Y^1 \times_2 Y^2 \times_3 \cdots \times_d Y^d\|_F^2 \\ \text{s.t.} & \mathcal{X} \in \mathbf{R}^{p_1 \times p_2 \times \cdots \times p_d}, \\ & Y^j \in \mathbf{R}^{n_j \times p_j} \text{ is a row assignment matrix}, \ j = 1, 2, \ldots, d. \end{array} \tag{1.12}$$

Note that our co-identification model could accommodate different evaluation and objective functions. Therefore, different co-clustering approaches previously developed in the literature could be considered as special cases of our approaches. Besides the norms L_1, L_2, L_∞, our model could use any Bregman divergence functions [48] instead;

in [48] the authors chose the appropriate Bregman divergence based on the underlying data generation process or noise model. For classical clustering, Euclidean distance and Pearson correlation are both reasonable distance measures, with Euclidean distance being more appropriate for log ratio data, and Pearson correlation working better for absolute-valued (e.g. Affymetrix) data [37, 49, 50]. If there is additional known information for the dataset, the weight information can be easily added to the objective function of our model. Our co-identification model is very general and could be used for evaluating different proximity measures, as well as helping in developing new principled functions for gene expression data clustering analysis.

1.3 Reformulation of tensor models

1.3.1 Low-n-rank tensor optimization

The difficulty of the low-rank tensor problem lies in dealing of the rank function $\mathrm{rank}(\cdot)$. Not only the rank function itself is difficult to deal with, but also determining the rank of a specific given tensor is already a difficult task, which is NP-hard in general [51]. To give an impression of the difficulty involved in computing tensor ranks, note that there is a particular $9 \times 9 \times 9$ tensor (cf. [52]) whose rank is only known to be in between 18 and 23. One way to deal with the difficulty is to convert the tensor optimization problems (1.2) and (1.10) into matrix optimization problems. A typical matricization technique is the so-called mode-n matricization [30]. Roughly speaking, given a tensor $\mathcal{A} \in \mathbf{R}^{n_1 \times n_2 \times \cdots \times n_m}$, its mode-$n$ matricization denoted by $A(n)$ is to arrange nth index of $\mathcal{A}$ to be the row index of the resulting matrix and merge other indices of $\mathcal{A}$ as the column index of $A(n)$. The precise definition of the mode-n matricization is given below.

Definition 1.3 For a given tensor $\mathcal{A} \in \mathbf{R}^{n_1 \times n_2 \times \cdots \times n_m}$, the matrix $A(n)$ is the associated mode-n matricization. In particular

$$A(n)_{i_n, j} := \mathcal{A}_{i_1, i_2, \ldots, i_m}, \ \forall\, 1 \le i_k \le n_k,\ 1 \le k \le m,$$

where

$$j = 1 + \sum_{\substack{k=1 \\ k \neq n}}^{m} (i_k - 1) J_k, \ \text{with } J_k = \prod_{\substack{\ell=1 \\ \ell \neq n}}^{k} n_\ell. \tag{1.13}$$

The so-called *n-rank* of $\mathcal{A}$ is defined by the vector

$$[\mathrm{rank}(A(1)), \mathrm{rank}(A(2)), \ldots, \mathrm{rank}(A(m))],$$

that is its nth component corresponds to the column rank of mode-n matrix $A(n)$.

The notion of n-rank has been widely used in the problems of tensor decomposition. Recently, Liu *et al.* [53] and Gandy *et al.* [54] considered the low-n-rank tensor completion problem, which were the first attempts to solve low-rank tensor optimization problems. Convexifying the robust low-rank tensor recovery problem was also studied by Tomioka *et al.* [55] and Goldfarb and Qin [56]. Other works on this topic include

[57–61]. Specifically, Tomioka *et al.* [57] analyzed the statistical performance of the convex relaxation of the low-n-rank minimization model. Signoretto *et al.* [58] compared the performance of the convex relaxation of the low-n-rank minimization model and the low-rank matrix completion model on applications in spectral image reconstruction. Mu *et al.* [59] and Jiang *et al.* [23] proposed a matricization technique to convert the tensor problem into a matrix problem, but minimizing the Tucker rank of the tensor was considered rather than the CP rank. Kressner *et al.* [60] proposed a Riemannian manifold optimization algorithm for finding a local optimum of the Tucker rank constrained optimization problem. Krishnamurthy and Singh [61] studied some adaptive sampling algorithms for low-rank tensor completion.

The big advantage of n-rank is that this rank is easy to compute. Along with this line, we can replace the CP rank in (1.2) by the average of n-rank and get the following formulation:

$$\begin{array}{ll} \min & \frac{1}{m}\sum_{i=1}^{m} \operatorname{rank}(\bar{X}(i)) + \rho\|\mathcal{Y}\|_1 \\ \text{s.t.} & \bar{\mathcal{X}} \otimes e + \mathcal{Y} + \mathcal{Z} = \mathcal{A} \\ & \|\mathcal{Z}\|_F \leq \delta. \end{array} \tag{1.14}$$

There has been a large amount of work that deals with the low-rank matrix optimization problems. Research in this area was mainly ignited by the recent emergence of compressed sensing [62, 63], matrix rank minimization and low-rank matrix completion problems [9–11]. The matrix rank minimization seeks a matrix with the lowest rank satisfying some linear constraints, i.e.

$$\min_{X \in \mathbf{R}^{n_1 \times n_2}} \operatorname{rank}(X), \text{ s.t. } \mathcal{C}(X) = b, \tag{1.15}$$

where $b \in \mathbf{R}^p$ and $\mathcal{C} : \mathbf{R}^{n_1 \times n_2} \to \mathbf{R}^p$ is a linear operator. The works of [9–11] show that under certain randomness hypothesis of the linear operator $\mathcal{C}$, the NP-hard problem (1.15) is equivalent to the following nuclear norm minimization problem, which is a convex programming problem, with high probability:

$$\min_{X \in \mathbf{R}^{n_1 \times n_2}} \|X\|_*, \text{ s.t. } \mathcal{C}(X) = b. \tag{1.16}$$

In other words, the optimal solution to the convex problem (1.16) is also the optimal solution to the original NP-hard problem (1.15).

Now since the objective in (1.14) is to minimize the sum of ranks of matrices and the nuclear norm has proved to be a reliable convex relaxation, we can further replace the rank functions in (1.14) by nuclear norms, which results in a convex optimization problem:

$$\begin{array}{ll} \min & \frac{1}{m}\sum_{i=1}^{m} \|\bar{X}(i)\|_* + \rho\|\mathcal{Y}\|_1 \\ \text{s.t.} & \bar{\mathcal{X}} \otimes e + \mathcal{Y} + \mathcal{Z} = \mathcal{A} \\ & \|\mathcal{Z}\|_F \leq \delta. \end{array} \tag{1.17}$$

1.3.2 Equivalent formulation of tensor PCA

A main drawback of the above approach is that it is not clear how low-n-rank of $\mathcal{X}$ is related to its CP rank. That is to say, even though solving (1.14) (or its convex relaxation) will result in a low-n-rank solution $\bar{\mathcal{X}}^*$, its CP rank may be still high. To address this issue, let us introduce another way to unfold a tensor.

Definition 1.4 For a given super-symmetric even-order tensor $\mathcal{F} \in \mathbf{S}^{n^{2d}}$, we define its square matricization, denoted by $\boldsymbol{M}(\mathcal{F}) \in \mathbf{R}^{n^d \times n^d}$, as the following:

$$\boldsymbol{M}(\mathcal{F})_{k\ell} := \mathcal{F}_{i_1 \cdots i_d i_{d+1} \cdots i_{2d}}, \quad 1 \leq i_1, \ldots, i_d, i_{d+1}, \ldots, i_{2d} \leq n,$$

where

$$k = \sum_{j=1}^{d} (i_j - 1)n^{d-j} + 1, \text{ and } \ell = \sum_{j=d+1}^{2d} (i_j - 1)n^{2d-j} + 1.$$

Similarly we introduce below the vectorization of a tensor.

Definition 1.5 The vectorization, $\boldsymbol{V}(\mathcal{F})$, of tensor $\mathcal{F} \in \mathbf{R}^{n^m}$ is defined as

$$\boldsymbol{V}(\mathcal{F})_k := \mathcal{F}_{i_1 \cdots i_m},$$

where

$$k = \sum_{j=1}^{m} (i_j - 1)n^{m-j} + 1, 1 \leq i_1, \ldots, i_m \leq n.$$

In the same vein, we can convert a vector or a matrix with appropriate dimensions to a tensor. In other words, the inverse of the operators $\boldsymbol{M}$ and $\boldsymbol{V}$ can be defined in the same manner. In the following, we denote $X = \boldsymbol{M}(\mathcal{X})$, and so

$$\mathrm{Tr}\,(X) = \sum_{\ell} X_{\ell,\ell} \text{ with } \ell = \sum_{j=1}^{d} (i_j - 1)n^{d-j} + 1.$$

If we assume $\mathcal{X}$ to be of rank one, then

$$\mathrm{Tr}\,(X) = \sum_{i_1,\ldots,i_d} \mathcal{X}_{i_1 \cdots i_d i_1 \cdots i_d} = \sum_{i_1,\ldots,i_d} \mathcal{X}_{i_1^2 \cdots i_d^2}.$$

In the above expression, $(i_1, \ldots, i_d)$ is a subset of $(1, 2, \ldots, n)$. Suppose that j appears k_j times in $(i_1, \ldots, i_d)$ with $j = 1, 2, \ldots, n$ and $\sum_{j=1}^{n} k_j = d$. Then for a fixed outcome $(k_1, k_2, \ldots, k_n)$, the total number of permutations $(i_1, \ldots, i_d)$ to achieve such an outcome is

$$\binom{d}{k_1}\binom{d-k_1}{k_2}\binom{d-k_1-k_2}{k_3}\cdots\binom{d-k_1-\cdots-k_{n-1}}{k_n} = \frac{d!}{\prod_{j=1}^{n} k_j!}.$$

Consequently,

$$\mathrm{Tr}\,(X) = \sum_{i_1,\ldots,i_d} \mathcal{X}_{i_1^2 \cdots i_d^2} = \sum_{k \in \mathbb{K}(n,d)} \frac{d!}{\prod_{j=1}^{n} k_j!} \mathcal{X}_{1^{2k_1} 2^{2k_2} \cdots n^{2k_n}}, \tag{1.18}$$

which is exactly the function of linear constraint in the tensor PCA problem (1.10). Thus, if we further denote $F = \boldsymbol{M}(\mathcal{F})$, then the objective in (1.10) is $\mathcal{F} \bullet \mathcal{X} = \mathrm{Tr}\,(FX)$, while the first constraint

$$\sum_{k \in \mathbb{K}(n,d)} \frac{d!}{\prod_{j=1}^{n} k_j!} \mathcal{X}_{1^{2k_1} 2^{2k_2} \cdots n^{2k_n}} = 1 \iff \mathrm{Tr}\,(X) = 1.$$

The hard constraint in (1.10) is $\mathrm{rank}(\mathcal{X}) = 1$. It is straightforward to see that if $\mathcal{X}$ is of rank one, then by letting $\mathcal{X} = \lambda \underbrace{x \otimes \cdots \otimes x}_{2d}$ for some $\lambda \in \{1, -1\}$ and $\mathcal{Y} = \underbrace{x \otimes \cdots \otimes x}_{d}$, we have $\boldsymbol{M}(\mathcal{X}) = \lambda \boldsymbol{V}(\mathcal{Y})\boldsymbol{V}(\mathcal{Y})^\top$, which is to say that matrix $\boldsymbol{M}(\mathcal{X})$ is of rank one too. In the following, we shall continue to show that the other way around is also true.

To proceed, we present a useful result below, whose proof can be found in [23].

Proposition 1.6 *Suppose* $\mathcal{A} \in \mathbf{R}^{n^d}$ *is an n-dimensional dth-order tensor. The following two statements are equivalent:*

(i) $\mathcal{A} \in \mathbf{S}^{n^d}$, *and* $\mathrm{rank}(\mathcal{A}) = 1$;
(ii) $\mathcal{A} \otimes \mathcal{A} \in \mathbf{S}^{n^{2d}}$.

Now we are ready to present the main result of this subsection.

Theorem 1.7 *Suppose* $\mathcal{X} \in \mathbf{S}^{n^{2d}}$ *and* $X = \boldsymbol{M}(\mathcal{X}) \in \mathbf{R}^{n^d \times n^d}$. *Then we have*

$$\mathrm{rank}(\mathcal{X}) = 1 \iff \mathrm{rank}(X) = 1.$$

Proof As remarked earlier, that $\mathrm{rank}(\mathcal{X}) = 1 \implies \mathrm{rank}(X) = 1$ is evident. To see this, suppose $\mathrm{rank}(\mathcal{X}) = 1$ and $\mathcal{X} = \underbrace{x \otimes \cdots \otimes x}_{2d}$ for some $x \in \mathbf{R}^n$. By constructing $\mathcal{Y} = \underbrace{x \otimes \cdots \otimes x}_{d}$, we have $X = \boldsymbol{M}(\mathcal{X}) = \boldsymbol{V}(\mathcal{Y})\boldsymbol{V}(\mathcal{Y})^\top$, which leads to $\mathrm{rank}(X) = 1$.

To prove the other implication, suppose that we have $\mathcal{X} \in \mathbf{S}^{n^{2d}}$ and $\boldsymbol{M}(\mathcal{X})$ is of rank one, i.e. $\boldsymbol{M}(\mathcal{X}) = yy^\top$ for some vector $y \in \mathbf{R}^{n^d}$. Then $\mathcal{X} = \boldsymbol{V}^{-1}(y) \otimes \boldsymbol{V}^{-1}(y)$, which combined with Proposition 1.6 implies $\boldsymbol{V}^{-1}(y)$ is super-symmetric and of rank one. Thus there exists $x \in \mathbf{R}^n$ such that $\boldsymbol{V}^{-1}(y) = \underbrace{x \otimes \cdots \otimes x}_{d}$ and $\mathcal{X} = \underbrace{x \otimes \cdots \otimes x}_{2d}$. □

Nuclear norm penalty relaxation

According to Theorem 1.7, we know that a super-symmetric tensor is of rank one, if and only if its matrix correspondence obtained via the matricization operation defined in Definition 1.4, is also of rank one. As a result, we can reformulate problem (1.10) equivalently as the following matrix optimization problem:

$$\begin{array}{ll} \max & \mathrm{Tr}\,(FX) \\ \text{s.t.} & \mathrm{Tr}\,(X) = 1,\ \boldsymbol{M}^{-1}(X) \in \mathbf{S}^{n^{2d}}, \\ & X \in \mathbf{S}^{n^d \times n^d},\ \mathrm{rank}(X) = 1, \end{array} \tag{1.19}$$

where $X = \boldsymbol{M}(\mathcal{X})$, $F = \boldsymbol{M}(\mathcal{F})$, and $\mathbf{S}^{n^d \times n^d}$ denotes the set of $n^d \times n^d$ symmetric matrices. Note that the constraints $\boldsymbol{M}^{-1}(X) \in \mathbf{S}^{n^{2d}}$ requires the tensor correspondence of X to be super-symmetric, which essentially correspond to $O(n^{2d})$ linear equality constraints. The rank constraint $\text{rank}(X) = 1$ makes the problem intractable. In fact, problem (1.19) is NP-hard in general, due to its equivalence to problem (1.7).

Motivated by the convex nuclear norm relaxation, one way to deal with the rank constraint in (1.19) is to introduce the nuclear norm term of X, which penalizes high-ranked Xs, in the objective function. This yields the following convex optimization formulation:

$$\begin{array}{ll} \max & \text{Tr}\,(FX) - \rho\|X\|_* \\ \text{s.t.} & \text{Tr}\,(X) = 1, \ \boldsymbol{M}^{-1}(X) \in \mathbf{S}^{n^{2d}}, \\ & X \in \mathbf{S}^{n^d \times n^d}, \end{array} \tag{1.20}$$

where $\rho > 0$ is a penalty parameter. It is easy to see that if the optimal solution of (1.20) (denoted by $\tilde{X}$) is of rank one, then $\|\tilde{X}\|_* = \text{Tr}\,(\tilde{X}) = 1$, which is a constant. In this case, the term $-\rho\|X\|_*$ added to the objective function is a constant, which leads to the fact the solution is also optimal with the constraint that X is rank one. In fact, problem (1.20) is the convex relaxation of the following problem:

$$\begin{array}{ll} \max & \text{Tr}\,(FX) - \rho\|X\|_* \\ \text{s.t.} & \text{Tr}\,(X) = 1, \ \boldsymbol{M}^{-1}(X) \in \mathbf{S}^{n^{2d}}, \\ & X \in \mathbf{S}^{n^d \times n^d}, \ \text{rank}(X) = 1, \end{array}$$

which is equivalent to the original problem (1.19) since $\rho\|X\|_* = \rho\text{Tr}\,(X) = \rho$.

After solving the convex optimization problem (1.20) and obtaining the optimal solution $\tilde{X}$, if $\text{rank}(\tilde{X}) = 1$, we can find $\tilde{x}$ such that $\boldsymbol{M}^{-1}(\tilde{X}) = \underbrace{\tilde{x} \otimes \cdots \otimes \tilde{x}}_{2d}$, according to Theorem 1.7. In this case, $\tilde{x}$ is the optimal solution to problem (1.7). The original tensor PCA problem, or the Z-eigenvalue problem (1.7), is thus solved to optimality. If $\text{rank}(\tilde{X})$ is not equal to 1, then we have to use some heuristics to find a rank-one approximation. In fact, as we will see from Table 1.1, for randomly created instances, $\text{rank}(\tilde{X})$ is almost always equal to 1, which indicates that our convex relaxation (1.20) serves a very good approximation to the original NP-hard problem (1.7).

Semidefinite programming relaxation

Note that the constraint

$$X \in \mathbf{S}^{n^d \times n^d}, \text{rank}(X) = 1$$

in (1.19) actually implies that X is positive semidefinite. To get a tractable convex problem, we drop the rank constraint and impose a semidefinite constraint to (1.19) and consider the following SDP relaxation:

$$\begin{array}{ll} \max & \text{Tr}\,(FX) \\ \text{s.t.} & \text{Tr}\,(X) = 1, \\ & \boldsymbol{M}^{-1}(X) \in \mathbf{S}^{n^{2d}}, \ X \succeq 0. \end{array} \tag{1.21}$$

Remark that replacing the rank-one constraint by SDP constraint is by now a common and standard practice; see, e.g., [64–66]. The next theorem shows that the SDP relaxation (1.21) is actually closely related to the nuclear norm penalty problem (1.20).

Theorem 1.8 *Let X^*_{SDR} and $X^*_{PNP}(\rho)$ be the optimal solutions of problems (1.21) and (1.20) respectively. Suppose $Eig^+(X)$ and $Eig^-(X)$ are the summations of non-negative eigenvalues and negative eigenvalues of X respectively, i.e.*

$$Eig^+(X) := \sum_{i:\, \lambda_i(X)\geq 0} \lambda_i(X), \quad Eig^-(X) := \sum_{i:\, \lambda_i(X)<0} \lambda_i(X).$$

It holds that

$$2(\rho - \upsilon)\left|Eig^-(X^*_{PNP}(\rho))\right| \leq \upsilon - F_0,$$

where $F_0 := \max_{1\leq i\leq n} \mathcal{F}_{i^{2d}}$ and υ is the optimal value of the following optimization problem

$$\begin{array}{ll} \max & \mathrm{Tr}\,(FX) \\ s.t. & \|X\|_* = 1, \\ & X \in \mathbf{S}^{n^d \times n^d}. \end{array} \tag{1.22}$$

As a result, $\lim_{\rho\to+\infty} \mathrm{Tr}\,(FX^*_{PNP}(\rho)) = \mathrm{Tr}\,(FX^*_{SDR})$.

Theorem 1.8 (proof of which can be found in [23]) shows that when ρ goes to infinity in (1.20), the optimal solution of the nuclear norm penalty problem (1.20) converges to the optimal solution of the SDP relaxation (1.21).

1.4 Solution methods

1.4.1 Directly resorting to some existing solver

Since both problems (1.20) and (1.21) are convex, we could use CVX [67] to solve them. Interestingly, we found from our extensive numerical tests that the optimal solution to problem (1.20) or problem (1.21) is a rank-one matrix almost all the time. For instance, consider an example from [29].

Example 1.9 *We consider a super-symmetric tensor $\mathcal{F} \in \mathbf{S}^{3^4}$ defined by*

$$\begin{array}{llll}
\mathcal{F}_{1111} = 0.2883, & \mathcal{F}_{1112} = -0.0031, & \mathcal{F}_{1113} = 0.1973, & \mathcal{F}_{1122} = -0.2485, \\
\mathcal{F}_{1123} = -0.2939, & \mathcal{F}_{1133} = 0.3847, & \mathcal{F}_{1222} = 0.2972, & \mathcal{F}_{1223} = 0.1862, \\
\mathcal{F}_{1233} = 0.0919, & \mathcal{F}_{1333} = -0.3619, & \mathcal{F}_{2222} = 0.1241, & \mathcal{F}_{2223} = -0.3420, \\
\mathcal{F}_{2233} = 0.2127, & \mathcal{F}_{2333} = 0.2727, & \mathcal{F}_{3333} = -0.3054. &
\end{array}$$

The task is to compute the largest Z-eigenvalue of $\mathcal{F}$.

Applying CVX [67] to solve problem (1.20) with $F = \boldsymbol{M}(\mathcal{F})$ and $\rho = 10$ yields a rank-one solution $\tilde{X} = aa^\top \in \mathbf{R}^{3^2\times 3^2}$, where

$$a = (0.4451, 0.1649, -0.4688, 0.1649, 0.0611, -0.1737, -0.4688, -0.1737, 0.4938)^\top.$$

Thus we get the matrix correspondence of a by reshaping a into a square matrix A:

$$A = [a(1:3), a(4:6), a(7:9)] = \begin{bmatrix} 0.4451 & 0.1649 & -0.4688 \\ 0.1649 & 0.0611 & -0.1737 \\ -0.4688 & -0.1737 & 0.4938 \end{bmatrix}.$$

It is easy to check that A is a rank-one matrix with the nonzero eigenvalue being 1. This further confirms our theory on the rank-one equivalence, i.e. Theorem 1.7. The eigenvector that corresponds to the nonzero eigenvalue of A is given by

$$\tilde{x} = (-0.6671, -0.2472, 0.7027)^\top,$$

which is the optimal solution to problem (1.7).

The next example is from a real magnetic resonance imaging (MRI) application studied by Ghosh *et al.* in [2]. Ghosh *et al.* studied a fiber detection problem in diffusion magnetic resonance imaging (MRI), where they tried to extract the geometric characteristics from an antipodally symmetric spherical function (ASSF), which can be described equivalently in the homogeneous polynomial basis constrained to the sphere. They showed that it is possible to extract the maxima and minima of an ASSF by computing the stationary points of a problem in the form of (1.7) with $d = 2$ and $n = 4$.

Example 1.10 *The objective function $\mathcal{F}(x, x, x, x)$ in this example is given by*

$$\begin{aligned} & 0.74694x_1^4 - 0.435103x_1^3x_2 + 0.454945x_1^2x_2^2 + 0.0657818x_1x_2^3 \\ + & \; x_2^4 + 0.37089x_1^3x_3 - 0.29883x_1^2x_2x_3 - 0.795157x_1x_2^2x_3 \\ + & \; 0.139751x_2^3x_3 + 1.24733x_1^2x_3^2 + 0.714359x_1x_2x_3^2 + 0.316264x_2^2x_3^2 \\ - & \; 0.397391x_1x_3^3 - 0.405544x_2x_3^3 + 0.794869x_3^4. \end{aligned}$$

Again, we used CVX to solve problem (1.20) with $F = \boldsymbol{M}(\mathcal{F})$ and $\rho = 10$, and a rank-one solution was found with $\tilde{X} = aa^\top$, with

$$a = (0.0001, 0.0116, 0.0004, 0.0116, 0.9984, 0.0382, 0.0004, 0.0382, 0.0015)^\top.$$

By reshaping vector a, we get the following expression of matrix A:

$$A = [a(1:3), a(4:6), a(7:9)] = \begin{bmatrix} 0.0001 & 0.0116 & 0.0004 \\ 0.0116 & 0.9984 & 0.0382 \\ 0.0004 & 0.0382 & 0.0015 \end{bmatrix}.$$

It is easy to check that A is a rank-one matrix with 1 being the nonzero eigenvalue. The eigenvector corresponding to the nonzero eigenvalue of A is given by

$$\tilde{x} = (0.0116, 0.9992, 0.0382)^\top,$$

which is also the optimal solution to the original problem (1.7).

We then conducted some numerical tests on randomly generated examples. We constructed fourth-order tensor $\mathcal{T}$ with its components drawn randomly from i.i.d. standard normal distribution. The super-symmetric tensor $\mathcal{F}$ in the tensor PCA problem was obtained by symmetrizing $\mathcal{T}$. All the numerical experiments in this chapter were conducted on an Intel Core i5-2520M 2.5 GHz computer with 4 GB of RAM, and all

Table 1.1 Frequency of nuclear norm penalty problem (1.20) having a rank-one solution

n	rank-one	CPU
3	100	0.21
4	100	0.56
5	100	1.31
6	100	6.16
7	100	47.84
8	100	166.61
9	100	703.82

Table 1.2 Frequency of SDP relaxation (1.21) having a rank-one solution

n	rank-one	CPU
3	100	0.14
4	100	0.25
5	100	0.55
6	100	1.16
7	100	2.37
8	100	4.82
9	100	8.89

the default settings of Matlab 2012b and CVX 1.22 were used for all the tests. We chose $d = 2$ and the dimension of $\mathcal{F}$ in the tensor PCA problem from $n = 3$ to $n = 9$. We chose $\rho = 10$. For each n, we tested 100 random instances. In Table 1.1, we report the number of instances that produced rank-one solutions. We also report the average CPU time (in seconds) using CVX to solve the problems.

Table 1.1 shows that, for these randomly created tensor PCA problems, the nuclear norm penalty problem (1.20) *always* gives a rank-one solution, and thus *always* solves the original problem (1.7) to optimality.

Encouraged by these results, it is expected that the SDP relaxation (1.21) will also be likely to give rank-one solutions. In fact, this is indeed the case as shown through the numerical results in Table 1.2. Like in Table 1.1, we tested 100 random instances for each n. In Table 1.2, we report the number of instances that produced rank-one solutions for $d = 2$. We also report the average CPU time (in seconds) using CVX to solve the problems. As we see from Table 1.2, for these randomly created tensor PCA problems, the SDP relaxation (1.21) *always* gives a rank-one solution, and thus *always* solves the original problem (1.7) to optimality.

1.4.2 First-order methods

The computational times reported in Tables 1.1 and 1.2 suggest that it can be time consuming to solve the convex problems (1.20) and (1.21) when the problem size is large (especially for the nuclear norm penalty problem (1.20)). In this subsection, we propose a first-order method called alternating direction method of multipliers (ADMM) for solving (1.20) and (1.21) that fully takes advantage of the structures of the problems. ADMM is closely related to some operator-splitting methods, known as Douglas–Rachford and Peaceman–Rachford methods, that were proposed in 1950s for solving variational problems arising from PDEs (see [68, 69]). These operator-splitting methods were extensively studied later in the literature for finding the zeros of the sum of monotone operators and for solving convex optimization problems (see [70–74]). The ADMM we will study in this section was shown to be equivalent to the Douglas–Rachford operator-splitting method applied to convex optimization problem (see [75]). ADMM was revisited recently as it was found to be very efficient for many sparse and low-rank optimization problems arising from the recent emergence of compressed sensing [76], compressive imaging [77, 78], robust PCA [79], sparse inverse covariance selection [80, 81], sparse PCA [82], and SDP [83], etc. For a more complete discussion and list of references on ADMM, we refer to the recent survey paper by Boyd *et al.* [84] and the references therein.

Generally speaking, ADMM solves the following convex optimization problem,

$$\begin{array}{ll} \min_{x\in\mathbf{R}^n, y\in\mathbf{R}^p} & f(x)+g(y) \\ \text{s,t.} & Ax+By=b \\ & x\in\mathcal{C},\ y\in\mathcal{D}, \end{array} \tag{1.23}$$

where f and g are convex functions, $A\in\mathbf{R}^{m\times n}$, $B\in\mathbf{R}^{m\times p}$, $b\in\mathbf{R}^m$, $\mathcal{C}$ and $\mathcal{D}$ are some simple convex sets. A typical iteration of ADMM for solving (1.23) can be described as follows:

$$\left\{\begin{array}{lll} x^{k+1} & := & \operatorname{argmin}_{x\in\mathcal{C}}\ \mathcal{L}_\mu(x,y^k;\lambda^k) \\ y^{k+1} & := & \operatorname{argmin}_{y\in\mathcal{D}}\ \mathcal{L}_\mu(x^{k+1},y;\lambda^k) \\ \lambda^{k+1} & := & \lambda^k-(Ax^{k+1}+By^{k+1}-b)/\mu, \end{array}\right. \tag{1.24}$$

where the augmented Lagrangian function $\mathcal{L}_\mu(x,y;\lambda)$ is defined as

$$\mathcal{L}_\mu(x,y;\lambda):=f(x)+g(y)-\langle\lambda,Ax+By-b\rangle+\frac{1}{2\mu}\|Ax+By-b\|^2,$$

λ is the Lagrange multiplier and $\mu>0$ is a penalty parameter. The following theorem gives the global convergence of (1.24) for solving (1.23), and this has been well studied in the literature (see, e.g., [71, 73]).

Theorem 1.11 *Assume both A and B are of full column rank, the sequence* $\{(x^k,y^k,\lambda^k)\}$ *generated by* (1.24) *globally converges to a pair of primal and dual optimal solutions* (x^*,y^*) *and* λ^* *of* (1.23) *from any starting point.*

Because both the nuclear norm penalty problem (1.20) and SDP relaxation (1.21) can be rewritten as the form of (1.23), we can apply ADMM to solve them.

ADMM for nuclear norm penalty problem (1.20)

Note that the nuclear norm penalty problem (1.20) can be rewritten equivalently (by introducing a new variable Y) as

$$\begin{array}{ll} \min & -\mathrm{Tr}(FY) + \rho\|Y\|_* \\ \text{s.t.} & X - Y = 0, \\ & X \in \mathcal{C}, \end{array} \tag{1.25}$$

where $\mathcal{C} := \{X \in \mathbf{S}^{n^d \times n^d} \mid \mathrm{Tr}(X) = 1,\ \boldsymbol{M}^{-1}(X) \in \mathbf{S}^{n^{2d}}\}$. Here we introduce the new variable Y and impose the constraint $X - Y = 0$ to split the two difficult parts in (1.20), i.e. the nuclear norm and the constraint set $\mathcal{C}$. It is easy to see that, in the reformulation (1.25), these two difficult parts are associated with different variables, and this allows us to apply ADMM to solve it efficiently. A typical iteration of ADMM for solving (1.25) can be described as

$$\left\{ \begin{array}{lll} X^{k+1} & := & \mathrm{argmin}_{X\in\mathcal{C}} -\mathrm{Tr}(FY^k) + \rho\|Y^k\|_* - \langle\Lambda^k, X - Y^k\rangle + \frac{1}{2\mu}\|X - Y^k\|_F^2 \\ Y^{k+1} & := & \mathrm{argmin}\ -\mathrm{Tr}(FY) + \rho\|Y\|_* - \langle\Lambda^k, X^{k+1} - Y\rangle + \frac{1}{2\mu}\|X^{k+1} - Y\|_F^2 \\ \Lambda^{k+1} & := & \Lambda^k - (X^{k+1} - Y^{k+1})/\mu, \end{array} \right. \tag{1.26}$$

where Λ is the Lagrange multiplier associated with the equality constraint in (1.25) and $\mu > 0$ is a penalty parameter. Following Theorem 1.11, we know that the sequence $\{(X^k, Y^k, \Lambda^k)\}$ generated by (1.26) globally converges to a pair of primal and dual optimal solutions (X^*, Y^*) and Λ^* of (1.25) from any starting point.

Next we show that the two subproblems in (1.26) are both easy to solve. The first subproblem in (1.26) can be equivalently written as

$$X^{k+1} := \operatorname*{argmin}_{X\in\mathcal{C}} \frac{1}{2}\|X - (Y^k + \mu\Lambda^k)\|_F^2, \tag{1.27}$$

i.e. the solution of the first subproblem in (1.26) corresponds to the projection of $Y^k + \mu\Lambda^k$ onto convex set $\mathcal{C}$. We will elaborate on how to compute this projection shortly.

The second subproblem in (1.26) can be reduced to:

$$Y^{k+1} := \operatorname*{argmin}_{Y} \quad \mu\rho\|Y\|_* + \frac{1}{2}\|Y - (X^{k+1} - \mu(\Lambda^k - F))\|_F^2. \tag{1.28}$$

This problem is known to have a closed-form solution that is given by the following so-called matrix shrinkage operation (see, e.g., [85]):

$$Y^{k+1} := U\mathrm{Diag}(\max\{\sigma - \mu\rho, 0\})V^\top,$$

where $U\mathrm{Diag}(\sigma)V^\top$ is the singular value decomposition of $X^{k+1} - \mu(\Lambda^k - F)$.

The projection

Now we show how to compute the projection in (1.27), i.e. for given Z, how to solve the following optimization problem:

$$\begin{array}{ll} \min & \|X - Z\|_F^2 \\ \text{s.t.} & \mathrm{Tr}\,(X) = 1, \\ & \boldsymbol{M}^{-1}(X) \in \mathbf{S}^{n^{2d}}. \end{array} \tag{1.29}$$

For the sake of discussion, in the following we consider the equivalent tensor representation of (1.29):

$$\begin{array}{ll} \min & \|\mathcal{X} - \mathcal{Z}\|_F^2 \\ \text{s.t.} & \sum\limits_{k \in \mathbb{K}(n,d)} \frac{d!}{\prod_{j=1}^n k_j!} \mathcal{X}_{1^{2k_1} 2^{2k_2} \ldots n^{2k_n}} = 1, \\ & \mathcal{X} \in \mathbf{S}^{n^{2d}}, \end{array} \tag{1.30}$$

where $\mathcal{X} = \boldsymbol{M}^{-1}(X)$, $\mathcal{Z} = \boldsymbol{M}^{-1}(Z)$, and the equality constraint is due to (1.18). Now we denote index set

$$\mathbf{I} = \left\{(i_1 \cdots i_{2d}) \in \pi(1^{2k_1} \cdots n^{2k_n}) \;\middle|\; k = (k_1, \ldots, k_n) \in \mathbb{K}(n, d)\right\}.$$

Then the first-order optimality conditions for problem (1.30) imply

$$\begin{cases} 2\left(|\pi(i_1 \cdots i_{2d})|\, \mathcal{X}_{i_1 \cdots i_{2d}} - \sum\limits_{\substack{j_1 \cdots j_{2d} \in \\ \pi(i_1 \cdots i_{2d})}} \mathcal{Z}_{j_1 \cdots j_{2d}}\right) = 0, & \text{if } (i_1 \cdots i_{2d}) \notin \mathbf{I}, \\ 2\left(\frac{(2d)!}{\prod_{j=1}^n (2k_j)!} \mathcal{X}_{1^{2k_1} \ldots n^{2k_n}} - \sum\limits_{\substack{j_1 \cdots j_{2d} \in \\ \pi(1^{2k_1} \ldots n^{2k_n})}} \mathcal{Z}_{j_1 \cdots j_{2d}}\right) - \lambda \frac{(d)!}{\prod_{j=1}^n (k_j)!} = 0, & \text{otherwise.} \end{cases}$$

Denote $\hat{\mathcal{Z}}$ to be the super-symmetric counterpart of tensor $\mathcal{Z}$, i.e.

$$\hat{\mathcal{Z}}_{i_1 \cdots i_{2d}} = \sum_{j_1 \cdots j_{2d} \in \pi(i_1 \cdots i_{2d})} \frac{\mathcal{Z}_{j_1 \cdots j_{2d}}}{|\pi(i_1 \cdots i_{2d})|}$$

and $\alpha(k, d) := \left(\frac{(d)!}{\prod_{j=1}^n (k_j)!}\right) / \left(\frac{(2d)!}{\prod_{j=1}^n (2k_j)!}\right)$. Then due to the first-order optimality conditions of (1.30), the optimal solution $\mathcal{X}^*$ of problem (1.30) satisfies

$$\begin{cases} \mathcal{X}^*_{i_1 \cdots i_{2d}} = \hat{\mathcal{Z}}_{i_1 \cdots i_{2d}}, & \text{if } (i_1 \cdots i_{2d}) \notin \mathbf{I}, \\ \mathcal{X}^*_{1^{2k_1} \ldots n^{2k_n}} = \frac{\lambda}{2} \alpha(k, d) + \hat{\mathcal{Z}}_{1^{2k_1} \ldots n^{2k_n}}, & \text{otherwise}. \end{cases} \tag{1.31}$$

Multiplying the second equality of (1.31) by $\frac{(d)!}{\prod_{j=1}^n (k_j)!}$ and summing the resulting equality over all $k = (k_1, \ldots, k_n)$ yield

$$\sum_{k \in \mathbb{K}(n,d)} \frac{(d)!}{\prod_{j=1}^n (k_j)!} \mathcal{X}^*_{1^{2k_1} \ldots n^{2k_n}}$$
$$= \frac{\lambda}{2} \sum_{k \in \mathbb{K}(n,d)} \frac{(d)!}{\prod_{j=1}^n (k_j)!} \alpha(k, d) + \sum_{k \in \mathbb{K}(n,d)} \frac{(d)!}{\prod_{j=1}^n (k_j)!} \hat{\mathcal{Z}}_{1^{2k_1} \ldots n^{2k_n}}.$$

It remains to determine λ. Noticing that $\mathcal{X}^*$ is a feasible solution for problem (1.30), we have $\sum_{k\in\mathbb{K}(n,d)} \frac{(d)!}{\prod_{j=1}^n (k_j)!} \mathcal{X}^*_{1^{2k_1}\dots n^{2k_n}} = 1$. As a result,

$$\lambda = 2\left(1 - \sum_{k\in\mathbb{K}(n,d)} \frac{(d)!}{\prod_{j=1}^n (k_j)!} \hat{\mathcal{Z}}_{1^{2k_1}\dots n^{2k_n}}\right) \Big/ \sum_{k\in\mathbb{K}(n,d)} \frac{(d)!}{\prod_{j=1}^n (k_j)!} \alpha(k,d),$$

and thus we derived $\mathcal{X}^*$ and $X^* = \boldsymbol{M}(\mathcal{X}^*)$ as the desired optimal solution for (1.29).

ADMM for SDP relaxation (1.21)

Note that the SDP relaxation problem (1.21) can be formulated as

$$\begin{array}{ll} \min & -\mathrm{Tr}\,(FY) \\ \text{s.t.} & \mathrm{Tr}\,(X) = 1, \quad \boldsymbol{M}^{-1}(X) \in \mathbf{S}^{n^{2d}} \\ & X - Y = 0, \quad Y \succeq 0. \end{array} \tag{1.32}$$

A typical iteration of ADMM for solving (1.32) is

$$\begin{cases} X^{k+1} & := \ \mathrm{argmin}_{X\in\mathcal{C}} -\mathrm{Tr}\,(FY^k) - \langle \Lambda^k, X - Y^k\rangle + \frac{1}{2\mu}\|X - Y^k\|_F^2 \\ Y^{k+1} & := \ \mathrm{argmin}_{Y\succeq 0} -\mathrm{Tr}\,(FY) - \langle \Lambda^k, X^{k+1} - Y\rangle + \frac{1}{2\mu}\|X^{k+1} - Y\|_F^2 \\ \Lambda^{k+1} & := \ \Lambda^k - (X^{k+1} - Y^{k+1})/\mu, \end{cases} \tag{1.33}$$

where $\mu > 0$ is a penalty parameter. Following Theorem 1.11, we know that the sequence $\{(X^k, Y^k, \Lambda^k)\}$ generated by (1.33) globally converges to a pair of primal and dual optimal solutions (X^*, Y^*) and Λ^* of (1.32) from any starting point.

It is easy to check that the two subproblems in (1.33) are both relatively easy to solve. Specifically, the solution of the first subproblem in (1.33) corresponds to the projection of $Y^k + \mu\Lambda^k$ onto $\mathcal{C}$. The solution of the second problem in (1.33) corresponds to the projection of $X^{k+1} + \mu F - \mu\Lambda^k$ onto the positive semidefinite cone $Y \succeq 0$, i.e.

$$Y^{k+1} := U\mathrm{Diag}\,(\max\{\sigma, 0\})U^\top,$$

where $U\mathrm{Diag}\,(\sigma)U^\top$ is the eigenvalue decomposition of matrix $X^{k+1} + \mu F - \mu\Lambda^k$.

1.4.3 The block optimization technique

In this subsection, we shall discuss a solution method tailored for a generic multi-block optimization model which has been frequently encountered in tensor-related applications:

$$\begin{array}{ll} \max & f(x^1, \dots, x^{i-1}, x^i, x^{i+1}, \dots, x^d) \\ \text{s.t.} & x^i \in \mathbb{G}_i \subseteq \mathbf{R}^{n_i}, \ i = 1, 2, \dots, d. \end{array} \tag{1.34}$$

Notice that both the tensor PCA problem (1.5) and the tensor co-clustering problem (1.11) can be treated as its special cases. A feature of the problem is that the variables $x^1, \dots, x^d$, each may be multi-dimensional and constrained, are only related through the objective function, but are independent in the constraints. This structure enables a well-known approach to the above problem, known as the *block coordinate descent method* (BCD) [86]. The nature of the method is to *cyclically* search along the blocks, say in the order of $x^1 \in \mathbb{G}_1$, $x^2 \in \mathbb{G}_2$, and so on, while fixing variables in the other blocks. More specifically, at iteration k, the blocks are updated by

$$x_k^i \in \arg\min_{x^i \in \mathbb{G}_i} f\left(x_k^1, \ldots, x_k^{i-1}, x^i, x_{k-1}^{i+1} \ldots, x_{k-1}^d\right).$$

We shall remark here that the well-known *alternating least square* (ALS) method in tensor decomposition is indeed the BCD method applied to the least squares problems. This method, though simple to implement, fails to converge to a stationary point (local optimum) if no additional condition is assumed. The existing analysis of BCD method requires the uniqueness of the minimizer for each subproblem [87, 88], or the quasi-convexity of f [89]. If the objective of (1.34) is strongly convex or a certain local error bound is satisfied around the solution set, then the BCD-type algorithm converges linearly [90–92]. When f is assumed to be only convex, some recent studies [93, 94] have established the $O(1/k)$ sublinear convergence.

When f is not convex and no extra condition is assumed, the issue of convergence was resolved in a very recent theoretical exploration [27] by a specific choice in the block improvement: one needs to attempt all the block improvements and then select the one that brings in the maximum improvement to actually implement the block variable updating. The enhanced search algorithm is termed the *maximum block improvement* (MBI) method. For reference the procedure is described in the following.

Algorithm MBI

Let $x_0^i \in \mathbb{G}_i$ for $i = 1, \ldots, d$.
for $k = 1, 2, \ldots,$ **do**
  for $i = 1, 2, \ldots, d$ **do**
    $y_k^i = \arg\min_{y^i \in \mathbb{G}_i} f\left(x_{k-1}^1, \ldots, x_{k-1}^{i-1}, y^i, x_{k-1}^{i+1}, \ldots, x_{k-1}^d\right);$
  end for
    Choose $i^* = \arg\min_{i=1,\ldots,d} f\left(x_{k-1}^1, \ldots, x_{k-1}^{i-1}, y_k^i, x_{k-1}^{i+1}, \ldots, x_{k-1}^d\right).$
    Set $x_k^{i^*} = y_k^{i^*}$ and $x_k^i = x_{k-1}^i$ for $i \neq i^*$.
end for

Compared to BCD, the MBI approach is more time consuming since it needs to check $d-1$ block updates in every iteration. However, the block updates can be parallelized to reduce the running time by a factor of $1/(d-1)$ approximately. Moreover, this method is tested to be highly effective empirically. Very recently the global and local linear convergence results of MBI have been established under certain conditions [95].

An interesting special case of multi-block optimization problem (1.34) is the tensor co-clustering problem (1.11), where $\mathcal{X}, Y^1, \ldots, Y^d$ are block variables. Moreover, the MBI scheme for this specific problem can be described as follows.

Generic Co-identification Algorithm

Input: $\mathcal{A} \in \mathbf{R}^{n_1 \times n_2 \times \cdots \times n_d}$ is a d-dimensional tensor, which holds the d-dimensional gene expression data set. Parameters $p_1, p_2, \ldots, p_d$, are all positive integers, $0 < p_i \leq n_i, 1 \leq i \leq d$.

Output: $p_1 \times p_2 \times \cdots \times p_d$ co-groups of $\mathcal{A}$.

Main Variables: A nonnegative integer k as the loop counter;
A $p_1 \times p_2 \times \cdots \times p_d$-tensor $\mathcal{X}$ with each entry as the artificial central point of each of the co-groups;
A $n_i \times p_i$-matrix Y_i as the assignment matrix with $\{0, 1\}$ as the value of each entry, $1 \leq i \leq d$.

Begin

0. *(Initialization).* $Y^0 = \mathcal{X}$. Choose a feasible solution $(Y_0^0, Y_0^1, Y_0^2, \ldots, Y_0^d)$ and compute the initial objective value $v_0 := f(Y_0^0, Y_0^1, Y_0^2, \ldots, Y_0^d)$. Set the loop counter $k := 0$.

1. *(Block Improvement).* For each $i = 0, 1, 2, \ldots, d$, solve

$$\begin{array}{lll} (G_i) & \max & f\left(Y_k^0, Y_k^1, \ldots, Y_k^{i-1}, Y^i, Y_k^{i+1}, \ldots, Y_k^d\right) \\ & \text{s.t.} & Y^i \in \mathbf{R}^{n_j \times p_j} \text{ is an assignment matrix,} \end{array}$$

and let

$$\begin{aligned} y_{k+1}^i &:= \arg\max f\left(Y_k^0, Y_k^1, \ldots, Y_k^{i-1}, Y^i, Y_k^{i+1}, \ldots, Y_k^d\right) \\ w_{k+1}^i &:= f\left(Y_k^0, Y_k^1, \ldots, Y_k^{i-1}, y_{k+1}^i, Y_k^{i+1}, \ldots, Y_k^d\right). \end{aligned}$$

2. *(Maximum Improvement).* Let $w_{k+1} := \max_{1 \leq i \leq d} w_{k+1}^i$ and $i^* = \arg\max_{1 \leq i \leq d} w_{k+1}^i$. Let

$$\begin{aligned} Y_{k+1}^i &:= Y_k^i, \ \forall\, i \in \{0, 1, 2, \ldots, d\} \backslash \{i^*\} \\ Y_{k+1}^{i^*} &:= y_{k+1}^{i^*} \\ v_{k+1} &:= w_{k+1}. \end{aligned}$$

3. *(Stopping Criterion).* If $|v_{k+1} - v_k| > \epsilon$, set $k := k + 1$, and go to Step 1; otherwise, set $V_{p_i} = v_{k+1}$.

End

1.5 Applications

High-throughput gene expression techniques are producing large-scale high-dimensional (e.g. 4D with genes vs. timepoints vs. conditions vs. tissues) genome-wide gene expression data. This induces increasing demands for more efficient and powerful computational methods for partitioning the data into biologically relevant groups.

Current clustering and co-clustering approaches have limitations in gene expression data analysis, which may be very time consuming and work for only low-dimensional gene expression datasets. In this section we discuss the bioinformatics applications of the tensor models we have discussed. The computational results presented in the section demonstrate the efficiency and effectiveness of the tensor models and the solution methods for analyzing large, high-dimensional gene expression data of biological and biomedical significance [35, 36].

1.5.1 Computational results on gene expression data

We apply our **generic co-identification algorithm** whereby

$$f(\mathcal{X}, Y^1, \ldots, Y^d) = \|\mathcal{A} - \mathcal{X} \times_1 Y^1 \times_2 \cdots \times_d Y^d\|_F^2$$

to gene expression datasets. One 2D dataset is the gene expression of a yeast cell cycle dataset with 2884 genes and 17 conditions, where the expression values are in the range 0 to 595. Another 2D dataset is the gene expression of a human B-cell lymphoma dataset with 4026 genes and 96 conditions, where the values are in the range -749 and 642. The detailed information about the datasets could be found in Cheng and Church [41], Tavazoie *et al.* [39] and Alizadeh *et al.* [96]. The 3D dataset is the *Arabidopsis thaliana* abiotic stress gene expression from [97, 98]. We extract a file which has 2395 genes, five conditions (cold, salt, drought, wound, and heat), with each condition containing six time points. Some synthetic datasets are also used to validate and evaluate our methods.

Our co-clustering method is implemented using C++. The experimental testing is performed on a regular PC (configuration: processor: Pentium dual-core CPU, T4200 @ 2.00GHz; memory: 3GB; operating system: 64-bit windows 7; compiler: Microsoft Visual C++ 2010). The running-time testing is conducted on a server (PowerEdge 2950III, 32GB Mem). The figures are generated using MATLAB R2010a.

Evaluating the co-clustering method

We tested our co-clustering algorithm using different initial values of the three matrices X, Y_1 and Y_2. The setup of the initial values of the three matrices includes using random values for the three matrices, using subsets of values in A to initialize X, limiting the number of 1s to be one in each row of matrices Y_1 and Y_2, and using the values of the matrices Y_1 and Y_2 to calculate the values of the matrix X. We found out that the initial values of the three matrices will not significantly affect the convergence of our algorithm (refer to Figure 1.2 for the final objective function values and the running times over 50 runs for the yeast dataset to generate 30×3 co-clusters).

We also tested our algorithm for different numbers of partitions of the rows and the columns, that is, different values of k_1 and k_2. For example, when $k_1 = 30$ and $k_2 = 3$, our program generates the co-clusters of the yeast cell dataset in 40.252 seconds with the final objective function value -7386.75, and when $k_1 = 100$ and $k_2 = 5$, our program generates the co-clusters of the yeast cell dataset in 90.138 seconds with the final objective function value -6737.86. The running time of our algorithm is comparable to the running time of the algorithms developed in [99].

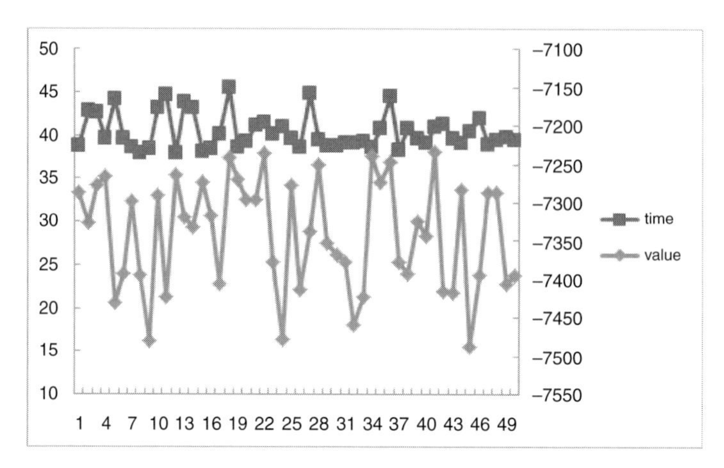

Figure 1.2 The final objective function values (the right axis) and the running time (the left axis, in seconds) of 50 runs of our algorithm with random initial values of the three matrices X, Y_1 and Y_2 on the yeast dataset to generate 30×3 co-clusters.

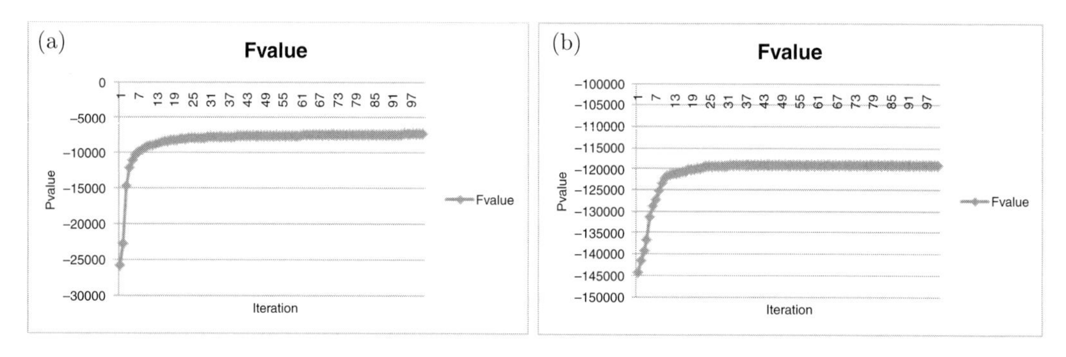

Figure 1.3 (a) The objective function value vs. iteration of our algorithm on the yeast dataset to generate 30×3 co-clusters. (b) The objective function value vs. iteration of our algorithm on the human dataset to generate 150×7 co-clusters.

Refer to Figure 1.3 for the objective function value versus iteration of our algorithm on the yeast cell dataset and the human lymphoma dataset. The average initial and final objective function values over 20 runs for the yeast dataset to generate 30×3 co-clusters are $-25\,818.1$ and -7323.42. The average initial and final objective function values over 20 runs for the human lymphoma dataset to generate 150×7 co-clusters are $-143\,958$ and $-119\,766$. There are 100 iterations of our implemented algorithm. We can see that our algorithm converges rapidly.

Determining the number of co-groups

To simplify the co-identification testing, we separate the determination of the number of co-groups from the co-identification analysis. We test the MBI approach for determining the number of co-groups. We first randomly generate some starting points, say the values of $(p_1, \ldots, p_d)$, and then we conduct a local improvement strategy, meaning that we try

Table 1.3 Testing of the maximum block improvement strategy on the 2D yeast dataset. The initial objective function value is −259 00; the first column: the initial p values; the second column: the new p values after the local search; the third column: the objective function values with the initial p values and with the new p values respectively; and the last column: the running time for the local search

Initial $p_{1,0}, p_{2,0}$	New p_1, p_2	Obj-value (Initial value −25 900)	Run time (seconds)
20,10	25,16	−7192.42, −6810.89	646.91
5,8	13,9	−9291.28, −7498.96	341.67
97,10	96,11	−6706.33, −6220.54	364.64
68,8	69,11	−6763.02, −6337.49	441.18
32,9	35,15	−6967.74, −6591.19	624.38
20,11	25,16	−7202.46, −6810.89	609.70
19,4	28,6	−7480.57, −7041.12	487.72
51,3	50,5	−7157.43, −6808.80	224.95
65,9	64,11	−6736.08, −6413.91	349.50
43,1	42,5	−7888.83, −6832.58	271.23
6,3	11,5	−8918.90, −7677.44	202.40
2,4	11,5	−15 973.50, −7677.44	280.21

to increase or decrease each p_i value until no more improvement is possible locally. We refer the reader to Table 1.3 for our testing on the effectiveness of the proposed local search strategy.

Efficiency analysis

Our approach is highly efficient and could be applied to 2D, 3D and higher-dimensional gene expression data. When testing on 3D datasets (genes vs. time points vs. conditions), the running time of our approach increases linearly with the number of genes, the number of time points, or the number of conditions. We conduct our running-time testing on the *Arabidopsis thaliana* abiotic stress 3D gene expression datasets from [98]. We use a file that has 2395 genes, five conditions (cold, salt, drought, wound, and heat), with each condition containing six time points. Especially when we keep the number of genes (2395) and increase the second dimension for the number of time points, or the third dimension for the number of the conditions, we increase the size of the dataset significantly, however, the running time of our algorithm still has only a linear increase (Figure 1.4). The performance of our algorithm is very robust. The testing results demonstrate the high efficiency of our algorithm. In contrast, other existing methods for 3D co-clustering such as *TriCluster* [45], the running time is exponential with the number of time points, or the number of conditions. Other existing methods usually do not work for gene expression data of four or higher dimensions.

Some identified exemplary co-clusters

Here we present some exemplary co-clusters identified by our co-clustering algorithm. We compare the co-clusters with those identified by other approaches. For all the figures

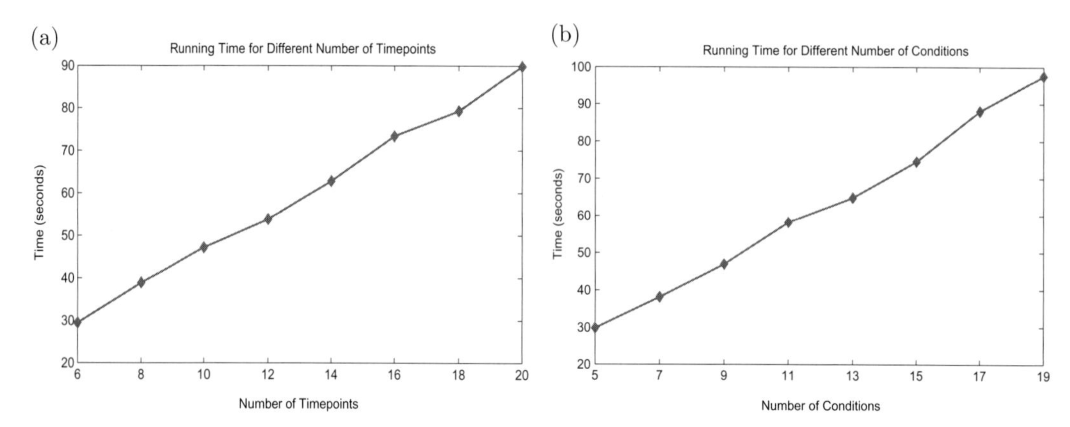

Figure 1.4 Evaluation of our approach on the 3D Arabidopsis dataset (2396 genes × 6 time points × 5 conditions). The parameters to control the number of co-groups for all these evaluations are set the same: $p_1 = 100$, $p_2 = 2$, $p_3 = 3$. For testing on different number of genes (this figure not shown due to space limit), the sizes of the eight groups are 300 × 6 × 5, 600 × 6 × 5, 900 × 6 × 5, 1200 × 6 × 5, 1500 × 6 × 5, 1800 × 6 × 5, 2100 × 6 × 5, 2400 × 6 × 5 (these datasets are truncated from the original dataset). (a) For testing on different number of time points, the sizes of the eight groups are 2396 × 6 × 5, 2396 × 8 × 5, 2396 × 10 × 5, 2396 × 12 × 5, 2396 × 14 × 5, 2396 × 16 × 5, 2396 × 18 × 5, 2396 × 20 × 5 (except for the first group, which is the original dataset, the other seven groups contain added repetitive time points). (b) For testing on different number of conditions, the sizes of the eight groups are 2396 × 6 × 5, 2396 × 6 × 7, 2396 × 6 × 9, 2396 × 6 × 11, 2396 × 6 × 13, 2396 × 6 × 15, 2396 × 6 × 17, 2396 × 6 × 19 (except for the first group, which is the original dataset, the other seven groups contain added repetitive conditions).

presented here, the x-axis represents the different number of conditions and the y-axis represents the values of the gene expression level.

We can see from the generated co-clusters that our algorithm can effectively identify groups of genes and groups of conditions that exhibit similar expression patterns. It can discover the same subset of genes that have different expression levels over different subsets of conditions, and can also discover different subsets of genes that have different expression levels over the same subset of conditions.

We use the mean squares residual score developed in [41] to evaluate the co-clusters generated by our algorithm. We identify 12 co-clusters with the best mean squares residual scores of the yeast cell dataset when $k_1 = 30$ and $k_2 = 3$. The list of the scores are 168.05, 182.04, 215.69, 335.72, 365.01, 378.37, 408.98, 410.03, 413.08, 416.63, 420.37, and 421.49. All 12 co-clusters have mean squares residual scores less than 450. They are meaningful co-clusters.

We test our algorithm using the 3D synthetic dataset from [98] which has six files with each file containing 1000 genes measured over 10 conditions with six time points for each condition. The co-clusters in Figure 1.5 show clear coherent patterns of the 3D dataset.

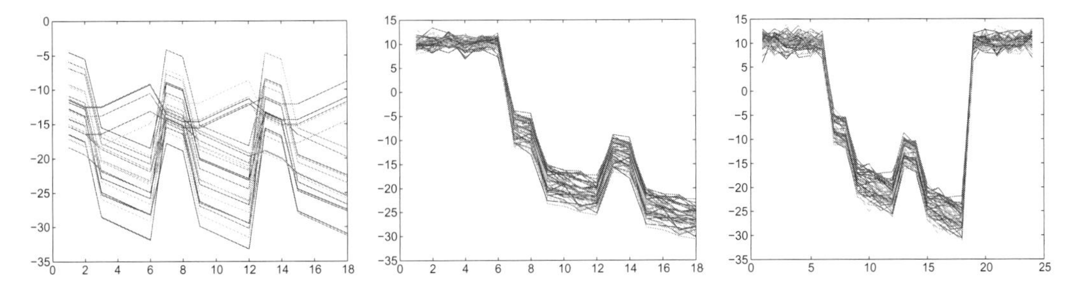

Figure 1.5 Co-clusters of the 3D dataset generated when the three parameters are $k_1 = 10$, $k_2 = 1$ and $k_3 = 3$. Each curve corresponds to the expression of one gene. The x-axis represents the different number of time points with every six time points in one condition, while the y-axis represents the values of the gene expression level.

Results from yeast gene expression data

We apply our co-identification approach to the analysis of the *Saccharomyces cerevisiae* gene expression data collected in [39]. These data contain the expression of 2884 genes under 17 conditions. From our co-identification analysis, we identify genes that are co-listed in two or more co-groups, such as YER068W, YDR103W, YGL130W, YJR129C, YLR425W, YOR383C, and YLL004W. This information could be used to predict the functions of unknown genes from the known functions of the genes in the same co-groups. This information could also lead to identification of previously undetected novel functions of genes. Specifically we have checked the function information (http://www.yeastgenome.org) of the following two genes, which are involved in more than one co-group.

Gene ORC3/YLL004W: subunit of the origin recognition complex, which directs DNA replication by binding to replication origins and is also involved in transcriptional silencing. We find out three pathways from KEGG Pathway Database (http://www.genome.jp/kegg/) in which this gene are involved.

Gene STE5/YDR103W: pheromone-response scaffold protein that controls the mating decision; binds Ste11p, Ste7p, and Fus3p kinases, forming a MAPK cascade complex that interacts with the plasma membrane and Ste4p-Ste18p; allosteric activator of Fus3p. Our approach identifies two co-groups in which Gene STE5/YDR103W is involved. The two co-groups are biologically significant with low p-values: co-group#41 (six genes: STE5 ADA2 AFG3 MOT2 PHO23 DSS4), zinc ion binding, with p-value 0.001735, co-group#62 (four genes: STE5 MOT2 ORC3 RAD52), pheromone response, mating-type determination, sex-specific proteins, with p-value 0.007773. (The p-value information is obtained from the website of Funcspec: http://funspec.med.utoronto.ca/.)

Clustering of lung cancer patients into groups with different survival outcomes

Lung cancer is the leading cause of cancer-related death worldwide (http://seer.cancer.gov/statfacts/html/lungb.html). Lung cancers are classified into two classes: small

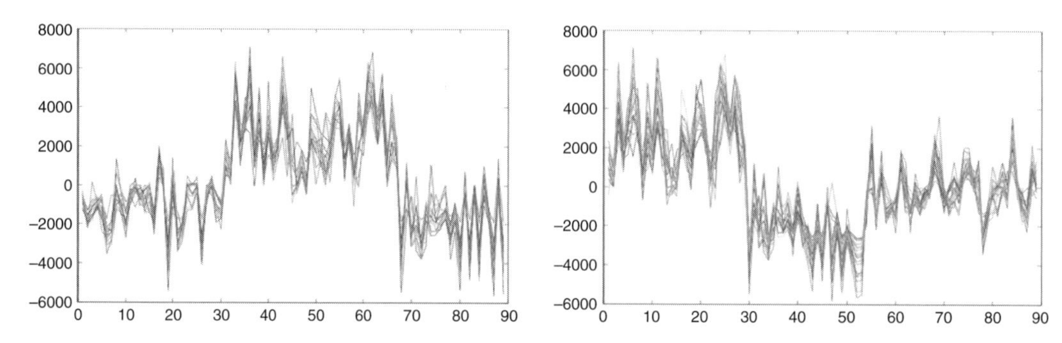

Figure 1.6 The gene expression of two groups of genes from MBI co-clustering approach.

cell lung cancer (SCLC) and non-small cell lung cancer (NSCLC). There are three main subtypes of NSCLC: squamous cell carcinoma (SQ), adenocarcinoma (ADCA), and large cell carcinoma (LC). Clinically, the non-small cell lung cancer (NSCLC) accounts for about the 85% of all lung cancers. Only 15% of all lung cancer patients are alive five years or more after diagnosis. One focus for the study of gene expression profiles of lung cancer patients is molecular typing and subtyping of lung tumors, and the other focus is identification of prognostic gene signatures using global gene expression profiling. Molecular diagnosis, prognosis, and classification of lung cancer types or subtypes are critical for the selection of appropriate therapies, to maximize efficacy and minimize toxicity [100]. The division of lung cancer into molecular subtypes with potentially different survival outcomes is purposeful in that each subtype has proposed treatment guidelines that include specific assays, targeted therapies, and clinical trials [101]. Global gene expression profiling study help to improve our understanding of the histological heterogeneity of non-small cell lung cancer and to identify novel potential biomarkers and gene signatures for classifying patients with significantly different survival outcomes [102]. We applied our MBI co-clustering method to the gene expression profiles for the Jacob ADCA data samples [102]. The following are the gene expression images of two selected groups of genes (Figure 1.6), which may be of particular interest as biomarkers.

1.6 Conclusions

The tensor data format has become an important source of *structured* big data in recent years, due to the exponential growth of its size in terms of the dimension. In this chapter we discussed several computational models related to tensors, including the tensor low-rank and sparse decomposition, the tensor principal component analysis, and the tensor co-clustering problem. These models have wide applications in scientific computation. To cope with the large-scale nature, we introduced various relaxation techniques, as well as the first-order optimization methods, tailored for solving the above mentioned tensor models. For tensor co-clustering, we introduced a block optimization scheme, and demonstrated its efficacy in the gene expression data analysis in bio-informatics. In conclusion, we believe that big data in tensor data analysis deserve special research attention.

References

[1] H. Wang and N. Ahuja, "Compact representation of multidimensional data using tensor rank-one decomposition," in *Proceedings of the Pattern Recognition, 17th International Conference on ICPR*, 2004.

[2] A. Ghosh, E. Tsigaridas, M. Descoteaux, P. Comon, B. Mourrain, and R. Deriche, "A polynomial based approach to extract the maxima of an antipodally symmetric spherical function and its application to extract fiber directions from the orientation distribution function in diffusion mri," in *Computational Diffusion MRI Workshop (CDMRI'08), New York*, 2008.

[3] L. Bloy and R. Verma, "On computing the underlying fiber directions from the diffusion orientation distribution function," in *Medical Image Computing and Computer-Assisted Intervention, MICCAI 2008*, D. Metaxas, L. Axel, G. Fichtinger and G. Szaäekeley, eds., 2008.

[4] L. Qi, G. Yu, and E. X. Wu, "Higher order positive semi-definite diffusion tensor imaging," *SIAM Journal on Imaging Sciences*, pp. 416–433, 2010.

[5] J. J. Hilling and A. Sudbery, "The geometric measure of multipartite entanglement and the singular values of a hypermatrix," *J. Math. Phys.*, vol. **51**, p. 072102, 2010.

[6] S. Hu and L. Qi, "Algebraic connectivity of an even uniform hypergraph," *Journal of Combinatorial Optimization*, vol. **24**, pp. 564–579, 2012.

[7] W. Li and M. Ng, "Existence and uniqueness of stationary probability vector of a transition probability tensor," Department of Mathematics, The Hong Kong Baptist University, Tech. Rep., 2011.

[8] M. Fazel, H. Hindi, and S. Boyd, "Rank minimization and applications in system theory." in *American Control Conference*, 2004, pp. 3273–3278.

[9] B. Recht, M. Fazel, and P. Parrilo, "Guaranteed minimum-rank solutions of linear matrix equations via nuclear norm minimization," *SIAM Review*, vol. **52**, no. 3, pp. 471–501, 2010.

[10] E. J. Candès and B. Recht, "Exact matrix completion via convex optimization," *Foundations of Computational Mathematics*, vol. **9**, pp. 717–772, 2009.

[11] E. J. Candès and T. Tao, "The power of convex relaxation: near-optimal matrix completion," *IEEE Trans. Inform. Theory*, vol. **56**, no. 5, pp. 2053–2080, 2009.

[12] E. J. Candès, X. Li, Y. Ma, and J. Wright, "Robust principal component analysis?" *Journal of ACM*, vol. **58**, no. 3, pp. 1–37, 2011.

[13] V. Chandrasekaran, S. Sanghavi, P. Parrilo, and A. Willsky, "Rank-sparsity incoherence for matrix decomposition," *SIAM Journal on Optimization*, vol. **21**, no. 2, pp. 572–596, 2011.

[14] A. d'Aspremont, L. E. Ghaoui, M. I. Jordan, and G. R. G. Lanckriet, "A direct formulation for sparse PCA using semidefinite programming," *SIAM Review*, vol. **49**, no. 3, pp. 434–448, 2007.

[15] S. Ma, D. Johnson, C. Ashby, *et al.*, "SPARCoC: a new framework for molecular pattern discovery and cancer gene identification," *PLoS ONE*, vol. **10**, no. 3, e0117135, 2015.

[16] X. Li, M. Ng, and X. Yuan, "Nuclear-norm-free variational models for background extraction from surveillance video," *Preprint*, 2013.

[17] F. L. Hitchcock, *The Expression of a Tensor or a Polyadic as a Sum of Products*. Institute of Technology, 1927.

[18] ——, "Multiple invariants and generalized rank of a p-way matrix or tensor," *Journal of Mathematical Physics*, vol. **7**, no. 1, pp. 39–79, 1927.

[19] J. D. Carroll and J. J. Chang, "Analysis of individual differences in multidimensional scaling via an n-way generalization of 'Eckart–Young' decomposition," *Psychometrika*, vol. **35**, no. 3, pp. 283–319, 1970.

[20] R. A. Harshman, *Foundations of the PARAFAC Procedure: Models and Conditions for an "Explanatory" Multimodal Factor Analysis*. Los Angeles: University of California at Los Angeles, 1970.

[21] L. Mackey, "Deflation methods for sparse PCA," in *Advances in Neural Information Processing Systems (NIPS)*, 2008.

[22] L. De Lathauwer, B. De Moor, and J. Vandewalle, "On the best rank-1 and rank-$(r_1, r_2, \ldots, r_n)$ approximation of higher-order tensors," *SIAM Journal on Matrix Analysis and Applications*, vol. **21**, no. 4, pp. 1324–1342, 2000.

[23] B. Jiang, S. Ma, and S. Zhang, "Tensor principal component analysis via convex optimization," *Mathematical Programming*, vol. **150**, pp. 423–457, 2015.

[24] L. Qi, "Eigenvalues of a real supersymmetric tensor," *Journal of Symbolic Computation*, vol. **40**, pp. 1302–1324, 2005.

[25] E. Kofidis and P. A. Regalia, "Tensor approximation and signal processing applications," in *Structured Matrices in Mathematics, Computer Science, and Engineering I*, V. Olshevsky, Ed., Contemporary Mathematics Series, American Mathematical Society, 2001.

[26] L. H. Lim, "Singular values and eigenvalues of tensors: a variational approach," in *Proceedings of the IEEE International Workshop on Computational Advances in Multi-Sensor Adaptive Processing (CAMSAP)*, 2005.

[27] B. Chen, S. He, Z. Li, and S. Zhang, "Maximum block improvement and polynomial optimization," *SIAM Journal on Optimization*, vol. **22**, pp. 87–107, 2012.

[28] L. Qi, F. Wang, and Y. Wang, "Z-eigenvalue methods for a global polynomial optimization problem," *Mathematical Programming, Series A*, vol. **118**, pp. 301–316, 2009.

[29] E. Kofidis and P. A. Regalia, "On the best rank-1 approximation of higher-order supersymmetric tensors," *SIAM Journal on Matrix Analysis and Applications*, vol. **23**, pp. 863–884, 2002.

[30] T. G. Kolda and B. W. Bader, "Tensor decompositions and applications," *SIAM Review*, vol. **51**, pp. 455–500, 2009.

[31] T. G. Kolda and J. R. Mayo, "Shifted power method for computing tensor eigenpairs," *SIAM J. Matrix Analysis*, vol. **32**, pp. 1095–1124, 2011.

[32] L. Suter, L. Babiss, and E. Wheeldon, "Toxicogenomics in predictive toxicology in drug development," *Chem. Biol.*, vol. **11**, pp. 161–171, 2004.

[33] Z. Magic, S. Radulovic, and M. Brankovic-Magic, "cdna microarrays: Identification of gene signatures and their application in clinical practice," *J. BUON*, vol. **12**, Suppl. 1, pp. S39–44, 2007.

[34] A. Cheung, "Molecular targets in gynaecological cancers," *Pathology*, vol. **39**, pp. 26–45, 2007.

[35] S. Zhang, K. Wang, B. Chen, and X. Huang, "A new framework for co-clustering of gene expression data," in *PRIB2011*, ser. Lecture Notes in Bio-Informatics, M. Loog *et al.*, Eds., Springer-Verlag, 2011, vol. **7036**, pp. 1–12.

[36] S. Zhang, K. Wang, C. Ashby, B. Chen, and X. Huang, "A unified adaptive co-identification framework for high-d expression data," in *PRIB2012*, ser. Lecture Notes in Bio-Informatics, Springer-Verlag, 2012, vol. **7632**, pp. 59–70.

[37] P. D'haeseleer, "How does gene expression clustering work?" *Nature Biotechnology*, vol. **23**, no. 12, pp. 1499–1502, 2005.

[38] M. Eisen, P. Spellman, P. Brown, and D. Botstein, "Cluster analysis and display of genome-wide expression patterns," *Proceedings of the National Academy of Sciences*, vol. **95**, no. 25, pp. 14 863–14 868, 1998.
[39] S. Tavazoie, J. Hughes, M. Campbell, R. Cho, and G. Church, "Systematic determination of genetic network architecture," *Nature Genetics*, vol. **22**, no. 3, pp. 281–285, 1999.
[40] P. Tamayo, D. Slonim, J. Mesirov, *et al.*, "Interpreting patterns of gene expression with self-organizing maps: methods and application to hematopoietic differentiation," *Proceedings of the National Academy of Sciences*, vol. **96**, no. 6, pp. 2907–2912, 1999.
[41] Y. Cheng and G. Church, "Biclustering of expression data," *Ismb*, vol. **8**, pp. 93–103, 2000.
[42] A. Prelić, S. Bleuler, P. Zimmermann, *et al.*, "A systematic comparison and evaluation of biclustering methods for gene expression data," *Bioinformatics*, vol. **22**, no. 9, pp. 1122–1129, 2006.
[43] M. Strauch, J. Supper, C. Spieth, *et al.*, "A two-step clustering for 3-d gene expression data reveals the main features of the arabidopsis stress response," *Journal of Integrative Bioinformatics*, vol. **4**, no. 1, p. 54, 2007.
[44] A. Li and D. Tuck, "An effective tri-clustering algorithm combining expression data with gene regulation information." *Gene Regulation and Systems Biology*, vol. **3**, pp. 49–64, 2008.
[45] L. Zhao and M. Zaki, "Tricluster: an effective algorithm for mining coherent clusters in 3d microarray data," in *Proceedings of the 2005 ACM SIGMOD International Conference on Management of Data*, ACM, 2005, pp. 694–705.
[46] D. Jiang, J. Pei, M. Ramanathan, C. Tang, and A. Zhang, "Mining coherent gene clusters from gene-sample-time microarray data," in *Proceedings of the Tenth ACM SIGKDD International Conference on Knowledge Discovery and Data Mining*, ACM, 2004, pp. 430–439.
[47] M. Deodhar, J. Ghosh, G. Gupta, H. Cho, and I. Dhillon, "Hunting for coherent co-clusters in high dimensional and noisy datasets," in *Data Mining Workshops, 2008. ICDMW'08. IEEE International Conference on*, IEEE, 2008, pp. 654–663.
[48] A. Banerjee, I. Dhillon, J. Ghosh, S. Merugu, and D. Modha, "A generalized maximum entropy approach to bregman co-clustering and matrix approximation," *Journal of Machine Learning Research*, vol. **8**, pp. 1919–1986, 2007.
[49] I. Costa, F. de Carvalho, and M. de Souto, "Comparative analysis of clustering methods for gene expression time course data," *Genetics and Molecular Biology*, vol. **27**, no. 4, pp. 623–631, 2004.
[50] F. Gibbons and F. Roth, "Judging the quality of gene expression-based clustering methods using gene annotation," *Genome research*, vol. **12**, no. 10, pp. 1574–1581, 2002.
[51] J. Håstad, "Tensor rank is NP-complete," *J. Algorithms*, vol. **11**, pp. 644–654, 1990.
[52] J. B. Kruskal, *Multiway Data Analysis*, North-Holland, Amsterdam, 1989, ch. Rank, Decomposition, and Uniqueness for 3-way and N-way Arrays, pp. 7–18.
[53] J. Liu, P. Musialski, P. Wonka, and J. Ye, "Tensor completion for estimating missing values in visual data," in *The Twelfth IEEE International Conference on Computer Vision*, 2009.
[54] S. Gandy, B. Recht, and I. Yamada, "Tensor completion and low-n-rank tensor recovery via convex optimization," *Inverse Problems*, vol. **27**, no. 2, p. 025010, 2011.
[55] R. Tomioka, K. Hayashi, and H. Kashima, "Estimation of low-rank tensors via convex optimization," preprint, 2011.
[56] D. Goldfarb and Z. Qin, "Robust low-rank tensor recovery: models and algorithms," preprint, 2013.

[57] R. Tomioka, T. Suzuki, and K. Hayashi, "Statistical performance of convex tensor decomposition," in *NIPS*, 2011.
[58] M. Signoretto, R. Van de Plas, B. De Moor, and J. Suykens, "Tensor versus matrix completion: a comparison with application to spectral data," *IEEE Signal Processing Letters*, vol. **18**, no. 7, pp. 403–406, 2011.
[59] C. Mu, B. Huang, J. Wright, and D. Goldfarb, "Square deal: lower bounds and improved relaxations for tensor recovery," preprint, 2013.
[60] D. Kressner, M. Steinlechner, and B. Vandereycken, "Low-rank tensor completion by Riemannian optimization," preprint, 2013.
[61] A. Krishnamurthy and A. Singh, "Low-rank matrix and tensor completion via adaptive sampling," preprint, 2013.
[62] E. J. Candès, J. Romberg, and T. Tao, "Robust uncertainty principles: Exact signal reconstruction from highly incomplete frequency information," *IEEE Transactions on Information Theory*, vol. **52**, pp. 489–509, 2006.
[63] D. Donoho, "Compressed sensing," *IEEE Transactions on Information Theory*, vol. **52**, pp. 1289–1306, 2006.
[64] F. Alizadeh, "Interior point methods in semidefinite programming with applications to combinatorial optimization," *SIAM Journal on Optimization*, vol. **5**, pp. 13–51, 1993.
[65] M. X. Goemans and D. P. Williamson, "Improved approximation algorithms for maximum cut and satisfiability problems using semidefinite programming," *J. Assoc. Comput. Mach.*, vol. **42**, no. 6, pp. 1115–1145, 1995.
[66] L. Vandenberghe and S. Boyd, "Semidefinite programming," *SIAM Rev.*, vol. **38**, no. 1, pp. 49–95, 1996.
[67] M. Grant and S. Boyd, "CVX: Matlab software for disciplined convex programming, version 1.21," http://cvxr.com/cvx, 2010.
[68] J. Douglas and H. H. Rachford, "On the numerical solution of the heat conduction problem in 2 and 3 space variables," *Transactions of the American Mathematical Society*, vol. **82**, pp. 421–439, 1956.
[69] D. H. Peaceman and H. H. Rachford, "The numerical solution of parabolic elliptic differential equations," *SIAM Journal on Applied Mathematics*, vol. **3**, pp. 28–41, 1955.
[70] P. L. Lions and B. Mercier, "Splitting algorithms for the sum of two nonlinear operators," *SIAM Journal on Numerical Analysis*, vol. **16**, pp. 964–979, 1979.
[71] M. Fortin and R. Glowinski, *Augmented Lagrangian Methods: Applications to the Numerical Solution of Boundary-Value Problems*, North-Holland Pub. Co., 1983.
[72] R. Glowinski and P. Le Tallec, *Augmented Lagrangian and Operator-Splitting Methods in Nonlinear Mechanics*, Philadelphia, Pennsylvania: SIAM, 1989.
[73] J. Eckstein, "Splitting methods for monotone operators with applications to parallel optimization," Ph.D. dissertation, Massachusetts Institute of Technology, 1989.
[74] J. Eckstein and D. P. Bertsekas, "On the Douglas–Rachford splitting method and the proximal point algorithm for maximal monotone operators," *Mathematical Programming*, vol. **55**, pp. 293–318, 1992.
[75] D. Gabay, "Applications of the method of multipliers to variational inequalities," in *Augmented Lagrangian Methods: Applications to the Solution of Boundary Value Problems*, M. Fortin and R. Glowinski, Eds., Amsterdam: North-Holland, 1983.
[76] J. Yang and Y. Zhang, "Alternating direction algorithms for ℓ_1 problems in compressive sensing," *SIAM Journal on Scientific Computing*, vol. **33**, no. 1, pp. 250–278, 2011.
[77] Y. Wang, J. Yang, W. Yin, and Y. Zhang, "A new alternating minimization algorithm for total variation image reconstruction," *SIAM Journal on Imaging Sciences*, vol. **1**, no. 3, pp. 248–272, 2008.

[78] T. Goldstein and S. Osher, "The split Bregman method for L1-regularized problems," *SIAM J. Imaging Sci.*, vol. **2**, pp. 323–343, 2009.
[79] M. Tao and X. Yuan, "Recovering low-rank and sparse components of matrices from incomplete and noisy observations," *SIAM J. Optim.*, vol. **21**, pp. 57–81, 2011.
[80] X. Yuan, "Alternating direction methods for sparse covariance selection," *Journal of Scientific Computing*, vol. **51**, pp. 261–273, 2012.
[81] K. Scheinberg, S. Ma, and D. Goldfarb, "Sparse inverse covariance selection via alternating linearization methods," in *NIPS*, 2010.
[82] S. Ma, "Alternating direction method of multipliers for sparse principal component analysis," *Journal of the Operations Research Society of China*, vol. **1**, no. 2, pp. 253–274, 2013.
[83] Z. Wen, D. Goldfarb, and W. Yin, "Alternating direction augmented Lagrangian methods for semidefinite programming," *Mathematical Programming Computation*, vol. **2**, pp. 203–230, 2010.
[84] S. Boyd, N. Parikh, E. Chu, B. Peleato, and J. Eckstein, "Distributed optimization and statistical learning via the alternating direction method of multipliers," *Foundations and Trends in Machine Learning*, vol. **3**, no. 1, pp. 1–122, 2011.
[85] S. Ma, D. Goldfarb, and L. Chen, "Fixed point and Bregman iterative methods for matrix rank minimization," *Mathematical Programming Series A*, vol. **128**, pp. 321–353, 2011.
[86] D. P. Bertsekas, *Nonlinear Programming*, 2nd Edn, Belmont, Massachusetts: Athena Scientific, 1999.
[87] P. Tseng, "Convergence of a block coordinate descent method for nondifferentiable minimization," *J. Optim. Theory Appl.*, vol. **109**, no. 3, pp. 475–494, 2001.
[88] D. P. Bertsekas and J. N. Tsitsiklis, *Parallel and Distributed Computation: Numerical Methods*, Prentice-Hall, Inc., Upper Saddle River, NJ, USA, 1989.
[89] L. Grippo and M. Sciandrone, "On the convergence of the block nonlinear Gauss–Seidel method under convex constraints," *Oper. Res. Lett.*, vol. **26**, no. 3, pp. 127–136, 2000.
[90] Z. Q. Luo and P. Tseng, "On the convergence of the coordinate descent method for convex differentiable minimization," *J. Optim. Theory Appl.*, vol. **72**, no. 1, pp. 7–35, 1992.
[91] Z.-Q. Luo and P. Tseng, "On the linear convergence of descent methods for convex essentially smooth minimization," *SIAM Journal on Control and Optimization*, vol. **30**, no. 2, pp. 408–425, 1992.
[92] ——, "Error bounds and convergence analysis of feasible descent methods: a general approach," *Annals of Operations Research*, vol. **46**, no. 1, pp. 157–178, 1993.
[93] A. Beck and L. Tetruashvili, "On the convergence of block coordinate descent type methods," *SIAM Journal on Optimization*, vol. **23**, no. 4, pp. 2037–2060, 2013.
[94] M. Hong, X. Wang, M. Razaviyayn, and Z.-Q. Luo, "Iteration complexity analysis of block coordinate descent methods," arXiv preprint arXiv:1310.6957, 2013.
[95] Z. Li, A. Uschmajew, and S. Zhang, "On convergence of the maximum block improvement method," to appear in *SIAM Journal on Optimization*, 2013.
[96] A. A. Alizadeh *et al.*, "Distinct types of diffuse large b-cell lymphoma identified by gene expression profiling," *Nature*, vol. **403**, no. 6769, pp. 503–511, 2000.
[97] J. Kilian, D. Whitehead, J. Horak, *et al.*, "The AtGenExpress global stress expression data set: protocols, evaluation and model data analysis of UV-b light, drought and cold stress responses," *The Plant Journal*, vol. **50**, no. 2, pp. 347–363, 2007.
[98] J. Supper, M. Strauch, D. Wanke, K. Harter, and A. Zell, "Edisa: extracting biclusters from multiple time-series of gene expression profiles," *BMC Bioinformatics*, vol. **8**, no. 1, pp. 334–347, 2007.

[99] H. Cho, I. S. Dhillon, Y. Guan, and S. Sra, "Minimum sum-squared residue co-clustering of gene expression data." in *Proceedings of The Fourth SIAM International Conference on Data Mining*, vol. **3**, SIAM, 2004, pp. 114–125.

[100] T. Li, H.-J. Kung, P. C. Mack, and D. R. Gandara, "Genotyping and genomic profiling of non–small-cell lung cancer: Implications for current and future therapies," *Journal of Clinical Oncology*, vol. **31**, no. 8, pp. 1039–1049, 2013.

[101] L. West, S. J. Vidwans, N. P. Campbell, *et al.*, "A novel classification of lung cancer into molecular subtypes," *PloS one*, vol. **7**, no. 2, p. e31906, 2012.

[102] K. Shedden *et al.*, "Gene expression–based survival prediction in lung adenocarcinoma: a multi-site, blinded validation study," *Nature Medicine*, vol. **14**, no. 8, pp. 822–827, 2008.

2 Sparsity-aware distributed learning

Symeon Chouvardas, Yannis Kopsinis, and Sergios Theodoridis

In this chapter, the problem of sparsity-aware distributed learning is studied. In particular, we consider the setup of an ad-hoc network, the nodes of which are tasked to estimate, in a collaborative way, a sparse parameter vector of interest. Both batch and online algorithms will be discussed. In the batch learning context, the distributed LASSO algorithm and a distributed greedy technique will be presented. Furthermore, an LMS-based sparsity promoting algorithm, revolving around the ℓ_1 norm, as well as a greedy distributed LMS will be discussed. Moreover, a set-theoretic sparsity promoting distributed technique will be examined. Finally, the performance of the presented algorithms will be validated in several scenarios.

2.1 Introduction

The volume of data captured worldwide is growing at an exponential rate posing certain challenges regarding their processing and analysis. Data mining, regression, and prediction/forecasting have played a leading role in learning insights and extracting useful information from raw data. The employment of such techniques covers a wide range of applications in several areas such as biomedical, econometrics, forecasting sales models, content preference, etc. The massive amount of data produced together with their increased complexity (new types of data emerge) as well as their involvement in the *Internet of Things* [1] paradigm call for further advances in already established machine learning techniques in order to cope with the new challenges.

Even though data tend to live in high-dimensional spaces, they often exhibit a high degree of redundancy; that is, their useful information can be represented by using a number of attributes much lower compared to their original dimensionality. Often, this redundancy can be effectively exploited by treating the data in a transformed domain, in which they can be represented by sparse models; that is, models comprising a few nonzero parameters. Besides, sparsity is an attribute that is met in a plethora of models, modeling natural signals, since nature tends to be parsimonious. Such sparse structures can be effectively exploited in big data applications in order to reduce processing demands. The advent of compressed sensing led to novel theoretical as well

Big Data over Networks, ed. Shuguang Cui, Alfred O. Hero III, Zhi-Quan Luo, and José M. F. Moura. Published by Cambridge University Press.

as algorithmic tools, which can be efficiently employed for *sparsity-aware learning*, e.g. [2–7].

In many cases, processing of large amount of data is not only cumbersome but might be proved to be infeasible due to lack of processing power and/or of storage capabilities. Naturally, one may think of splitting the full data analysis problem into subtasks in order to feed a number of distributed processing units. This philosophy has been followed in several methods in the framework of MapReduce/Hadoop [8, 9]. This involves the fusion of the subproblem outcomes in a central processor, which needs to communicate with all the subunits. An alternative path is to resort to fully distributed/decentralized solutions based on recent advances, which stem from the signal processing and machine learning communities [10–12]. The latter approach offers certain advantages. First, the existence of a fusion center is avoided and one solely relies on in-network processing in ad-hoc topologies. This leads to increased reliability and robustness of the network system, because it is not affected by possible fusion center failures. Overall, besides the computational and storage ease per processing unit, which can be attained by distributed processing, another important attribute is that of privacy. In particular, the processing of data is performed locally avoiding the need for sensitive information exchange [13].

There are mainly two major roadmaps for data processing. The first, and more common one, is referred to as *batch* processing. According to this, the full amount of data is considered as a whole and it is processed as a single entity, often iteratively, for optimizing the respective task. The alternative approach of data processing is the *online* one. According to this, the data are processed sequentially, one per iteration step, until all data have been considered or a convergence criterion has been met. Clearly, big data pose certain limitations in the batch processing framework. First, whenever a new datum is available, the whole process needs to be repeated from scratch. Second, the vast load of data may exceed the available memory and computational resources. Although batch processing can be facilitated via the use of distributed techniques, as was discussed above, in many cases, online processing seems to be the sensible choice for handling massive data volumes.

In online learning, data need not be stored and the estimate is updated each time a new observation/measurement becomes available. On top of that, online algorithms usually yield to reduced computational complexity per iteration. This is crucial in streaming scenarios, where data arrive rapidly and are either processed immediately or they are lost. Moreover, a notable characteristic of online methods is that they are capable of tracking possible variations, which take place in the environment. Finally, there is another important factor which concerns performance accuracy. It must be kept in mind that the ultimate goal in any machine learning task is to optimize the generalization performance of the respective prediction model. This is defined by the so-called *expected loss* and not by the *empirical loss*, as this is expressed in terms of a finite number of data points, as that involved in any batch processing technique. In big data applications, and given a fixed budget of computational resources, one can use many more data points if a simple online algorithm is employed, compared to more expensive batch processing algorithmic schemes. This offers the luxury to an online algorithm to achieve a better overall performance, see, e.g., [14, 15].

The goal of this chapter is to present, in a unified way, sparsity-aware distributed processing techniques focussing on the parameter estimation/regression tasks. Algorithms that belong to either of the two frameworks will be discussed. Typical examples of batch processing are the ℓ_1-norm regularized algorithms [12, 16], greedy algorithms, [17], and iterative thresholding-based methods, [18]. Some representative examples for the online mode of operation include the ℓ_1 regularized least mean squares (LMS) [19], set-theoretic sparsity promoting algorithms [20, 21], and a greedy online approach [17].

The chapter is organized as follows. The batch approach is discussed in Section 2.2, whereas the online one is presented in Section 2.3. In Section 2.2.2, the celebrated LASSO task in a distributed reformulation is discussed. In Section 2.2.3, the greedy pursuit rationale is derived under the decentralized framework and possible algorithmic solutions are given. A brief discussion of alternative distributed batch approaches is given in Section 2.2.4. The first adaptive algorithm, which is presented in this chapter (Section 2.3.2), is an ℓ_1-norm constrained LMS, which belongs to the so-called diffusion algorithmic family, and a greedy LMS based approach is given in Section 2.3.3. The set-theoretic philosophy in online learning is exposed in Section 2.3.4 adjusted to the distributed and sparse estimation requirements. Finally, Section 2.4 deals with the performance evaluation of the presented algorithmic schemes in several scenarios.

2.2 Batch distributed sparsity promoting algorithms

2.2.1 Problem formulation

Consider a collection of agents, which are capable of performing computations locally. Moreover, each agent has a number of neighbors; these are agents, which it is connected to and with which it can exchange information. All agents share a common goal; that of estimating an unknown parameter vector, $\boldsymbol{h}_* \in \mathbb{R}^m$, which is considered to be sparse, i.e. most of its coefficients are zeros. The network consisting of these agents can be modeled as an undirected graph $\mathcal{G}(\mathcal{N}, \mathcal{E})$, where $\mathcal{N} = \{1, \dots, K\}$ stands for the set of all the nodes and $\mathcal{E}$ is the set of pairs of agents, which are neighbors. Of special interest are the strongly connected networks, in which there exists at least one (possibly multihop) path connecting every two nodes of the network. Such a network is illustrated in Figure 2.1.

Each node, $k \in \mathcal{N}$, has access to a set of training/observation data points $(y_{k,i}, \boldsymbol{x}_{k,i})$, $y_{k,i} \in \mathbb{R}$, $\boldsymbol{x}_{k,i} \in \mathbb{R}^m$, $i = 1, \dots, N$, which are related via the following linear system:

$$\boldsymbol{y}_k = \boldsymbol{X}_k \boldsymbol{h}_* + \boldsymbol{\eta}_k, \ \forall k \in \mathcal{N}, \tag{2.1}$$

where $\boldsymbol{X}_k \in \mathbb{R}^{N \times m}$ is the sensing (input) matrix comprising the input vectors, $[\boldsymbol{x}_{k,1}^T \dots, \boldsymbol{x}_{k,N}^T]^T$, $\boldsymbol{y}_k = [y_1, \dots, y_N]^T \in \mathbb{R}^N$, and $\boldsymbol{\eta}_k \in \mathbb{R}^N$ is the vector of the noise process samples. As will be further discussed in the next section, the number of measurements, N, is allowed to be lower than m. The vector to be estimated is assumed to be at most s-sparse, i.e. $\|\boldsymbol{h}_*\|_0 \leq s \ll m$, where $\|\cdot\|_0$ denotes the ℓ_0 (pseudo) norm.

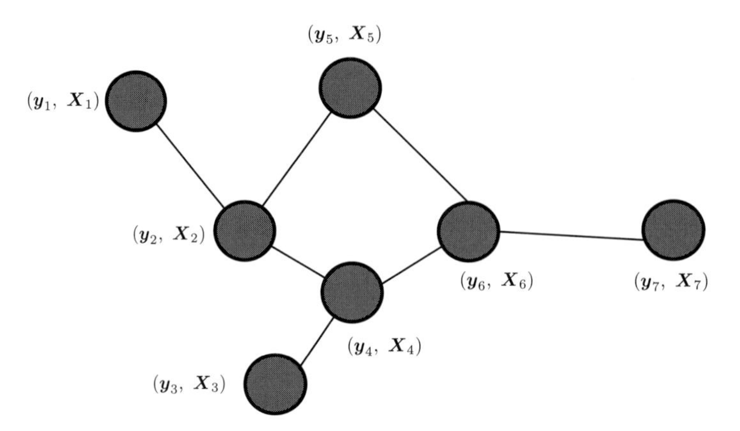

Figure 2.1 Illustration of an ad-hoc network.

As (2.1) suggests, it is assumed that all observations are generated via a single unknown parameter vector, $\boldsymbol{h}_*$, and the common task is its estimation. Furthermore, each agent has access to partial information, since it possess a subset of the whole training data set. In such a case, cooperation among the agents can be beneficial in terms of improved performance. Intuitively, each node benefits, via cooperating with its neighbours, which in turn have already benefited via their cooperation with their own neighbours. In this way, even though an agent does not have at its disposal the whole training data set, it becomes capable of estimating the global optimum, i.e. the solution that would be obtained if one had full access to it. At the heart of the decentralized processing lies the derivation of algorithms and cooperation protocols, which lead the distributed agents to achieve a global objective, by relying solely on local measurements and on in-network processing.

2.2.2 LASSO and its distributed learning formulation

In this section, an ℓ_1-norm regularized least squares (LS) algorithm, which operates in a distributed/decentralized fashion, will be presented. For the sake of clarifying the involved concepts, let us first concentrate on the non-distributed case and consider a single agent that has access to all data pairs. The goal is to exploit the fact that the unknown vector, $\boldsymbol{h}_*$, is sparse in order to estimate it by using fewer measurements than the dimensionality of the problem. It is well established that this can be effectively achieved by employing an ℓ_1-norm regularization term, e.g. [2, 7, 22]. Such a celebrated example is the least absolute shrinkage selection operator (LASSO), which solves the following optimization problem:

$$\widehat{\boldsymbol{h}} = \arg\min_{\boldsymbol{h}} \|\boldsymbol{y} - \boldsymbol{X}\boldsymbol{h}\|^2 + \lambda\|\boldsymbol{h}\|_1, \tag{2.2}$$

where $\|\cdot\|_1$ stands for the ℓ_1 norm and λ is a user-defined regularization parameter, which controls the trade-off between the sparsity imposing norm and the error counting

LS loss function. Note that since (2.2) refers to a single node, subscript k has been suppressed. Several techniques have been proposed for solving (2.2), e.g. [7, 23–25].

Let us now recast (2.2) for the multiagent scenario, involving data from all agents, i.e.,

$$\widehat{\boldsymbol{h}} = \arg\min_{\boldsymbol{h}} \sum_{k\in\mathcal{N}} \|\boldsymbol{y}_k - \boldsymbol{X}_k\boldsymbol{h}\|^2 + \lambda\|\boldsymbol{h}\|_1. \tag{2.3}$$

The optimization task (2.3) is referred to as distributed LASSO (DLasso) albeit all the measurement subsets $\boldsymbol{X}_k,\ k = 1, \dots, K$ need to be available on the same processing unit. In [12], (2.3) has been reformulated in order to fit the needs of decentralized processing, and it is stated as:

$$\begin{aligned} \widehat{\boldsymbol{h}}_k &= \arg\min_{\boldsymbol{h}_k} \|\boldsymbol{y}_k - \boldsymbol{X}_k\boldsymbol{h}_k\|^2 + \frac{2\lambda}{K}\|\boldsymbol{h}_k\|_1,\ \forall k \in \mathcal{N} \\ &\text{s.t. } \boldsymbol{h}_k = \boldsymbol{h}_l,\ \forall l \in \mathcal{N}_k, \forall k \in \mathcal{N}, \end{aligned} \tag{2.4}$$

where $\mathcal{N}_k$ is the neighborhood of node k, i.e. the nodes with which it can exchange information. A very interesting property, proved in [12], is that if the network is strongly connected, then the problems (2.3) and (2.4) are equivalent. In other words, each node computes the global solution, relying only on locally obtained information combined with that received within the respective neighborhood. A key point in (2.4) is the imposed constraint regarding the equality of the estimates of the unknown parameter vector, which are computed at each node. This gives rise to the so-called *consensus* cooperation protocol, which has been adopted in several distributed learning algorithms, e.g. distributed support vector machines [26], and distributed clustering [27], to name just a few.

The optimization problem (2.4), can be solved by employing the alternating direction method of multipliers (ADMM) [28]. The consensus cooperation protocol adopted and implemented via the ADMM can ensure that the nodes converge to the global solution. The solution via the ADMM involves the reformulation of the constraints and the introduction of local auxiliary vectors, denoted by $\boldsymbol{\gamma}_k^{(i)}$. Moreover, the constraints are treated via embedding Lagrange multipliers, $\boldsymbol{p}_k^{(i)},\ \boldsymbol{v}_k^{(i)}$, under the quadratically augmented Lagrangian framework. The steps of the DLasso algorithm, which stem from the ADMM solver, are summarized in Table 2.1; $\boldsymbol{h}_k^{(i)}$ stands for the estimate obtained at node k and at iteration step i, with $c > 0$ being a penalty parameter, which is associated to the consensus constraints. Moreover, $\mathcal{S}$ is a soft-thresholding operator, defined as

$$\begin{aligned} &\mathcal{S} : \mathbb{R}^m \times \mathbb{R} \mapsto \mathbb{R}^m \\ &[\mathcal{S}(\boldsymbol{h}, x)]_j := \operatorname{sign}([\boldsymbol{h}]_j)\max(0, |[\boldsymbol{h}]_j| - x), \end{aligned}$$

where $[\cdot]_j$ stands for the jth coefficient of the respective vector and $\operatorname{sign}(\cdot)$ is the sign function. Finally, $\boldsymbol{I}_m$ is the $m \times m$ identity matrix and $|\mathcal{N}_k|$ stands for the cardinality of the neighborhood set $\mathcal{N}_k$.

As it has been shown in [12], the DLasso algorithm converges, asymptotically, to the solution obtained by solving (2.3). This property could be exploited, potentially,

Table 2.1 The DLasso algorithm

Each node initializes to zero $\boldsymbol{h}_k^{(0)}$, $\boldsymbol{p}_k^{(0)}$, $\boldsymbol{\gamma}_k^{(0)}$, $\boldsymbol{v}_k^{(0)}$	
FOR $i = 1, \ldots,$ **DO**	
Node k transmits $\boldsymbol{h}_k^{(i-1)}$ to each node which belongs to $\mathcal{N}_k$	
$\boldsymbol{p}_k^{(i)} = \boldsymbol{p}_k^{(i-1)} + c\sum_{l\in\mathcal{N}_k}\left(\boldsymbol{h}_k^{(i-1)} - \boldsymbol{h}_l^{(i-1)}\right)$	$O(\lvert\mathcal{N}_k\rvert m)$
$\boldsymbol{v}_k^{(i)} = \boldsymbol{v}_k^{(i-1)} + c\left(\boldsymbol{h}_k^{(i-1)} - \boldsymbol{\gamma}_k^{(i-1)}\right)$	$O(m)$
$\boldsymbol{h}_k^{(i)} = \frac{1}{c(2\lvert\mathcal{N}_k\rvert + 1)}\mathcal{S}\Big(c\boldsymbol{\gamma}_k^{(i-1)} - \boldsymbol{p}_k^{(i-1)} - \boldsymbol{v}_k^{(i-1)} + \sum_{l\in\mathcal{N}_k}\left(\boldsymbol{h}_k^{(i-1)} - \boldsymbol{h}_l^{(i-1)}\right), \frac{\lambda}{K}\Big)$	$O(m)$
$\boldsymbol{\gamma}_k^{(i)} = \left(c\boldsymbol{I}_m + \boldsymbol{X}_k^T\boldsymbol{X}_k\right)^{-1}\left(\boldsymbol{X}_k^T\boldsymbol{y}_k + c\boldsymbol{h}_k^{(i)} + \boldsymbol{v}_k^{(i)}\right)$	$O(Nm)$
END	

in big data applications, since the original problem can be divided into a number of subproblems and consider each subproblem as a separate agent; then the task can be solved by adopting the previously discussed distributed optimization rationale.

Remark 2.1

Computational complexity and communication cost

The DLasso algorithm requires the inversion of an $m \times m$ matrix. Nevertheless, this computation can be done offline, since the matrix to be inverted, does not carry iteration-dependent information. Besides that, matrix vector multiplications take place, which require $O(Nm)$.

Regarding the communication cost, as can be seen from Table 2.1, each node transmits at each iteration m scalars, via the $m \times 1$ vector $\boldsymbol{h}_k^{(i-1)}$. However, this communication cost can be greatly reduced when the unknown vector, $\boldsymbol{h}_k^{(i-1)}$, is highly sparse. In this case, only the nonzero values and their positions need to be transmitted.

2.2.3 Sparsity-aware learning: the greedy point of view

In the previous section, LASSO was discussed, as a typical ℓ_1-norm based sparsity promoting path. An alternative to ℓ_1 regularization, for treating sparsity-inspired models, is the greedy viewpoint, [22, 29–34]. In the sequel, we focus on a particular family of greedy algorithms, which estimate the unknown parameter by applying iteratively the following two-step approach.

- **Support set selection** First, based on the input-output available observations and the tentative estimate of the unknown vector, a signal that acts as a proxy for the true one is computed. This proxy signal (vector) has the property to admit larger in amplitude values at indices pointing to the nonzero components of $\boldsymbol{h}_*$. In the sequel, the s largest in amplitude coefficients of the proxy are identified and they are selected in order to form the so-called *support set*.

- **Greedy update** An updated estimate of the unknown vector is computed, by performing a least squares fit restricted on the identified support set.

To keep the discussion simple, let us first assume that the signal proxy is computed in a non-cooperative way; that is, each node exploits solely its locally obtained observations. Following the rationale of the majority of greedy algorithms (see, e.g., [22, 29–34]) the signal proxy is chosen so that

$$X_k^T\left(y_k - X_k\widehat{h}_k\right) \approx X_k^T X_k\left(h_* - \widehat{h}_k\right). \tag{2.5}$$

In particular, the s ([31]) or the $2s$ ([32]) largest in amplitude coefficients are selected. However, in a distributed mode of operation these widely acclaimed options for proxy signals appear to be problematic. This is because the proxy signal (2.5) essentially estimates the residual, $(h_* - \widehat{h}_k)$. As is readily seen from (2.5), noting that when X_k is appropriately chosen, then $X_k^T X_k \approx I_m$. Accordingly, this proxy carries node-dependent information (via $\widehat{h}_k$) and it may pose problems to the support set consensus; that is, the nodes' "agreement" to the same support set. In distributed learning, it is only the vector h_* that commonly affects all the nodes (see also (2.1)). Thus, it is preferable to find a proxy which is close to h_* rather than to $(h_* - \widehat{h}_k)$. This is an important point for the analysis to follow, since such an option leads to the required consensus among the nodes.

Towards this direction in [17], a similar methodology to the one presented in [35], is adopted, which selects the s dominant coefficients of the following proxy:

$$\widehat{h}_k + X_k^T\left(y_k - X_k\widehat{h}_k\right) \approx h_*. \tag{2.6}$$

The intuition behind this proxy selection is that the left-hand side (lhs) of (2.6) is a gradient descent iteration of the cost function

$$\|y_k - X_k h_k\|^2, \tag{2.7}$$

e.g. [7]. It has been verified experimentally (for example [35] for the centralized case) that, for such a choice, few iterations are enough to recover the support.

In decentralized/distributed learning, the support selection and greedy update steps can be further *enriched* by exploiting the information that is available at the nodes of the network. More precisely, the signal proxies can be computed in a cooperative fashion and instead of exploiting only local quantities of interest, spatially received data will take part in the formation of the support set.

The core stages of the greedy-based distributed algorithm, proposed in [17], are given next.

- **Information enhancement for support set selection** Each node exchanges information, regarding its own observations, with its neighbours, in order to fuse them together and construct the signal proxy.
- **Cooperative estimation restricted on the support set** At each node, a least squares estimation is performed restricted on the identified support sets. The nodes in the neighborhood exchange their estimates and an information fusion under a certain protocol takes place.

Table 2.2 The distributed hard thresholding algorithm (DiHaT)

Algorithm description	Complexity
$\boldsymbol{h}_k^{(0)} = \boldsymbol{0}_m, \quad \overline{\boldsymbol{y}}_{k,0} = \boldsymbol{y}_k, \quad \overline{\boldsymbol{X}}_{k,0} = \boldsymbol{X}_k \quad \mathcal{S}_{k,0} = \emptyset$	{Initialization}
FOR $i = 1, \ldots,$ **DO**	
1: $\overline{\boldsymbol{y}}_{k,i} = \sum_{l \in \mathcal{N}_k} a_{l,k} \overline{\boldsymbol{y}}_{l,i-1}$	$O(\|\mathcal{N}_k\| N)$
2: $\overline{\boldsymbol{X}}_{k,i} = \sum_{l \in \mathcal{N}_k} a_{l,k} \overline{\boldsymbol{X}}_{l,i-1}$	$O(\|\mathcal{N}_k\| Nm)$
3: $\mathcal{S}_{k,i} = \text{supp}_s \left(\boldsymbol{h}_k^{(i-1)} + \overline{\boldsymbol{X}}_{k,i}^T (\overline{\boldsymbol{y}}_{k,i} - \overline{\boldsymbol{X}}_{k,i} \boldsymbol{h}_k^{(i-1)}) \right)$	$\mathcal{O}(Nm)$
4: $\widehat{\boldsymbol{h}}_k^{(i)} = \arg\min_{\boldsymbol{h}} \left\{ \\|\overline{\boldsymbol{y}}_{k,i} - \overline{\boldsymbol{X}}_{k,i} \boldsymbol{h}\\|_{\ell_2}^2 \right\}, \ \text{supp}(\boldsymbol{h}) \subset \mathcal{S}_{k,i}$	$\mathcal{O}(Ns^2)$
5: $\tilde{\boldsymbol{h}}_k^{(i)} = \sum_{l \in \mathcal{N}_k} b_{l,k} \widehat{\boldsymbol{h}}_l^{(i)}$	$O(\|\mathcal{N}_k\| m)$
6: $\boldsymbol{h}_k^{(i)} = \text{supp}_s \left(\tilde{\boldsymbol{h}}_k^{(i)} \right)$	$O(m)$
END	

- **Pruning step** Neighboring nodes may have identified different support sets, due to the different observations they have access to. Hence, the result of the previously mentioned fusion may not give an s-sparse estimate. This is a consequence of the fact that vector estimates having different support sets are aggregated. To this end, pruning ensures that, at each iteration, an s-sparse vector is produced.

Let us now turn our focus to explaining the basic principles concerning the cooperation (information fusion) among the nodes. Each node assigns a weight (combination coefficient) to the estimates and/or the observations received from the neighborhood and they are convexly combined. A well-known example of combination coefficients results by the Metropolis rule, where

$$a_{l,k} = \begin{cases} \frac{1}{\max\{|\mathcal{N}_k|, |\mathcal{N}_l|\}}, & \text{if } l \in \mathcal{N}_k, \text{ and } l \neq k, \\ 1 - \sum_{l \in \mathcal{N}_k \setminus k} a_{l,k}, & \text{if } l = k, \\ 0, & \text{otherwise.} \end{cases}$$

Note that $a_{l,k} > 0$, if $l \in \mathcal{N}_k$, $a_{l,k} = 0$, if $l \notin \mathcal{N}_k$, and $\sum_{l \in \mathcal{N}_k} a_{l,k} = 1, \forall k \in \mathcal{N}$. Furthermore, it is assumed that each node is a neighbor of itself, i.e. $a_{k,k} > 0, \ \forall k \in \mathcal{N}$.

The algorithmic steps of the greedy-based algorithm, which is hereafter referred to as distributed hard thresholding pursuit (DiHaT), are given in Table 2.2. In steps 1 and 2, the nodes exchange their input–output observations and fuse them using combination weights $a_{l,k}, \ \forall k \in \mathcal{N}, \ \forall l \in \mathcal{N}_k$. It can be shown that, if these coefficients are chosen properly, e.g. stem from the Metropolis rule, then $\overline{\boldsymbol{X}}_{k,i}$ and $\overline{\boldsymbol{y}}_{k,i}$ tend asymptotically to the average values, i.e. $\frac{1}{K} \sum_{k \in \mathcal{N}} \boldsymbol{X}_k$ and $\frac{1}{K} \sum_{k \in \mathcal{N}} \boldsymbol{y}_k$, respectively (see, e.g., [36]). This improves significantly the performance of the algorithm, since as the number of iteration increases, the information related to the support set is accumulated; that is, the support set identification and the parameter estimation procedure will contain information which comes from the entire network. In step 3, the s largest in amplitude coefficients of the signal proxy are selected, and step 4 performs a least squares operation restricted on the support set, as identified in the previous step. Next (step 5), the nodes exchange their estimates and fuse them in a similar way as in steps 1 and 2. Notice that the combination coefficients may be different from the weights used for the combination of the data observations (denoted as $b_{l,k}$). Finally, since the nodes have access to different measurements, the estimated support sets among the neighborhood, in general, may be different. This is more likely to happen in the first iterations, in which the estimates of the nodes have not converged yet. This implies that in step 5 there is no guarantee that the produced estimate will be s-sparse. For this reason, in step 6 a thresholding operation takes place and the final estimate at each node is s-sparse.

Regarding the theoretical properties of the DiHaT, it is shown in [17] that the algorithm converges to the true unknown vector, under certain assumptions regarding the noise vectors and a restricted isometry constant for the average value of the input matrices.

Remark 2.2

Computational complexity and communication cost

The complexity of each step of the DiHaT algorithm is presented in Table 2.2. Regarding the communication costs of this algorithm, each node transmits an $m \times N$ matrix, an $N \times 1$ vector, and $2s \times 1$ coefficients, from an s-sparse estimate and the positions of the nonzero coefficients. It should be pointed out that the complexity and the communications cost can be significantly reduced, if the nodes compute the signal proxy relying only on local information, i.e. if they do not perform steps 1 and 2. In that case, the DiHaT and the DLasso algorithms are of similar complexity. Comparative experiments between the two schemes will be presented in Section 2.4.

2.2.4 Other distributed sparse recovery algorithms

Another ℓ_1 regularized algorithm has been proposed in [16]. More specifically, the problem to solve is the so-called *distributed basis pursuit*. A decentralized ad-hoc network is considered and the problem takes the following form:

$$\begin{aligned} &\min \frac{1}{K} \sum_{k \in \mathcal{N}_k} \|\boldsymbol{h}_k\|_1 \\ &\text{s.t. } \boldsymbol{X}_k \boldsymbol{h}_k = \boldsymbol{y}_k, \quad k = 1, \ldots, K \\ &\boldsymbol{h}_k = \boldsymbol{h}_l, \ \forall l \in \mathcal{N}_k, \forall k \in \mathcal{N}. \end{aligned}$$

The optimization is solved in a similar way to the one presented in Section 2.2.2; that is, via the ADMM. Compared to the DLasso, a major difference is that distributed basis pursuit is applicable on networks that correspond to a bipartite graph. In contrast, DLasso and all the batch algorithms, which have been described so far, operate to more general network cases, under the assumption of full graph connectivity.

In [18], a distributed compressed sensing algorithm based on the iterative hard thresholding (ITH) technique, is developed. In a nutshell, ITH algorithms perform a gradient step based on the input–output data. In the sequel, a hard thresholding operator, which keeps only the s-largest in amplitude coefficients and forces all the rest to zero, takes place. The problem is first examined in static networks. More specifically, global computations take place, based on a broadcast operation. Then, the case where the network is time varying, and, consequently, global computations are not feasible, is considered. In that case, consensus averaging iterations substitute the previously described global computations. Two notable differences are found compared to the previously discussed algorithms. First, a hierarchical network is considered. That is, there exists a node that has a distinct role, compared to the rest. In particular, this node averages via consensus iterations the gradients computed at the nodes, and, then, the final estimation via a hard thresholding step is performed. In contrast, the algorithms discussed in previous subsections perform a single consensus averaging step at each iteration, all the nodes perform the same algorithmic task per iteration and, therefore, the computational load is equally distributed across the whole node set. Note that instead of hard thresholding the closely related soft thresholding philosophy has also been addressed in [37].

2.3 Online sparsity-aware distributed learning

2.3.1 Problem description

When dealing with big data problems, the dimensions of the input vector and the available measurements, are assumed to be very large. In the previous sections, we focused on batch distributed processing techniques, where the complexity per time iteration is proportional to N and all observations are assumed to be available prior to the processing/optimization task. It is obvious that, for large N, batch algorithms cannot perform the distributed estimation task in an efficient way, since using all the measurements demands a large amount of computational and storage resources. In order to overcome this drawback, the employment of distributed online learning techniques is considered next.

Formally, the goal in distributed online learning is to estimate an unknown vector, $\boldsymbol{h}_* \in \mathbb{R}^m$, exploiting, theoretically, an *infinite* number of *sequentially* obtained observations collected at the K nodes of an ad-hoc network, e.g. [38, 39]. Each node k, at each (discrete) time instance, has access to the training data pair, $(y_k(n), \boldsymbol{x}_k(n)) \in \mathbb{R} \times \mathbb{R}^m$, which are related via the linear model

$$y_k(n) = \boldsymbol{x}_k^T(n)\boldsymbol{h}_* + v_k(n), \ \forall k \in \mathcal{N}, \ \forall n \in \mathbb{Z}, \tag{2.8}$$

where the term $v_k(n)$ stands for the additive noise process at each node.

The following modes of cooperation have been proposed in the context of distributed adaptive learning.

(1) **Adapt then combine (ATC) [39, 40]** According to this strategy, each node computes an intermediate estimate, by exploiting locally sensed measurements. Following this step, each network agent receives these estimates from the neighboring nodes and combines them, in order to come up with the final estimate, for the current time instant.

(2) **Combine then adapt (CTA) [38, 41]** According to this scenario, the combination step precedes the adaptation one.
(3) **Consensus based [42, 43]** According to this strategy, the computations are made in parallel and there is no clear distinction between the combine and the adapt steps.

The ATC and the CTA cooperation strategies characterizes the family of the so-called *diffusion* algorithms. In the next sections, distributed online algorithms will be presented, starting from an ℓ_1 regularized diffusion, LMS [19]. In the sequel, a greedy-based adaptive algorithm will be discussed [17], and, finally, a CTA projection-based scheme will be presented [20].

2.3.2 LMS based sparsity-promoting algorithm

The goal in this section is to present a sparsity-promoting adaptive algorithm for distributed learning, which is based on the least mean squares (LMS) algorithm, e.g. [14, 44, 45].

In order to grasp the reasoning and the basic concepts regarding the algorithm developed in [19], we will first discuss its centralized version; that is, we will assume that all the information available locally in each node is also accessible by a central node. The cost function to be minimized with respect to $\boldsymbol{h}$ is the following:

$$J^{\text{glob}}(\boldsymbol{h}) = \frac{1}{2}\sum_{k\in\mathcal{N}} \mathbb{E}\left[(y_k(n) - \boldsymbol{x}_k^T(n)\boldsymbol{h})^2\right] + \lambda f(\boldsymbol{h}), \tag{2.9}$$

where $\mathbb{E}[\cdot]$ denotes the expectation operator, $f(\boldsymbol{h}) : \mathbb{R}^m \mapsto \mathbb{R}$ is a regularization function, which promotes sparsity, and λ is a regularization parameter, similar to the one employed in Section 2.2.2. Notice that, if we omit the sparsity enforcing term, i.e. $\lambda f(\boldsymbol{h})$, (2.9) reduces to the cost function of the classical LMS algorithm. Motivated by the need for decentralized processing, in which each node exploits locally sensed information and information received from the neighborhood, the following cost is proposed in order to approximate (2.9):

$$J^{\text{dist}}(\boldsymbol{h}) = \frac{1}{2}\sum_{l\in\mathcal{N}_k} c_{l,k}\mathbb{E}\left[(y_l(n) - \boldsymbol{x}_l^T(n)\boldsymbol{h})^2\right] + \frac{1}{2}\sum_{l\in\mathcal{N}_k\setminus\{k\}} b_{l,k}\|\boldsymbol{h} - \boldsymbol{\psi}_l\|^2 + \lambda f(\boldsymbol{h}), \tag{2.10}$$

where $\boldsymbol{\psi}_l$ is a local estimate of the unknown vector, obtained at node l, and $c_{l,k}$, $b_{l,k}$ are combination weights, defined similarly to those in Section 2.2.3. A subgradient of the cost function in (2.10) is denoted as $J^{'dist}(\boldsymbol{h})$ and given by

$$J^{'\text{dist}}(\boldsymbol{h}) = \sum_{l\in\mathcal{N}_k} c_{l,k}(\boldsymbol{R}_l\boldsymbol{h} - \boldsymbol{p}_l) + \sum_{l\in\mathcal{N}_k\setminus\{k\}} b_{l,k}(\boldsymbol{h} - \boldsymbol{\psi}_l) + \lambda f'(\boldsymbol{h}), \tag{2.11}$$

where $\boldsymbol{R}_k := \mathbb{E}[\boldsymbol{x}_k(n)\boldsymbol{x}_k^T(n)]$ is the so-called autocorrelation matrix, $\boldsymbol{p}_k := \mathbb{E}[\boldsymbol{x}_k(n)\boldsymbol{y}_k(n)]$ is the crosscorrelation vector and $f'(\boldsymbol{h})$ is a subgradient of $f(\boldsymbol{h})$. The subgradient scheme can be employed for the iterative minimization of (2.10). Hence,

Table 2.3 The SpaDiLMS algorithm

	Each node initializes to zero $\boldsymbol{h}_k(0)$	
FOR $n = 1, \ldots,$ **DO**		
1:	$\boldsymbol{\psi}_k(n) = \boldsymbol{h}_k(n-1) + \mu_k \sum_{l \in \mathcal{N}_k} c_{l,k}(y_l(n) - \boldsymbol{x}_l^T(n)\boldsymbol{h}_k(n-1))\boldsymbol{x}_l(n) - \mu_k \lambda \partial f(\boldsymbol{h}_k(n-1))$	$O(\lvert\mathcal{N}_k\rvert m)$
2:	$\boldsymbol{h}_k(n) = \sum_{l \in \mathcal{N}_k} a_{l,k} \boldsymbol{\psi}_l(n)$	$O(\lvert\mathcal{N}_k\rvert m)$
END		

the estimate obtained at node k at time instance n is given by

$$\boldsymbol{h}_k(n) = \boldsymbol{h}_k(n-1) + \mu_k \sum_{l \in \mathcal{N}_k} c_{l,k}(\boldsymbol{p}_l - \boldsymbol{R}_l \boldsymbol{h}_k(n-1)) + \mu_k \sum_{l \in \mathcal{N}_k} b_{l,k}(\boldsymbol{\psi}_l - \boldsymbol{h}_k(n-1)) - \mu_k \lambda f'(\boldsymbol{h}_k(n-1)). \tag{2.12}$$

The update (2.12) can be rewritten as follows:

$$\boldsymbol{\psi}_k(n) = \boldsymbol{h}_k(n-1) + \mu_k \sum_{l \in \mathcal{N}_k} c_{l,k}(\boldsymbol{p}_l - \boldsymbol{R}_l \boldsymbol{h}_k(n-1)) - \mu_k \lambda f'(\boldsymbol{h}_k(n-1)) \tag{2.13}$$

$$\boldsymbol{h}_k(n) = \boldsymbol{\psi}_k(n) + \mu_k \sum_{l \in \mathcal{N}_k} b_{l,k}(\boldsymbol{\psi}_l - \boldsymbol{h}_k(n-1)). \tag{2.14}$$

As suggested in [19], $\boldsymbol{\psi}_l$ is substituted by $\boldsymbol{\psi}_l(n)$ and $\boldsymbol{h}_k(n-1)$ by $\boldsymbol{\psi}_k(n)$, which are given by (2.13). According to that, (2.14) becomes:

$$\boldsymbol{h}_k(n) = \sum_{l \in \mathcal{N}_k} a_{l,k} \boldsymbol{\psi}_l(n), \tag{2.15}$$

clearly following an ATC strategy. The weights $a_{l,k}$ are defined as:

$$a_{l,k} = \mu_k b_{l,k}, \quad l \neq k$$
$$a_{k,k} = 1 - \mu_k \sum_{l \in \mathcal{N}_k} b_{l,k}, \quad l = k.$$

So far, we have considered that we have access to the input/output statistics, i.e. the second-order moments $\boldsymbol{R}_l$, $\boldsymbol{p}_l$, which is unrealistic in many cases, e.g. [46]. Following stochastic approximation arguments, a way to overcome this problem is to rely on the following instantaneous approximations, $\boldsymbol{R}_k \approx \boldsymbol{x}_k(n)\boldsymbol{x}_k^T(n)$ and $\boldsymbol{p}_k \approx \boldsymbol{x}_k(n)y_k(n)$. Taking the previous two facts into consideration, a proper reformulation of (2.14), gives birth to the so-called sparse diffusion LMS (SpaDiLMS). The steps of this algorithm are summarized in Table 2.3.

In step 1 (adaptation step), each node performs an iteration, employing measurements received by the neighborhood and sparsity is enforced via the regularization function $f(\cdot)$. Methodologies for choosing this function will be discussed later on. In step 2 (combination step), the nodes fuse the estimates, computed in the previous step, with respect to the combination coefficients.

Two strategies, those first employed in [47], have been proposed in [19] for choosing the regularization function. For the first one: $f(\boldsymbol{h}) = \|\boldsymbol{h}\|_1$ with $f'(\boldsymbol{h}) = \text{sign}(\boldsymbol{h})$. The entries of the sign function are given by:

$$j = 1, \ldots, m, \quad \text{sign}(h_j) = \begin{cases} h_j/|h_j|, & h_j \neq 0 \\ 0, & h_j = 0. \end{cases} \tag{2.16}$$

By choosing the regularization function in such a way, the so-called *zero-attracting* diffusion LMS results. The second option is to use the following rule, which, for a very small positive ε, approximates the ℓ_0 norm, i.e. $f(\boldsymbol{h}) = \sum_{j=1}^{m} \frac{|h_j|}{\varepsilon + |h_j|} \approx \|\boldsymbol{h}\|_0$. The subgradient of this cost function is taking to be equal to

$$f' = \text{diag}\left\{ \frac{|h_1|}{\varepsilon + |h_1|}, \ldots, \frac{|h_m|}{\varepsilon + |h_m|} \right\},$$

and the resulting algorithm is known as the *reweighted zero–attracting diffusion LMS.*

Remark 2.3 The theoretical properties of the SpaDiLMS are thoughtfully studied in [19]. It should be mentioned that the regularization function introduces bias, i.e. in contrast to the classical diffusion LMS, the algorithm does not converge in the mean to the true unknown vector.

Remark 2.4

Computational complexity and communication cost

A notable advantage of the gradient/subgradient-based algorithms is that their computational complexity is relatively low. This attribute is also inherited in the SpaDiLMS, where the adaptation step exhibits a linear dependence on the dimensionality of the systems, i.e. $O(|\mathcal{N}_k|m)$. Regarding the communications cost, each node transmits an $m \times 1$ estimate vector, the $m \times 1$ input vector and a scalar corresponding to $y_k(n)$.

2.3.3 The GreeDi LMS algorithm

The main idea behind the GreeDi LMS is summarized as follows. The nodes of the network compute a signal proxy, locally, yet in a cooperative way; this proxy estimates the positions of the nonzero coefficients of the unknown vector. In the sequel, the ATC diffusion LMS, restricted on the support identified in each node, is employed in order to update the estimates of each node.

For simplicity, let us first discuss the non-cooperative case. As we have already seen in the batch learning version (Section 2.2.3), the proxy signal is constructed via the available measurements at each node; i.e. the local measurements and the information received by the neighborhood. Nevertheless, in online learning, the observations are received sequentially, one per time step and, consequently, a different route has to be followed. Recall the definition of the signal proxy, given in (2.6). A first approach is to make the following modifications; $\boldsymbol{X}_k^T \boldsymbol{y}_k$ with $\boldsymbol{x}_k(n) y_k(n)$ and $\boldsymbol{X}_k^T \boldsymbol{X}_k$ with $\boldsymbol{x}_k(n) \boldsymbol{x}_k^T(n)$. A drawback of this choice is that the proxy is constructed exploiting just a single pair of data, which in practice carries insufficient information. Another viewpoint, which will be followed here, is to rely on approximations of the expected value of the previous quantities, i.e. $\mathbb{E}[\boldsymbol{x}_k(n)y_k(n)] := \boldsymbol{p}_k$, $\mathbb{E}[\boldsymbol{x}_k(n)\boldsymbol{x}_k^T(n)] := \boldsymbol{R}_k$. Since the statistics are usually unknown and might exhibit time variations, they can be replaced by the following approximations:

$$\mathbb{E}[\boldsymbol{x}_k(n)y_k(n)] \approx \sum_{i=1}^{n} \beta^{n-i}\boldsymbol{x}_k(i)y_k(i)$$

$$\mathbb{E}[\boldsymbol{x}_k(n)\boldsymbol{x}_k^T(n)] \approx \sum_{i=1}^{n} \beta^{n-i}\boldsymbol{x}_k(i)\boldsymbol{x}_k^T(i).$$

Parameter $\beta \in (0, 1]$ is the so-called forgetting factor, incorporated in order to give less weight to past values, so as to cope with cases where the unknown vector and the statistics undergo changes.

The modified proxy, which is suitable for online operation, takes the following form:

$$\boldsymbol{h}_k(n-1) + \tilde{\mu}_k(\overline{\boldsymbol{p}}_k(n) - \overline{\boldsymbol{R}}_k(n)\boldsymbol{h}_k(n-1)) \approx \boldsymbol{h}_*, \tag{2.17}$$

where $\boldsymbol{h}_k(n-1)$, $\overline{\boldsymbol{p}}_k(n)$ and $\overline{\boldsymbol{R}}_k(n)$ are defined in Table 2.4. The step size $\tilde{\mu}_k$ accelerates the convergence speed. The proposed proxy constitutes an exponentially weighted extension of its batch form (discussed in Section 2.2.3). Notice that (2.17) defines a gradient descent iteration using the approximate statistics $\overline{\boldsymbol{p}}_k(n)$ and $\overline{\boldsymbol{R}}_k(n)$. Assuming that we have at our disposal the true mean values, i.e. $\overline{\boldsymbol{p}}_k := \sum_{l\in\mathcal{N}_k} a_{l,k}\boldsymbol{p}_l$ and $\overline{\boldsymbol{R}}_k := \sum_{l\in\mathcal{N}_k} a_{l,k}\boldsymbol{R}_l$, then the recursion (2.17) converges to the true solution, see e.g. [11]. In practice, the previously mentioned approximate values are used, since the nodes have no access to the true statistics.

Table 2.4 summarizes the core steps of the GreeDi LMS. Steps 1–6 constitute the *distributed greedy support set selection* process. It is worth pointing out that steps 1–4 create, in a cooperative manner, the basic elements needed for the proxy update and steps 5–6 select the s-largest components. Notice that D is a parameter that determines if the estimate in the proxy computation steps will be normalized (step 5) or not (step 6). More precisely, if the norm of the estimate takes extensively large values, which are determined by D, then in the signal proxy a normalized version of it is employed, i.e. $\boldsymbol{h}_k(n-1)/\|\boldsymbol{h}_k(n-1)\|$.[1] It should be pointed out that the *distributed greedy update* is established via the adapt–combine LMS (steps 7–8) of [39]. The diffusion steps 2, 4, and 8 bring the local estimates closer to the global estimates based on the entire network [39]. Finally, because we are operating in a decentralized network, where the nodes perform local computations, some of them will achieve support set convergence faster than others. Therefore, at a node level, we pay more attention to the s-dominant positions via the introduction of a pruning step directly after the adaptation in the combined greedy update (step 9), where $\tilde{S}^c_{k,n}$ stands for the complementary of the respective set.

Remark 2.5

Computational complexity and communication cost

The computational complexity of GreeDi LMS is linear except from the update of $\boldsymbol{R}_k(n)$, which requires $\mathcal{O}(m^2)$ operations per iteration. However, exploitation of the

[1] This treatment merely facilitates the theoretical convergence proof of the algorithm, rather than been crucial to the performance of the algorithm in practice.

Table 2.4 The Greedy Diffusion LMS algorithm

Algorithm description	Complexity
$\boldsymbol{h}_k(0) = \mathbf{0}_m$, $\boldsymbol{p}_k(0) = \mathbf{0}_m$, $\boldsymbol{R}_k(0) = \boldsymbol{O}_{m\times m}$	
$0 < \beta \leq 1$	
$0 < D$	
FOR $n := 1, 2, \ldots$ **DO**	
1: $\boldsymbol{p}_k(n) = \frac{n}{n+1}\beta \boldsymbol{p}_k(n-1) + \frac{\boldsymbol{x}_k(n) y_k(n)}{n+1}$	$O(m)$
2: $\overline{\boldsymbol{p}}_k(n) = \sum_{l\in\mathcal{N}_k} a_{l,k}\boldsymbol{p}_l(n)$	$O(\lvert\mathcal{N}_k\rvert m)$
3: $\boldsymbol{R}_k(n) = \frac{n}{n+1}\beta \boldsymbol{R}_k(n-1) + \frac{\boldsymbol{x}_k(n)\boldsymbol{x}_k^T(n)}{n+1}$	$O(m^2)$
4: $\overline{\boldsymbol{R}}_k(n) = \sum_{l\in\mathcal{N}_k} a_{l,k}\boldsymbol{R}_l(n)$	$O(\lvert\mathcal{N}_k\rvert m^2)$
If $\lVert\boldsymbol{h}_k(n-1)\rVert \leq D$	
5: $\widehat{\mathcal{S}}_{k,n} = \operatorname{supp}_s\big(\boldsymbol{h}_k(n-1) + \tilde{\mu}_k\big(\overline{\boldsymbol{p}}_k(n) - \overline{\boldsymbol{R}}_k(n)\boldsymbol{h}_k(n-1)\big)\big)$	$O(m)$
Else	
6: $\widehat{\mathcal{S}}_{k,n} = \operatorname{supp}_s\left(\frac{\boldsymbol{h}_k(n-1)}{\lVert\boldsymbol{h}_k(n-1)\rVert} + \tilde{\mu}_k\left(\overline{\boldsymbol{p}}_k(n) - \overline{\boldsymbol{R}}_k(n)\frac{\boldsymbol{h}_k(n-1)}{\lVert\boldsymbol{h}_k(n-1)\rVert}\right)\right)$	$O(m)$
7: $\boldsymbol{\psi}_k(n) = \boldsymbol{h}_{k\lvert\widehat{\mathcal{S}}_{k,n}}(n-1) + \mu_k \boldsymbol{x}_{k\lvert\widehat{\mathcal{S}}_{k,n}}(n)\big[y_k(n) - \boldsymbol{x}^T_{k\lvert\widehat{\mathcal{S}}_{k,n}}(n)\boldsymbol{h}_{k\lvert\widehat{\mathcal{S}}_{k,n}}(n-1)\big]$	$O(s)$
8: $\widetilde{\boldsymbol{h}}_k(n) = \sum_{l\in\mathcal{N}_k} b_{l,k}\boldsymbol{\psi}_{l\lvert\widehat{\mathcal{S}}_{l,n}}(n)$	$O(\lvert\mathcal{N}_k\rvert s)$
9: $\widetilde{\mathcal{S}}_{k,n} = \operatorname{supp}_s(\widetilde{\boldsymbol{h}}_k(n))$, $\boldsymbol{h}_{k\lvert\widetilde{\mathcal{S}}^c_{k,n}}(n) = \mathbf{0}$	$O(m)$
END	

shift structure in the regressor vector, which is the case in many applications, e.g. echo cancellation, channel equalization, etc., allows us to drop the computational cost to $\mathcal{O}(m)$, e.g. [14, 45, 46]. As regards the communication cost of the algorithm, each node transmits an $2s \times 1$ vector (the estimate and the positions of the nonzero coefficients) an $m \times 1$ vector (the approximated crosscorelation vector) and an $m \times m$ matrix (the approximated autocorellation matrix). Complexity as well as the number of transmitted coefficients, can be significantly decreased if, in steps 5 and 6, each node exploits only local information.

2.3.4 Set-theoretic sparsity-aware distributed learning

The main idea of the algorithms that have been discussed so far is to employ a loss function that measures the deviation between the desired and the actual prediction provided by the model, and then employ a minimization procedure.

In this section, a different route is followed. More specifically, the sparse distributed adaptive projected subgradient method (SpaDiAPSM), proposed in [20], will be discussed. This algorithm follows the set-theoretic estimation rationale [48–50]; that is,

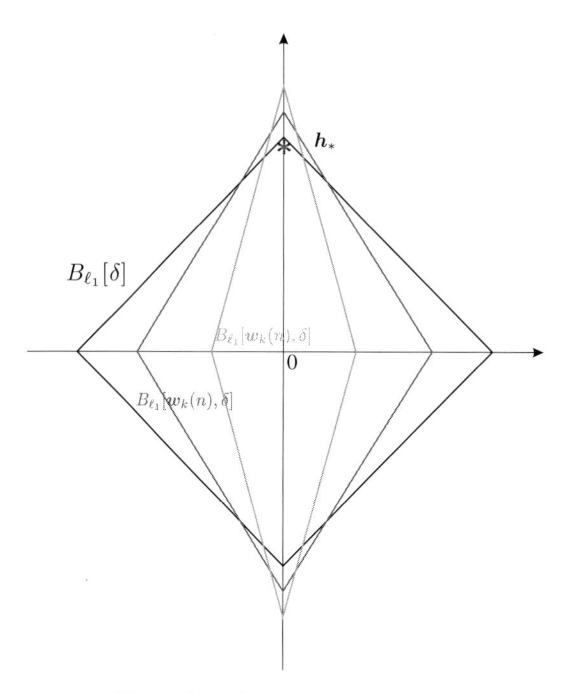

Figure 2.2 Illustration of two weighted and an unweighted ℓ_1 balls.

for each pair of measured input/output data, a closed convex set is constructed, where the unknown vector lies with high probability. Moreover, in several applications/cases, a-priori knowledge concerning characteristics of the unknown parameter vector may be available. A notable example is the sparsity characteristic. This kind of a-priori knowledge can be embedded into the algorithm in the form of constraints, where each constraint also defines a corresponding closed convex set of points. Any vector belonging to the common intersection of all the sets, which is referred to as feasibility set, is considered to be an acceptable solution. Therefore, the objective of the adaptive set-theoretic algorithms is to converge to a point that belongs to the feasibility set.

A popular (convex) set, which is used to construct property sets around the training data, takes the form of a *hyperslab* around $(\boldsymbol{x}_k(n), y_k(n))$, defined as

$$S_{k,n} := \left\{ \boldsymbol{h} \in \mathbb{R}^m : \left| \boldsymbol{x}_k^T(n)\boldsymbol{h} - y_k(n) \right| \leq \epsilon_k \right\}, \quad \forall n \in \mathcal{N}, \tag{2.18}$$

for some user-defined tolerance $\epsilon_k \geq 0$. The parameter ϵ_k is a threshold, which takes into consideration the noise as well as model and/or sensing inaccuracies and determines the width of the hyperslabs. An example of two hyperslabs is shown in Figure 2.2. In the context of sparsity-aware learning, popular sparsity promoting constraint sets are ℓ_1 balls or weighted ℓ_1 balls [5, 51]. The constraint sets employed in the SpaDiAPSM algorithm are weighted ℓ_1 balls, since as it has been shown in [51] they lead to an enhanced performance compared to the conventional ℓ_1 balls. Given a vector of weights

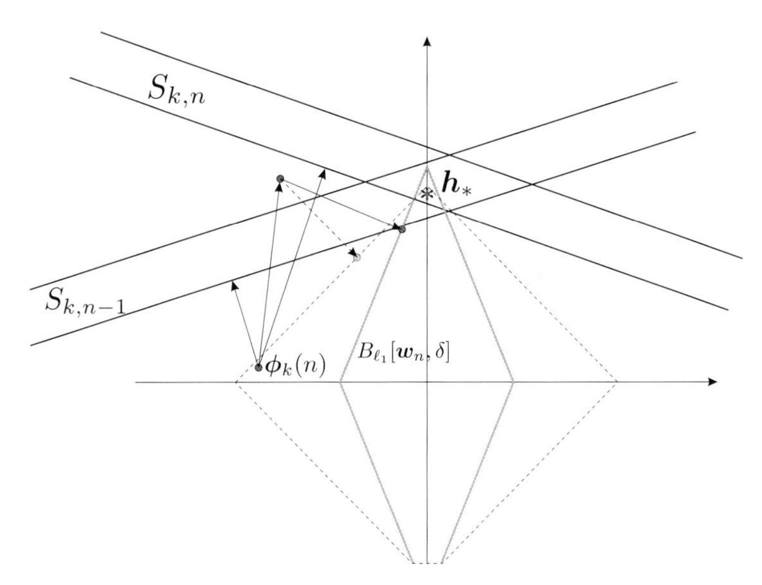

Figure 2.3 Geometrical illustration of the SpaDiAPSM algorithm. The result of the combination step is projected onto the two most recently constructed hyperslabs, and the new estimate is computed.

$\boldsymbol{w}_k(n) = [w_1^{(k,n)}, \ldots, w_m^{(k,n)}]^T$, where $w_j^{(k,n)} > 0, \forall j = 1, \ldots, m$, and a positive radius, δ, the weighted ℓ_1 ball is defined as:

$$B_{\ell_1}[\boldsymbol{w}_k(n), \delta] := \left\{ \boldsymbol{h} \in \mathbb{R}^m : \sum_{j=1}^{m} w_j^{(k,n)} |h_j| \leq \delta \right\}. \tag{2.19}$$

It is worth pointing out that the classical ℓ_1 ball occurs if $\boldsymbol{w}_n = \mathbf{1}_m$. The geometry of these sets is illustrated in Figure 2.3.

A popular strategy of constructing the weights associated to the weighted ℓ_1 ball (see for example [20, 51]) is the following: $w_j^{(k,n)} = 1/(|h_j^{(k,n)}| + \tilde{\epsilon}_{k,n}),\ j = 1, \ldots, m$, where $\tilde{\epsilon}_{k,n}$ is a sequence of positive numbers used in order to avoid divisions by zero. As shown in [51], if one chooses the weights according to the previously mentioned strategy and if $\delta \geq \|\boldsymbol{h}_*\|_0$, then it holds that $\boldsymbol{h}_* \in B_{\ell_1}[\boldsymbol{w}_k(n), \delta]$. The algorithm that will be presented in this section is based on the set-theoretic adaptive projected subgradient method (APSM), e.g. [52–55]. In a nutshell, the task of identifying a point in the intersection of the convex sets, can be accomplished by iteratively performing two successive steps. First, the currently available estimate is projected, concurrently, onto the q most-recently obtained/chosen hyperslabs and then an intermediate estimate is obtained via the convex combination of these projections. Second this estimate is projected onto the constraint set, i.e. the weighted ℓ_1 ball.

The projection onto a hyperslab, is defined as [55, 56]:

$$\forall \boldsymbol{h} \in \mathbb{R}^m, \quad P_{S_{k,n}}(\boldsymbol{h}) := \boldsymbol{h} + \beta_k(n)\boldsymbol{x}_k(n), \tag{2.20}$$

Table 2.5 Sparse distributed adative projected subgradient method (SpaDiAPSM)

Algorithm description	Complexity
Each node initializes to zero $\boldsymbol{h}_k(0)$	
FOR $n = 1, \ldots,$ **DO**	
Node k transmits $\boldsymbol{h}_k(n)$ to each node which belongs to $\mathcal{N}_k$	
1: $\boldsymbol{\phi}_k(n) = \sum_{l \in \mathcal{N}_k} a_{l,k} \boldsymbol{h}_l(n)$	$O(\lvert\mathcal{N}_k\rvert m)$
2: $\boldsymbol{h}'_k(n) = \left(\boldsymbol{\phi}_k(n) + \mu_{k,n} 1/q \left(\sum_{j=n-q+1}^{n} P_{S_{k,j}}(\boldsymbol{\phi}_k(n)) - \boldsymbol{\phi}_k(n)\right)\right)$	$O(qm)$
3: $\boldsymbol{h}_k(n+1) = P_{B_{\ell_1}[\boldsymbol{w}_k(n), \delta]}(\boldsymbol{h}'_k(n))$	$O(m \log m)$
END	

where

$$\beta_k(n) = \begin{cases} \dfrac{y_k(n) - \boldsymbol{x}_k^T(n)\boldsymbol{h} + \epsilon_k}{\|\boldsymbol{x}_k(n)\|^2}, & \text{if } y_k(n) - \boldsymbol{x}_k^T(n)\boldsymbol{h} < -\epsilon_k, \\ 0, & \text{if } |y_k(n) - \boldsymbol{x}_k^T(n)\boldsymbol{h}| \le \epsilon_k, \\ \dfrac{y_k(n) - \boldsymbol{x}_k^T(n)\boldsymbol{h} - \epsilon_k}{\|\boldsymbol{x}_k(n)\|^2}, & \text{if} y_k(n) - \boldsymbol{x}_k^T(n)\boldsymbol{h} > \epsilon_k. \end{cases}$$

In contrast, the projection of a point onto $B_{\ell_1}[\boldsymbol{w}_k(n), \delta]$ does not admit a closed-form solution; albeit it is computed in a finite number of steps and it turns out to be equivalent with a soft thresholding operation [51, Theorem 1].

Based on the previous findings, a corresponding distributed scheme can be derived; the combine–adapt diffusion strategy has been adopted. The main steps of the algorithm, are summarized in Table 2.5 and the geometry of the algorithm is illustrated in Figure 2.3. In step 1, at time instance n each node fuses the information, which consists of the most recently obtained estimates of the neighborhood. In the sequel (step 2), exploiting the newly available measurement, the hyperslab $S_{k,n}$ is formed. The aggregate $\boldsymbol{\phi}_k(n)$ is projected, onto the q most-recent hyperslabs, constructed locally at the kth node and a simple convex combination scheme leads to the intermediate estimate. The larger the q is, the faster the convergence speed of the algorithm becomes with a simultaneous increment in computational complexity. Convergence is guaranteed for step–size $\mu_{k,n} \in (0, 2)$. However, convergence speed can be accelerated via a properly defined extrapolation term, [55], given by

$$\mathcal{M}_{k,n} := \begin{cases} \dfrac{1/q \sum_{j=n-q+1}^{n} \|P_{S_{k,j}}(\boldsymbol{\phi}_k(n)) - \boldsymbol{\phi}_k(n)\|^2}{\left\|1/q \sum_{j=n-q+1}^{n} P_{S_j}(\boldsymbol{\phi}_k(n)) - \boldsymbol{\phi}_{k,n}\right\|^2}, & \\ & \text{if } 1/q \sum_{j=n-q+1}^{n} P_{S_{k,j}}(\boldsymbol{\phi}_k(n)) \neq \boldsymbol{\phi}_k(n) \\ 1, & \text{otherwise.} \end{cases}$$

In this case, the step size is allowed to belong to the interval $(0, 2\mathcal{M}_{k,n})$. Finally, in the third step, the intermediate estimate is projected onto the sparsity constraint set, i.e. the weighted ℓ_1 ball.

Variations of the APSM

In order to keep the discussion simple, the APSM was presented in its very basic form. However, the specific algorithmic framework is robust in allowing, for example, the use of subdimensional projections for complexity reduction [57], as well as the incorporation of more advanced sparsity promoting practices. Accordingly, the projections onto weighted ℓ_1 balls can be replaced by a generalized thresholding operator, which embodies a number of well appreciated thresholding operators and penalty functions, which may also be nonconvex; examples are the hard thresholding and the smoothly clipped absolute deviation (SCAD) penalty [58, 59]. Moreover, for enhanced performance, the classical Euclidean projections onto the involved sets can be replaced by variable metric projections, e.g. [20, 60]. The main use of the variable metric projections, is to assign different weights to each coefficient of the updated estimate, which are made proportional to the magnitude of the particular estimated coefficient. This rationale is also followed in the so-caled *proportionate* adaptive schemes, e.g. [61, 62]. In other words, variable metric projections offer complementary means for exploiting the a priori information regarding the sparsity characteristic of the unknown vector. The variable projection extension of the APSM scheme is briefly presented next.

Let $\boldsymbol{G}_k(n)$ be the $m \times m$ diagonal matrix with entries $g_j^{(k,n)} = \left(\frac{1-\alpha}{2m} + (1+\alpha)\frac{|h_j^{(k,n)}|}{2\|\boldsymbol{h}_k(n)\|_1}\right)^{-1}$, where $h_j^{(k,n)}$ stands for the jth coefficient of the estimate $\boldsymbol{h}_k(n)$. The parameter $\alpha \in [-1, 1]$ determines to what extend the sparsity will be taken into account [62]. Setting $\alpha = 0$ is a commonly chosen option, which is also adopted here. Notice that the coefficients of the matrix $\boldsymbol{G}_k(n)$ are in line with the previously described philosophy, of assigning larger weights to the more significant coefficients.

The variable metric projections, determined by $\boldsymbol{G}_k(n)$, onto the respective hyperslabs are given by:

$$\forall \boldsymbol{h} \in \mathbb{R}^m, \quad P_{S_{k,n}}^{(\boldsymbol{G}_k(n))}(\boldsymbol{h}) := \boldsymbol{h} + \beta_k(n)\boldsymbol{G}_k(n)\boldsymbol{x}_k(n), \tag{2.21}$$

where

$$\beta_k(n) = \begin{cases} \dfrac{y_k(n) - \boldsymbol{x}_k^T(n)\boldsymbol{h} + \epsilon_k}{\boldsymbol{x}_k^T(n)\boldsymbol{G}_k(n)\boldsymbol{x}_k(n)}, & \text{if } y_k(n) - \boldsymbol{x}_k^T(n)\boldsymbol{h} < -\epsilon_k, \\ 0, & \text{if } |y_k(n) - \boldsymbol{x}_k^T(n)\boldsymbol{h}| \leq \epsilon_k, \\ \dfrac{y_k(n) - \boldsymbol{x}_k^T(n)\boldsymbol{h} - \epsilon_k}{\boldsymbol{x}_k^T(n)\boldsymbol{G}_k(n)\boldsymbol{x}_k(n)}, & \text{if } y_k(n) - \boldsymbol{x}_k^T(n)\boldsymbol{h} > \epsilon_k, \end{cases}$$

and the algorithm, in this case, follows similar steps as in the one given in Table 2.5, albeit the Euclidean projections are substituted by the variable metric projections. It is worth pointing out, that the use of these modified projections accelerates the convergence of the algorithm, as it has been experimentally verified in [20].

Finally, cases where the input–output relation is not linear can also be accommodated in the APSM framework with the aid of Volterra series expansion [63].

Remark 2.6 Regarding the theoretical analysis of the algorithm, it has been shown in [20] that if the hyperslabs share a common intersection and if the constraint sets, e.g. the weight ℓ_1 balls, are common at each node, then the scheme enjoys a number of theoretical properties. The most important ones are: (a) monotonicity, i.e. the distance of the estimates from the intersection of the hyperslabs with the weighted ℓ_1 balls is a non-increasing function, and (b) the nodes converge to the *same* estimate. In practice, it has been shown in [20] that if the assumption about the common constraint sets per node is relaxed, the performance of the algorithm is not affected.

Remark 2.7 Data reliability is an emerging issue arising in the big data era. The larger the volume of data, the more likely it is to become contaminated by erroneous entries, the so called *outliers*. A major source of such corrupted data is the failure of sensors and/or recording systems. Even though such errors appear rarely, algorithms that are unable to deal with outliers, such as those solely relying on LS minimization, are likely to diverge from the true solution even in the presence of a single gross error. More importantly, in cases where manual data inputs are involved, e.g. in product recommendations/ratings, completing questionnaires, mouse clicks on the internet, etc., erroneous or even deliberately fake data can appear in large amounts among healthy and useful data entries. So, it is a definite plus for an algorithm to be able to deal with outliers. Although such an algorithm has not been proposed yet for distributed learning, it should be emphasized that the set-theoretic framework can easily cope with such scenarios, see, e.g. [41, 64]. Other robust approaches can be found in, e.g. [65–67].

Remark 2.8

Computational complexity and communication cost
As can be seen in Table 2.5, the complexity of the algorithm is of order $O(qm)$ springing from the projection operators and $O(m\log_2 m)$ occurring from the projection onto the weighted ℓ_1 ball. The logarithmic term in the last projections results from the required sorting operation. However, by properly employing a divide-and-conquer approach, sorting could be avoided leading to linear overall complexity. Linear overall complexity can be more efficiently achieved by using easy to implement sparsity promoting generalized thresholds [58, 59]. Moreover, computational savings in the hyperslab projections can be achieved by replacing metric projections with subdimensional projections [68]. The communication cost of the algorithm is relatively low, since each node transmits only m coefficients coming from the computed estimate.

2.4 Simulation examples

This section deals with numerical simulations and performance evaluation of some of the algorithms that were discussed throughout this chapter. First, the performance of the algorithms, operating in a batch mode, is validated in the context of the distributed

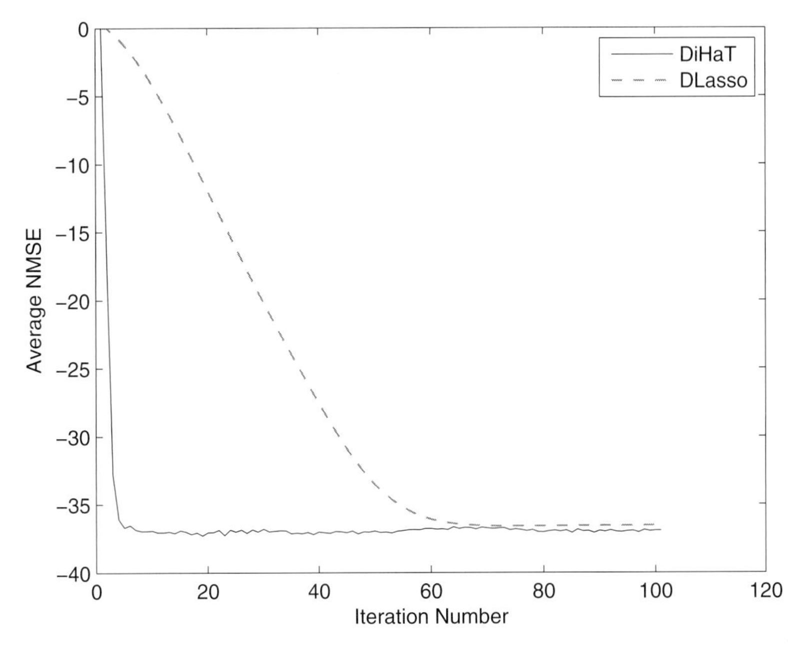

Figure 2.4 Average normalized MSD curves for the first experiment (batch operation).

parameter estimation task. Secondly, the online scenario is considered. In particular, the SpaDiLMS and the SpaDiAPSM are validated in an adaptive distributed network.

2.4.1 Performance evaluation of batch methods

In the first experiment, we compare the performance of the DiHaT and the DLasso algorithms. A network comprising $K = 20$ nodes is considered, the dimension of the unknown vector equals $m = 2000$ and the number of measurements at each node is $N = 1500$. The entries of the input matrices $\boldsymbol{X}_k,\ \forall k \in \mathcal{N}$ follow a Gaussian distribution with zero mean and variance 1. The noise is generated according to the Gaussian distribution and the signal-to-noise ratio (SNR) at each node equals to 20 dB. Furthermore, the sparsity level of the unknown vector equals to 200, i.e. $\|\boldsymbol{h}_*\|_0 = 200$. The performance metric is the average normalized mean square deviation, which equals, at each iteration, $\text{MSD}(i) = \frac{1}{K}\sum_{k\in\mathcal{N}} \frac{\|\boldsymbol{h}_k^{(i)} - \boldsymbol{h}_*\|^2}{\|\boldsymbol{h}_*\|^2}$. For the DiHaT algorithm, it is assumed that the sparsity level of the unknown vector is known, i.e. in each iteration, we keep the 200 largest in amplitude coefficients of the signal proxy. Furthermore, the weights $a_{l,k},\ b_{l,k}$ are chosen with respect to the Metropolis rule, which was discussed in Section 2.2.3. The free parameters, which are employed in the DLasso algorithm, are set so as the algorithm to attain an optimum trade-off between convergence speed and steady state error floor. More specifically, for the first experiment we set $\lambda = 2 \cdot 10^{-4}$ and $c = 0.3$. From Figure 2.4 it can be readily seen that the DiHaT algorithm outperforms the DLasso, since it converges faster to a slightly lower error floor.

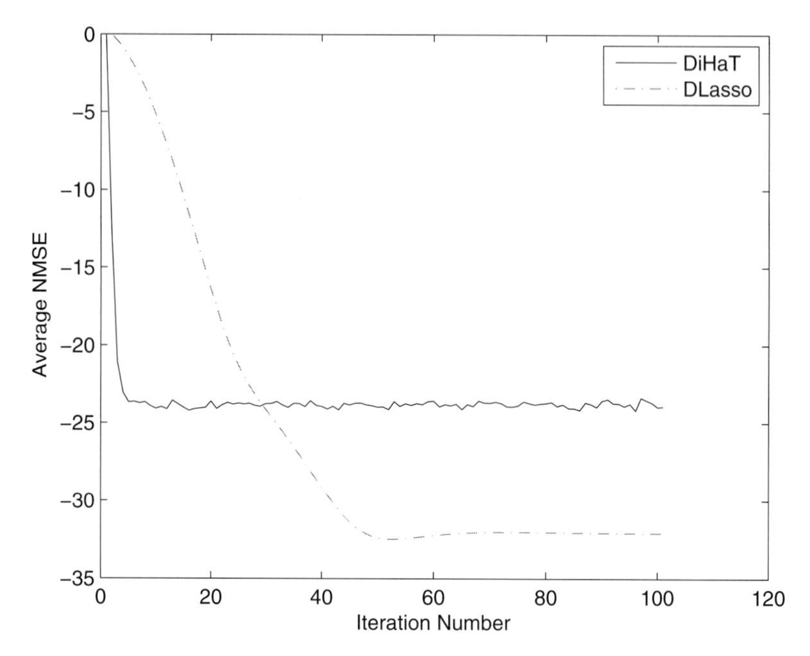

Figure 2.5 Average normalized MSD curves for the second experiment (batch operation).

In the second experiment, the parameters are the same as in the first one, but we increase the sparsity level of the unknown vector to $\|\boldsymbol{h}_*\|_0 = 500$. Figure 2.5 illustrates that the DLasso algorithm converges to an error floor lower than that attained by DiHaT. Moreover, the latter algorithm converges faster than the former one. Intuitively, the increased error floor in the higher sparsity level case is due to the fact that the algorithm fails to identify the true support correctly.

2.4.2 Performance evaluation of online methods

In the first example, the ability of online methods to handle large volumes of data without needing to store them locally will be demonstrated. Moreover, the computational complexity per iteration will be linear with respect to the dimensionality of the unknown vector, i.e. $\mathcal{O}(m)$. The network used comprises $K = 10$ nodes, as shown in Figure 2.6. Furthermore, the number of connections between nodes was intentionally kept small in order to demonstrate the ability of the algorithms to operate reliable under such a scenario. In particular, only 12 connections were considered out 40 that one would have if each node was directly communicating with all the rest. Note that the small number of connections lead to reduced computational complexity in the combine stage of the algorithms, i.e. step 2 of Table 2.3 and step 1 of Table 2.5 for the SpaDiLMS and the SpaDiAPSM respectively.

The dimensionality of the unknown vector is $m = 10\,000$ and each node receives a stream of $N = 5 \cdot 10^5$ data pairs which are processed on-the-fly. Assuming double precision floating-point number format, the values above sum up to a total of more than

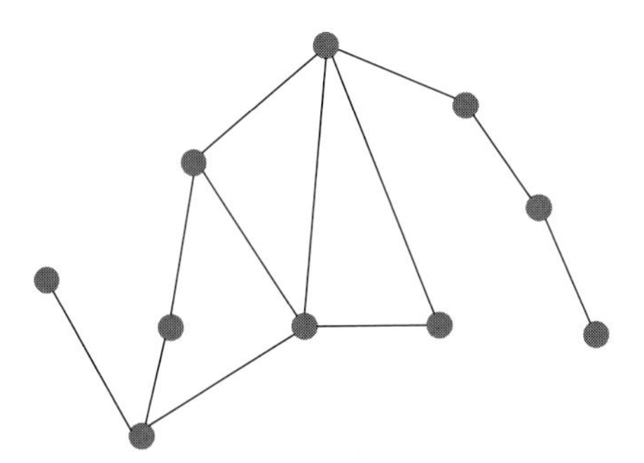

Figure 2.6 The network adopted in the evaluation of the online methods.

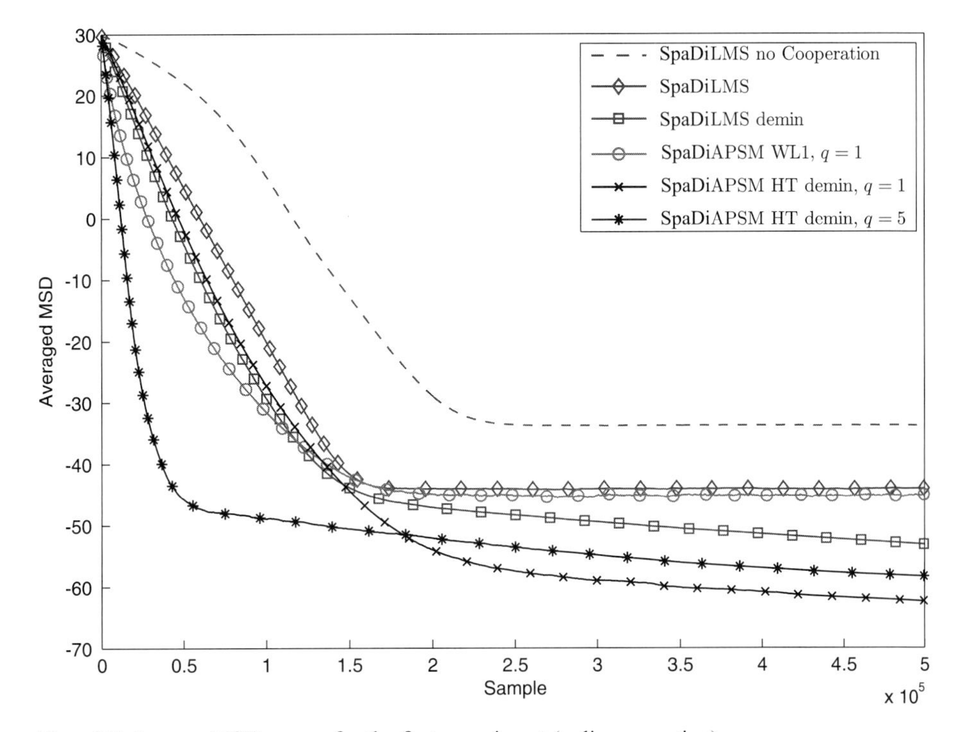

Figure 2.7 Average MSD curves for the first experiment (online operation).

0.3 Tb of data, which are processed in the whole network, something that is not feasible in batch operation. The components of the input vectors are randomly drawn from a uniform distribution, and the SNR is chosen to be different in each node with the best and the worse node having SNR equal to 25 dB and 15 dB respectively. The rest of the nodes admit SNR values linearly spread between these two values. Moreover, the unknown vector is sparse, having $\|\boldsymbol{h}_*\|_0 = 1000$ nonzero values.

The results are shown in Figure 2.7, where, in contrast to the batch case, as a performance measure the non-normalized MSD is used in order to comply with [19, 20]. The curves indicated with diamonds and circles correspond to the SpaDiLMS and the SpaDiAPSM with projections to the weighted ℓ_1 (SpaDiAPSM-WL1) with $q = 1$, $\epsilon = 0.1\sigma_k$, with σ_k being equal to the noise standard deviation in the kth node. The projections performed are variable with the parameter α being set equal to -0.2. With respect to the step size parameters used LMS and SpaDiAPSM-WL1, they where set equal to $6 \cdot 10^{-5}$ and $0.5 \cdot \mathcal{M}_{k,n}$, respectively. The fine tuning of the sparsity promoting regularization parameter λ of SpaDiLMS is hard to be realized in such large data scales so its iterative estimation proposed in [19], eq. (80), is adopted. According to this, SpaDiLMS requires knowledge of the sparsity level (or at least an estimate of it). The same is true for the SpaDiAPSM-WL1 as well, so both algorithms are provided with this piece of information. A somewhat better performance with respect to both convergence speed and error floor is achieved by SpaDiAPSM-WL1. The cooperation of the nodes lead to a significant performance boost something that is demonstrated with the aid of the dashed performance curve. This curve corresponds to the SpaDiLMS operating in a non-cooperative mode, where each node independently processes its own data.

When large volumes of data can be accessed, a scenario typical in big data era, lower error floors can be attained by using diminishing step size parameters $\mu_k(n)$, $n = 1 \ldots \mathcal{N}$. As an example here, step size of the form $\mu_k(n) = \gamma_0(1 + \gamma_0\zeta n)^{-\hat{c}}$, [69, 70], has been examined in order to show the potential of such approach in the distributed online learning set up. The curve denoted with squares is the diminishing step size SpaDiLMS, where $\mu_k(n)$ is the diminishing step size. Clearly, improvements on both convergence speed and error floor over the constant μ_k case (curve with diamonds) can be observed. APSM can also be benefited with diminishing μ_k. The corresponding curve is the one indicated with x-crosses. In this latter curve, instead of projections onto weighted ℓ_1 balls for sparsity promotion (curve with circle), in each iteration hard thresholding (HT) was employed, forcing the 6000 smaller in magnitude components to set to zero whereas the rest of them remain unaltered. Note that HT is an instance of a large family of thresholding operators, which are theoretically supported in APSM [58, 59], and other more advances and computationally complex options could lead to better performance. Different convergence behaviors regarding convergence speed and error floor can be observed depending on the values chosen for γ_0, ζ, and $\hat{c}$ parameters in $\mu_k(n)$, and it is an open problem to optimize the diminishing step size rule with respect to performance. Indicatively, in the SpaDiAPSM-WL1 and SpaDiLMS examples mentioned above the values used were $\gamma_0 = 1$, $\zeta = 0.05$ and $\hat{c} = 10$. Moreover, similarly to the non-distributed case, the convergence achieved with APSM can be speeded up with larger q values, e.g. $q = 5$, as shown in the curve indicated with asterisks. In this latter case, wider hyperslabs have been used with $\epsilon = 1\sigma_k$.

In the next simulation example depicted in Figure 2.8 the tracking ability of the online methods is demonstrated in a setting where the network is the same as in Figure 2.7 and the unknown vector has m and $\|\boldsymbol{h}_*\|_0$, which are set equal to 1000 and 100 respectively. After 5000 instances (received samples) the unknown vector is changed abruptly and, particularly, both its support and the values of the nonzero components are changing in a random way. Again the worse performance correspond to the non-cooperative operation

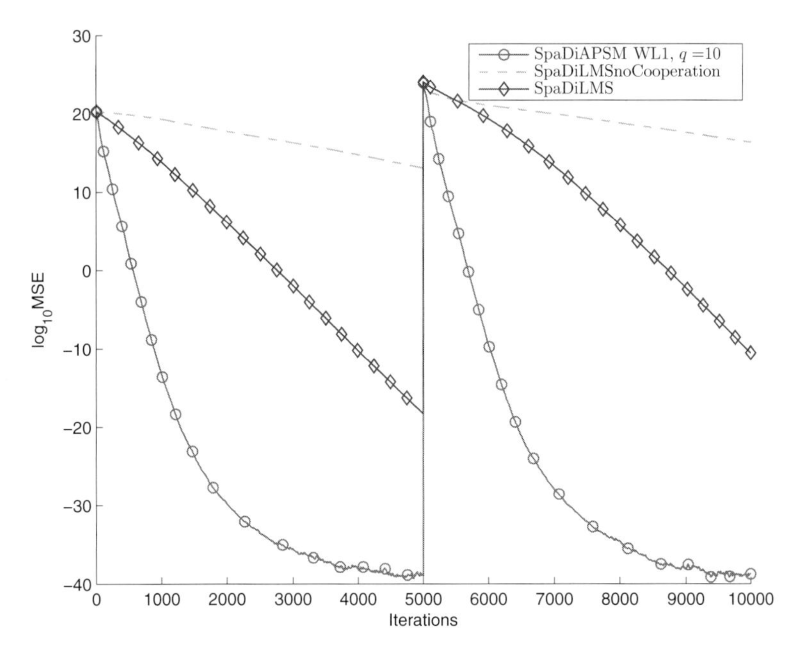

Figure 2.8 Average MSD curves for the second experiment (online operation) with the unknown vector exhibiting an abrupt change.

(dashed curve). When cooperation is in place, neither SpaDiLMS have the power to converge before the vector change (curve marked with diamonds) nor the SpaDiAPSM with $q = 1$ which, in this case, exhibits a performance similar to SpaDiLMS so it is not shown. However, a larger q, e.g. the $q = 10$ case corresponding to curve marked with circles, enhances the tracking potential of the algorithm significantly.

References

[1] L. Atzori, A. Iera, and G. Morabito, "The internet of things: a survey," *Computer networks*, vol. **54**, no. 15, pp. 2787–2805, 2010.

[2] E. Candès and T. Tao, "Near optimal signal recovery from random projections: Universal encoding strategies," *IEEE Transactions on Information Theory*, vol. **52**, no. 12, pp. 5406–5425, 2006.

[3] E. Candès, J. Romberg, and T. Tao, "Robust uncertainty principles: Exact signal reconstruction from highly incomplete Fourier information," *IEEE Transactions on Information Theory*, vol. **52**, no. 2, pp. 489–509, 2006.

[4] D. Donoho, "Compressed sensing," *Information Theory, IEEE Transactions on*, vol. **52**, no. 4, pp. 1289–1306, April 2006.

[5] E. J. Candès, M. B. Wakin, and S. P. Boyd, "Enhancing sparsity by reweighted ℓ_1 minimization," *The Journal of Fourier Analysis and Applications*, vol. **14**, no. 5, pp. 877–905, 2008.

[6] M. F. Duarte and Y. Eldar, "Structured compressed sensing: From theory to applications," *IEEE Transactions on Signal Processing*, vol. **59**, no. 9, pp. 4053–4085, 2011.

[7] S. Theodoridis, Y. Kopsinis, and K. Slavakis, "Sparsity-aware learning and compressed sensing: An overview," 2014, to appear in the *E-Reference Signal Processing*, Elsevier.
[8] C.-T. Chu, S. K. Kim, Y.-A. Lin, Y. Yu, G. Bradski, A. Y. Ng, and K. Olukotun, "Map-reduce for machine learning on multicore," in *NIPS*, vol. **6**, 2006, pp. 281–288.
[9] J. Lin, "Mapreduce is good enough? If all you have is a hammer, throw away everything that's not a nail!" *Big Data*, vol. **1**, no. 1, pp. 28–37, 2013.
[10] I. F. Akyildiz, W. Su, Y. Sankarasubramaniam, and E. Cayirci, "Wireless sensor networks: a survey," *Computer Networks*, vol. **38**, no. 4, pp. 393–422, 2002.
[11] A. H. Sayed, "Diffusion adaptation over networks," arXiv preprint arXiv:1205.4220, 2012.
[12] G. Mateos, J. Bazerque, and G. Giannakis, "Distributed sparse linear regression," *IEEE Trans. Signal Process.*, vol. **58**, no. 10, pp. 5262–5276, 2010.
[13] C. Clifton, M. Kantarcioglu, J. Vaidya, X. Lin, and M. Y. Zhu, "Tools for privacy preserving distributed data mining," *ACM SIGKDD Explorations*, vol. **4**, no. 2, pp. 28–34, 2002.
[14] S. Theodoridis, *Machine Learning: A Bayesian and Optimization Perspective*, Academic Press, 2015.
[15] L. Bottou and O. Bousquet, "The tradeoffs of large scale learning." in *Advances in Neural Information Processing System*, vol. **20**, 2007, pp. 161–168.
[16] J. F. Mota, J. Xavier, P. M. Aguiar, and M. Puschel, "Distributed basis pursuit," *Signal Processing, IEEE Transactions on*, vol. **60**, no. 4, pp. 1942–1956, 2012.
[17] S. Chouvardas, G. Mileounis, N. Kalouptsidis, and S. Theodoridis, "Greedy sparsity-promoting algorithms for distributed learning," *IEEE Transactions on Signal Processing*, vol. **63**, no. 6, pp. 1419–1432, 2015.
[18] S. Patterson, Y. C. Eldar, and I. Keidar, "Distributed compressed sensing for static and time-varying networks," arXiv preprint arXiv:1308.6086, 2013.
[19] P. Di Lorenzo and A. Sayed, "Sparse distributed learning based on diffusion adaptation," *IEEE Transactions on Signal Processing*, vol. **61**, no. 6, pp. 1419–1433, 2013.
[20] S. Chouvardas, K. Slavakis, Y. Kopsinis, and S. Theodoridis, "A sparsity promoting adaptive algorithm for distributed learning," *IEEE Transactions on Signal Processing*, vol. **60**, no. 10, pp. 5412 –5425, oct. 2012.
[21] S. Ono, M. Yamagishi, and I. Yamada, "A sparse system identification by using adaptively-weighted total variation via a primal-dual splitting approach," in *IEEE International Conference on Acoustics, Speech and Signal Processing (ICASSP)*. IEEE, 2013, pp. 6029–6033.
[22] D. L. Donoho, Y. Tsaig, I. Drori, and J.-L. Starck, "Sparse solution of underdetermined systems of linear equations by stagewise orthogonal matching pursuit," *IEEE Transactions on Information Theory*, vol. **58**, no. 2, pp. 1094–1121, 2012.
[23] S. S. Chen, D. L. Donoho, and M. A. Saunders, "Atomic decomposition by basis pursuit," *SIAM journal on Scientific Computing*, vol. **20**, no. 1, pp. 33–61, 1998.
[24] R. Tibshirani, "Regression shrinkage and selection via the lasso," *Journal of the Royal Statistical Society, Series B (Methodological)*, vol. **58**, pp. 267–288, 1996.
[25] A. M. Bruckstein, D. L. Donoho, and M. Elad, "From sparse solutions of systems of equations to sparse modeling of signals and images," *SIAM Review*, vol. **51**, no. 1, pp. 34–81, 2009.
[26] P. A. Forero, A. Cano, and G. B. Giannakis, "Consensus-based distributed support vector machines," *The Journal of Machine Learning Research*, vol. **99**, pp. 1663–1707, 2010.
[27] ——, "Distributed clustering using wireless sensor networks," *Selected Topics in Signal Processing, IEEE Journal of*, vol. **5**, no. 4, pp. 707–724, 2011.
[28] D. P. Bertsekas and J. N. Tsitsiklis, *Parallel and Distributed Computation: Numerical Methods*, Athena-Scientific, second edition, 1999.

[29] J. A. Tropp, "Greed is good: algorithmic results for sparse approximation," *IEEE Transactions on Information Theory*, vol. **50**, no. 10, pp. 2231–2242, 2004.
[30] T. Peleg, Y. Eldar, and M. Elad, "Exploiting statistical dependencies in sparse representations for signal recovery," *IEEE Transactions on Signal Processing*, vol. **60**, no. 5, pp. 2286–2303, 2012.
[31] W. Dai and O. Milenkovic, "Subspace pursuit for compressive sensing signal reconstruction," *IEEE Trans. on Information Theory*, vol. **55**, no. 5, pp. 2230–2249, 2009.
[32] D. Needell and J. Tropp, "Cosamp: Iterative signal recovery from incomplete and inaccurate samples," *Applied and Computational Harmonic Analysis*, vol. **26**, no. 3, pp. 301–321, 2009.
[33] D. Needell and R. Vershynin, "Uniform uncertainty principle and signal recovery via regularized orthogonal matching pursuit," *Found. Comput. Math*, vol. **9**, no. 3, pp. 317–334, 2009.
[34] H. Huang and A. Makur, "Backtracking-based matching pursuit method for sparse signal reconstruction," *IEEE Signal Processing Letters*, vol. **18**, no. 7, pp. 391–394, 2011.
[35] S. Foucart, "Hard thresholding pursuit: an algorithm for compressive sensing," *SIAM Journal on Numerical Analysis*, vol. **49**, no. 6, pp. 2543–2563, 2011.
[36] L. Xiao, S. Boyd, and S. Lall, "A scheme for robust distributed sensor fusion based on average consensus," in *International Symposium on Information Processing in Sensor Networks IPSN*, IEEE, 2005, pp. 63–70.
[37] C. Ravazzi, S. M. Fosson, and E. Magli, "Distributed soft thresholding for sparse signal recovery," arXiv preprint arXiv:1301.2130, 2013.
[38] C. Lopes and A. Sayed, "Diffusion least-mean squares over adaptive networks: Formulation and performance analysis," *IEEE Trans. Signal Process.*, vol. **56**, no. 7, pp. 3122–3136, 2008.
[39] F. Cattivelli and A. Sayed, "Diffusion LMS strategies for distributed estimation," *IEEE Trans. Signal Process.*, vol. **58**, no. 3, pp. 1035 –1048, 2010.
[40] R. L. Cavalcante, I. Yamada, and B. Mulgrew, "An adaptive projected subgradient approach to learning in diffusion networks," *IEEE Transactions on Signal Processing*, vol. **57**, no. 7, pp. 2762–2774, 2009.
[41] S. Chouvardas, K. Slavakis, and S. Theodoridis, "Adaptive robust distributed learning in diffusion sensor networks," *IEEE Trans. Signal Process.*, vol. **59**, no. 10, pp. 4692–4707, 2011.
[42] I. Schizas, G. Mateos, and G. Giannakis, "Distributed LMS for consensus-based in-network adaptive processing," *IEEE Trans. Signal Process.*, vol. **57**, no. 6, pp. 2365 –2382, 2009.
[43] G. Mateos, I. D. Schizas, and G. B. Giannakis, "Performance analysis of the consensus-based distributed LMS algorithm," *EURASIP Journal on Advances in Signal Processing*, vol. **2009**, p. 68, 2009.
[44] A. Sayed, *Fundamentals of Adaptive Filtering*, John Wiley & Sons, New Jersey, 2003.
[45] S. Haykin, *Adaptive Filter Theory*, Prentice Hall, 1996.
[46] A. Sayed, *Adaptive Filters*, John Wiley and Sons, 2008.
[47] Y. Chen, Y. Gu, and A. O. Hero, "Sparse lms for system identification," in *IEEE International Conference on Acoustics, Speech and Signal Processing (ICASSP) 2009*, IEEE, 2009, pp. 3125–3128.
[48] P. L. Combettes, "The foundations of set theoretic estimation," *Proceedings of the IEEE*, vol. **81**, no. 2, pp. 182–208, 1993.

[49] H. Stark and Y. Yang, *Vector Space Projections: a Numerical Approach to Signal and Image Processing, Neural Nets, and Optics*, John Wiley & Sons, Inc., 1998.

[50] R. T. Rockafellar, *Convex Analysis*, Princeton University Press, 1997, vol. **28**.

[51] Y. Kopsinis, K. Slavakis, and S. Theodoridis, "Online sparse system identification and signal reconstruction using projections onto weighted balls," *IEEE Transactions on Signal Processing*, vol. **59**, no. 3, pp. 936–952, 2011.

[52] I. Yamada and N. Ogura, *Adaptive Projected Subgradient Method for Asymptotic Minimization of Sequence of Nonnegative Convex Functions*, Taylor & Francis, 2005.

[53] K. Slavakis, I. Yamada, and N. Ogura, "The adaptive projected subgradient method over the fixed point set of strongly attracting nonexpansive mappings," *Numerical Functional Analysis and Optimization, 27*, vol. **7**, no. 8, pp. 905–930, 2006.

[54] K. Slavakis and I. Yamada, "The adaptive projected subgradient method constrained by families of quasi-nonexpansive mappings and its application to online learning," *SIAM Journal on Optimization*, vol. **23**, no. 1, pp. 126–152, 2013.

[55] S. Theodoridis, K. Slavakis, and I. Yamada, "Adaptive learning in a world of projections," *IEEE Signal Processing Magazine*, vol. **28**, no. 1, pp. 97–123, 2011.

[56] S. P. Boyd and L. Vandenberghe, *Convex Optimization*, Cambridge University Press, 2004.

[57] Y. Kopsinis, K. Slavakis, S. Theodoridis, and S. McLaughlin, "Reduced complexity online sparse signal reconstruction using projections onto weighted ℓ_1 balls," in *Digital Signal Processing (DSP), 2011 17th International Conference on*, July 2011, pp. 1–8.

[58] K. Slavakis, Y. Kopsinis, S. Theodoridis, and S. McLaughlin, "Generalized thresholding and online sparsity-aware learning in a union of subspaces," *Signal Processing, IEEE Transactions on*, vol. **61**, no. 15, pp. 3760–3773, August 2013.

[59] Y. Kopsinis, K. Slavakis, S. Theodoridis, and S. McLaughlin, "Generalized thresholding sparsity-aware algorithm for low complexity online learning," in *Acoustics, Speech and Signal Processing (ICASSP), 2012 IEEE International Conference on*, March 2012, pp. 3277–3280.

[60] M. Yukawa and I. Yamada, "A unified view of adaptive variable-metric projection algorithms," *EURASIP Journal on Advances in Signal Processing*, vol. **2009**, p. 34, 2009.

[61] D. L. Duttweiler, "Proportionate NLMS adaptation in echo cancelers," *IEEE Trans. Speech Audio Processing*, vol. **8**, pp. 508–518, 2000.

[62] J. Benesty and S. L. Gay, "An improved PNLMS algorithm," in *Acoustics, Speech, and Signal Processing (ICASSP), 2002 IEEE International Conference on*, vol. **2**, May 2002, pp. II-1881–II-1884.

[63] K. Slavakis, Y. Kopsinis, S. Theodoridis, G. B. Giannakis, and V. Kekatos, "Generalized iterative thresholding for sparsity-aware online volterra system identification," in *Wireless Communication Systems (ISWCS 2013), Proceedings of the Tenth International Symposium on*, VDE, 2013, pp. 1–5.

[64] M. Bhotto and A. Antoniou, "Robust set-membership affine-projection adaptive-filtering algorithm," *Signal Processing, IEEE Transactions on*, vol. **60**, no. 1, pp. 73 –81, January 2012.

[65] V. Kekatos and G. Giannakis, "From sparse signals to sparse residuals for robust sensing," *Signal Processing, IEEE Transactions on*, vol. **59**, no. 7, pp. 3355–3368, July 2011.

[66] M. Rabbat and R. Nowak, "Distributed optimization in sensor networks," in *Proceedings of the 3rd International Symposium on Information Processing in Sensor Networks*, ACM, 2004, pp. 20–27.

[67] G. Papageorgiou, P. Bouboulis, and S. Theodoridis, "Robust kernel-based regression using orthogonal matching pursuit," in *IEEE International Workshop on Machine Learning for Signal Processing (MLSP)*, IEEE, 2013, pp. 1–6.

[68] Y. Kopsinis, K. Slavakis, S. Theodoridis, and S. McLaughlin, "Reduced complexity online sparse signal reconstruction using projections onto weighted ℓ_1 balls," in *Digital Signal Processing (DSP), 2011 17th International Conference on*, IEEE, 2011, pp. 1–8.

[69] L. Bottou, "Large-scale machine learning with stochastic gradient descent," in *Proceedings of the 19th International Conference on Computational Statistics (COMPSTAT'2010)*, Y. Lechevallier and G. Saporta, Eds., Paris, France: Springer, August 2010, pp. 177–187. [Online]. Available: http://leon.bottou.org/papers/bottou-2010

[70] W. Xu, "Towards optimal one pass large scale learning with averaged stochastic gradient descent," arXiv preprint arXiv:1107.2490, 2011. [Online]. Available: http://arxiv.org/abs/1107.2490

3 Optimization algorithms for big data with application in wireless networks

Mingyi Hong, Wei-Cheng Liao, Ruoyu Sun, and Zhi-Quan Luo

This chapter proposes the use of modern first-order large-scale optimization techniques to manage a cloud-based densely deployed next-generation wireless network. In the first part of the chapter we survey a few popular first-order methods for large-scale optimization, including the block coordinate descent (BCD) method, the block successive upper-bound minimization (BSUM) method and the alternating direction method of multipliers (ADMM). In the second part of the chapter, we show that many difficult problems in managing large wireless networks can be solved efficiently and in a parallel manner, by modern first-order optimization methods. Extensive numerical results are provided to demonstrate the benefit of the proposed approach.

3.1 Introduction

3.1.1 Motivation

The ever-increasing demand for rapid access to large amounts of data *anywhere anytime* has been the driving force in the current development of next-generation wireless network infrastructure. It is projected that within 10 years, the wireless cellular network will offer up to 1000× throughput performance over the current 4G technology [1]. By that time the network should also be able to deliver a fiber-like user experience, boasting 10 Gb/s individual transmission rate for data-intensive cloud-based applications.

Achieving this lofty goal requires revolutionary infrastructure and highly sophisticated resource management solutions. A promising network architecture to meet this requirement is the so-called cloud-based radio access network (RAN), where a large number of networked base stations (BSs) are deployed for wireless access, while powerful cloud centers are used at the back end to perform centralized network management [1–4]. Intuitively, a large number of networked access nodes, when intelligently provisioned, will offer significantly improved spectrum efficiency, real-time load balancing and hotspot coverage. In practice, the optimal network provisioning is extremely challenging, and its success depends on smart joint backhaul provisioning, physical layer transmit/receive schemes, BS/user cooperation and so on.

Big Data over Networks, ed. Shuguang Cui, Alfred O. Hero III, Zhi-Quan Luo, and José M. F. Moura. Published by Cambridge University Press.

This chapter proposes the use of modern first-order large-scale optimization techniques to manage a cloud-based densely deployed next-generation wireless network. We show that many difficult problems in this domain can be solved efficiently and in a parallel manner, by advanced optimization algorithms such as the block successive upper-bound minimization (BSUM) method and the alternating direction methods of multipliers (ADMM) method.

3.1.2 The organization of the chapter

To begin with, we introduce a few well-known first-order optimization algorithms. Our focus is on algorithms suitable for solving problems with certain block-structure, where the optimization variables can be divided into (possibly overlapping) blocks. Next we show that this type of block-structured problem turns out to be crucial in modeling many network provisioning problems arising in next-generation network design. A few detailed examples are provided to demonstrate the applicability of the first-order optimization algorithms in large-scale data delivery and network provisioning. Numerical examples are given at the end to demonstrate the efficiency of the algorithms studied throughout the article.

3.2 First-order algorithms for big data

In this chapter we consider algorithms that can solve the block-structured optimization problems of the following form

$$\underset{x}{\text{minimize}} \quad f(x_1, x_2, \ldots, x_n), \qquad \text{s.t. } (x_1, x_2, \ldots, x_n) \in \mathcal{X}, \tag{3.1}$$

where $f(\cdot)$ is a continuous function (possibly nonconvex and nonsmooth), $\mathcal{X}$ is a closed convex set, and each $x_i \in \mathbb{R}^{m_i}$ is a block variable, $i = 1, 2, \ldots, n$. Later we will see that this type of problem appears frequently in many network provisioning problems that arise in next-generation network design.

3.2.1 The block coordinate descent algorithm

In practice, solving (3.1) directly can be very challenging, due to either its nonconvexity, nonsmoothness, or the sheer problem size. However, consider the special case of (3.1) where the constraint set has a Cartesian product structure: $\mathcal{X} = \prod_{i=1}^{n} \mathcal{X}_i$, and the nonsmooth part of the objective is separable among the variables. A well-known technique for such special case is the so-called block coordinate descent (BCD) method whereby, at every iteration, a single block of variables is optimized while the remaining blocks are held fixed. More specifically, we consider the following special case of problem (3.1)

$$\begin{aligned} \underset{x}{\text{minimize}} \quad & f(x) = h_0(x_1, x_2, \ldots, x_n) + \sum_{i=1}^{n} h_i(x_i), \\ \text{s.t.} \quad & x_i, \in \mathcal{X}_i, \; i = 1, \ldots, n, \end{aligned} \tag{3.2}$$

where $h_0(\cdot)$ is a smooth function (possibly nonconvex), and $h_i(\cdot), i = 1, \ldots, n$ are convex functions (possibly nonsmooth). When following the classic Gauss–Seidel (G-S) update rule, at iteration t, the block $i = (t \bmod n) + 1$ is updated by

$$x_i^{(t)} \in \arg\min_{y_i \in \mathcal{X}_i} \quad f\left(x_1^{(t-1)}, \ldots, x_{i-1}^{(t-1)}, y_i, x_{i+1}^{(t-1)}, \ldots, x_n^{(t-1)}\right) \tag{3.3}$$

while the remaining blocks are kept unchanged, i.e. $x_k^{(t)} = x_k^{(t-1)}$ for all $k \neq i$. Since each step involves solving a simple subproblem of small size, the BCD method can be quite effective for solving large-scale problems, provided that certain regularity conditions are met. For instance, the existing analysis of the BCD method [5–7] requires the uniqueness of the minimizer for the subproblems (3.3), or the quasi-convexity of f. Below is a summary of the convergence results of the BCD method for solving (3.2).

Theorem 3.1 *Assume that the level set $X^0 = \{f(x) \leq f(x^0)\}$ is compact. Then the sequence $\{x^{(t)} = (x_1^{(t)}, \ldots, x_n^{(t)})\}$ generated by the BCD method is well-defined and bounded. Further, we have the following.*

(1) If $f(x_1, \ldots, x_n)$ is pseudoconvex in (x_k, x_i) for every $(i, k) \in \{1, \ldots, n\}$, then every cluster point of $\{x^{(t)}\}$ is a stationary point of f.

(2) If $f(x_1, \ldots, x_n)$ has at most one minimum in x_k for $k = 2, \ldots, n-1$, then every cluster point z of $\{x^{(t)}\}_{t=(n-1)\mathrm{mod} n}$ is a stationary point of f.

This result is adapted from [5, Theorem 4.1], where the "regularity" of f therein is implied by the smooth plus separable nonsmooth objective of problem (3.2). Further, the "stationary solutions" here are the solutions that satisfying the first-order optimality condition; see [5] for the precise definition.

When $f(\cdot)$ is a convex function, it is possible to characterize the *rate of convergence* for BCD-type algorithm. For example, when the objective function is strongly convex, the BCD algorithm converges globally linearly [8], that is

$$f(x^{(t+1)}) - f(x^*) \leq c\left(f(x^{(t)}) - f(x^*)\right) \tag{3.4}$$

for some constant $0 < c < 1$. When the objective function is smooth but not strongly convex, Luo and Tseng have shown that the BCD method with the G-S rule converges linearly, provided that a certain local error bound is satisfied around the solution set [8–10]. For more general convex problems, several recent studies have established the $\mathcal{O}(1/t)$ iteration complexity for various BCD-type algorithms [11–14]. In these works, it is shown that when the problem satisfies certain regularity conditions, and when the coordinates are selected according to certain probability distribution, then the bound of the following type is true:

$$\mathbb{E}\left[f(x^{(t)}) - f(x^*)\right] \leq \frac{d}{t}, \tag{3.5}$$

where the expectation is taken over the randomization of the choice of the coordinates, and $d > 0$ is some constant. When the coordinates are updated according to the traditional G-S rule, a few recent works [15–17] have proven the $\mathcal{O}(1/t)$ rate for the G-S BCD algorithm when applied to certain special convex problems. Some recent works [18, 19] propose BCD-based algorithms with parallel block update rules. These algorithms are designed for both convex and nonconvex problems, and the built-in parallelism offers a significant speed up in computation when multiple computing nodes are available.

It is important to note that without the assumptions such as the uniqueness of the minimizers of the subproblems or the separability of the constraint set, the BCD method may get stuck at a non-stationary point of the problem (see [20] and [21] for well-known examples). Unfortunately, sometimes these assumptions can be restrictive in practice. We will show how to generalize the BCD method when these assumptions are not satisfied in the following subsections.

3.2.2 The ADMM algorithm

In many contemporary applications involving big data, the objective function of (3.1) is convex separable, and the block variables are linearly coupled in the constraint:

$$\begin{aligned} \text{minimize} \quad & f(x) = f_1(x_1) + f_2(x_2) + \cdots + f_n(x_n) \\ \text{s.t.} \quad & Ex = E_1 x_1 + E_2 x_2 + \cdots + E_n x_n = q, \\ & x_i \in \mathcal{X}_i, \ i = 1, \ldots, n \end{aligned} \tag{3.6}$$

where $E = (E_1, \ldots, E_n)$ is the partition of matrix E corresponding to the block variables $x_1, \ldots, x_n$.

Directly applying the BCD method to problem (3.6) may fail to find any (local) optimal solution. For instance, the following simple quadratic problem has an optimal objective of 0, but the BCD method can get stuck at the non-interesting point $(1, -1)$:

$$\text{minimize} \quad x_1^2 + x_2^2, \quad \text{s.t.} \quad x_1 + x_2 = 0.$$

In the ADMM method, instead of maintaining feasibility all the time, the constraint $Ex = q$ is dualized using the Lagrange multiplier y and a quadratic penalty term is added. The resulting *augmented Lagrangian function* is of the form:

$$L(x; y) = f(x) + \langle y, q - Ex \rangle + \frac{\rho}{2} \|q - Ex\|^2, \tag{3.7}$$

where $\rho > 0$ is a constant and $\langle \cdot, \cdot \rangle$ denotes the inner product operator. The ADMM method updates the primal block variables $x_1, \ldots, x_n$ similarly to BCD to minimize $L(x; y)$, which often leads to simple subproblems with closed-form solutions. These updates are followed by a gradient ascent update of the dual variable y. Equation (3.8) summarizes the ADMM method.

Alternating direction method of multipliers (ADMM)

At each iteration $t \geq 1$:

$$\begin{cases} x_i^{(t+1)} = \arg\min_{x_i} L\left(x_1^{(t+1)}, \ldots, x_{i-1}^{(t+1)}, x_i, x_{i+1}^{(t)}, \ldots, x_n^{(t)}; y^{(t)}\right), \ i = 1, 2, \ldots, n, \\ y^{(t+1)} = y^{(t)} + \alpha\left(q - Ex^{(t+1)}\right) = y^{(t)} + \alpha\left(q - \sum_{i=1}^{n} E_i x_i^{(t+1)}\right), \end{cases} \tag{3.8}$$

where $\alpha > 0$ is the step size for the dual update.

Although the ADMM algorithm was introduced as early as 1976 by Gabay, Mercier, Glowinski, and Marrocco [22, 23], it became popular only recently due to its applications in modern large-scale optimization problems arising from machine learning and computer vision [24–28]. In practice, the algorithm is often computationally very efficient and exhibits much faster convergence than other traditional algorithms such as the dual ascent algorithm [29–31] or the method of multipliers [32].

When there are only two block variables ($n = 2$), the ADMM converges under very mild conditions; see the following basic result from [7, Proposition 4.2].

Theorem 3.2 *Suppose that $n = 2$ and $\alpha = \rho$. Assume that the optimal solution set of problem* (3.6) *is non-empty, and $E_1^T E_1$ and $E_2^T E_2$ are invertible. Then the sequence of $\{x^{(t)}, y^{(t)}\}$ generated by the ADMM algorithm is bounded and every limit point of $\{x^{(t)}, y^{(t)}\}$ is an optimal primal-dual solution of problem* (3.6).

Several recent works [33, 34] have shown that the ADMM method converges with the rate of $\mathcal{O}(\frac{1}{t})$. Moreover, references [35–37] have shown that the ADMM converges linearly when the objective function is strongly convex and there are only two blocks of variables. Unfortunately, the understanding of the algorithm for the case of $n \geq 3$ is still very limited. In fact, the convergence of the ADMM method for the case of $n \geq 3$ has been an open question since the late 1980s, precluding its direct application to many important problems such as the robust PCA [38]. Recent advances in extending the convergence analysis of ADMM to multiple-block case can be found for example in [39–42].

3.2.3 The BSUM method

If the per-block subproblem (3.3) is *nonconvex*, the BCD algorithm cannot be used due to the difficulty in solving each of the subproblems. To broaden the applicability of the BCD method, a block successive upper-bound minimization (BSUM) is proposed in [43], in which a sequence of approximate versions (e.g. upper bounds) of the objective function is minimized. It is shown that in many applications it is possible to construct subproblems with simple solutions.

Specifically, at each iteration t of the BSUM method, one chooses an index set $I^{(t)} \subseteq \{1, 2, \ldots, n\}$ and performs the following update

$$\begin{cases} x_i^{(t+1)} = \operatorname{argmin}_{x_i \in \mathcal{X}_i} g_i(x_i; x^{(t)}) + h_i(x_i), \ \forall\, i \in I^{(t)} \\ x_i^{(t+1)} = x_i^{(t)}, \ \forall\, i \notin I^{(t)}, \end{cases} \tag{3.9}$$

where $g_i(x_i; z)$ is an approximation of the smooth function $h_0(x_i, z_{-i})$ at a given z which satisfies the following assumption.

Assumption A.

$$g_i(x_i; x) = h_0(x), \quad \forall\, x \in \mathcal{X}, \ \forall\, i \tag{A1}$$

$$g_i(x_i; z) \geq h_0(x_i, z_{-i}), \quad \forall\, x_i \in \mathcal{X}_i, \ \forall\, z \in \mathcal{X}, \forall\, i \tag{A2}$$

$$\nabla g_i(z_i; z) = \nabla_i h_0(z), \quad \forall\ z_i \in \mathcal{X}_i,\ \forall\, i \tag{A3}$$

$$g_i(x_i; z) \text{ is continuous in } (x_i, z), \quad \forall\, i \tag{A4}$$

$$g_i(x_i; z) \text{ is strictly convex in } x_i, \quad \forall\, i. \tag{A5}$$

The assumptions (A1) and (A2) imply that the approximation function is a *global upper bound* of $h_0(x)$; while the assumption (A3) guarantees that the first-order behavior of the objective function and the approximation function are the same.

The BSUM method has wide application in various engineering domains. Many well-known existing algorithms for solving *both* convex and nonconvex problems are in fact special cases of BSUM. Examples include the proximal gradient method [44], the alternating least square (ALS) method for tensor decomposition [45], the weighted minimum mean square error (WMMSE) algorithm in wireless communication [46], the EM algorithm in statistics [47], the convex concave procedure (CCP) [48], the majorization minimization method (MM) [49] and the nonnegative matrix factorization [50] for machine learning. One related method is the inner approximation algorithm (IAA) developed by Marks and Wright in [51]. Its convergence analysis is quite restrictive: it is applicable only to smooth problems with a single block variable. Moreover, convergence to a stationary solution is established only under the unreasonable assumption that the *whole iterate sequence* converges (see [51, Theorem 1]).

Below we present a general convergence theorem for the BSUM method.

Theorem 3.3 *The following hold true.*

(a) Suppose that the function $g_i(x_i; y)$ is quasi-convex in x_i and Assumptions (A1)–(A4) *hold. Further assume that the subproblem* (3.9) *has a unique solution for all $x^{(t-1)} \in \mathcal{X}$. Then every limit point z of the iterates generated by the BSUM algorithm is a stationary point of* (3.2).

(b) Suppose the level set $\mathcal{X}^{(0)} = \{x \mid f(x) \le f(x^{(0)})\}$ is compact and Assumptions (A1)–(A4) *hold. Further assume that the subproblem* (3.9) *has a unique solution for any point $x^{(t-1)} \in \mathcal{X}$ for at least $n-1$ blocks. Then the iterates generated by the BSUM algorithm converge to the set of stationary points.*

This result is adapted from [43, Theorem 2], where again the "regularity" of f is implied by the smooth plus nonsmooth structure of the objective in (3.2). The convergence of BSUM algorithm can also be established under other assumptions. For example, it is possible to drop the uniqueness requirement in the solution of subproblems provided we update the block that provides the maximum amount of improvement; see [43, 52].

To close this section, we remind the readers that the main strength of all the first-order algorithms discussed in this chapter lies in the simplicity of solving their subproblems. Therefore, when applying these algorithms to solve practical problems, it is often desirable to find the right problem structure that leads to easy updates. In the next section we will show how this can be done for a wide class of network provisioning problems.

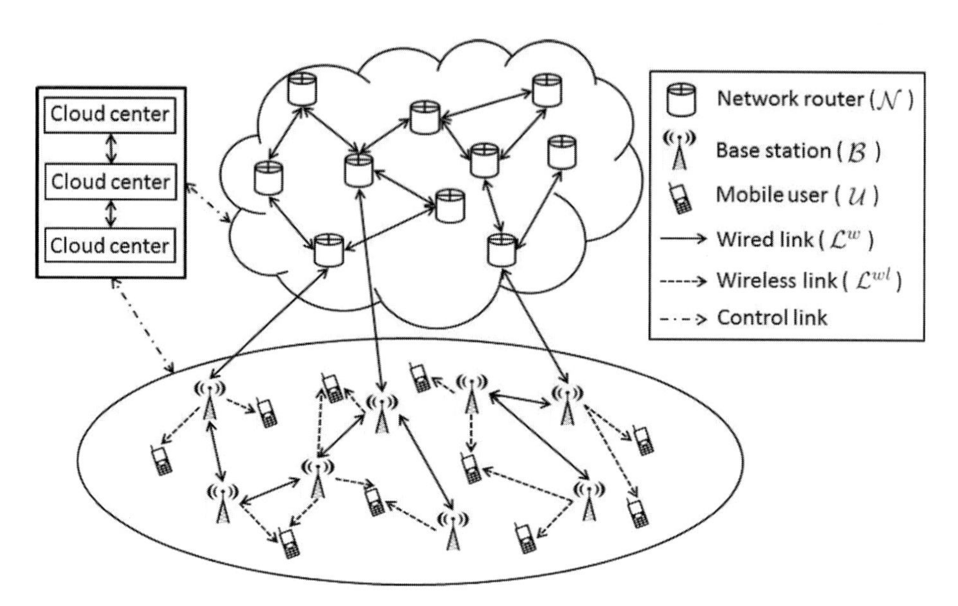

Figure 3.1 Illustration of the considered system.

3.3 Application to network provisioning problem

In this section, we provide a few concrete examples to demonstrate the applicability of the first-order algorithms such as BSUM and ADMM for large-scale network provisioning problems. We start with describing the general network setting.

3.3.1 The setting

We first describe the generic network model to be studied in the subsequent discussion; see Figure 3.1 for an illustration. For simplicity, we consider the *downlink* direction in which the traffic flows from the network to the users.

Consider the next-generation access network consisting both the wired backhaul network, which delivers the data flow from the core network to the BSs, and the wireless radio access network (RAN) that transfers the data wirelessly to the users. The wireless RAN consists of a set of BSs $\mathcal{B}$, a set of mobile users $\mathcal{U}$ and a set of wireless links:

$$\mathcal{L}^{\text{wl}} \triangleq \{(s_\ell, d_\ell) \in \mathcal{B} \times \mathcal{U}\}. \tag{3.10}$$

Here we have used s_ℓ and d_ℓ to denote the BS–user pair that uniquely defines a link ℓ. Also suppose that each node in the system has a single antenna, and use $h_{d_\ell s_k} \in \mathbb{C}$, or simply $h_{\ell k} \in \mathbb{C}$, to denote the channel between BS s_k and user d_ℓ. Using this notation, the wireless link ℓ is said to be interfered by the set of wireless links $\mathcal{I}(\ell) \triangleq \{k \in \mathcal{L}^{\text{wl}} \mid h_{\ell k} \neq 0\}$. See Figure 3.2 for an illustration of this simple network setting.

For a wireless link $\ell \in \mathcal{L}^{\text{wl}}$, BS s_ℓ uses a linear precoder $v_\ell \in \mathbb{C}$ to transmit to user d_ℓ. Use $\mathbf{v}$ to collect the precoders from all the BSs. Then the transmit rate achievable over

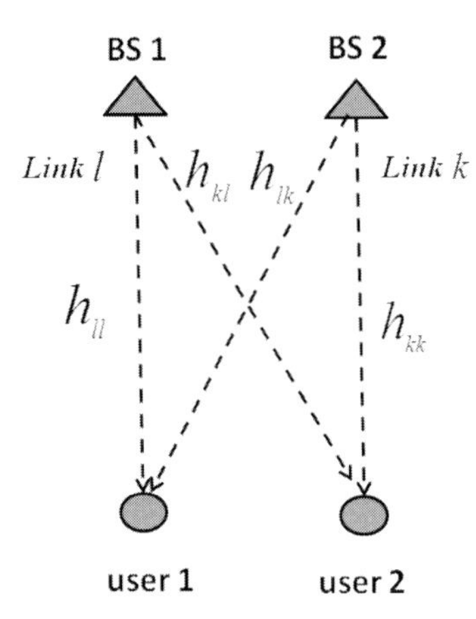

Figure 3.2 Illustration of the wireless part of the network.

the link ℓ is given by

$$r_\ell(\mathbf{v}) \triangleq \log\left(1 + \frac{|h_{\ell\ell}|^2|v_\ell|^2}{\sum_{n\in\mathcal{I}(\ell)\backslash\{\ell\}} |h_{\ell n}|^2|v_n|^2 + \sigma_\ell^2}\right), \tag{3.11}$$

where σ_ℓ^2 is the variance of AWGN noise at mobile user d_ℓ.

To describe the entire access network, let $\mathcal{V}$ denote the set of nodes in the network, including a set of network routers $\mathcal{N}$, a set of BSs $\mathcal{B}$, and a set of mobile users $\mathcal{U}$. Let $\mathcal{L}$ denote the set of directed links that connect the nodes of $\mathcal{V}$. The set $\mathcal{L}$ consists both wireless links and wired links: the former is defined in (3.10), and the latter connects the nodes in the backhaul and is given by

$$\mathcal{L}^{\text{w}} \triangleq \{(s_\ell, d_\ell) \in \mathcal{L} \mid \forall\, s_\ell, d_\ell \in \mathcal{N} \cup \mathcal{B}\}. \tag{3.12}$$

Suppose a set of $\mathcal{M}$ flows is to be delivered from the network to the users, and each $m \in \mathcal{M}$ has a source $s(m) \in \mathcal{V}$ and a sink $d(m) \in \mathcal{V}$. Owing to the fact that only the downlink direction is considered, $d(m)$ must be one of the users. We use $r(m) \geq 0$ and $r_\ell(m) \geq 0$ to denote the rate of flow m and the rate of flow m on link $\ell \in \mathcal{L}$, respectively. Define

$$\mathbf{r} \triangleq \{r(m), r_\ell(m)\}_{m\in\mathcal{M},\ell\in\mathcal{L}}.$$

We use the following notation for the set of links going into and coming out of a node v respectively

$$\text{In}(v) \triangleq \{\ell \in \mathcal{L} \mid\ d_\ell = v\} \quad \text{and} \quad \text{Out}(v) \triangleq \{\ell \in \mathcal{L} \mid\ s_\ell = v\}. \tag{3.13}$$

It is important to note that using the above network model, we implicitly allow a mobile user to be served by more than one BSs; see Figure 3.3 for an illustration.

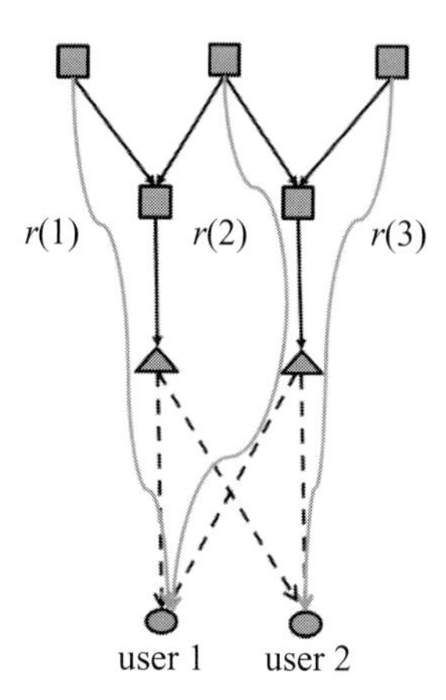

Figure 3.3 Illustration of the data flow through the network.

Below we describe a few link-level constraints that regulate the data flows.

(1) **Wired link capacity constraint** Assume each wired link $\ell \in \mathcal{L}^{\mathrm{w}}$ has a fixed capacity C_ℓ. The total flow rate on link ℓ is constrained by

$$\sum_{m \in \mathcal{M}} r_\ell(m) \le C_\ell, \ \forall\, \ell \in \mathcal{L}^{\mathrm{w}}. \tag{3.14}$$

(2) **Wireless link capacity constraint** The total flow rates on a given wireless link $\ell \in \mathcal{L}^{\mathrm{wl}}$ should not exceed the capacity (3.11):

$$\sum_{m \in \mathcal{M}} r_\ell(m) \le r_\ell(\mathbf{v}) = \log\left(1 + \frac{|h_{\ell\ell}|^2 |v_\ell|^2}{\sum_{n \in \mathcal{I}(\ell)\setminus\{\ell\}} |h_{\ell n}|^2 |v_n|^2 + \sigma_\ell^2}\right), \ \forall\, \ell \in \mathcal{L}^{\mathrm{wl}}. \tag{3.15}$$

(3) **Flow conservation constraint** For any node $v \in \mathcal{V}$, the total incoming flow should be equal to the total outgoing flow:

$$\sum_{\ell \in \mathrm{In}(v)} r_\ell(m) + 1_{s(m)}(v) r(m) = \sum_{\ell \in \mathrm{Out}(v)} r_\ell(m) + 1_{d(m)}(v) r(m),$$
$$\forall\, m \in \mathcal{M}, \ \forall\, v \in \mathcal{V} \tag{3.16}$$

where the notation $1_{\mathcal{A}}(x)$ denotes the indicator function for a set $\mathcal{A}$, i.e. $1_{\mathcal{A}}(x) = 1$ if $x \in \mathcal{A}$, and $1_{\mathcal{A}}(x) = 0$ otherwise.

(4) **BS power budget constraint** The transmit power used by each BS $b \in \mathcal{B}$ should not exceed given budget $\bar{p}_b \geq 0$:

$$\sum_{\ell \in \text{Out}(b) \bigcap \mathcal{L}^{\text{wl}}} |v_\ell|^2 \leq \bar{p}_b, \ \forall \, b \in \mathcal{B}. \tag{3.17}$$

We are interested in data delivery problem formulated in the following utility maximization form

$$\begin{aligned} \max_{\mathbf{v},\mathbf{r}} \quad & U\left(\{r(m)\}_{m \in \mathcal{M}}\right) \\ \text{st} \quad & (3.14), (3.15), (3.16), (3.17) \\ & r(m) \geq 0, \ \forall \, m \in \mathcal{M} \end{aligned} \tag{3.18}$$

where $U(\cdot)$ is the system utility function which measures the performance of the entire network. We note that the system utility maximization problem (3.18) is described in a fairly simple manner to facilitate presentation. The solutions described here can be applied to more complicated formulations that involve nodes with multiple transmit/receive antennas as well as nodes capable of operating on multiple frequency channels. We refer the readers to [2, 53] for extended discussions.

Next we describe a decomposition-based optimization approach to solve the utility maximization problem (3.18). To gain some insights into the problem, we first consider the idealized scenario in which the capacity of the backhaul links is infinite. In this case the problem reduces to a resource management problem for the wireless access network only. We will show that for a large-family of utility functions, problem (3.18) can be solved effectively by BSUM algorithm. Using the insights obtained from this special case, we then generalize the approach to the full-fledged network provisioning problem with limited backhaul capacity.

3.3.2 Network with an uncapacitated backhaul

In this section, we consider a simplified network that has infinite backhaul capacity, and each user gets precisely a single flow; see Figure 3.4 for an illustration. In this case $\mathcal{M} = \mathcal{U}$, and each wireless link $\ell \in \mathcal{L}^{\text{wl}}$ carries a single flow, denoted as $r_\ell(\mathbf{v})$. The utility maximization problem (3.18) reduces to

$$\begin{aligned} \max_{\mathbf{v},\mathbf{r}} \quad & U\left(\{r_\ell(\mathbf{v})\}_{\ell \in \mathcal{L}^{\text{wl}}}\right) \\ \text{s.t.} \quad & \sum_{\ell \in \text{Out}(b) \bigcap \mathcal{L}^{wl}} |v_\ell|^2 \leq \bar{p}_b, \ \forall \, b \in \mathcal{B}, \\ & r_\ell(\mathbf{v}) \leq \log\left(1 + \frac{|h_{\ell\ell}|^2 |v_\ell|^2}{\sum_{n \in \mathcal{I}(\ell) \setminus \{\ell\}} |h_{\ell n}|^2 |v_n|^2 + \sigma_\ell^2}\right), \ \forall \, \ell \in \mathcal{L}^{\text{wl}}. \end{aligned} \tag{3.19}$$

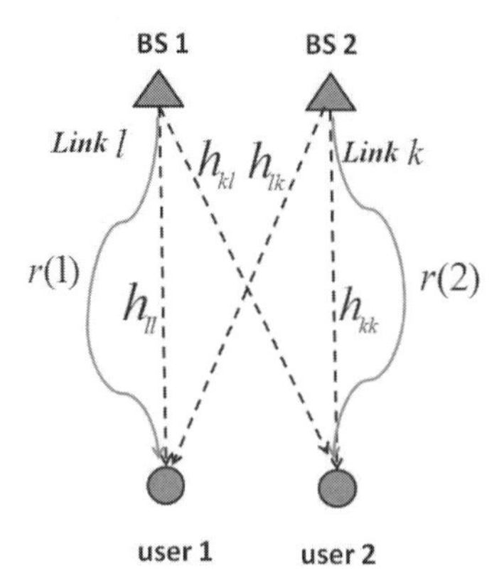

Figure 3.4 Illustration of the simplified problem with infinite backhaul capacity. The routing in backhaul is ignored here as any routing scheme is feasible.

The sum rate maximization problem

For illustration purpose, in the following we specialize the utility function to be the well-known sum rate utility. The problem becomes

$$
\begin{aligned}
\max_{\mathbf{v}} \quad & U(\mathbf{v}) = \sum_{\ell \in \mathcal{L}^{\mathrm{wl}}} \log \left(1 + \frac{|h_{\ell\ell}|^2 |v_\ell|^2}{\sum_{n \in \mathcal{I}(\ell) \setminus \{\ell\}} |h_{\ell n}|^2 |v_n|^2 + \sigma_\ell^2} \right) \\
\text{s.t.} \quad & \sum_{\ell \in \mathrm{Out}(b) \bigcap \mathcal{L}^{\mathrm{wl}}} |v_\ell|^2 \le \bar{p}_b, \ \forall\, b \in \mathcal{B}.
\end{aligned}
\tag{3.20}
$$

This problem is precisely the block-structured problem discussed in Section 3.1. More specifically, it falls into the category of problem (3.2), where the precoder for a given BS b, $\{v_\ell\}_{\ell \in \mathrm{Out}(b) \bigcap \mathcal{L}^{\mathrm{wl}}}$, corresponds to a block variable x_i in (3.2).

The difficulty in solving problem (3.20) is quite obvious now: the variables v_ns are coupled in a nonlinear way in the objective through mutual interference, making the problem highly nonconvex. One may resort to general purpose algorithm such as gradient projection, but its dependence on stepsize as well as the requirement to perform projection make it difficult to implement for large-scale problems. What we propose here is to use the BSUM approach discussed in Section 3.2.3, in which approximate versions of the original problem are successively solved to progressively obtain improved solutions; see Figure 3.5 for the illustration of the algorithm.[1] Clearly the key here is to find an appropriate lower bound of the objective function at any given point $\hat{\mathbf{v}}$, so that the resulting subproblem can be solved cheaply.

[1] Note that here the problem is formulated in a *maximization form*, rather than the *minimization form* considered in Section 3.2. So the BSUM algorithm successively constructs and solves *lower bounds* as opposed to the *upper bounds* stated in Section 3.2.3.

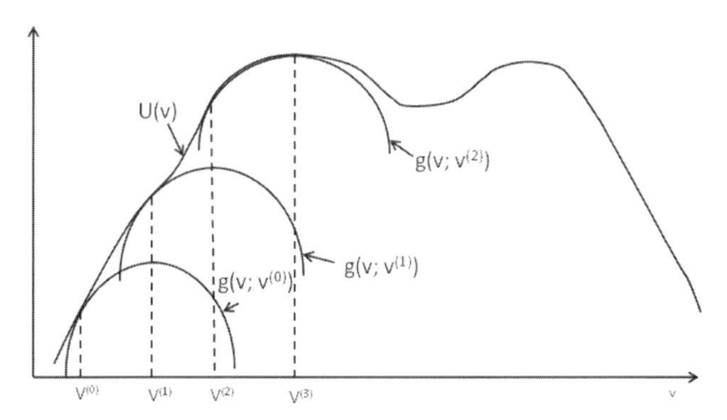

Figure 3.5 Illustration of the BSUM algorithm. Superscript (t) denotes the iteration number, $g(v; v^{(t)})$ denotes the approximate function (lower bound) constructed at iteration (t); $v^{(t)}$s are the iterates generated by the algorithm.

To this end, let us first introduce two useful quantities. For a given collection of precoder $\mathbf{v}$ and a given link ℓ, let us define $c_\ell(\mathbf{v})$ and $e_\ell(\mathbf{v})$ as

$$c_\ell(\mathbf{v}) = \sigma_\ell^2 + \sum_{n \in \mathcal{I}(\ell)} |h_{\ell n}|^2 |v_n|^2 \tag{3.21}$$

$$e_\ell(\mathbf{v}) = 1 - c_\ell^{-1}(\mathbf{v}) |h_{\ell\ell}|^2 |v_\ell|^2. \tag{3.22}$$

Here $c_\ell(\mathbf{v})$ can be interpreted as the total signal plus interference power received by user d_ℓ, while $e_\ell(\mathbf{v})$ is the minimum mean square error for decoding user d_ℓ's message (see, e.g., [46] for a more detailed explanation).

Our first lemma finds a lower bound $f_\ell(\mathbf{v}; \hat{\mathbf{v}})$ for the rate $r_\ell(\mathbf{v})$ at a given point $\hat{\mathbf{v}}$.

Lemma 3.4 *For any feasible solution $\hat{\mathbf{v}}$, define*

$$\hat{e}_\ell \triangleq 1 - c_\ell^{-1}(\hat{\mathbf{v}}) |h_{\ell\ell}|^2 |\hat{v}_\ell|^2. \tag{3.23}$$

Then we have

$$r_\ell(\mathbf{v}) \geq r_\ell(\hat{\mathbf{v}}) + \frac{1}{\hat{e}_\ell} \left(v_\ell^H h_{\ell\ell}^H c_\ell^{-1}(\mathbf{v}) h_{\ell\ell} v_\ell - 1 \right) + 1 \triangleq f_\ell(\mathbf{v}; \hat{\mathbf{v}}). \tag{3.24}$$

To see why this result is true, let us first express $r_\ell(\mathbf{v})$ as follows

$$r_\ell(\mathbf{v}) = \log \left(1 + v_\ell^H h_{\ell\ell}^H \left(\sum_{n \in \mathcal{I}(\ell) \setminus \{\ell\}} |h_{\ell n}|^2 |v_n|^2 + \sigma_\ell^2 \right)^{-1} h_{\ell\ell} v_\ell \right). \tag{3.25}$$

Applying the following inversion lemma

$$1 + ab^{-1}c = (1 - a(b + ac)^{-1}c)^{-1} \tag{3.26}$$

we have

$$r_\ell(\mathbf{v}) = -\log \left(1 - v_\ell^H h_{\ell\ell}^H c_\ell^{-1}(\mathbf{v}) h_{\ell\ell} v_\ell \right). \tag{3.27}$$

Now it is clear that the right-hand side (RHS) is a convex function on $1 - v_\ell^H h_{\ell\ell}^H c_\ell^{-1}(\mathbf{v}) h_{\ell\ell} v_\ell$. Therefore we can linearize the RHS at any given $\hat{\mathbf{v}}$ and use the defining property of a convex function to obtain

$$\begin{aligned} r_\ell(\mathbf{v}) &\geq r_\ell(\hat{\mathbf{v}}) - \frac{\left(1 - v_\ell^H h_{\ell\ell}^H c_\ell^{-1}(\mathbf{v}) h_{\ell\ell} v_\ell\right) - \left(1 - \hat{v}_\ell^H h_{\ell\ell}^H c_\ell^{-1}(\hat{\mathbf{v}}) h_{\ell\ell} \hat{v}_\ell\right)}{1 - \hat{v}_\ell^H h_{\ell\ell}^H c_\ell^{-1}(\hat{\mathbf{v}}) h_{\ell\ell} \hat{v}_\ell} \\ &= r_\ell(\hat{\mathbf{v}}) + \frac{1}{\hat{e}_\ell}\left(v_\ell^H h_{\ell\ell}^H c_\ell^{-1}(\mathbf{v}) h_{\ell\ell} v_\ell - 1\right) + 1 \\ &\triangleq f_\ell(\mathbf{v}; \hat{\mathbf{v}}). \end{aligned} \tag{3.28}$$

Summing over all the links, we obtain

$$U(\mathbf{v}) = \sum_\ell r_\ell(\mathbf{v}) \geq \sum_\ell f_\ell(\mathbf{v}; \hat{\mathbf{v}}) \triangleq f(\mathbf{v}; \hat{\mathbf{v}}). \tag{3.29}$$

Unfortunately, the lower bound $f(\mathbf{v}; \hat{\mathbf{v}})$ obtained is not so useful yet, as it still couples all the variables and is again a nonconvex function w.r.t. the optimization variable $\mathbf{v}$. What we will do next is to further construct a *concave* lower bound for $f(\mathbf{v}; \hat{\mathbf{v}})$, which in turn is a concave lower bound for $U(\mathbf{v})$.

Lemma 3.5 *For any given* $\hat{\mathbf{v}}$ *we have*

$$v_\ell^H h_{\ell\ell}^H c_\ell^{-1}(\mathbf{v}) h_{\ell\ell} v_\ell \geq \hat{u}_\ell^H h_{\ell\ell} v_\ell + v_\ell^H h_{\ell\ell} \hat{u}_\ell - \hat{u}_\ell^H c_\ell(\mathbf{v}) u_\ell \tag{3.30}$$

where we have defined

$$\hat{u}_\ell \triangleq c_\ell^{-1}(\hat{\mathbf{v}}) h_{\ell\ell} \hat{v}_\ell. \tag{3.31}$$

The proof of this result is again very simple. First note that the function $h(x, y) = \frac{x^2}{y}$ is jointly convex on (x, y) over the domain $y > 0$. Thus using the property of the convex function, for any given tuple $(\hat{x}, \hat{y})$ with $\hat{y} > 0$ we have

$$h(x, y) \geq h(\hat{x}, \hat{y}) + \frac{\partial h(\hat{x}, \hat{y})}{\partial x}(x - \hat{x}) + \frac{\partial h(\hat{x}, \hat{y})}{\partial y}(y - \hat{y}). \tag{3.32}$$

Now applying this inequality to $v_\ell^H h_{\ell\ell}^H c_\ell^{-1}(\mathbf{v}) h_{\ell\ell} v_\ell$ with the identification that $y = c_\ell^{-1}(\mathbf{v})$ and $x = v_\ell^H h_{\ell\ell}^H$, we obtain the desired result

$$\begin{aligned} &v_\ell^H h_{\ell\ell}^H c_\ell^{-1}(\mathbf{v}) h_{\ell\ell} v_\ell \\ &\quad \geq \hat{v}_\ell^H h_{\ell\ell}^H c_\ell^{-1}(\hat{\mathbf{v}}) h_{\ell\ell} \hat{v}_\ell + \hat{u}_\ell^H h_{\ell\ell}(v_\ell - \hat{v}_\ell) + (v_\ell - \hat{v}_\ell)^H h_{\ell\ell}^H \hat{u}_\ell - \hat{u}_\ell^H (c_\ell(\mathbf{v}) - c_\ell(\hat{\mathbf{v}})) \hat{u}_\ell \\ &\quad = \hat{u}_\ell^H h_{\ell\ell} v_\ell + v_\ell^H h_{\ell\ell} \hat{u}_\ell - \hat{u}_\ell^H c_\ell(\mathbf{v}) \hat{u}_\ell. \end{aligned}$$

As a result, combining Lemma 3.4 and Lemma 3.5, we have

$$r_\ell(\mathbf{v}) \geq f_\ell(\mathbf{v}; \hat{\mathbf{v}}) \geq \underbrace{\frac{1}{\hat{e}_\ell}\left(\hat{u}_\ell^H h_{\ell\ell} v_\ell + v_\ell^H h_{\ell\ell}^H \hat{u}_\ell - \hat{u}_\ell^H c_\ell(\mathbf{v}) \hat{u}_\ell\right)}_{\text{concave quadratic function on } \mathbf{v}} + \underbrace{1 + r_\ell(\hat{\mathbf{v}}) - \frac{1}{\hat{e}_\ell}}_{\text{constants}}. \tag{3.33}$$

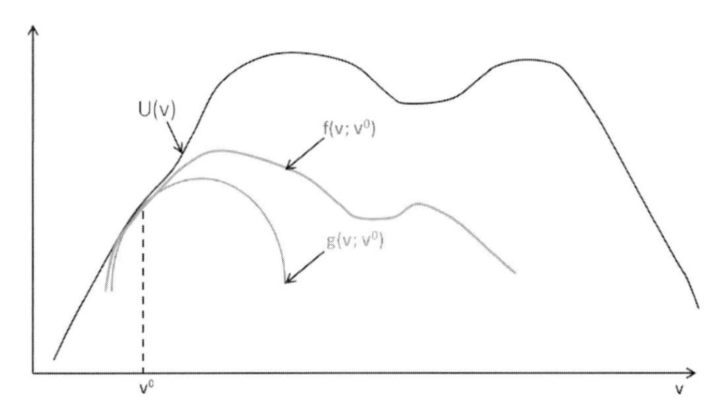

Figure 3.6 Illustration of the two-layer convex approximation process. $f(\mathbf{v}; \hat{\mathbf{v}})$ is the first-layer approximation; $g(\mathbf{v}; \hat{\mathbf{v}})$ is the second-layer approximation.

Using this bound, the objective of the sum rate maximization problem can be bounded below by

$$\begin{aligned}
U(\mathbf{r}) &= \sum_{\ell} r_\ell(\mathbf{v}) \geq \sum_{\ell} f_\ell(\mathbf{v}; \hat{\mathbf{v}}) \\
&\geq \sum_{\ell} \left(\underbrace{\frac{1}{\hat{e}_\ell} \left(\hat{u}_\ell^H h_{\ell\ell} v_\ell + v_\ell^H h_{\ell\ell}^H \hat{u}_\ell - \hat{u}_\ell^H c_\ell(\mathbf{v}) \hat{u}_\ell \right)}_{\text{concave quadratic function on } \mathbf{v}} + \underbrace{1 + r_\ell(\hat{\mathbf{v}}) - \frac{1}{\hat{e}_\ell}}_{\text{constants}} \right) \\
&= \underbrace{\sum_{\ell} \left(\frac{1}{\hat{e}_\ell} \left(\hat{u}_\ell^H h_{\ell\ell} v_\ell + v_\ell^H h_{\ell\ell}^H \hat{u}_\ell \right) - \sum_{n \in \mathcal{I}(\ell)} \frac{1}{\hat{e}_n} \left(|\hat{u}_n|^2 |h_{n\ell}^H|^2 |v_\ell|^2 \right) \right)}_{\text{concave quadratic on } v_\ell} + \text{constant} \\
&\triangleq \sum_{\ell} g_\ell(v_\ell; \hat{\mathbf{v}}) \triangleq g(\mathbf{v}; \hat{\mathbf{v}})
\end{aligned} \tag{3.34}$$

The above two-layer convex approximation process is illustrated in Figure 3.6. From (3.34) it is clear that the lower bound function $g(\mathbf{v}; \hat{\mathbf{v}})$ is not only a concave quadratic function on $\mathbf{v}$, but is also completely separable among v_ℓs. Utilizing this favorable structure of the lower bound, we can successively minimize the lower bound function $g(\mathbf{v}; \hat{\mathbf{v}})$ by solving $|\mathcal{B}|$ independent subproblems of the following form, one for each BS b:

$$\begin{aligned}
\max \quad & \sum_{\ell \in \text{Out}(b) \bigcap \mathcal{L}^{wl}} g_\ell(v_\ell; \hat{\mathbf{v}}) \\
\text{s.t.} \quad & \sum_{\ell \in \text{Out}(b) \bigcap \mathcal{L}^{wl}} |v_\ell|^2 \leq \bar{p}_b.
\end{aligned} \tag{3.35}$$

The solution of problem (3.35) can be written down in closed form

$$v_\ell^* = \left(\sum_{n \in \mathcal{I}(\ell)} |\hat{u}_n|^2 |h_{n\ell}^H|^2 \frac{1}{\hat{e}_n} + \lambda_b \right)^{-1} h_{\ell\ell}^H \hat{u}_\ell \hat{e}_\ell^{-1}, \ \forall\, \ell \in \text{Out}(b) \bigcap \mathcal{L}^{\text{wl}}, \tag{3.36}$$

where $\lambda_b \geq 0$ is the Lagrangian multiplier used to guarantee the total power constraint for BS b.

The overall algorithm is summarized in the following box, where we use the superscript (t) to denote the iteration index. It can be easily checked that the lower bound $g(\mathbf{v}; \hat{\mathbf{v}})$ satisfies Assumptions (A1)–(A5). By appealing to Theorem 3.3, this algorithm is guaranteed to converge to a stationary solution of problem (3.3.2).

(1) Update $\{\hat{u}_\ell\}$: $\hat{u}_\ell \leftarrow c_\ell^{-1}(\mathbf{v}^{(t)}) h_{\ell\ell} v_\ell^{(t)}$.
(2) Update $\{\hat{e}_\ell\}$: $\hat{e}_\ell \leftarrow 1 - c_\ell^{-1}(\mathbf{v}^{(t)}) |h_{\ell\ell}|^2 \left(v_\ell^{(t)}\right)^2$.
(3) Update $\{v_\ell\}$:

$$v_\ell^{(t+1)} \leftarrow \left(\sum_{n \in \mathcal{I}(\ell)} |\hat{u}_n|^2 |h_{n\ell}^H|^2 \frac{1}{\hat{e}_n} + \lambda_b \right)^{-1} h_{\ell\ell}^H \hat{u}_\ell \hat{e}_\ell^{-1},$$
$$\forall\, \ell \in \text{Out}(b) \bigcap \mathcal{L}^{wl}\ \forall\, b \in \mathcal{B}.$$

(4) Let $t = t + 1$, go to step (1).

It is important to note that by exploring the hidden convexity of the rate function $r_\ell(\mathbf{v})$, problem (3.20) can be solved with simple closed-form updates, and the computation can be further carried out in parallel by all the BSs. The algorithm described above is the so-called weighted minimum mean square error (WMMSE) algorithm, which is originally developed using certain equivalence argument between problem (3.20) and certain weighted MSE minimization problem [46, 54]. The preceding derivation based on BSUM is first given in [53], which provides an interesting alternative interpretation of the algorithm. We remark that the above analysis and the WMMSE algorithm can be easily generalized to networks with multi-antenna wireless nodes, or to problems having different (possibly nonsmooth) utility functions [55].

The min rate maximization problem

In this section we briefly discuss how the bounds derived in the previous section can be utilized to solve another popular problem – the min rate maximization problem – which results in a fair rate allocation. In particular, we are interested in maximizing the minimum rate achieved by all the users $u \in \mathcal{U}$:

$$\begin{aligned} \max_{\mathbf{v}} \quad & \min_{u} \sum_{\ell: u = d_\ell} \log \left(1 + \frac{|h_{\ell\ell}|^2 |v_\ell|^2}{\sum_{n \in \mathcal{I}(\ell) \setminus \{\ell\}} |h_{\ell n}|^2 |v_n|^2 + \sigma_\ell^2} \right) \\ \text{s.t.} \quad & \sum_{\ell \in \text{Out}(b) \bigcap \mathcal{L}^{wl}} |v_\ell|^2 \leq \bar{p}_b, \ \forall\, b \in \mathcal{B}. \end{aligned} \tag{3.37}$$

First we introduce a new variable $r \geq 0$ and transform the above problem to the following equivalent form

$$\begin{aligned} \max_{\mathbf{v},r} \quad & r \\ \text{s.t.} \quad & \sum_{\ell:u=d_\ell} \log\left(1 + \frac{|h_{\ell\ell}|^2|v_\ell|^2}{\sum_{n\in\mathcal{I}(\ell)\setminus\{\ell\}} |h_{\ell n}|^2|v_n|^2 + \sigma_\ell^2}\right) \geq r, \ \forall\, u \in \mathcal{U} \\ & \sum_{\ell\in \mathrm{Out}(b)\bigcap \mathcal{L}^{wl}} |v_\ell|^2 \leq \bar{p}_b, \ \forall\, b \in \mathcal{B}. \end{aligned} \tag{3.38}$$

The transformed problem now has a simple linear objective, but a set of difficult rate constraints. Once again we can utilize the bounds developed in the previous section, but to successively approximate the *feasible set* instead. That is, for a given $\hat{\mathbf{v}}$, we first compute $\hat{u}_\ell$ and $\hat{e}_\ell$ just as before, and then solve the following convex problem

$$\begin{aligned} \max_{\mathbf{v},r} \quad & r \\ \text{s.t.} \quad & \sum_{\ell:u=d_\ell} g_\ell(\mathbf{v};\hat{\mathbf{v}}) \geq r, \ \forall\, u \in \mathcal{U} \\ & \sum_{\ell\in \mathrm{Out}(b)\bigcap \mathcal{L}^{wl}} |v_\ell|^2 \leq \bar{p}_b, \ \forall\, b \in \mathcal{B}. \end{aligned} \tag{3.39}$$

Unlike the original problem (3.38), this problem is now a convex problem as it has a convex, albeit smaller, feasible set. The overall algorithm is presented in the following box.

(1) Update $\{\hat{u}_\ell\}$: $\hat{u}_\ell \leftarrow c_\ell^{-1}(\mathbf{v}^{(t)})h_{\ell\ell}v_\ell^{(t)}$.
(2) Update $\{\hat{e}_\ell\}$: $\hat{e}_\ell \leftarrow 1 - c_\ell^{-1}(\mathbf{v}^{(t)})|h_{\ell\ell}|^2\left(v_\ell^{(t)}\right)^2$.
(3) Update $\{v_\ell\}$: obtain $\mathbf{v}_\ell^{(t+1)}$ by solving problem (3.39).
(4) Let $t = t+1$, go to step (1).

We note that the above algorithm is *not* a special case of BSUM, because it is the feasible set that has been approximated here. Therefore the previous analysis of BSUM in Section 3.2.3 does not apply. Fortunately by carefully studying the optimality conditions of the resulting subproblems, one can still show that the iterates $\{\mathbf{v}^{(t)}\}$ converge to the set of stationary solutions of problem (3.37); see [56] for detailed analysis.

At this point it should be noted that the subproblem for solving $\mathbf{v}$ is convex but does not have closed-form solution. Therefore general purpose solvers need to be used repeatedly for this subproblem, which can be computationally expensive when the problem size becomes large (i.e. large number of BSs, flows, users, etc.). Later when we discuss the general network provisioning problem, we will revisit this issue and design an efficient algorithm for solving the related subproblem.

3.3.3 Network with a capacitated backhaul

Now we are ready to solve the problem posed in Section 3.3.1 in the setting of large-scale cloud-based RAN. Without loss of generality, we let $v_\ell \in \mathbb{R}$ for all ℓ. We focus on the following per-flow min rate maximization problem

$$\max_{\mathbf{v},\mathbf{r}} \quad r \tag{3.40a}$$

$$\text{s.t.} \quad \mathbf{r} \geq 0, r(m) \geq r, \ m \in \mathcal{M} \tag{3.40b}$$

$$\sum_{m\in\mathcal{M}} r_\ell(m) \leq C_\ell, \ \forall\, \ell \in \mathcal{L}^{\mathrm{w}} \tag{3.40c}$$

$$\sum_{m\in\mathcal{M}} r_\ell(m) \leq r_\ell(\mathbf{v}) = \log\left(1 + \frac{|h_{\ell\ell}|^2 v_\ell^2}{\sum_{n\in\mathcal{I}(\ell)\setminus\{\ell\}} |h_{\ell n}|^2 v_n^2 + \sigma_\ell^2}\right), \ \forall\, \ell \in \mathcal{L}^{\mathrm{wl}} \tag{3.40d}$$

$$\sum_{\ell\in\mathrm{In}(v)} r_\ell(m) + 1_{s(m)}(v) r(m) = \sum_{\ell\in\mathrm{Out}(v)} r_\ell(m) + 1_{d(m)}(v) r(m), \quad \forall\, m \in \mathcal{M}, \ \forall\, v \in \mathcal{V}, \tag{3.40e}$$

$$\sum_{\ell\in\mathrm{Out}(b)\bigcap\mathcal{L}^{wl}} v_\ell^2 \leq \bar{p}_b, \ \forall\, b \in \mathcal{B}. \tag{3.40f}$$

Here with a little abuse of notation, we have defined

$$\mathbf{r} \triangleq [r, \{r(m), r_\ell(m) \mid \ell \in \mathcal{L}\}_{m\in\mathcal{M}}]^T .$$

The constraints (3.40c)–(3.40f) are, respectively, the wired link capacity constraints, the wireless link capacity constraints, the flow conservation constraints and the BS power budget constraint introduced in Section 3.3.1.

The N-MaxMin algorithm

In practice, problem (3.40) needs to be solved frequently to determine the dynamic resource and flow allocation. However, this is very challenging because:

- the problem is nonconvex due to the wireless rate constraints (3.40d);
- the design variables $\mathbf{v}$ and $\mathbf{r}$ are tightly coupled through the rate expressions; and
- the size of the problem can be huge.

To obtain an effective algorithm, our first step is again to approximate the rate $r_\ell(\mathbf{v})$ using its lower bound. To this end, let us simplify the expression for $g_\ell(\mathbf{v}; \hat{\mathbf{v}})$ in (3.34) by the following:

$$g_\ell(\mathbf{v}; \hat{\mathbf{v}}) = \hat{c}_{1,\ell} + \hat{c}_{2,\ell} v_\ell - \sum_{n\in\mathcal{I}(\ell)} \hat{c}_{3,\ell n} v_n^2, \tag{3.41}$$

where the constants $(\hat{c}_{1,\ell}, \hat{c}_{2,\ell}, \hat{c}_{3,\ell n})$ are given by

$$\hat{c}_{1,\ell} = 1 + r_\ell(\hat{\mathbf{v}}) - \frac{1}{\hat{e}_\ell}\left(1 + \sigma_\ell^2 \hat{u}_\ell^2\right) \tag{3.42}$$

$$\hat{c}_{2,\ell} = \frac{2}{\hat{e}_\ell}\hat{u}_\ell |h_{\ell\ell}| \tag{3.43}$$

$$\hat{c}_{3,\ell n} = \frac{1}{\hat{e}_\ell}\hat{u}_\ell|^2 h_{\ell n}|^2. \tag{3.44}$$

Then at any given point $\hat{\mathbf{v}}$, we can approximate problem (3.40) by

$$\max_{\mathbf{v},\mathbf{r}} \quad r \tag{3.45a}$$

$$\text{s.t.} \quad (3.40b), (3.40c), (3.40e), (3.40f), \tag{3.45b}$$

$$\sum_{m\in\mathcal{M}} r_\ell(m) \le g_\ell(\mathbf{v}; \hat{\mathbf{v}}) = \hat{c}_{1,\ell} + \hat{c}_{2,\ell} v_\ell - \sum_{n\in\mathcal{I}(\ell)} \hat{c}_{3,\ell n} v_n^2, \ \forall\, \ell \in \mathcal{L}^{\text{wl}}. \tag{3.45c}$$

Similarly as solving the max-min problem in Section 3.3.2, the above problem is again convex and can be solved by using general-purpose solvers. The resulting algorithm, termed the network max-min WMMSE (N-MaxMin) algorithm, is given in the following box. Again one can show that this algorithm converges to the set of stationary solutions of the network provisioning problem (3.40); see [2] for detailed analysis.

(1) Update $\{\hat{u}_\ell\}$: $\hat{u}_\ell \leftarrow c_\ell^{-1}(\mathbf{v}^{(t)})|h_{\ell\ell}|v_\ell^{(t)}$.
(2) Update $\{\hat{e}_\ell\}$: $\hat{e}_\ell \leftarrow 1 - c_\ell^{-1}(\mathbf{v}^{(t)})|h_{\ell\ell}|^2\left(v_\ell^{(t)}\right)^2$.
(3) Update $(\mathbf{v}, \mathbf{r})$: obtain $(\mathbf{v}^{(t+1)}, \mathbf{r}^{(t+1)})$ by solving problem (3.45).
(4) Let $t = t + 1$, go to step (1).

Once again, the computation of $\hat{u}_\ell$s and $\hat{e}_\ell$s is in closed form. The main computational complexity is in step (3) where $(\mathbf{v}, \mathbf{r})$ are updated. When the number of variables and constraints are large, the efficiency of the entire algorithm critically depends on the implementation of this step. How this can be done is the topic that we address in the following section.

An ADMM approach for updating $(\mathbf{v}, \mathbf{r})$

We propose to use the ADMM algorithm for solving problem (3.45). ADMM is chosen because it allows us to implement a highly parallelizable algorithm that fits ideally to the cloud-based architecture of the next-generation wireless networks.

In order to apply the ADMM, the first step is to formulate problem (3.45) into the form of (3.6). Our main approach is to properly split the variables in the coupling constraints (3.40e) and (3.45c), so that these constraints decompose nicely over the variables.

Let us first look at the flow conservation constraint (3.40e), restated below for convenience:

$$\sum_{\ell\in\text{In}(v)} r_\ell(m) + 1_{s(m)}(v) r(m) = \sum_{\ell\in\text{Out}(v)} r_\ell(m) + 1_{d(m)}(v) r(m), \ \forall\, v, m.$$

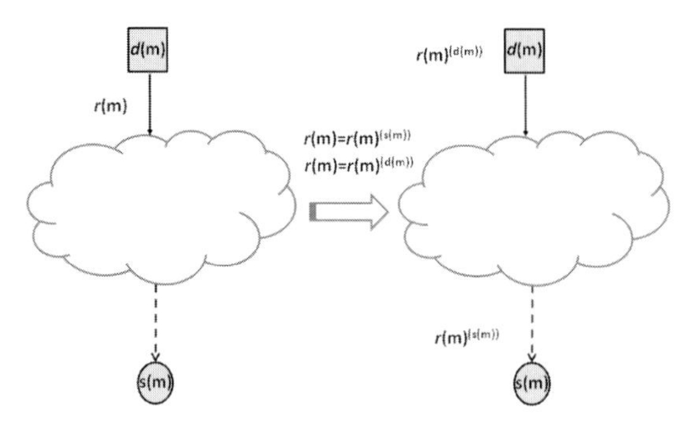

Figure 3.7 Illustration of flow rate splitting.

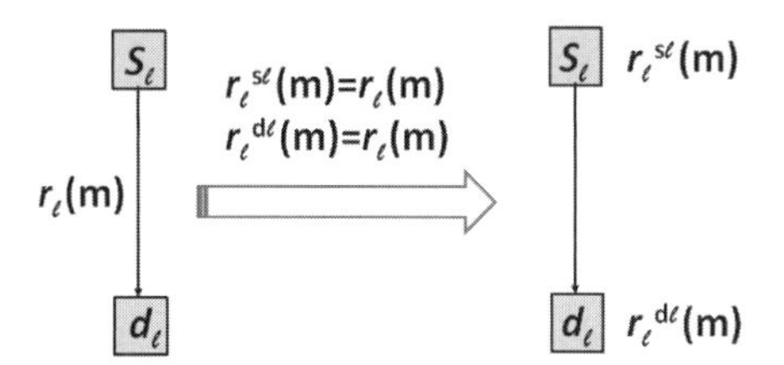

Figure 3.8 Illustration of link rate splitting.

At first sight it appears that the link rate variables $\{r_\ell(m)\}$ are all tightly coupled together, making (3.40e) a very difficult constraint to satisfy. However, a careful study reveals that each link rate $r_\ell(m)$ appears exactly twice, in the constraints for node s_ℓ and d_ℓ. It follows that if we introduce two copies of $r_\ell(m)$ (denoted as $\hat{r}_\ell^{s_\ell}(m)$ and $\hat{r}_\ell^{d_\ell}(m)$) and use $\hat{r}_\ell^{s_\ell}(m)$ (resp. use $\hat{r}_\ell^{d_\ell}(m)$) in the constraint for node s_ℓ (resp. for node d_ℓ), then each of these new auxiliary variables only appears in a *single* flow conservation constraint. The same is true for the flow rate variable $r(m)$: they appear only twice, in those constraints defined by the source node $s(m)$ and the destination node $d(m)$. Similarly we introduce two variables $\hat{r}(m)^{s(m)}$ and $\hat{r}(m)^{d(m)}$ for each flow m, and use them in the constraints for the source and the destination node of flow m, respectively. See Figure 3.7 and Figure 3.8 for illustrations of the above splitting process.

Mathematically, we have introduced the following auxiliary variables:

$$\hat{r}_\ell^{s_\ell}(m) = r_\ell(m),\ \hat{r}_\ell^{d_\ell}(m) = r_\ell(m),\ \forall\, \ell \in \mathcal{L}, m \in \mathcal{M}; \tag{3.46a}$$

$$\hat{r}(m)^{s(m)} = r(m),\ \hat{r}(m)^{d(m)} = r(m),\ \forall\, m \in \mathcal{M}. \tag{3.46b}$$

We have also modified the flow rate conservation constraints to

$$\sum_{\ell\in \mathrm{In}(v)} \hat{r}_\ell^v(m) + 1_{s(m)}(v)\hat{r}(m)^v = \sum_{l\in \mathrm{Out}(v)} \hat{r}_\ell^v(m) + 1_{d(m)}(v)\hat{r}(m)^v,\ \forall\, m, v. \tag{3.47}$$

To facilitate analysis, we also split r by introducing $\hat{r}$ that satisfies $r = \hat{r}$. Let us collect the new variables and define

$$\hat{\mathbf{r}} \triangleq \left[\, \hat{r}, \left\{\hat{r}(m)^{s(m)}, \hat{r}_\ell^{s_\ell}(m) \mid l \in \mathcal{L}\right\}_{m=1}^{M}, \left\{\hat{r}(m)^{d(m)}, \hat{r}_l^{d_\ell}(m) \mid l \in \mathcal{L}\right\}_{m=1}^{M} \right]^T.$$

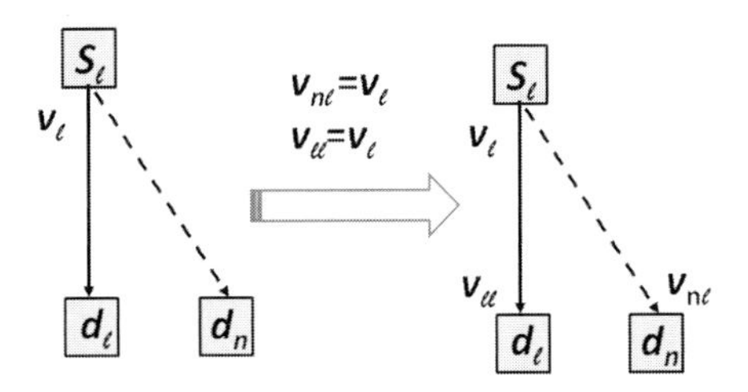

Figure 3.9 Illustration of precoder splitting. The solid line represents the link ℓ; the dotted line represents the link n which is interfered by ℓ.

Next let us look at the rate constraint (3.45c), restated below for convenience

$$\sum_{m\in\mathcal{M}} r_\ell(m) \le \hat{c}_{1,\ell} + \hat{c}_{2,\ell} v_\ell - \sum_{n\in\mathcal{I}(\ell)} \hat{c}_{3,ln} v_n^2, \ \forall\, \ell \in \mathcal{L}^{\rm wl}. \tag{3.48}$$

Again these constraints are coupled because each variable v_n appears in multiple constraints, i.e. constrains for the links that are interfered by n. To decouple the constraints, we introduce several copies of the transmit precoders for each v_n, denoted by $\hat{v}_{\ell n}$, one for each link ℓ interfered by n (i.e. for nodes ℓs satisfying $n \in \mathcal{I}(\ell)$). By doing so, each variable $\hat{v}_{\ell n}$ appears only in a single constraint. See Figure 3.9 for an illustration of the splitting process.

Formally, we have introduced a set of new variables

$$\hat{v}_{\ell n} = v_n, \ \forall\, \ell \text{ such that } n \in \mathcal{I}(\ell), \forall\, n \in \mathcal{L}^{\rm wl}. \tag{3.49}$$

We have also modified the rate constraints to

$$\sum_{m=1}^{M} r_\ell(m) \le \hat{c}_{1,\ell} + \hat{c}_{2,\ell} \hat{v}_{\ell\ell} - \sum_{n\in\mathcal{I}(\ell)} \hat{c}_{3,\ell n} \hat{v}_{\ell n}^2, \ \forall\, \ell \in \mathcal{L}^{\rm wl}. \tag{3.50}$$

For notational convenience, define

$$\hat{\mathbf{v}} \triangleq \left[[\hat{v}_{n\ell}, \forall\, n \in \bar{\mathcal{I}}(\ell)], \forall\, \ell \in \mathcal{L}^{\rm wl}\right]^T,$$

with

$$\bar{\mathcal{I}}(\ell) \triangleq \{n \mid \ell \in \mathcal{I}(n)\}$$

being the set of wireless links interfered by ℓ.

By utilizing the new variables introduced so far, problem (3.45) is equivalently expressed as

$$\begin{aligned} \max \quad & (r+\hat{r})/2 \\ \text{s.t.} \quad & (3.40\text{b}),\ (3.40\text{c}),\ (3.40\text{f}), \\ & (3.47),\ (3.50), \\ & (3.46),\ (3.49),\ \text{and } r=\hat{r}. \end{aligned} \tag{3.51}$$

We will see shortly that the above equivalent formulation decomposes the constraints (except the linear equality constraints $r = \hat{r}$, (3.46) and (3.49)) between the variable sets

Table 3.1 The ADMM based algorithm for (3.51)

Algorithm 1

(1) **Initialize** all primal variables $\mathbf{r}^{(0)}, \hat{\mathbf{r}}^{(0)}, \mathbf{v}^{(0)}, \hat{\mathbf{v}}^{(0)}$ (not necessarily a feasible solution for (3.51)), and all dual variables $\boldsymbol{\delta}^{(0)}, \boldsymbol{\theta}^{(0)}$; set $t = 0$.
(2) **Repeat**.
(3) Solve the following problem and obtain $\mathbf{r}^{(t+1)}, \hat{\mathbf{v}}^{(t+1)}$:

$$\begin{aligned} \max_{\mathbf{r},\, \hat{\mathbf{v}}} \quad & L_{\rho_1,\rho_2}(\mathbf{r}, \hat{\mathbf{v}}, \hat{\mathbf{r}}^{(t)}, \mathbf{v}^{(t)}; \boldsymbol{\delta}^{(t)}, \boldsymbol{\theta}^{(t)}) \\ \text{s.t.} \quad & (3.40\text{c}),\ (3.40\text{b}),\ \text{and } (3.50). \end{aligned} \tag{3.52}$$

(4) Solve the following problem and obtain $\hat{\mathbf{r}}^{(t+1)}, \mathbf{v}^{(t+1)}$:

$$\begin{aligned} \max_{\hat{\mathbf{r}},\, \mathbf{v}} \quad & L_{\rho_1,\rho_2}(\mathbf{r}^{(t+1)}, \hat{\mathbf{v}}^{(t+1)}, \hat{\mathbf{r}}, \mathbf{v}; \boldsymbol{\delta}^{(t)}, \boldsymbol{\theta}^{(t)}) \\ \text{s.t.} \quad & (3.17) \text{ and } (3.47). \end{aligned} \tag{3.53}$$

(5) Update the Lagrange dual multipliers $\boldsymbol{\delta}^{(t+1)}$ and $\boldsymbol{\theta}^{(t+1)}$ by

$$\begin{aligned} \boldsymbol{\delta}^{(t+1)} &= \boldsymbol{\delta}^{(t)} - \rho_1(\hat{\mathbf{r}}^{(t+1)} - \mathbf{C}\mathbf{r}^{(t+1)}), \\ \boldsymbol{\theta}^{(t+1)} &= \boldsymbol{\theta}^{(t)} - \rho_2(\mathbf{D}\mathbf{v}^{(t+1)} - \hat{\mathbf{v}}^{(t+1)}). \end{aligned} \tag{3.54}$$

(6) $t = t + 1$.
(7) **Until** Desired stopping criterion is met.

$(\mathbf{r}, \hat{\mathbf{v}})$ and $(\hat{\mathbf{r}}, \mathbf{v})$. In particular, we can write the linear equalities $r = \hat{r}$, (3.46) and (3.49) as $\mathbf{C}\mathbf{r} = \hat{\mathbf{r}},\ \mathbf{D}\mathbf{v} = \hat{\mathbf{v}}$ with

$$\mathbf{C} = \begin{bmatrix} 1 & 0 & 0 & 0 & 0 \\ 0 & \mathbf{I} & \mathbf{0} & \mathbf{I} & \mathbf{0} \\ 0 & \mathbf{0} & \mathbf{I} & \mathbf{0} & \mathbf{I} \end{bmatrix}^T; \quad \mathbf{D} = \text{blkdg}\left[\{\mathbf{1}_\ell\}_{\ell \in \mathcal{L}^{\text{wl}}}\right],$$

where $\mathbf{1}_\ell$ is an all one column vector of size equal to $|\bar{\mathcal{I}}(\ell)|$.

Now ADMM can be used to solve (3.51), where the linear equality constraints $\mathbf{C}\mathbf{r} = \hat{\mathbf{r}},\ \mathbf{D}\mathbf{v} = \hat{\mathbf{v}}$ are to be relaxed and penalized in the augmented Lagrangian. To write down the ADMM iteration, let us use δ, $\{\delta_\ell^{s_\ell}(m), \delta_\ell^{d_\ell}(m)\}$, $\{\delta_m^{s(m)}, \delta_m^{d(m)}\}$, and $\{\theta_{n\ell}\}$ to denote the Lagrangian multipliers for equality constraints $r = \hat{r}$, (3.46a), (3.46b), and (3.49), respectively. Collect the multipliers in the vectors $\boldsymbol{\delta}$ and $\boldsymbol{\theta}$. Let $\rho_1 > 0$ and $\rho_2 > 0$ denote the dual stepsizes. Then the partial augmented Lagrangian for problem (3.51) is given by

$$\begin{aligned} & L_{\rho_1,\rho_2}(\mathbf{r}, \hat{\mathbf{v}}, \hat{\mathbf{r}}, \mathbf{v}; \boldsymbol{\delta}, \boldsymbol{\theta}) \\ &= (r + \hat{r})/2 + \underbrace{\left[\boldsymbol{\delta}^T(\hat{\mathbf{r}} - \mathbf{C}\mathbf{r}) - \frac{\rho_1}{2}\|\hat{\mathbf{r}} - \mathbf{C}\mathbf{r}\|^2\right]}_{\text{relaxing (3.46)}} + \underbrace{\left[\boldsymbol{\theta}^T(\mathbf{D}\mathbf{v} - \hat{\mathbf{v}}) - \frac{\rho_2}{2}\|\mathbf{D}\mathbf{v} - \hat{\mathbf{v}}\|^2\right]}_{\text{relaxing (3.49)}}. \end{aligned} \tag{3.55}$$

The resulting algorithm, named Algorithm 1, is described in Table 3.1. The convergence of this algorithm to the optimal solutions of problem (3.51) (hence the original subproblem (3.45) for $(\mathbf{v}, \mathbf{r})$) is readily implied by the standard analysis of ADMM (cf. Theorem 3.2). In the appendix, we provide guidelines on solving the two primal subproblems (3.52) and (3.53). Our focus is given to demonstrating the distinctive feature of these subproblems, that they naturally decompose into a series of independent small subproblems, which can be solved easily and in parallel.

Implementation of Algorithm 1 in cloud-based RAN

We reiterate here that each step of the N-MaxMin algorithm is in closed form. Further, the computation can be completely distributed to each node or link of the network. For the detailed discussion on the distributed implementation we refer the readers to [2, Sec. III]. However, in a cloud-based RAN, it is more desirable that a few cloud centers handle the computation *centrally*, each of them taking care of a subset of nodes located in a specific geographical zone. The key question here is whether the proposed algorithm can also be used in this scenario. Below we show that a properly modified version of Algorithm 1 does the trick. For simplicity, we will only focus on the backhaul network, but the extension for incorporating the wireless links follows the same idea.

Let us revisit the variable splitting in (3.46), which is introduced to decompose the flow conservation constraints (3.16) into each node. We assume that the set of nodes $\mathcal{V}$ is partitioned into $\mathcal{Z}$ non-overlapping zones, and $v \in \mathcal{Z}_i$ if node v is within the ith zone. We modify the variable splitting procedure (3.46a) as follows:

$$\hat{r}_l^{s_l}(m) = r_l(m),\ \hat{r}_l^{d_l}(m) = r_l(m), \quad \forall\, s_l \in \mathcal{Z}_i, d_l \in \mathcal{Z}_j,\ i \neq j,\ m = 1 \sim M, \tag{3.56a}$$

$$\hat{r}_l(m) = r_l(m), \quad \forall\, s_l \in \mathcal{Z}_i, d_l \in \mathcal{Z}_j,\ i = j,\ m = 1 \sim M. \tag{3.56b}$$

That is, we only split the link rates on the bordering links.

Given this new variable splitting method, the flow conservation constraints for each node within zone i becomes

$$\begin{aligned}&\sum_{l \in \mathrm{In}(v)} \left(1_{\{\bigcup_{k\neq i} \mathcal{Z}_k\}}(s_l)\hat{r}_l^{s_l}(m) + 1_{\{\mathcal{Z}_i\}}(s_l)\hat{r}_l(m)\right) + 1_{s(m)}(v)\hat{r}(m)^v \\ &\quad = \sum_{l \in \mathrm{Out}(v)} \left(1_{\{\bigcup_{k\neq i} \mathcal{Z}_k\}}(d_l)\hat{r}_l^{d_l}(m) + 1_{\{\mathcal{Z}_i\}}(d_l)\hat{r}_l(m)\right) + 1_{d(m)}(v)\hat{r}(m)^v, \\ &\qquad\qquad \forall\, v \in \mathcal{Z}_i, \forall\, m = 1 \sim M.\end{aligned} \tag{3.57}$$

We can observe that the variables are now decoupled over each zone of nodes instead of each node.

With this new variable splitting we can again apply the ADMM. The resulting algorithm has closed-form updates except for the step related to the flow conservation constraints (i.e. the corresponding subproblem (3.63)). This step now is decomposable into each zone. To describe the subproblem in detail, let us first introduce the following sets of links

$$\mathrm{BD}_i \triangleq \{l \in \mathcal{L} \mid \forall\, s_l \in \mathcal{Z}_i,\ d_l \in \mathcal{Z}_k,\ \text{or}\ \forall\, d_l \in \mathcal{Z}_i,\ s_l \in \mathcal{Z}_k,\ k \neq i\} \quad \text{(bordering links)}$$

$$\mathrm{IT}_i \triangleq \{l \in \mathcal{L} \mid \forall s_l, d_l \in \mathcal{Z}_i\} \quad \text{(interior links).}$$

Then the per-zone subproblem can be explicitly expressed as the following quadratic problem with a set of linear constraints:

$$\begin{aligned} \min \ & \sum_{l \in \mathrm{BD}_i} \left(\hat{r}_l^v(m) - r_l(m) - \frac{\delta_l^v(m)}{\rho_1} \right)^2 \\ & + \sum_{l \in \mathrm{IT}_i} \left(\hat{r}_l(m) - r_l(m) - \frac{\delta_l^v(m)}{\rho_1} \right)^2 + 1_{\{S(m), D(m)\}}(v) \left(\hat{r}_m^v - r_m - \frac{\delta_m^v}{\rho_1} \right)^2 \\ \text{s.t.} \ & (3.57). \end{aligned} \tag{3.58}$$

Although problem (3.58) does not have an easy closed-form solution, it can be solved efficiently in a centralized way via well-known network optimization algorithms such as the relax code [57]. Moreover, when each zone has a single node, the above modified algorithm reduces to the original Algorithm 1.

We emphasize that this modified approach is particularly suitable for the cloud based RAN architecture, where the computation is distributed to each cloud center. One additional benefit offered by this zone-based algorithm is that the splitting procedure is performed less frequently, leading to far fewer number of slack variables. Therefore compared with original Algorithm 1, the modified approach also enjoys faster convergence speed (measured by the number of iterations). This will be demonstrated in the subsequent numerical experiments.

3.4 Numerical results

In this section, we report some numerical results on the performance of the proposed algorithms. Most of the numerical experiments are conducted on a network with 57 BSs and 11 network routers; see Figure 3.10 for an illustration of the network. The detailed specification of the network is given below.

(1) **The backhaul network** Each link $\ell \in \mathcal{L}\mathrm{w}$ is bidirectional, and the capacities for both directions are the same. Detailed link capacities are given below:
  - links between routers and those between gateway BSs and the routers: 1 (Gnats/s);
  - 1-hop to the gateways: 100 (Mnats/s);
  - 2-hop to the gateways: [10,50] (Mnats/s);
  - 3-hop to the gateways: [2,5] (Mnats/s);
  - more than 4-hop to the gateways: 0 (nats/s).

(2) **The wireless access network** We use a slightly more general model in which the BSs can also operate on $K = 3$ subchannels, each with 1 MHz bandwidth. The power budget for each BS is chosen equally by $p = \bar{p}_s$, $\forall\, s \in \mathcal{B}$, and $\sigma_l = 1$, $\forall\, l \in \mathcal{L}$. The wireless links follow the Rayleigh distribution with $CN(0, (200/\mathrm{dist})^3)$, where "dist" represents the distance between BS and the corresponding user. The source (destination) node of each commodity is randomly selected from network routers (mobile users), and all simulation results are averaged over 100 randomly

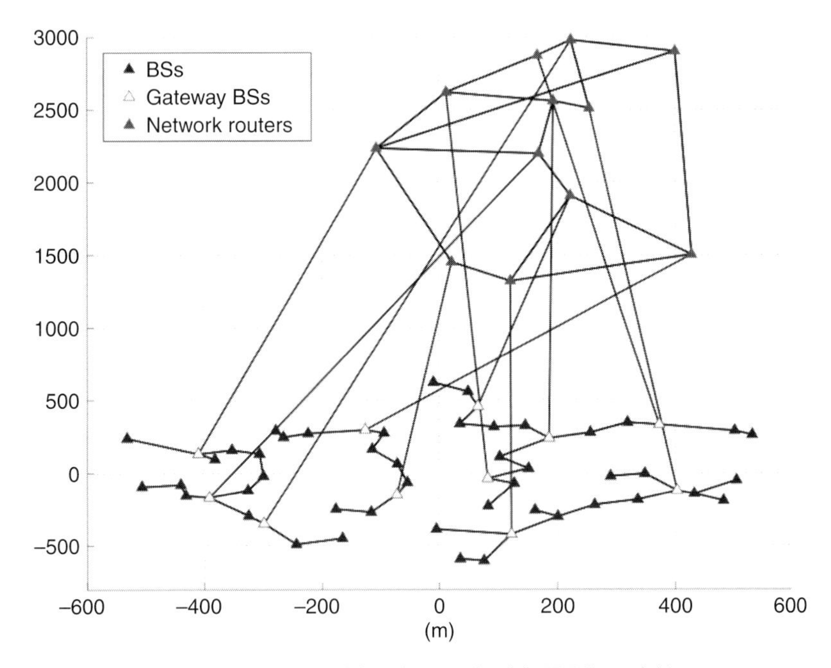

Figure 3.10 Illustration of the considered network with 57 BSs and 11 routers.

selected end-to-end commodity pairs. Below we refer to one round of the N-MaxMin iteration as an *outer iteration*, and one round of Algorithm 1 in Table 3.1 for solving $(\mathbf{r}, \mathbf{v})$ as an *inner iteration*.

3.4.1 Scenario 1: Performance comparison with heuristic algorithms

In the first experiment, we assume that each mobile user can be served by the BSs within 300 m radius. Further, each user is interfered by all BSs in the network. For this problem, the parameters of N-MaxMin algorithm are set to be $\rho_1 = 0.1$ and $\rho_2 = 0.1, 0.05$, and 0.01 for, respectively, $p = 0$ dB, 10 dB, and 20 dB. The termination criteria are

$$\frac{(r^{(t+1)} + \hat{r}^{(t+1)}) - (r^{(t)} + \hat{r}^{(t)})}{r^{(t)} + \hat{r}^{(t)}} < 10^{-3}$$
$$\max\{\|\mathbf{Cr}^{(t)} - \hat{\mathbf{r}}^{(t)}\|_\infty, \|(\mathbf{Dv}^{(t)})^2 - (\hat{\mathbf{v}}^{(t)})^2\|_\infty\}\} < 5 \times 10^{-4},$$

where $(\cdot)^2$ represents elementwise square operation.

For comparison purposes, the following two heuristic algorithms are considered.

- **Heuristic 1 (greedy approach)**
  We assume that each mobile user is served by a single BS on a specific frequency tone. For each user, we pick the BS and channel pair that has the strongest channel as its serving BS and channel. After BS–user association is determined, each BS uniformly allocates its power budget to the available frequency tones as well as to the served users on each tone. With the obtained power allocation and BS–user association, the

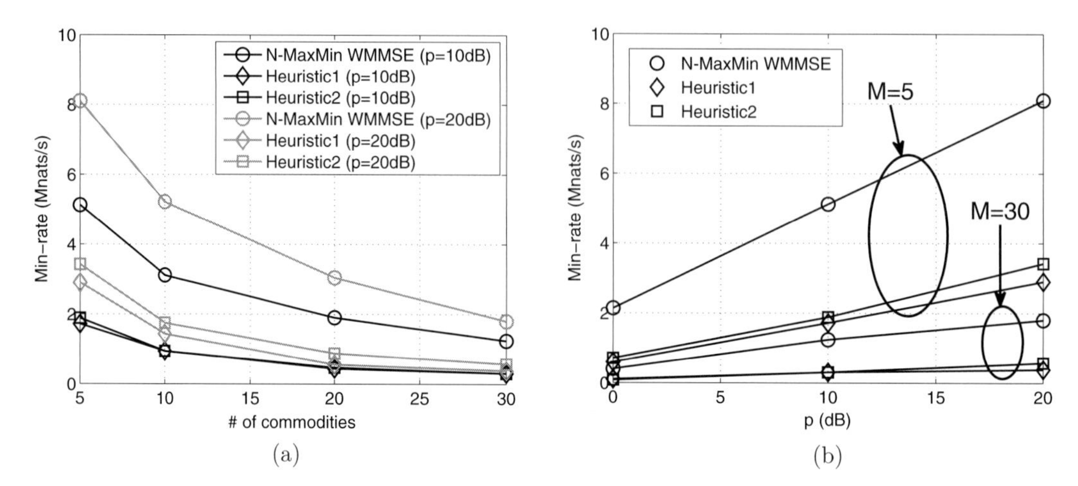

Figure 3.11 The min rate achieved by N-MaxMin algorithm and the two heuristic algorithms for different numbers of commodities and different power budgets.

capacity of all wireless links are available and fixed. Therefore the min rate of all commodities can be maximized by solving a wireline routing problem.

- **Heuristic 2 (orthogonal wireless transmission)**
  For the second heuristic algorithm, each BS uniformly allocates its power budget to each subchannel. To obtain a tractable problem formulation, we further assume that each active wireless link is interference free. Hence, each wireless link rate constraint now becomes convex. To impose this interference free constraint, additional variables $\beta_l \in \{0, 1\},\ \forall\, l \in \mathcal{L}^{wl}$ are introduced, where $\beta_l = 1$ if wireless link l is active, otherwise $\beta_l = 0$. In this way, there is no interference on wireless link l if $\sum_{n\in I(l)} \beta_n = 1$. To summarize, we solve the following optimization problem:

$$
\begin{aligned}
&\max\ r \\
&\text{s.t.} \sum_{m=1}^{M} r_l(m) \le \beta_l \log\left(1 + \frac{|h_l|^2 \bar{p}_{s_l}/K}{\sigma_{d_l}^2}\right), \\
&\sum_{n\in I(l)} \beta_n = 1,\ \beta_l \in \{0, 1\},\ \forall\, l, n \in \mathcal{L}^{wl}, \\
&(3.40\text{b}),\ (3.40\text{c}),\ \text{and}\ (3.40\text{e}).
\end{aligned}
$$

Since the integer constraints on $\{\beta_l \mid \forall\, l \in \mathcal{L}^{wl}\}$ are also intractable, we relax it to $\beta_l = [0, 1]$. In this way the problem becomes a large-scale LP, whose solution represents an upper bound value of this heuristic.

In Figure 3.11, we show the min rate performance of different algorithms for different numbers of commodities and power budget. We observe that the minimum rates achieved by the N-MaxMin algorithm are more than twice of those achieved by the heuristic algorithms.

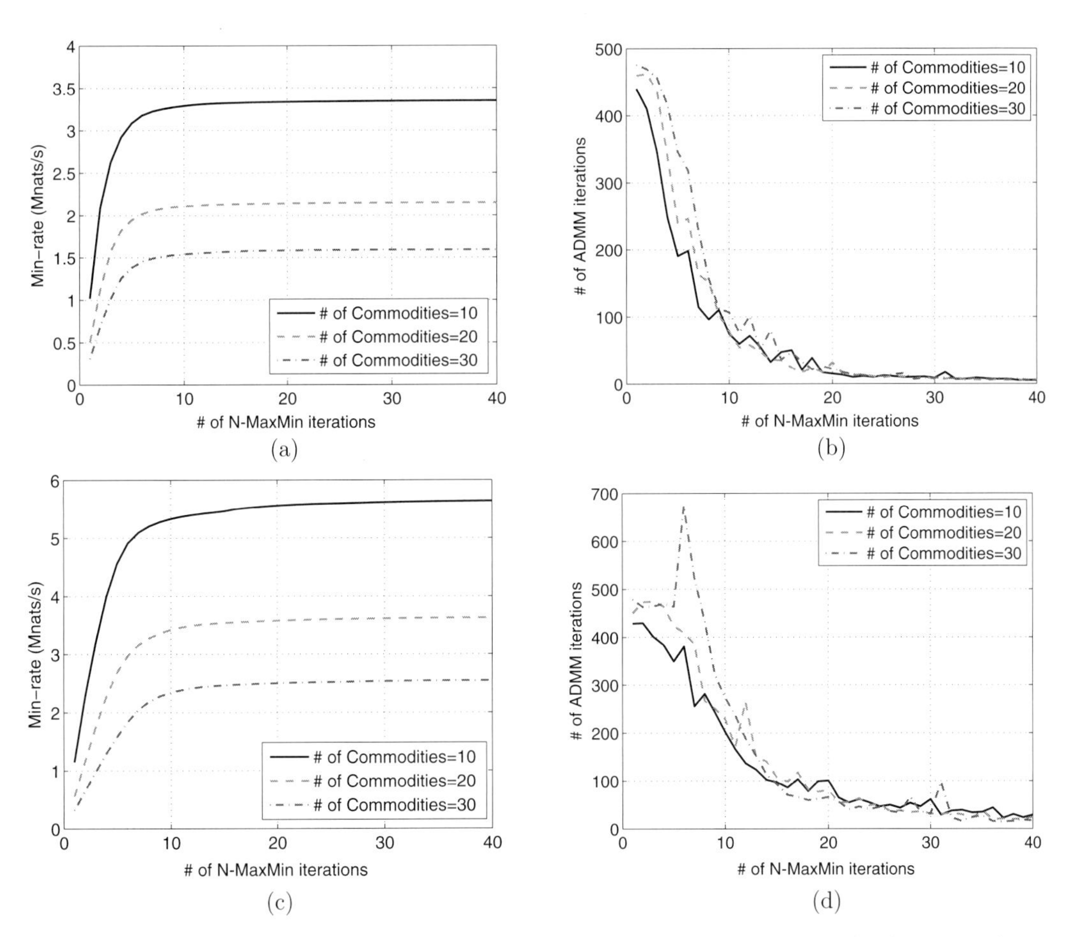

Figure 3.12 The min rate performance and the required number of iterations for the proposed N-MaxMin algorithm. In [(a)(b)] $p = 10$ dB and in [(c)(d)] $p = 20$ dB. In [(a)(c)], the obtained min rate versus the iterations of N-MaxMin is plotted. In [(b)(d)], the required number of inner ADMM iterations is plotted against the iteration for the outer N-MaxMin algorithm.

3.4.2 Scenario 2: The efficiency of N-MaxMin WMMSE algorithm

In the second set of numerical experiments, we evaluate the proposed N-MaxMin algorithm using different number of commodity pairs and different power budgets at the BSs. Here we use the same settings as in the previous experiment, except that all mobile users are interfered by the BSs within a distance of 800 m, and that we set $\rho_2 = 0.005$ (resp. $\rho_2 = 0.001$) when $p = 10$ dB (resp. $p = 20$ dB). The min rate performance for the N-MaxMin algorithm and the required number of inner iterations are plotted in Figure 3.12. Owing to the fact that the obtained $\{\mathbf{r}, \mathbf{v}\}$ is far from the stationary solution in the first few outer iterations, there is no need to complete the inner Algorithm 1 at the very beginning. Hence, we limit the number of inner iterations to be no more than 500 for the first five outer iterations. After the early termination of the inner algorithm, we use the obtained $\mathbf{v}$ to update $\{\hat{u}_l\}_l$ and $\{\hat{e}_l\}_l$.

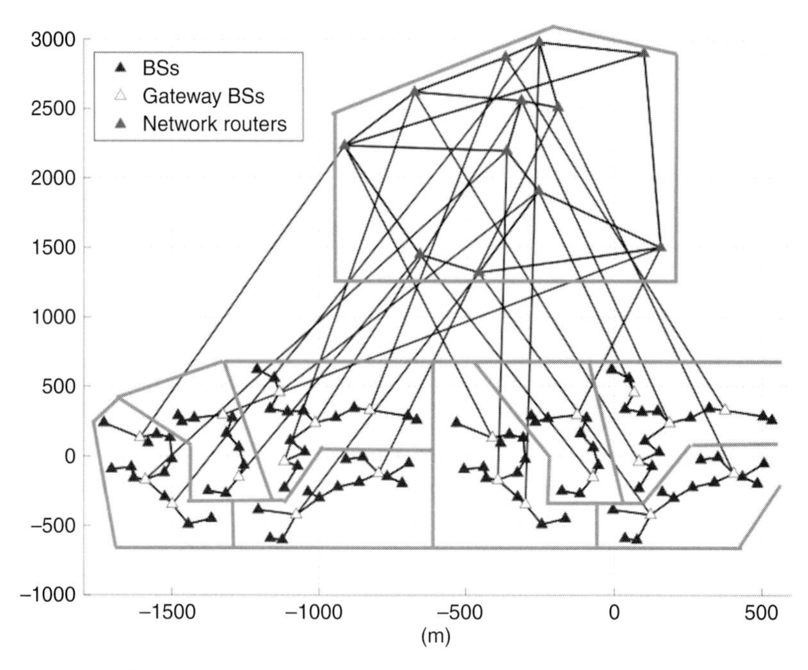

Figure 3.13 The considered network consists of 114 BSs and 11 routers with the locations and the connectivity between these nodes. Each computation core is responsible for one group of nodes shown in the figure.

In Figure 3.12(a), (b), we see that when $p = 10$ dB, the min rate converges at about the tenth outer iteration when the number of commodities is up to 30, while less than 500 inner iterations are needed per outer iteration. Moreover, after the tenth outer iteration, the number of inner ADMM iterations reaches below 100. In Figure 3.12(c), (d), the case with $p = 20$ dB is considered. Clearly the required number of outer iterations is slightly larger than that in the case of $p = 10$ dB, since the objective value and the feasible set are both larger. However, in all cases the algorithm still converges fairly quickly.

3.4.3 Scenario 3: Multi-commodity routing problem with parallel implementation

In this set of numerical experiments, we demonstrate how parallel implementation can speed up the inner Algorithm 1 considerably. To illustrate the benefit of parallelization, we consider a larger network (see Figure 3.13) which is derived by merging two identical BS networks shown in Figure 3.10. The new network consists of 126 nodes (12 network routers and 114 BSs).

For simplicity, we removed all the wireless links, so constraints (3.40d) and (3.40f) of problem (3.40) are absent. This reduces problem (3.40) to a network flow problem (a very large linear program).

We implement Algorithm 1 using the Open MPI package, and compare its efficiency with the commercial LP solver, Gurobi [58]. For the Open MPI implementation, we use nine computation cores for each set of network nodes as illustrated in Figure 3.13.

Table 3.2 Comparison of computation time used by different implementations of the ADMM approach for the routing only problem. The size of the problems solved are specified using a range of metrics (total number of commodities, variables and constraints)

# of Commodities	50	100	300
# of Variables	1.4×10^4	2.9×10^4	8.7×10^4
# of Constraints	2.1×10^4	4.2×10^4	1.3×10^5
Time (s) (Sequential)	1.04	2.03	8.53
Time (s) (Parallel)	0.20	0.37	1.10
Time (s) (Gurobi)	0.20	0.64	2.51

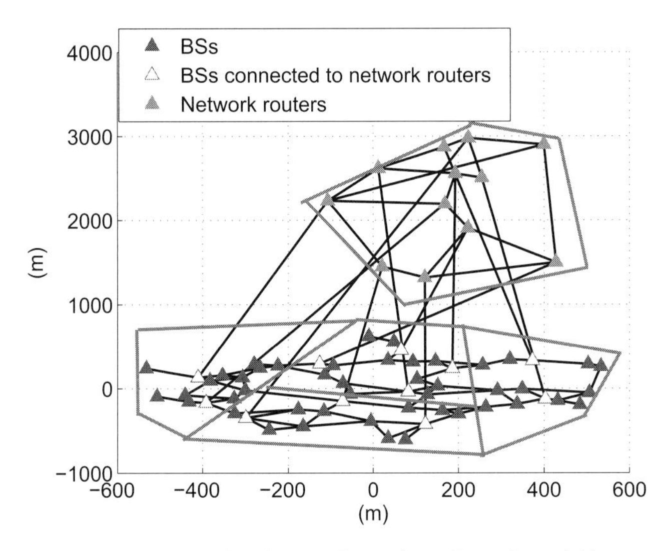

Figure 3.14 The considered network consists of 57 BSs and 11 routers with the locations and the connectivity between these nodes. Each zone of nodes is labeled by one group of nodes shown in the figure.

We choose $\rho_1 = 0.01$ and let the BSs serve as the destination nodes for commodities. Table 3.2 compares the computation time required for different implementation of Algorithm 1 and that of Gurobi. We observe that parallel implementation of the ADMM approach leads to more than five-fold improvement in computation time.[2] We also note that when the problem size increases, the performance of Gurobi becomes worse than that achieved by the parallel implementation of Algorithm 1. Thus, the proposed algorithm (implemented in parallel) appears to scale nicely to large problem sizes.

[2] All the computation is performed on a SunFire X4600 server with AMD Opteron 8356 2.3 GHz CPUs.

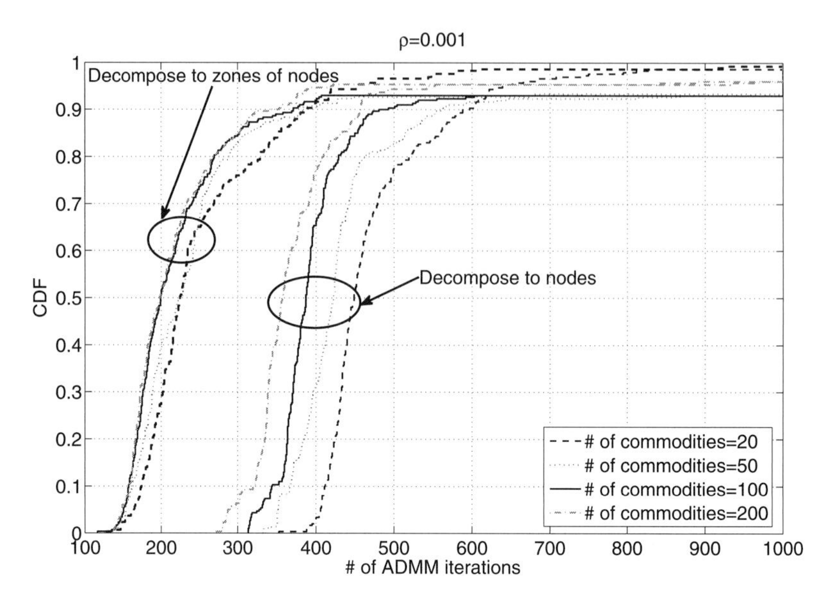

Figure 3.15 The CDF of the required number of iterations for two versions of Algorithm 1, one decomposes to the nodes and the other to the zones.

3.4.4 Scenario 4: Performance evaluation for Algorithm 1 with zones of nodes

In the last set of numerical experiments, the advantage of applying the modified Algorithm 1 with predefined zones of nodes will be demonstrated (cf. Section 3.3.3). In Figure 3.14, we provide the considered mesh network with 57 BSs and 11 network routers where the light gray lines label each zone of the nodes. The parameter ρ_1 of ADMM is set to be 0.001. In Figure 3.15, the CDF of the required number of iterations is illustrated for Algorithm 1 that decomposes to nodes or zones. One can easily observe that using the modified Algorithm 1 with predetermined zones of nodes, the number of ADMM iterations can be greatly decreased. This is because fewer slack variables are introduced for the zone-based implementation.

3.5 Appendix

In the appendix, we provide guidelines for solving subproblems (3.52) and (3.53).

Solving subproblem (3.52)

A closer look at this problem reveals that it naturally decomposes over the following three sets of variables

$$\{r, \{r(m)\}_{m\in\mathcal{M}}\},\ \left\{\{r_\ell(m)\}_{m\in\mathcal{M}} \mid \forall\, \ell \in \mathcal{L}^{\mathrm{w}}\right\},\ \hat{\mathbf{v}} \cup \{\{r_\ell(m)\}_{m\in\mathcal{M}} \mid \forall\, \ell \in \mathcal{L}^{\mathrm{wl}}\}.$$

Note that the first two subblocks only have to do with the wired links, while the last subblock corresponds to the variables over the wireless link. In the following we provide explicit forms of these subproblems. We refer the interested readers to [2, Appendix B] for detailed expressions for their solutions.

(i) Subproblem for $\{r, \{r(m)\}_{m\in\mathcal{M}}\}$ This subproblem updates the current minimum flow rate among all commodities. It takes the following form:

$$\max \frac{r}{2} - \frac{\rho_1}{2}\left(\hat{r} - r - \frac{\delta}{\rho_1}\right)^2 - \frac{\rho_1}{2}\sum_{m\in\mathcal{M}}\sum_{v\in\{s(m),d(m)\}}\left(\hat{r}(m)^v - r(m) - \frac{\delta_m^v}{\rho_1}\right)^2$$
$$\text{s.t. } r \geq 0,\ r(m) \geq r,\ m \in \mathcal{M}. \tag{3.59}$$

This problem is a quadratic problem with simple nonnegativity constraints. By checking the first-order optimality condition, its solution can be written down in (semi)closed form.

(ii) Subproblem for $\{\{r_\ell(m)\}_{m\in\mathcal{M}} \mid \forall\, \ell \in \mathcal{L}^{\mathrm{w}}\}$ The problem takes the following form:

$$\min \sum_{\ell\in\mathcal{L}^{\mathrm{w}}}\sum_{m\in\mathcal{M}}\sum_{v\in\{s_\ell,d_\ell\}}\left(\hat{r}_\ell^v(m) - r_\ell(m) - \frac{\delta_\ell^v(m)}{\rho_1}\right)^2$$
$$\text{s.t. } \sum_{m\in\mathcal{M}} r_\ell(m) \leq C_\ell,\ r_\ell(m) \geq 0,\ m \in \mathcal{M},\ \ell \in \mathcal{L}^{\mathrm{w}}.$$

Obviously it can be further decomposed into $|\mathcal{L}^{\mathrm{w}}|$ subproblems, expressed below, *one for each link* $\ell \in \mathcal{L}^{\mathrm{w}}$

$$\min \sum_{m\in\mathcal{M}}\sum_{v\in\{s_\ell,d_\ell\}}\left(\hat{r}_\ell^v(m) - r_\ell(m) - \frac{\delta_\ell^v(m)}{\rho_1}\right)^2$$
$$\text{s.t. } \sum_{m\in\mathcal{M}} r_\ell(m) \leq C_\ell,\ r_\ell(m) \geq 0,\ m \in \mathcal{M}. \tag{3.60}$$

This problem is a quadratic problem with a single linear inequality constraint and a number of simple nonnegativity constraints. Its solution can be again written down in closed form.

(iii) Subproblem for $\hat{\mathbf{v}} \cup \{\{r_\ell(m)\}_{m\in\mathcal{M}} \mid \forall\, \ell \in \mathcal{L}^{\mathrm{wl}}\}$ This problem is given by

$$\min \frac{\rho_1}{2}\sum_{\ell\in\mathcal{L}^{\mathrm{wl}}}\sum_{m\in\mathcal{M}}\sum_{v\in\{s_\ell,d_\ell\}}\left(\hat{r}_\ell^v(m) - r_\ell(m) - \frac{\delta_\ell^v(m)}{\rho_1}\right)^2 + \frac{\rho_2}{2}\sum_{n\in\mathcal{I}(\ell)}\left(v_n - \hat{v}_{\ell n} - \frac{\theta_{\ell n}}{\rho_2}\right)^2$$
$$\text{s.t. } r_\ell(m) \geq 0,\ m \in \mathcal{M},\ \text{and (3.50)}.$$

First note that the objective as well as the constraints in (3.50) are separable among the wireless links. Further due to variable splitting, each variable $\hat{v}_{\ell n}$ and $r_\ell(m)$ only appears in a *single* constraint in (3.50). Consequently the above problem can be decomposed into $|\mathcal{L}^{\mathrm{wl}}|$ subproblems, one for each wireless link $\ell \in \mathcal{L}^{\mathrm{wl}}$:

$$\begin{aligned} \min \quad & \frac{\rho_1}{2} \sum_{m\in\mathcal{M}, v\in\{s_\ell, d_\ell\}} \left(\hat{r}_\ell^v(m) - r_\ell(m) - \frac{\delta_\ell^v(m)}{\rho_1} \right)^2 + \frac{\rho_2}{2} \sum_{n\in\mathcal{I}(\ell)} \left(v_n - \hat{v}_{\ell n} - \frac{\theta_{\ell n}}{\rho_2} \right)^2 \\ \text{s.t.} \quad & r_\ell(m) \geq 0, \; m \in \mathcal{M}, \\ & \sum_{m\in\mathcal{M}} r_\ell(m) \leq \hat{c}_{1,\ell} + \hat{c}_{2,\ell}\hat{v}_{\ell\ell} - \sum_{n\in\mathcal{I}(\ell)} \hat{c}_{3,\ell n}\hat{v}_{\ell n}^2. \end{aligned} \tag{3.61}$$

Each of these problems is a quadratic problem with a single quadratic constraint and a number of nonnegativity constraints, therefore has closed-form solutions.

Solving subproblem (3.53)

This problem can be decomposed into two parts: one optimizes $\hat{\mathbf{r}}$ subject to the flow rate conservation constraint, and the other optimizes $\mathbf{v}$. The respective forms of the subproblem will be shown shortly. Again we refer the interested readers to [2, Appendix B] for exact solutions for these problems.

(i) Subproblem for $\hat{\mathbf{r}}$ This subproblem decomposes into two independent problems. The first one optimizes $\hat{r}$

$$\arg\max \frac{\hat{r}}{2} - \frac{\rho_1}{2}\left(\hat{r} - r - \frac{\delta}{\rho_1}\right)^2 = r + \frac{1+2\delta}{2\rho_1}. \tag{3.62}$$

The second one optimizes $\{\hat{r}(m)^{s(m)}, \hat{r}(m)^{d(m)}, \hat{r}_l^{s_l}(m), \hat{r}_l^{d_l}(m)\}$, subject to the conservation constraints of flow rate. Observe that the set of flow conservation constraints (3.47) decomposes among *each node* and *each commodity*. Further the introduction of the auxiliary variables made sure that each variable in $\hat{\mathbf{r}}$ only appears in a *single* constraint in (3.47). As such, the second problem further decomposes into a number of simpler problems, one for each commodity-node tuple (m, v)

$$\min \sum_{\ell\in\mathrm{In}(v)\cup\mathrm{Out}(v)} \left(\hat{r}_\ell^v(m) - r_\ell(m) - \frac{\delta_\ell^v(m)}{\rho_1} \right)^2 + 1_{\{s(m),d(m)\}}(v) \left(\hat{r}(m)^v - r(m) - \frac{\delta_m^v}{\rho_1} \right)^2 \tag{3.63}$$

$$\text{s.t.} \sum_{\ell\in\mathrm{In}(v)} \hat{r}_\ell^v(m) + 1_{s(m)}(v)\hat{r}(m)^v = \sum_{\ell\in\mathrm{Out}(v)} \hat{r}_\ell^v(m) + 1_{d(m)}(v)\hat{r}(m)^v.$$

Since problem (3.63) is a quadratic problem with a single equality constraint, it admits a closed-form solution.

(ii) Subproblem for $\mathbf{v}$ The variable $\mathbf{v}$ is constrained by the per-BS power constraint, therefore its related subproblem naturally decomposes over the BSs. For BS $s \in \mathcal{B}$, the problem is given by

$$\begin{aligned} \min \quad & \sum_{\ell \in \operatorname{Out}(s) \bigcap \mathcal{L}^{\mathrm{wl}},\, n \in \bar{\mathcal{I}}(\ell)} \left(v_\ell - \hat{v}_{n\ell} - \frac{\theta_{n\ell}}{\rho_2} \right)^2 \\ \text{s.t.} \quad & \sum_{\ell \in \operatorname{Out}(s) \bigcap \mathcal{L}^{\mathrm{wl}}} v_\ell^2 \le \bar{p}_s, \end{aligned} \tag{3.64}$$

which is a simple quadratic problem with a single quadratic constraint. Its solution can be again obtained analytically.

References

[1] Huawei, "5G: A technology vision," Huawei Technologies Inc., White paper, 2013.

[2] W.-C. Liao, M. Hong, H. Farmanbar, *et al.*, "Min flow rate maximization for software defined radio access networks," *IEEE Journal on Selected Areas in Communication*, vol. **32**, no. 6, pp. 1282–1294, 2014.

[3] J. Andrews, "Seven ways that HetNets are a cellular paradigm shift," *IEEE Communications Magazine*, vol. **51**, no. 3, pp. 136–144, March 2013.

[4] S.-H. Park, O. Simeone, O. Sahin, and S. Shamai, "Joint precoding and multivariate backhaul compression for the downlink of cloud radio access networks," *IEEE Transactions on Signal Processing*, vol. **61**, no. 22, pp. 5646–5658, November 2013.

[5] P. Tseng, "Convergence of a block coordinate descent method for nondifferentiable minimization," *Journal of Optimization Theory and Applications*, vol. **103**, no. 9, pp. 475–494, 2001.

[6] D. P. Bertsekas and J. N. Tsitsiklis, *Neuro-Dynamic Programming*, Belmont, MA: Athena Scientific, 1996.

[7] ——, *Parallel and Distributed Computation: Numerical Methods*, 2nd edn, Belmont, MA: Athena Scientific, 1997.

[8] Z.-Q. Luo and P. Tseng, "Error bounds and convergence analysis of feasible descent methods: a general approach," *Annals of Operations Research*, vol. **46–47**, pp. 157–178, 1993.

[9] ——, "On the convergence of the coordinate descent method for convex differentiable minimization," *Journal of Optimization Theory and Application*, vol. **72**, no. 1, pp. 7–35, 1992.

[10] ——, "On the linear convergence of descent methods for convex essentially smooth minimization," *SIAM Journal on Control and Optimization*, vol. **30**, no. 2, pp. 408–425, 1992.

[11] Y. Nesterov, "Efficiency of coordiate descent methods on huge-scale optimization problems," *SIAM Journal on Optimization*, vol. **22**, no. 2, pp. 341–362, 2012.

[12] P. Richtarik and M. Takac, "Iteration complexity of randomized block-coordinate descent methods for minimizing a composite function," *Mathematical Programming*, pp. 1–38, 2012.

[13] S. Shalev-Shwartz and A. Tewari, "Stochastic methods for ℓ_1 regularized loss minimization," *Journal of Machine Learning Research*, vol. **12**, pp. 1865–1892, 2011.

[14] Z. Lu and X. Lin, "On the complexity analysis of randomized block-coordinate descent methods," *Mathematical Programming*, 2013, accepted.

[15] A. Saha and A. Tewari, "On the nonasymptotic convergence of cyclic coordinate descent method," *SIAM Journal on Optimization*, vol. **23**, no. 1, pp. 576–601, 2013.

[16] A. Beck and L. Tetruashvili, "On the convergence of block coordinate descent type methods," *SIAM Journal on Optimization*, vol. **23**, no. 4, pp. 2037–2060, 2013.

[17] M. Hong, X. Wang, M. Razaviyayn, and Z.-Q. Luo, "Iteration complexity analysis of block coordinate descent methods," preprint, 2013, available online arXiv:1310.6957.

[18] F. Facchinei, S. Sagratella, and G. Scutari, "Flexible parallel algorithms for big data optimization," in *2014 IEEE International Conference on Acoustics, Speech and Signal Processing (ICASSP)*, 2014.

[19] G. Scutari, F. Facchinei, P. Song, D. P. Palomar, and J.-S. Pang, "Decomposition by partial linearization: Parallel optimization of multi-agent systems," *IEEE Transactions on Signal Processing*, vol. **63**, no. 3, pp. 641–656, 2014.

[20] M. J. D. Powell, "On search directions for minimization algorithms," *Mathematical Programming*, vol. **4**, pp. 193–201, 1973.

[21] M. V. Solodov, "On the convergence of constrained parallel variable distribution algorithms," *SIAM Journal on Optimization*, vol. **8**, no. 1, pp. 187–196, 1998.

[22] R. Glowinski and A. Marroco, "Sur l'approximation, par elements finis d'ordre un,et la resolution, par penalisation-dualite, d'une classe de problemes de dirichlet non lineares," *Revue Franqaise d'Automatique, Informatique et Recherche Opirationelle*, vol. **9**, pp. 41–76, 1975.

[23] D. Gabay and B. Mercier, "A dual algorithm for the solution of nonlinear variational problems via finite element approximation," *Computers & Mathematics with Applications*, vol. **2**, pp. 17–40, 1976.

[24] S. Boyd, N. Parikh, E. Chu, B. Peleato, and J. Eckstein, "Distributed optimization and statistical learning via the alternating direction method of multipliers," *Foundations and Trends in Machine Learning*, vol. **3**, 2011.

[25] W. Yin, S. Osher, D. Goldfarb, and J. Darbon, "Bregman iterative algorithms for l1-minimization with applications to compressed sensing," *SIAM Journal on Imgaging Science*, vol. **1**, no. 1, pp. 143–168, March 2008.

[26] J. Yang, Y. Zhang, and W. Yin, "An efficient TVL1 algorithm for deblurring multichannel images corrupted by impulsive noise," *SIAM Journal on Scientific Computing*, vol. **31**, no. 4, pp. 2842–2865, 2009.

[27] X. Zhang, M. Burger, and S. Osher, "A unified primal-dual algorithm framework based on Bregman iteration," *Journal of Scientific Computing*, vol. **46**, no. 1, pp. 20–46, 2011.

[28] K. Scheinberg, S. Ma, and D. Goldfarb, "Sparse inverse covariance selection via alternating linearization methods," in *Twenty-Fourth Annual Conference on Neural Information Processing Systems (NIPS)*, 2010.

[29] D. P. Bertsekas, *Nonlinear Programming*, 2nd edn, Belmont, MA: Athena Scientific, 1999.

[30] S. Boyd and L. Vandenberghe, *Convex Optimization*, Cambridge University Press, 2004.

[31] A. Nedic and A. Ozdaglar, "Cooperative distributed multi-agent optimization," in *Convex Optimization in Signal Processing and Communications*, Cambridge University Press, 2009.

[32] D. P. Bertsekas, *Constrained Optimization and Lagrange Multiplier Method*, Belmont, MA: Academic Press, 1982.

[33] B. He and X. Yuan, "On the O(1/n) convergence rate of the Douglas–Rachford alternating direction method," *SIAM Journal on Numerical Analysis*, vol. **50**, no. 2, pp. 700–709, 2012.

[34] R. Monteiro and B. Svaiter, "Iteration-complexity of block-decomposition algorithms and the alternating direction method of multipliers," *SIAM Journal on Optimization*, vol. **23**, no. 1, pp. 475–507, 2013.

[35] T. Goldstein, B. O'Donoghue, and S. Setzer, "Fast alternating direction optimization methods," *UCLA CAM technical report*, 2012.
[36] D. Boley, "Linear convergence of ADMM on a model problem," *SIAM Journal on Optimization*, vol. **23**, pp. 2183–2207, 2013.
[37] W. Deng and W. Yin, "On the global linear convergence of alternating direction methods," preprint, 2012.
[38] Z. Zhou, X. Li, J. Wright, E. Candes, and Y. Ma, "Stable principal component pursuit," *Proceedings of 2010 IEEE International Symposium on Information Theory*, 2010.
[39] M. Hong and Z.-Q. Luo, "On the linear convergence of the alternating direction method of multipliers," arXiv preprint arXiv:1208.3922, 2012.
[40] X. Wang, M. Hong, S. Ma, and Z.-Q. Luo, "Solving multiple-block separable convex minimization problems using two-block alternating direction method of multipliers," preprint, 2013.
[41] C. Chen, B. He, X. Yuan, and Y. Ye, "The direct extension of ADMM for multi-block convex minimization problems is not necessarily convergent," preprint, 2013.
[42] B. He, M. Tao, and X. Yuan, "Alternating direction method with Gaussian back substitution for separable convex programming," *SIAM Journal on Optimization*, vol. **22**, pp. 313–340, 2012.
[43] M. Razaviyayn, M. Hong, and Z.-Q. Luo, "A unified convergence analysis of block successive minimization methods for nonsmooth optimization," *SIAM Journal on Optimization*, vol. **23**, no. 2, pp. 1126–1153, 2013.
[44] P. Combettes and J.-C. Pesquet, "Proximal splitting methods in signal processing," in *Fixed-Point Algorithms for Inverse Problems in Science and Engineering*, ser. Springer Optimization and Its Applications, New York: Springer, 2011, pp. 185–212.
[45] C. Navasca, L. D. Lathauwer, and S. Kindermann, "Swamp reducing technique for tensor decomposition," *Proceedings 16th European Signal Processing Conference (EUSIPCO)*, August 2008.
[46] Q. Shi, M. Razaviyayn, Z.-Q. Luo, and C. He, "An iteratively weighted MMSE approach to distributed sum-utility maximization for a MIMO interfering broadcast channel," *IEEE Transactions on Signal Processing*, vol. **59**, no. 9, pp. 4331–4340, 2011.
[47] A. P. Dempster, N. M. Laird, and D. B. Rubin, "Maximum likelihood from incomplete data via the EM algorithm," *Journal of the Royal Statistical Society Series B*, vol. **39**, pp. 1–38, 1977.
[48] A. L. Yuille and A. Rangarajan, "The concave-convex procedure," *Neural Computation*, vol. **15**, no. 4, pp. 915–936, Apr. 2003.
[49] D. Hunter and K. Lange, "Quantile regression via an mm algorithm," *Journal of Computational and Graphical Statistics*, vol. **9**, pp. 60–77, 2000.
[50] D. D. Lee and H. S. Seung, "Algorithms for non-negative matrix factorization," in *Neural Information Processing Systems (NIPS)*, 2000, pp. 556–562.
[51] B. R. Marks and G. P. Wright, "A general inner approximation algorithm for nonconvex mathematical programs," *Operations Research*, vol. **26**, pp. 681–683, July–August 1978.
[52] B. Chen, S. He, Z. Li, and S. Zhang, "Maximum block improvement and polynomial optimization," *SIAM Journal on Optimization*, vol. **22**, no. 1, pp. 87–107, 2012.
[53] M. Hong, Q. Li, and Y.-F. Liu, "Decomposition by successive convex approximation: a unifying approach for linear transceiver design in interfering heterogeneous networks," manuscript, 2013, available online arXiv:1210.1507.

[54] S. S. Christensen, R. Agarwal, E. D. Carvalho, and J. M. Cioffi, "Weighted sum-rate maximization using weighted MMSE for MIMO-BC beamforming design," *IEEE Transactions on Wireless Communications*, vol. **7**, no. 12, pp. 4792–4799, 2008.

[55] M. Hong, R. Sun, H. Baligh, and Z.-Q. Luo, "Joint base station clustering and beamformer design for partial coordinated transmission in heterogenous networks," *IEEE Journal on Selected Areas in Communications.*, vol. **31**, no. 2, pp. 226–240, 2013.

[56] M. Razaviyayn, M. Hong, and Z.-Q. Luo, "Linear transceiver design for a MIMO interfering broadcast channel achieving max-min fairness," *Signal Processing*, vol. **93**, no. 12, pp. 3327–3340, 2013.

[57] D. P. Bertsekas, P. Hosein, and P. Tseng, "Relaxation methods for network flow problems with convex arc costs," *SIAM Journal on Control and Optimization*, vol. **25**, no. 5, pp. 1219–1243, September 1987.

[58] Gurobi, "Gurobi optimizer reference manual," 2013.

4 A unified distributed algorithm for non-cooperative games

Jong-Shi Pang and Meisam Razaviyayn

This chapter presents a unified framework for the design and analysis of distributed algorithms for computing first-order stationary solutions of non-cooperative games with non-differentiable player objective functions. These games are closely associated with multi-agent optimization wherein a large number of selfish players compete non-cooperatively to optimize their individual objectives under various constraints. Unlike centralized algorithms that require a certain system mechanism to coordinate the players' actions, distributed algorithms have the advantage that the players, either individually or in subgroups, can each make their best responses without full information of their rivals' actions. These distributed algorithms by nature are particularly suited for solving huge-size games where the large number of players in the game makes the coordination of the players almost impossible. The distributed algorithms are distinguished by several features: parallel versus sequential implementations, scheduled versus randomized player selections, synchronized versus asynchronous transfer of information, and individual versus multiple player updates. Covering many variations of distributed algorithms, the unified algorithm employs convex surrogate functions to handle nonsmooth nonconvex functions and a (possibly multi-valued) choice function to dictate the players' turns to update their strategies. There are two general approaches to establish the convergence of such algorithms: contraction versus potential based, each requiring different properties of the players' objective functions. We present the details of the convergence analysis based on these two approaches and discuss randomized extensions of the algorithms that require less coordination and hence are more suitable for big data problems.

4.1 Introduction

Introduced by John von Neumann [1], modern-day game theory has developed into a very fruitful research discipline with applications in many fields. There are two major classifications of a game, cooperative versus non-cooperative. This chapter pertains to one aspect of non-cooperative games for potential applications to big data, namely, the computation of a "solution" to such a game by a distributed algorithm. In a (basic) non-cooperative game, there are finitely many selfish players/agents each optimizing

Big Data over Networks, ed. Shuguang Cui, Alfred O. Hero III, Zhi-Quan Luo, and José M. F. Moura. Published by Cambridge University Press.

a rival-dependent objective by choosing feasible strategies satisfying certain private constraints. Providing a solution concept to such a game, a Nash equilibrium (NE) [2, 3] is by definition a tuple of strategies, one for each player, such that no player will be better off by unilaterally deviating from his/her equilibrium strategy while the rivals keep executing their equilibrium strategies. The game paradigm and the Nash solution concept have found recent applications in many engineering fields; two of these are dynamic spectrum management in signal processing [4–10]; and pricing and distribution in electricity markets [11–15]. In the absence of full information available to the players, distributed algorithms are particularly useful in large-size games because they allow the players to choose their best responses without extensive coordination with their rivals. While such algorithms have not been particularly popular in the domain of electricity markets, they have received much attention in the signal processing area due to the lack of information of the cross interferences among the channels which are modeled as the players of a game. Two recent surveys on Nash equilibria via the variational approach are [16, 17], where many references can be found.

With the recent advances in storage, communication, and processing of large amounts of data, the resulting game models arising from different applications are of big size. For an application of game theory models to big data analytics and decision making in the field of geosciences and remote sensing, see [18]. Another source of game theory application to big data engineering arises from modern heterogeneous wireless networks where many low-power small transmitters such as macro/micro/pico/femto base stations are densely deployed to provide coverage extension for users with low signal reception. The close proximity of the transmitters results in substantial interference in the network which, if not properly managed, can cause communication failure. Owing to the lack of multi-user interference knowledge, this problem can be modeled as a "big size" non-cooperative game with large number of variables [19]. A further (potential) application of such big games is in solving huge size inference problems on modern computing platforms. These huge-size inference problems appear in a wide range of applications such as feature selection and classification in bioinformatics [20, 21] and text processing [22]. Although these problems are mostly formulated as optimization problems at the present time, their solution could benefit from a game-theoretic perspective when there is a lack of a centralized coordination scheme in the inference process and it becomes essential to distribute the computation among competing groups. Owing to the gigantic dimension of the problem, one should use modern computing platforms with multi-thread/multi-core/cluster structures to solve such inference problems. Consequently, parallel distributed algorithms, with low coordination among players and processing nodes, are essential for analyzing such big games.

As a way of surveying the state of the art of distributed algorithms for computing Nash equilibria of non-cooperative games, this chapter presents a unified treatment of such algorithms, introducing flexibilities and exploring extensions of the algorithms that are more suitable to big data problems and modern high performance multi-core computing platforms. Our unification is accomplished by the use of convex functional approximations to deal with possibly nonconvex player objective functions (for minimization) and a set-valued player selection map to identify the sequences and subsets of players in updating their responses. While there is a growing literature on distributed (and other) algorithms for solving the so-called generalized Nash games in which the players' constraints, called *coupled constraints*, also contain the rivals' decision variables (see e.g. [16, 23, 24]), we focus herein on the basic version of the Nash game as described in the opening paragraph. As explained in several references (e.g. [25, 26] for optimization and [27] for games), one way to extend the algorithms for decoupled constraints to coupled constraints is to employ a sequential pricing scheme that converts the problems with coupled constraints to a sequence of problems with private constraints only, resulting in a doubly iterative loop wherein the outer loop updates the prices (or the multipliers of the coupled constraints) and each inner loop updates the players' strategies at the current prices. A major departure of our presentation from many papers in the literature is that we employ surrogate objectives in defining the sequence of subproblems to be solved at each iteration; this substitution of objectives is needed to handle games with non-convex objective functions so that the results from the subproblems are the players' "surrogate best responses" to their rivals' current iterates that are available to the players when they do their updates. For such nonconvex games, our goal is to compute a so-called *quasi-Nash equilibrium* [27] that is in essence a strategy tuple satisfying the first-order stationarity conditions of each players' optimization problems.

Although the employment of surrogate functions in nonconvex optimization dates back to the early days of the subject (see e.g. [28]), the systematic study of such methods has received renewed interests recently and attracted many researchers due to a host of applications in various modern practical problems; see [29, 30] and the examples therein. In particular, the recent works [30, 31] study the convergence of such methods in multi-block optimization of convex albeit nonsmooth functions; the references [32, 33] study the iteration complexity of this method for single and multi-variable block optimization problems. The recent papers [34, 35] utilize the same idea to develop a parallel/Jacobi-type algorithms for big data optimization problems. The extensions of the surrogate idea to linearly constrained optimization and stochastic optimization problems can be found in [36] and [32, 37], respectively. In this paper, we study the use of surrogate objective functions for solving games with nonconvex objectives in a *block method* wherein non-overlapping groups of players update their strategies by solving subgames in parallel.

While the distributed algorithms are simple to describe, their (theoretical) convergence requires certain properties of the players' objective functions. Extending the pioneer work of Ortega and Rheinboldt [38] on the convergence of iterative methods for systems of nonlinear equations and the subsequent treatment for the linear complementarity problem [39, Chapter 5], an extensive analysis of the convergence of such methods for solving the finite-dimensional variational inequalities (VIs) [17] can be found in the papers [40, 41]. Particularly worthy of note is the latter paper [41] that deals with a partitioned VI defined on a Cartesian product of sets of lower dimensions; the results therein are directly applicable to certain distributed algorithms for computing Nash equilibria. Also of relevance is the monograph [42] that presents a comprehensive and rigorous treatment of parallel and distributed numerical methods for optimization problems and VIs. Although these early references have laid a good foundation for analyzing the convergence of distributed algorithms, they are not tailored to game problems that contain special features not generally taken into account in the more general context. As can be seen from these references, particularly [39, Chapter 5], there are in general three approaches to establish the convergence of iterative algorithms: contraction of the iterates, a potential descent argument, and monotonicity of the iterates. The applicability of the last approach to games is highly restrictive; thus we will not consider it in this chapter and focus only on the contraction and potential approaches and their randomized extension which is more amenable to big data problems. Originated from the analysis of a special resource allocation problem in dynamic spectrum management [4], extended to several enhanced versions of this problem [5, 9, 43, 44], and given a general treatment in [17], a spectral radius condition on the second derivatives of the players' objective functions has played a fundamental role in the contraction-based convergence proof of the distributed algorithms. When a potential function exists for the game, the paper [24] has analyzed the convergence of Gauss–Seidel-type algorithms, even with shared constraints that couple the players' variables. To date, the two approaches, contraction and potential, provide the-state-of-the-art convergence analysis of the distributed algorithms for solving games. It remains a challenge to invent a third approach for such an analysis, thereby allowing the asymptotic behavior of these algorithms to be better understood. A need indeed exists because, in practical implementations, these algorithms exhibit convergence even though the existing theory cannot predict it. We hope that this chapter will provide the foundation to expand the theory so that the empirical convergence that is currently not understood can be explained.

Referring to the above cited references where applications are discussed, this chapter is written with an emphasis on the algorithms and their convergence. In addition to providing a unified treatment of these algorithms, the chapter contributes to the existing literature by expanding the algorithms with added flexibility in their implementation, allowing a mixture of Gauss–Seidel and Jacobi updates in the iterations, introducing randomized choices of players to do their updates at each iteration, and presenting a unified convergence theory based on two principal approaches. While no new application is presented here, as the need for large-scale distributed computing in a competitive environment increases, the game-theoretic framework will grow in importance; having a unified perspective of the iterative solution of the resulting games can be expected to benefit today's world of big data computations.

4.2 The nonsmooth, nonconvex game

Consider an n-player non-cooperative game $\mathcal{G}$ wherein each player $i = 1, \ldots, n$, anticipating the rivals' strategy tuple $x^{-i} \triangleq (x^j)_{i \neq j=1}^n \in \mathcal{X}^{-i} \triangleq \prod_{i \neq j=1}^n \mathcal{X}^j$, solves the optimization problem:

$$\underset{x^i \in \mathcal{X}^i}{\text{minimize}}\ \theta_i(x^i, x^{-i}). \tag{4.1}$$

Here $\mathcal{X}^i \subseteq \mathbb{R}^{n_i}$ is a closed convex set and $\theta_i : \mathbf{\Omega} \to \mathbb{R}$ is a continuous (possibly nonsmooth and nonconvex) function defined on the open convex set $\mathbf{\Omega} \triangleq \prod_{i=1}^n \Omega^i$, where each Ω^i is an open convex set containing $\mathcal{X}^i$. Let $\boldsymbol{\mathcal{X}} \triangleq \prod_{i=1}^n \mathcal{X}^i$ be the set of feasible strategies of all the players, which is a closed convex set in $\mathbb{R}^N$, where $N \triangleq \sum_{i=1}^n n_i$. In spite of the convexity of the constraints, no convexity or differentiability is imposed on the objective functions. This is a major departure from the extensive literature on the game $\mathcal{G}$ where $\theta_i(\bullet, x^{-i})$ is both convex and differentiable; see e.g. [17]. The first complication in this nonsmooth and nonconvex situation is that well-known optimality conditions for (4.1) that depend on the differentiability of the objective function are not readily applicable. Thus, we focus on the computation of a quasi-Nash equilibrium as defined below. This definition is not in terms of the (partial) gradient vectors of the objective functions, which may not exist, but only in terms of their directional derivatives, which we assume exist.

Definition 4.1 Assuming that for each i, $\theta_i(\bullet, x^{-i})$ is directionally differentiable with directional derivative $\theta_i(\bullet, x^{-i})'(x^i; d^i)$ at x^i in the direction d^i, we say that a tuple $x^* \triangleq \left(x^{*;i}\right)_{i=1}^n \in \boldsymbol{\mathcal{X}}$ is a *quasi-Nash equilibrium* (QNE) of the game $\mathcal{G}$ if for all $i = 1, \ldots, n$,

$$\theta_i(\bullet, x^{*;-i})'(x^{*;i}; x^i - x^{*;i}) \geq 0, \qquad \text{for all } x^i \in \mathcal{X}^i. \tag{4.2}$$

If $\theta_i(\bullet, x^{-i})$ is differentiable, then the above inequality becomes

$$\nabla_{x^i}\theta_i(x^*)^T(x^i - x^{*;i}) \geq 0, \qquad \text{for all } x^i \in \mathcal{X}^i.$$

Hence in this differentiable case, x^* is a QNE if and only if x^* is a solution of the variational inequality [45] defined by the pair $(\Theta, \boldsymbol{\mathcal{X}})$, where $\Theta(x) \triangleq (\nabla_{x^i}\theta_i(x))_{i=1}^n$.

Example of non-existence of a QNE

Owing to the nonconvexity and non-differentiability of the objective functions, the above game $\mathcal{G}$ may not have a QNE in general (let alone a Nash equilibrium). For example, consider a two-player game with

$$\theta_1(x_1, x_2) = x_1x_2 - \frac{1}{2}|x_1|, \quad \text{and} \quad \theta_2(x_1, x_2) = -x_1x_2$$

where the feasible strategy sets for the players are the same and given by: $-1 \leq x_1 \leq 1$ and $-1 \leq x_2 \leq 1$, respectively. It is not hard to check that a QNE of the game does not exist even though the feasible sets are compact and convex. □

Separately, the existence of a NE (for player-convex but not necessarily differentiable objectives) and that of a QNE (for player-differentiable but not necessarily convex objectives) have been studied in two recent papers, [17] for the former and [5] for the latter. Nevertheless, these results are not applicable to the case studied here, namely, when $\theta_i(\bullet, x^{-i})$ is neither convex nor differentiable. In what follows, we present a class of (simplified) games for which existence of a QNE can be proved using the approach of *surrogate objectives*; for this approach to be successful, a "derivative consistency" condition is required that links the two families of objectives. We keep this discussion simple by assuming that the set $\mathcal{X}$ is bounded, in addition to being closed and convex. Furthermore, a structural form of the players' objective functions is needed as specified in the statement of the following existence result.

Proposition 4.2 *Suppose that $\mathcal{X}$ is compact and convex, and that for every $i = 1, \ldots, n$, the function $\theta_i(x^i, x^{-i}) = f_i(x^i, x^{-i}) + g_i(x^i, x^{-i})$, where $f_i(\bullet)$ is twice continuously differentiable and $g_i(\bullet, x^{-i})$ is convex for every fixed but arbitrary $x^{-i} \in \mathcal{X}^{-i}$. Then the game $\mathcal{G}$ has a QNE.*

Proof Evidently, there exists a constant $\alpha > 0$ such that the function

$$\widetilde{\theta}_i(x^i; y) \triangleq f_i(x^i, y^{-i}) + g_i(x^i, y^{-i}) + \frac{\alpha}{2}\|x^i - y^i\|^2$$

is strongly convex in $x^i \in \mathcal{X}^i$ for any fixed $y \in \mathcal{X}$ and for all $i = 1, 2, \ldots, n$. Define the self-map $\varphi : \mathcal{X} \mapsto \mathcal{X}$ with $\varphi(y) \triangleq (\widehat{x}_i(y))_{i=1}^n$ and for $i = 1, \ldots, n$, $\widehat{x}_i(y) \triangleq \operatorname{argmin}_{x^i \in \mathcal{X}^i} \widetilde{\theta}_i(x^i; y)$, which is well defined by the compactness of $\mathcal{X}^i$. By the strong convexity of the mapping $\widetilde{\theta}_i(\bullet; y)$, it follows that $\widehat{x}_i$ is a continuous self-map from $\mathcal{X}$ onto itself. It therefore has a fixed point by Brouwer's fixed-point theorem; i.e. there exists $x^* \in \mathcal{X}$ so that $\varphi(x^*) = x^*$. It is not difficult to verify that x^* is a QNE of the game with the objective functions θ_i, in view of the following equality:

$$\widetilde{\theta}_i(\bullet; x^*)'(x^{*;i}; x^i - x^{*;i}) = \theta_i(\bullet, x^{*;-i})'(x^{*;i}; x^i - x^{*;i}),$$

which is the derivative consistency condition mentioned above. □

Remark 4.1 Without the functions f_i, Proposition 4.2 reduces to the classical result for the existence of a Nash equilibrium when each $\theta_i(\bullet, x^{-i})$ is convex. This proposition extends the classical result to the situation where each $\theta_i(\bullet, x^{-i})$ differs from a convex function by a C^2 function f_i with bounded second self-partial derivatives; i.e. $\|\nabla^2_{x^i x^i} f_i(x)\|$ is bounded on the domain $\mathcal{X}$. Under this assumption, the resulting function $\theta_i(\bullet, x^{-i}) = f_i(\bullet, x^{-i}) + g_i(\bullet, x^{-i})$ is of the dc type (where dc stands for "difference of convex"). In the accompanying paper [46], the reader can find a comprehensive study of a dc optimization problem that is applicable to each individual player's problem (4.1) in the game $\mathcal{G}$. □

Through Proposition 4.2, we see that there are three key elements that have played an important role in providing the existence of a QNE to the nonsmooth and nonconvex game $\mathcal{G}$: the idea of surrogate objectives, a derivative consistency condition, and a structural property of the objectives. These three attributes will continue to play a major role throughout the rest of this chapter.

4.3 The unified algorithm

In the design of a unified iterative distributed algorithm for computing a QNE of the game $\mathcal{G}$, we employ, at each iteration ν, a selection tuple $\boldsymbol{\sigma}^{\nu}$ of players and surrogate objectives $\widehat{\boldsymbol{\theta}}^{\nu}$ as follows.

(a) The players' selection tuple $\boldsymbol{\sigma}^{\nu} \triangleq \{\sigma_1^{\nu}, \ldots, \sigma_{\kappa_\nu}^{\nu}\}$ consists of κ_ν pairwise disjoint subsets of the players' labels, for some integer $\kappa_\nu > 0$, (thus, each $\sigma_j^{\nu} \subseteq \{1, \ldots, n\}$ and $\sigma_j^{\nu} \cap \sigma_k^{\nu} = \emptyset$ for $j \neq k$). We do not require that the (disjoint) union $\mathcal{N}_\nu \triangleq \bigcup_{k=1}^{\kappa_\nu} \sigma_k^{\nu}$ is equal to the full set $\{1, \ldots, n\}$ although some mild conditions will subsequently be imposed on these tuples of index subsets in order to establish the convergence of the algorithm. We refer to each subset σ_k^{ν} as a (non-cooperating) block/group of players whose strategies are updated as a group at iteration ν; players in the same block will update their strategies by solving a subgame that could be a single optimization problem if σ_k^{ν} happens to be a singleton. By attaching the superscript ν to σ_k^{ν}, we allow in principle that the blocks of players could vary from one iteration to the next, resulting in varying block updates. In the context of a game defined on a sparse network with a large number of nodes, where the players are identified with the nodes of the network, each group σ_k^{ν} may consist of nodes/players among whom some information sharing is feasible (e.g. they are neighbors) while these players still compete selfishly to optimize their individual objectives; thus it is reasonable for the players in the same group σ_k^{ν} to update their responses to their rivals current strategies by solving a subgame in which the strategies of the players from the other groups σ_ℓ^{ν} are fixed. During the solution of such a subgame, there is no exchange of information between the players in different groups while there is intergroup information sharing among the players in the same group.

(b) With the tuple $x^{\nu} \triangleq (x^{\nu;i})_{i=1}^{n}$ given and with $x^{\sigma_k^{\nu}} \triangleq (x^{\ell} : \ell \in \sigma_k^{\nu}) \in \mathcal{X}^{\sigma_k^{\nu}} \triangleq \prod_{\ell \in \sigma_k^{\nu}} \mathcal{X}^{\ell}$ denoting the players in the block σ_k^{ν} for $k = 1, \ldots, \kappa_\nu$, $\widehat{\boldsymbol{\theta}}^{\nu}(\bullet; x^{\nu})$ denotes the family of bivariate surrogate objective functions $\{\widehat{\theta}_i^{\sigma_k^{\nu}}(x^{\sigma_k^{\nu}}; x^{\nu}) \,:\, i \in \sigma_k^{\nu}\}_{k=1}^{\kappa_\nu}$ that will be used in place of the given objectives θ_i in defining "convex" subgames that are computationally easier to solve than the original game $\mathcal{G}$. For this purpose, we will subsequently introduce desirable properties on these surrogate functions that will facilitate the update of the iterates and their convergence. The subscript i and superscript σ_k^{ν} in $\widehat{\theta}_i^{\sigma_k^{\nu}}$ signify that this function could be dependent on both the individual player i and the player group σ_k^{ν}.

At the beginning of iteration ν, the triple $\{x^{\nu}, \boldsymbol{\sigma}^{\nu}, \widehat{\boldsymbol{\theta}}^{\nu}(\bullet; x^{\nu})\}$ is given. In parallel, the players in $\mathcal{N}_\nu$ update their strategies as follows: for $k = 1, \ldots, \kappa_\nu$, the players in the block σ_k^{ν} solve a subgame, denoted $\mathcal{G}^{\sigma_k^{\nu}}(x^{\nu})$, where each player is optimization problem is (for $i \in \sigma_k^{\nu}$):

$$\underset{x^i \in \mathcal{X}^i}{\text{minimize}}\ \widehat{\theta}_i^{\sigma_k^{\nu}} \left(\underbrace{x^i, x^{\sigma_k^{\nu};-i}}_{\text{subgame variables}} \;;\; \underbrace{x^{\nu}}_{\substack{\text{input to subgame} \\ \text{at iteration } \nu}} \right), \tag{4.3}$$

where $x^{\sigma_k^{\nu};-i} \triangleq (x^{\ell} : \ell \in \sigma_k^{\nu} \setminus \{i\})$. (Not reflected in the notation, it is understood that each of the subgames $\{\mathcal{G}^{\sigma_k^{\nu}}(x^{\nu})\}_{k=1}^{\kappa_\nu}$ utilizes the surrogate objectives $\widehat{\boldsymbol{\theta}}^{\nu}(\bullet; x^{\nu})$.) Let $\widehat{x}^{\nu;\sigma_k^{\nu}}$ be a solution strategy obtained by players in the set σ_k^{ν} after solving the above subgame. One assumption we will make in this setting is that each function $\widehat{\theta}_i^{\sigma_k^{\nu}}(\bullet, x^{\sigma_k^{\nu};-i}; x^{\nu})$ is convex (and thus continuous) on Ω^i so that each iterate $\widehat{x}^{\nu;\sigma_k^\nu}$ is actually a (standard) Nash equilibrium of the subgame $\mathcal{G}^{\sigma_k^\nu}(x^\nu)$, whose existence is an immediate consequence of the compactness and convexity of the set $\mathcal{X}^i$. We then let the new iterate be $x^{\nu+1;\sigma_k^\nu} \triangleq x^{\nu;\sigma_k^\nu} + \tau_{\sigma_k^\nu}(\widehat{x}^{\nu;\sigma_k^\nu} - x^{\nu;\sigma_k^\nu})$ for some scalar $\tau_{\sigma_k^\nu} \in (0, 1]$. (For simplicity, we take the step size $\tau_{\sigma_k^\nu}$ to depend only on the block σ_k^ν and not on its individual members.) If $\mathcal{N}_\nu$ is a proper subset of $\{1, \ldots, n\}$, then not all players update their strategies in iteration ν. At the end of the iteration, these updated components $x^{\nu+1;\mathcal{N}_\nu} \triangleq \{x^{\nu+1;\sigma_k^\nu}\}_{k=1}^{\kappa_\nu}$, along with the remaining components that are not updated, are concatenated to form the new iterate $x^{\nu+1}$. Note that the update of the variables $x^{\nu+1;\sigma_k^\nu}$ and $x^{\nu+1;\sigma_\ell^\nu}$ for $k \neq \ell$ can be carried out independently of each other. Thus the overall algorithm admits a distributed implementation in which groups of players update their responses simultaneously. The details of how each subgame $\mathcal{G}^{\sigma_k^\nu}(x^\nu)$ is solved are left unspecified.

An illustration

Consider a 10-player game with the grouping: $\boldsymbol{\sigma}^\nu = \{\{1, 2\}, \{3, 4, 5\}, \{6, 7, 8, 9\}\}$ so that $\kappa_\nu = 3$ and $\mathcal{N}_\nu = \{1, \cdots, 9\}$, leaving out the tenth-player. Players 1 and 2 update their strategies by solving a subgame $\mathcal{G}^{\{1,2\}}(x^\nu)$ defined by the surrogate objectives $\widehat{\theta}_1^{\{1,2\}}(\bullet; x^\nu)$ and $\widehat{\theta}_2^{\{1,2\}}(\bullet; x^\nu)$. In parallel, players 3, 4, and 5 update their strategies by solving a subgame $\mathcal{G}^{\{3,4,5\}}(x^\nu)$ using the surrogate objective functions $\{\widehat{\theta}_3^{\{3,4,5\}}(\bullet; x^\nu), \widehat{\theta}_4^{\{3,4,5\}}(\bullet; x^\nu), \widehat{\theta}_5^{\{3,4,5\}}(\bullet; x^\nu)\}$; similarly for players 6 through 9; the tenth player is not performing an update in the current iteration ν according to the given grouping. □

4.3.1 Special cases

By specializing the pair $\{\boldsymbol{\sigma}^\nu, \widehat{\boldsymbol{\theta}}^\nu(\bullet; x^\nu)\}$ of player blocks and surrogate objectives, we obtain a host of distributed algorithms for computing a QNE to the game $\mathcal{G}$. We first present two commonly used specializations of the player choices in each iteration.

- **Block Jacobi**: $\mathcal{N}_\nu = \{1, \ldots, n\}$. In each iteration ν of this scheme, all players update their strategies simultaneously through the blocks σ_k^ν which may be of different sizes; multi-player subgames need to be solved if some of the blocks have more than one players. The case where $\kappa_\nu = n$ and thus each $\sigma_k^\nu = \{k\}$ is a singleton for all $k = 1, \ldots, n$ is the classical point Jacobi iteration where each player i updates his/her strategy by solving the single optimization problem:

$$\underset{x^i \in \mathcal{X}^i}{\text{minimize}} \ \widehat{\theta}_i(x^i; x^\nu).$$

- **Block Gauss–Seidel**: $\kappa_\nu = 1$ for all ν. In each iteration ν of this scheme, only the players in the block σ_1^ν update their strategies that immediately become the inputs to the new iterate $x^{\nu+1}$ while all other players $j \notin \sigma_1^\nu$ keep their strategies at the current iterate $x^{\nu,j}$. The point version of this scheme has each σ_1^ν being a singleton.

We next present several classes of surrogate objective functions that can handle various kinds of nonconvex objective functions θ_i. We begin with the vanilla case where convexity is present in the latter functions.

- For $i \in \sigma_k^\nu$, let $\widehat{\theta}_i^{\sigma_k^\nu}(x^{\sigma_k^\nu}; z) \triangleq \theta_i(x^{\sigma_k^\nu}, z^{-\sigma_k^\nu}) + \frac{c_i}{2} \| x^i - z^i \|^2$ for some positive scalar c_i; the latter quadratic term is used to "regularize" the function $\widehat{\theta}_i^{\sigma_k^\nu}(\bullet, x^{\sigma_k^\nu;-i}; z)$ so that it is strongly convex for given $(x^{\sigma_k^\nu;-i}; z)$ when player i's objective function $\theta_i(\bullet, x^{-i})$ is convex for given x^{-i}; see [17]. Other regularization terms can be used; one example is the elliptic term: $\frac{1}{2}(x^i - z^i)^T A^{\sigma_k^\nu;i}(x^i - z^i)$ for some symmetric positive definite matrix $A^{\sigma_k^\nu;i}$.
- Consider the case of the objective functions θ_i satisfying the structural assumption in Proposition 4.2. That is, assume that each $\theta_i(\bullet, x^{-i}) = g_i(\bullet, x^{-i}) + f_i(\bullet, x^{-i})$, where $g_i(\bullet, x^{-i})$ is convex and $f_i(\bullet, x^{-i})$ is differentiable. (Subsequently, we need twice differentiability of f_i when we discuss the convergence of the method.) At a given tuple z, we first introduce the linear approximation:
$$f_i(x^{\sigma_k^\nu}, z^{-\sigma_k^\nu}) \approx f_i(z) + \sum_{j \in \sigma_k^\nu} \nabla_{z^j} f_i(z)^T (x^j - z^j),$$
and then let, for $i \in \sigma_k^\nu$,
$$\widehat{\theta}_i^{\sigma_k^\nu}(x^{\sigma_k^\nu}; z) \triangleq g_i(x^{\sigma_k^\nu}, z^{-\sigma_k^\nu}) + f_i(z) + \sum_{j \in \sigma_k^\nu} \nabla_{z^j} f_i(z)^T (x^j - z^j) + \frac{c_i}{2} \| x^i - z^i \|^2, \tag{4.4}$$
which is a convex approximation in $x^{\sigma_k^\nu}$ for fixed z. This choice leads to a block generalization of the point version of the partial linearization scheme in [26] for a multi-agent optimization problem with decoupled constraints (see the extension in [25] that contains convex, coupled constraints for an optimization problem). To date, a partial linearization scheme has not been applied to games even with convex player objectives.
- The above two choices lead to non-quadratic surrogate functions $\widehat{\theta}_i^{\sigma_k^\nu}(\bullet; z)$ when the given functions $\theta_i(\bullet, x^{-i})$ and $g_i(\bullet, x^{-i})$ are non-quadratic. These non-quadratic choices have the benefit of retaining as much as possible the favorable properties of the players' given objectives. Yet, for various reasons, it may be beneficial to employ quadratic approximations to facilitate and accelerate the solution of the subproblems. For instance, if one is interested in a distributed Newton-type method in smooth games, then one may consider using the following approximation function (assuming that $\nabla\theta_i$ exists):
$$\begin{aligned}\widehat{\theta}_i^{\sigma_k^\nu}(x^{\sigma_k^\nu}; z) \triangleq\ & \theta_i(z) + \sum_{j \in \sigma_k^\nu} \nabla_{x^j} \theta_i(z)^T (x^j - z^j) \\ & + \frac{1}{2} \sum_{j,j' \in \sigma_k^\nu} (x^{j'} - z^{j'})^T B^{\sigma_k^\nu;j,j'} (x^j - z^j),\end{aligned} \tag{4.5}$$
where $B^{\sigma_k^\nu;j,j'}$ is a symmetric positive definite matrix that serves as an approximation of the partial Hessian matrix of $\theta_i(\bullet, z^{-\sigma_k^\nu})$ with respect to the x^j and $x^{j'}$ variables.

The early paper [41] has discussed extensively the special case where $\kappa_\nu = 1$ and $\sigma_1^\nu = \{1, \ldots, n\}$ for all ν that updates all players' strategies by solving a single optimization problem with a strictly convex quadratic objective.

- Distributed projection algorithms. By letting $B^{\sigma_k^\nu;\ell,\ell'}$ be the identity matrix for $\ell = \ell'$ and the zero matrix otherwise, the approximation function (4.5) leads to the following "explicit" update formulae: for $i \in \sigma_k^\nu$ (again assuming that $\nabla\theta_i$ exists),

$$\widehat{x}^{\,\nu;i} \triangleq \Pi_{\mathcal{X}^i}\left(x^{\nu;i} - \nabla_{x^i}\theta_i(x^\nu)\right) \quad \text{and} \quad x^{\,\nu+1;i} \triangleq x^{\,\nu;i} + \tau_{\sigma_k^\nu}(\widehat{x}^{\,\nu;i} - x^{\nu;i}),$$

where $\Pi_S(\bullet)$ denotes the Euclidean projection operator onto a closed convex set S. For this particular choice of the matrices $B^{\sigma_k^\nu;\ell,\ell'}$, the updates of the iterates are distributed down to the players' levels in each iteration ν with each player implementing a projection step followed by a relaxation, whereas in the general case, the updates of the iterates are distributed at the blocks' levels with the players in each block solving a subgame.

Another specialization of the above framework is the successive convex approximation method for solving a single optimization problem.

- Successive convex approximation. Specialization of the above framework to the single objective optimization setup, where $\theta_1(x) = \theta_2(x) = \cdots = \theta_n(x) = \theta(x)$, yields the general *successive convex approximation* (also known as *block successive upper-bound minimization* [30] or *majorization minimization*) framework. This approach, which has been widely used for solving various practical problems, updates the variable blocks by successively minimizing a sequence of convex local approximations of the objective function $\theta(\bullet)$. For more details on this specialization, the interested readers are referred to [28, 32] for single block optimization, [30, 34] for multi-block optimization, and [37] for its stochastic counterpart.

4.4 Convergence analysis: contraction approach

In order to establish the convergence of the sequence of iterates obtained from the unified scheme using a contraction argument, we need to impose additional structures on the player selection groups in $\boldsymbol{\sigma}^\nu$ and the surrogate objective functions in $\widehat{\boldsymbol{\theta}}^{\,\nu}$. These are explained below.

On the player groups $\boldsymbol{\sigma}^\nu$

We assume that there is a fixed family $\{\widehat{\boldsymbol{\sigma}}^{\,t}\}_{t=1}^T$, for some $T > 0$, of index subsets of the players' labels with $\widehat{\boldsymbol{\sigma}}^{\,t} = \{\sigma_1^t, \ldots, \sigma_{\kappa_t}^t\}$ for some $\kappa_t > 0$ such that the union $\bigcup_{t=1}^T \bigcup_{k=1}^{\kappa_t} \sigma_k^t$ partitions the players' label set $\{1, \ldots, n\}$; i.e. this union is equal to the latter set and $\sigma_j^t \cap \sigma_k^{t'} = \emptyset$ for all $(j, t) \neq (k, t')$. Thus each player label i belongs to a σ_k^t for a unique pair (k, t) with $k \in \{1, \ldots, \kappa_t\}$ and $t \in \{1, \ldots, T\}$. Corresponding to these index sets is the family $\widehat{\boldsymbol{\theta}} = \{\widehat{\theta}_i\}_{i=1}^n$ of bivariate surrogate functions with each $\widehat{\theta}_i$ being a function of the pair $(x^{\sigma_k^t}; z)$; moreover for a given but arbitrary pair $(x^{\sigma_k^t;-i}; z)$, the function $\widehat{\theta}_i(\bullet, x^{\sigma_k^t;-i}; z)$ is convex in the argument x^i. For each pair (k, t) with $k \in \{1, \ldots, \kappa_t\}$ and $t \in \{1, \ldots, T\}$,

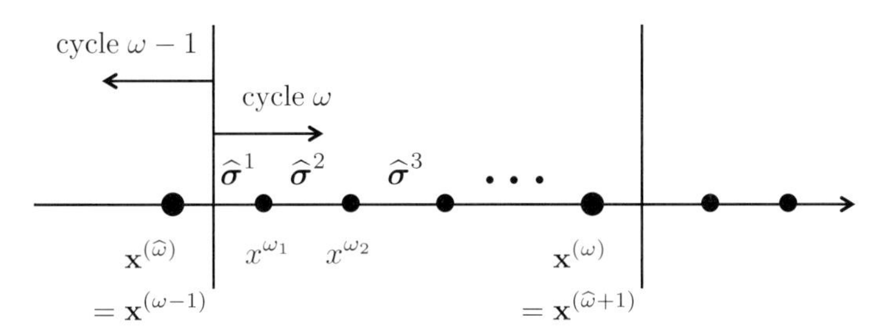

Figure 4.1 Illustration of the consecutive cycles $\widehat{\omega} = \omega - 1$ and ω.

let $\mathcal{G}^t_k(z)$ denote the subgame consisting of the players $i \in \sigma^t_k$ with objective functions $\widehat{\theta}_i(\bullet; z)$ for certain (known) iterate z to be specified. Given the partition $\{\widehat{\boldsymbol{\sigma}}^t\}_{t=1}^T$ of players' labels and the family $\widehat{\boldsymbol{\theta}}$ of surrogate functions, we may reconcile the change in notation here from the notation in the general discussion in Section 4.3 by letting $\kappa_\nu = \kappa_t$ and $\sigma^\nu_k = \sigma^t_k$ for $\nu \equiv t$ modulo T and for all $k = 1, \ldots, \kappa_t$; thus, for each $i = 1, \ldots, n$, $\widehat{\theta}_i^{\sigma^\nu_k} = \widehat{\theta}_i$ at such an iteration ν. Finally, we take each $\tau_{\sigma^\nu_k} = 1$.

To summarize, under the above setting, each player i and the members in σ^t_k will update their strategy tuple exactly once every T iterations through the solution of the subgame $\mathcal{G}^t_k(x^{\rm cr})$, where $x^{\rm cr}$ is the current strategy tuple of all the players when this update takes place. In addition to the notation x^ν, where ν denotes the count of the current iteration, we let $\omega = 1, 2, \ldots$ denote the cycle count and $\mathbf{x}^{(\omega)}$ denote the iterate at the beginning of cycle ω with components $x^{\omega_t;\sigma^t_k}$, where $\omega_t \triangleq (\omega - 1)T + t$ is the iteration count in which the players in σ^t_k update their strategies, for $t = 1, \ldots, T$ and $k = 1, \ldots, \kappa_t$. Thus,

$$\mathbf{x}^{(\omega);i} = x^{\omega_t;i} = x^{\omega_t+1;i} = \cdots = x^{\omega T+t-1;i}, \qquad \text{for } i \in \sigma^t_k, \tag{4.6}$$

which equate player i's strategy $\mathbf{x}^{(\omega);i}$ in cycle ω with the same player's strategy in iterations $(\omega - 1)T + t$ through $\omega T + t - 1$. See Figure 4.1 for an illustration. Overall, the iterations employ a mixture of the Jacobi and Gauss–Seidel ideas where some groups of players update their strategies simultaneously employing the latest updates of all players' strategies to the extent possible. Throughout the iterations, the same family $\widehat{\boldsymbol{\theta}}$ of bivariate surrogate objectives is employed in each cycle that is updated via the second argument of each of the functions.

A further illustration

Consider a 12-player game with $T = 3$ and with $\widehat{\boldsymbol{\sigma}}^1 = \{\{1, 2\}, \{3, 5, 6\}\}$, $\widehat{\boldsymbol{\sigma}}^2 = \{\{4, 7, 8\}\}$, and $\widehat{\boldsymbol{\sigma}}^3 = \{\{9\}, \{10, 11, 12\}\}$. Starting with $x^0 = (x^{0;i})_{i=1}^{12} = \mathbf{x}^{(0)}$, we obtain after one iteration $x^1 = (x^{1;\{1,2\}}, x^{1;\{3,5,6\}}, x^{0;4}, x^{0;\{7\text{thru}12\}})$. The subvectors $x^{1;\{1,2\}}$ and $x^{1;\{3,5,6\}}$ mean that the players 1 and 2 update their strategies by solving a two-player subgame and simultaneously the players 3, 5, and 6 update their strategies by solving a three-player subgame. The remaining players 4, 7 through 12 do not update their strategy in this first iteration. The next two iterations yield, respectively, $x^2 = (x^{1;\{1,2\}}, x^{1;\{3,5,6\}}, x^{2;\{4,7,8\}}, x^{0;\{9\text{thru}12\}})$ and $x^3 = (x^{1;\{1,2\}}, x^{1;\{3,5,6\}}, x^{2;\{4,7,8\}}, x^{3;\{9\}},$

$x^{3;\{10,11,12\}}$). The update of x^2 employs x^1 in defining the player objectives $\widehat{\theta}_4(\bullet; x^1)$, $\widehat{\theta}_7(\bullet; x^1)$, and $\widehat{\theta}_8(\bullet; x^1)$. Similarly, the update of x^3 employs x^2. After three iterations, we have completed a full cycle where all players have updated their strategies exactly once, obtaining the new iterate $\mathbf{x}^{(1)} = x^3$. The next cycle of updates is then initiated according to the same partition $\{\widehat{\boldsymbol{\sigma}}^1, \widehat{\boldsymbol{\sigma}}^2, \widehat{\boldsymbol{\sigma}}^3\}$ and employs the same family of bivariate surrogate functions $\{\widehat{\theta}_i\}_{i=1}^{12}$. At the start of an update, the second argument of the prevalent objective functions is substituted by the most recent iterate computed thus far. □

On the surrogate objectives $\widehat{\theta}$

As suggested by Proposition 4.2, some structural assumptions are needed on the objective functions θ_i for the nonsmooth, nonconvex game $\mathcal{G}$ to have a QNE. Such assumptions persist in order for the unified algorithm to produce a sequence of iterates that contract to a QNE of the game. Specifically, we focus on a similar structure on the surrogate function $\widehat{\theta}_i(x^{\sigma_k^t}; z) = g_i(x^i) + \widehat{f}_i(x^{\sigma_k^t}; z)$, where g_i is convex and $\widehat{f}_i$ is twice continuously differentiable with bounded mixed second derivatives. It turns out that the dependence of the convex function g_i on x^i only is essential for the convergence analysis, in both the contraction approach below and the potential approach later. Thus the state of the art of such an analysis for a game with nonsmooth, nonconvex player objectives is still restricted and there is room for improvement.

A key matrix

We assume that for every $i \in \{1, \ldots, n\}$ belonging to the player group σ_k^t with (k, t) being uniquely dependent on i and with $k \in \{1, \ldots, \kappa_t\}$ and $t \in \{1, \ldots, T\}$, there exists a constant $\gamma_{ii} > 0$ such that

$$(x^i - y^i)^T \nabla^2_{u^i u^i} \widehat{f}_i(u^{\sigma_k^t}; z)(x^i - y^i) \geq \gamma_{ii} \, \| \, x^i - y^i \, \|^2,$$

for all x^i and y^i in $\mathcal{X}^i$; all $u^{\sigma_k^t}$ in $\mathcal{X}^{\sigma_k^t}$, and all z in $\boldsymbol{\mathcal{X}}$, where $\nabla^2_{u^i u^i} \widehat{f}_i(u^{\sigma_k^t}; z)$ is the Jacobian matrix of $\nabla_{u^i} f_i(u^{\sigma_k^t}; z)$ with respect to the u^i variables:

$$\left[\nabla^2_{u^i u^i} \widehat{f}_i(u^{\sigma_k^t}; z) \right]_{rs} = \frac{\partial}{\partial u^i_s} \left(\frac{\partial \widehat{f}_i(u^{\sigma_k^t}; z)}{\partial u^i_r} \right) \qquad \forall \, r, s = 1, \ldots, n_i.$$

Defining the matrix $\nabla^2_{u^j u^i} \widehat{f}_i(u^{\sigma_k^t}; z)$ similarly for all $i \neq j$ belonging to the same player group σ_k^t, we let $\gamma_{ij} \triangleq \sup_{u^{\sigma_k^t} \in \mathcal{X}^{\sigma_k^t}; z \in \boldsymbol{\mathcal{X}}} \| \nabla^2_{u^j u^i} \widehat{f}_i(u^{\sigma_k^t}; z) \|$, which we assume is a finite scalar. We let $\gamma_{ij} = 0$ if i and j belong to two different player groups. Similarly, for all i and ℓ in $\{1, \ldots, n\}$ with $i \in \sigma_k^t$, let $\widetilde{\gamma}_{i\ell} \triangleq \sup_{u^{\sigma_k^t} \in \mathcal{X}^{\sigma_k^t}; z \in \boldsymbol{\mathcal{X}}} \| \nabla^2_{z^\ell u^i} \widehat{f}_i(u^{\sigma_k^t}; z) \|$, which we also assume is finite. We collect the above scalars in two matrices $\Gamma \triangleq [\, \gamma_{ij} \,]_{i,j=1}^n$ and $\widetilde{\boldsymbol{\Gamma}} \triangleq [\, \widetilde{\gamma}_{i\ell} \,]_{i,\ell=1}^n$ that will play a central role in the convergence proof of the unified iterative method. Let $\overline{\boldsymbol{\Gamma}}$ denote the *comparison matrix* of $\boldsymbol{\Gamma}$ so that

$$\overline{\boldsymbol{\Gamma}}_{ij} \triangleq \begin{cases} \gamma_{ii} & \text{if } i = j \\ -|\, \gamma_{ij} \,| & \text{if } i \neq j. \end{cases}$$

Let the matrix $\widetilde{\boldsymbol{\Gamma}} = \widetilde{\mathbf{L}} + \widetilde{\mathbf{D}} + \widetilde{\mathbf{U}}$ be written as the sum of its strictly lower triangular, diagonal, strictly upper triangular blocks, respectively.

The key assumption we make is that the matrix $\mathbf{M} \triangleq \overline{\boldsymbol{\Gamma}} - \widetilde{\boldsymbol{\Gamma}}$, which itself is a Z-matrix of order n (the number of players in the game $\mathcal{G}$), is also a P-matrix, i.e. $\mathbf{M}$ is a Minkowski matrix. For properties of these matrices, see [39, Chapter 3]. The fundamental role of these matrix-theoretic properties in the convergence analysis of distributed methods is well known; see e.g. [17, 39]. There are many characterizations of a Minkowski matrix; for our purpose, the following facts will be used [39]: (a) the block lower triangular matrix $\overline{\boldsymbol{\Gamma}} - \widetilde{\mathbf{L}}$ is invertible and has a nonnegative inverse; (b) the spectral radius of the (nonnegative) matrix $\widehat{\mathbf{M}} \triangleq [\overline{\boldsymbol{\Gamma}} - \widetilde{\mathbf{L}}]^{-1} (\widetilde{\mathbf{D}} + \widetilde{\mathbf{U}})$ is less than unity; and (c) statements (a) and (b) are equivalent and each is equivalent to the existence of positive scalars d_i for $i = 1, \ldots, n$ such that $\gamma_{ii}\, d_i > \sum_{i \neq j \in \sigma_k^t} \gamma_{ij}\, d_j + \sum_{\ell=1}^{n} \widetilde{\gamma}_{i\ell}\, d_\ell, \quad \forall i = 1, \ldots, n$, equivalently,

$$(\gamma_{ii} - \widetilde{\gamma}_{ii})\, d_i > \sum_{i \neq j \in \sigma_k^t} (\gamma_{ij} + \widetilde{\gamma}_{ij})\, d_j + \sum_{\ell \notin \sigma_k^t} \widetilde{\gamma}_{i\ell}\, d_\ell, \qquad \forall i = 1, \ldots, n. \tag{4.7}$$

See also Lemma 4.11 for another property.

The entries of $\mathbf{M}$ are defined by the second partial derivatives of the surrogate objectives $\widehat{f}_i(x^{\sigma_k^t}; z)$. The (convex) functions g_i are not involved in the definition of the matrices $\mathbf{M}$ and $\widehat{\mathbf{M}}$. As noted before, the player separability of these functions $g_i(x^i)$ (whose differentiability is not assumed) is needed in the convergence of the unified iterative algorithm. Presently, extension of the proof to more general classes of non-differentiable surrogate objective functions $\widehat{\theta}_i(\bullet; z)$ appears difficult.

We further illustrate the condition (4.7) with the player objective function $\theta_i(x) = g_i(x^i) + f_i(x)$, where g_i is convex, f_i is twice continuous differentiable and $f_i(\bullet, x^{-i})$ is strongly convex in x^i for fixed x^{-i}. Letting

$$\widehat{f}_i(x^{\sigma_k^t}; z) \triangleq f_i(x^{\sigma_k^t}, z^{-\sigma_k^t}), \qquad \text{for } i \in \sigma_k^t,$$

the condition (4.7) is satisfied if positive constants d_i exist such that

$$d_i \inf_{x \in \mathcal{X}} \left(\text{smallest eigenvalue of } \nabla^2_{x^i x^i} f(x)\right) > \sum_{j \neq i} d_j \sup_{x \in \mathcal{X}} \| \nabla^2_{x^j x^i} \nabla f_i(x) \|, \qquad \forall i,$$

which is a generalized diagonal dominance property of the Jacobian matrix of the vector function $F(x) \triangleq (\nabla_{x^i} f_i(x))_{i=1}^n$. Notice that this property is independent of the player groups $\{\widehat{\boldsymbol{\sigma}}^t\}_{t=1}^T$. Beginning with the analysis of the iterative waterfilling algorithm for multi-user power control in digital subscriber lines [4] and continuing to the survey [17], the generalized diagonal dominance property of the Jacobian matrix of F has played a central role in the convergence of the (block) Jacobi and Gauss–Seidel methods for solving Nash games. Roughly speaking, this property essentially stipulates that each player's objective is impacted more by his/her own strategies than by the totality of the rivals' strategies. This sort of dominance property is intuitive and has always been a key condition in the contraction approach for convergence of distributed algorithms. Over the basic methods such as the well-known Jacobi and Gauss–Seidel schemes, the extension offered by the unified algorithm lies in the flexibility of applying the parallel and sequential modes, allowing for the parallel updates carried out by proper subsets of the players in an iteration using the current information until all of them have updated their strategies once within a cycle consisting of a given number (T) of iterations. Most interestingly, the main matrix-theoretic condition for the convergence of the unified algorithm remains the same as in the well-known basic schemes.

The convergence result

For ease of reference, we summarize the assumptions needed for the convergence of the unified algorithm based on a contraction argument.

($\mathbf{A}_\sigma$) The player groups at iteration $\nu = t \pmod T$ are given by the family $\widehat{\boldsymbol{\sigma}}^{\,t} = \{\sigma_k^t\}_{k=1}^{K_t}$.

($\mathbf{A}_f^1$) The surrogate objectives of the players are given by the family $\{\,\widehat{\theta}_i\,\}_{i=1}^n$, where, for $i \in \sigma_k^t$, $\widehat{\theta}_i(x^{\sigma_k^t}; z) = g_i(x^i) + \widehat{f}_i(x^{\sigma_k^t}; z)$. We assume that g_i is convex and $\widehat{f}_i$ is C^2.

($\mathbf{A}_f^2$) each function $\widehat{f}_i(\bullet, u^{\sigma_k^t;-i}; z)$ is strongly convex with a strong convexity modulus γ_{ii} independent of $(u^{\sigma_k^t;-i}, z) \in \mathcal{X}^{\sigma_k^t;-i} \times \boldsymbol{\mathcal{X}}$, where $i \in \sigma_k^t$.

($\mathbf{A}_f^3$) each function $\widehat{f}_i$ has bounded mixed second partial derivatives; specifically, $\|\nabla^2_{u^j u^i} \widehat{f}_i(u^{\sigma_k^t}; z)\|$ and $\|\nabla^2_{z^\ell u^i} \widehat{f}_i(u^{\sigma_k^t}; z)\|$ are bounded above on $\mathcal{X}^{\sigma_k^t} \times \boldsymbol{\mathcal{X}}$ by γ_{ij} and $\widetilde{\gamma}_{i\ell}$, respectively, for all $i \neq j$ in σ_k^t and $\ell = 1, \ldots, n$.

($\mathbf{A}_f^4$) positive scalars d_i exist for $i = 1, \ldots, n$ such that (4.7) holds.

Assumptions (A_f^1)–(A_f^4) all refer to the surrogate objective functions $\widehat{\theta}_i$. Needless to say, we need to connect these objectives with the original objectives θ_i; this connection is specified by the condition of directional derivative consistency (4.8) in the theorem below whose proof is given in the Appendix.

Theorem 4.3 *Under the player selection rule (A_σ) and the assumptions (A_f^1)–(A_f^4), the sequence of iterates $\{x^\nu\}$ generated by the unified algorithm is well defined and contracts to a limit x^∞. Moreover, if*

$$\theta_i(\bullet, x^{\infty,-i})'(x^{\infty,i}; x^i - x^{\infty,i}) \geq \widehat{\theta}_i(\bullet, x^{\infty,\sigma_k^t;-i}; x^\infty)'(x^{\infty,i}; x^i - x^{\infty,i}), \tag{4.8}$$

for every $x^i \in \mathcal{X}^i$, then x^∞ is a QNE of the game $\mathcal{G}$. □

It should be pointed out that the condition (4.8) can be made to hold by properly choosing the surrogate functions. For example, it is not hard to check that the surrogate functions in (4.4) and (4.5) satisfy condition (4.8). In a nutshell, this is the main idea underlying the treatment of nonconvexity: replace it by convex surrogates, impose a key matrix-theoretic condition on these surrogates, and link them to the original objectives via a mild condition. This idea persists in the potential approach to be presented in Section 4.5.

Another noteworthy remark is that while Theorem 4.3 ensures the convergence of the entire sequence $\{x^\nu\}$ to a unique limit, which must be a QNE if it satisfies a directional derivative consistency condition, this result does not guarantee that the game $\mathcal{G}$ has a unique QNE because (4.8) is a pointwise condition at the limit of the generated sequence $\{x^\nu\}$.

4.4.1 Probabilistic player choices

Next, we employ the contraction approach to analyze the convergence of the above algorithm for *set-randomized* player selection rules. Unlike the previous case of a deterministic rule, where each player updates his/her strategy tuple exactly once every T iterations through the solution of the subgame $\mathcal{G}_k^t(x^\nu)$, in the case of a randomized rule, there is no restriction on the frequency each player updates his/her strategy profile (in particular, (s)he may update repeatedly for several consecutive iterations), nor is there a fixed number (T) of iterations within which each player will have updated his/her strategy exactly once (in particular, (s)he may update only occasionally). The only stipulation is that, at every iteration, there is a positive probability for a player to carry out the update. In what follows, we describe the setting of the randomized player selection rule and present the convergence result. The condition on the surrogate functions is similar to (4.8).

On the player groups $\boldsymbol{\sigma}^\nu$

Playing a similar role as the family $\{\widehat{\boldsymbol{\sigma}}^t\}_{t=1}^T$ of index subsets of the players' labels is the family $\{\sigma_k\}_{k=1}^K$ that partitions the set $\{1, \ldots, n\}$ of such labels; i.e. $\sigma_k \bigcap \sigma_{k'} = \emptyset$ for all $k \neq k'$, and $\bigcup_{k=1}^K \sigma_k = \{1, 2, \ldots, n\}$. Thus each player i belongs to one and only one of these subsets. At each iteration ν of the set-randomized algorithm, the subfamily $\boldsymbol{\sigma}^\nu \subseteq \{\sigma_1, \ldots, \sigma_K\}$ is chosen randomly and independently from the previous iterations, so that

$$\Pr(\sigma_k \in \boldsymbol{\sigma}^\nu \mid \mathcal{F}^\nu) = p_k > 0,$$

where $\mathcal{F}^\nu$ is the natural history/filtration of the trajectory of the iterates up to iteration ν. In other words, the above assumption states that there is positive probability p_k, which is the same at all iterations ν, for the subset σ_k of players to be chosen to update their strategies. In actual implementation, i.e. in a particular realization of the randomization, it is possible for the same group σ_k to be picked in two consecutive iterations; if this happens, the players in σ_k will update their strategies in these iterations using updated surrogate functions (see below) that are derived from the other players' current strategies. These kinds of consecutive updates are not possible in the previous deterministic player selection choice.

On the surrogate objectives $\widehat{\boldsymbol{\theta}}^\nu$

The surrogate objectives have the same structure and satisfy the same assumptions as in the deterministic rule. The only change occurs in the notation of player groups: from σ_k^t in the tth (inner) iteration within a cycle to σ_k in (the outer) iteration ν. Specifically, for $i \in \sigma_k$, where σ_k is the (unique) player group containing i, $\widehat{\theta}_i(x^{\sigma_k}; z) = g_i(x^i) + \widehat{f}_i(x^{\sigma_k}; z)$ with g_i being convex and $\widehat{f}_i$ being C^2; moreover, the second derivatives of the functions $\widehat{f}_i$ satisfy the same matrix-theoretic condition as before. In particular, the constants γ_{ii}

for $i \in \sigma_k$, γ_{ij} for $i \neq j$ belonging to the same σ_k, and $\gamma^t_{i\ell}$ for $i \in \sigma_k$ and $\ell = 1, \ldots, n$ are defined similarly and the condition (4.7) is assumed to be satisfied by some positive scalars d_i for $i = 1, \ldots, n$.

Given the triple $\{x^\nu, \boldsymbol{\sigma}^\nu, \widehat{\boldsymbol{\theta}}^\nu(\bullet; x^\nu)\}$ the players in different groups, if chosen to perform the updates, solve their corresponding subgames in parallel similar to (4.3). This set-randomized player selection rule could be particularly beneficial in large-size games since it is designed to require minimal coordination among the players.

The convergence result

Similar to Theorem 4.3, the following theorem can be proved with the set-randomized choice of player groups. See the Appendix for the detailed proof.

Theorem 4.4 *Assume that the directional derivative consistency condition* (4.8) *holds at every feasible tuple* $x^\infty \in \boldsymbol{\mathcal{X}}$. *Under the set-randomized player selection rule and the assumptions* (A^1_f–A^4_f), *the sequence of iterates* $\{x^\nu\}$ *generated by the unified algorithm is well defined and converges to a QNE of the game* $\mathcal{G}$ *almost surely.* □

4.5 Convergence analysis: potential approach

In this section, we establish the convergence of the proposed unified method in the presence of a potential function of the game. Two types of potentials are considered in our analysis.

Preliminary discussion

We say that a family of functions $\{\theta_i(x)\}_{i=1}^n$ on the set $\boldsymbol{\mathcal{X}}$ admits

- an *exact potential function* $P : \boldsymbol{\Omega} \to \mathbb{R}$ if P is continuous such that for all i, all $x^{-i} \in \Omega^{-i}$, and all y^i and $z^i \in \Omega^i$,

$$P(y^i, x^{-i}) - P(z^i, x^{-i}) = \theta_i(y^i, x^{-i}) - \theta_i(z^i, x^{-i}); \tag{4.9}$$

- a *generalized potential function* $P : \boldsymbol{\Omega} \to \mathbb{R}$ if P is continuous such that for all i, all $x^{-i} \in \Omega^{-i}$, and all y^i and $z^i \in \Omega^i$,

$$\begin{aligned} &\theta_i(y^i, x^{-i}) > \theta_i(z^i, x^{-i}) \\ &\quad\Rightarrow P(y^i, x^{-i}) - P(z^i, x^{-i}) \geq \xi_i(\theta_i(y^i, x^{-i}) - \theta_i(z^i, x^{-i})), \end{aligned}$$

  for some *forcing functions* $\xi_i : \mathbb{R}_+ \to \mathbb{R}_+$, i.e. $\lim_{\nu\to\infty} \xi_i(t_\nu) = 0 \Rightarrow \lim_{\nu\to\infty} t_\nu = 0$.

If such a function P exists, we call the game $\mathcal{G}$ defined by the family $\{\theta_i(x)\}$ an *exact (generalized) potential game*, respectively. Widely studied in economics [47], potential games constitute an important class of non-cooperative games for which the theory and methods of optimization can be profitably employed; see the recent reference [24] which motivates the above definition of a generalized potential game. It is easy to see that if the family of original objectives $\{\theta_i\}$ has an exact potential function, then so does the family of modified functions $\{\widetilde{\theta}_i\}$ where each $\widetilde{\theta}_i(x) = \theta_i(x) + \varphi_i(x^i)$ for some continuous player-dependent-only functions $\varphi_i(x^i)$.

In what follows, we summarize some properties of the family of objective functions $\{\theta_i\}$ when it admits an exact potential function, under the stipulation that each $\theta_i(\bullet, x^{-i})$ is convex; see Proposition 4.5 whose proof is given in the Appendix. Recalling the well-known concept of the subdifferential of a convex function [48], we define the multifunction:

$$\boldsymbol{\Theta}(x) \triangleq \prod_{i=1}^{n} \partial_{x^i}\theta_i(x), \qquad x \in \boldsymbol{\mathcal{X}},$$

where $\partial_{x^i}\theta_i(x)$ is the subdifferential of the function $\theta_i(\bullet, x^{-i})$ at x^i. We further recall the following definition [49, Definition 12.24]. A general multifunction $\Phi : \mathbb{R}^N \to \mathbb{R}^N$ is *cyclically monotone* if for any choice of points $\{u^0, u^1, \ldots, u^m\} \subset \mathbb{R}^N$ (for arbitrary $m \geq 1$) and elements $v^i \in \Phi(u^i)$, one has

$$\sum_{i=0}^{m} (u^{i+1} - u^i)^T v^i \leq 0, \text{ with } u^{m+1} = u^0.$$

We also say that Φ is *maximally cyclically monotone* if it is cyclically monotone and its graph cannot be enlarged without destroying this property. Every cyclically monotone map must be monotone (the case $m = 1$) but not conversely when $N \geq 1$. The subdifferential of a convex function defined on a convex open set is maximally cyclically monotone.

Proposition 4.5 *Suppose that each function $\theta_i(\bullet, x^{-i})$ is convex on the open convex set Ω^i. Associated with the multifunction Θ, consider the following four statements:*

(a) $\Theta(x)$ is maximally cyclically monotone on $\Omega \triangleq \prod_{i=1}^{n}\Omega^i$;
(b) there exists a convex (and thus continuous) function $\psi(x)$ such that $\partial\psi(x) = \boldsymbol{\Theta}(x)$ for all $x \in \Omega$;
(c) there exist a convex (and thus continuous) function $\psi(x)$ on Ω and continuous functions $A_i(x^{-i})$ on Ω^{-i} such that $\theta_i(x) = \psi(x) + A_i(x^{-i})$ for all $x \in \Omega$ and all $i = 1, \ldots, n$; and
(d) the game $\mathcal{G}$ is an exact potential game with a convex potential $P(x)$.

It holds that (a) $\Leftrightarrow$ (b) $\Rightarrow$ (c) $\Leftrightarrow$ (d). □

Proposition 4.5 is the analog of a well-known symmetry, or integrability result [45, Theorem 1.3.1] in the case where each $\theta_i(\bullet, x^{-i})$ is continuously differentiable (but not necessarily convex). In this case, we have $\Theta(x) = (\nabla_{x^i}\theta_i(x))_{i=1}^n$. If the Jacobian matrix $J\Theta(x)$ is symmetric for all $x \in \Omega$, then there exists a differentiable function $\psi(x)$ on Ω such that $\nabla\psi(x) = \Theta(x)$ on Ω; in fact, $\psi(x)$ is given as a line integral of $\Theta(x)$; see the cited reference for details. When such a function ψ exists, we have

$$\begin{aligned}
\psi(y^i, x^{-i}) - \psi(z^i, x^{-i}) &= \int_0^1 \nabla_{x^i}\psi(y^i + s(z^i - x^i), x^{-i})^T(z^i - x^i)\, ds \\
&= \int_0^1 \nabla_{x^i}\theta_i(y^i + s(z^i - x^i), x^{-i})^T(z^i - x^i)\, ds \\
&= \theta_i(y^i, x^{-i}) - \theta_i(z^i, x^{-i}),
\end{aligned}$$

which shows that ψ is an exact potential function for the family $\{\theta_i\}$ in the sense defined above.

With a proof given in the Appendix, the next result provides further property of an exact potential function; in particular, part (b) of the result shows that if each function $\theta_i(x) = f_i(x) + g_i(x^i)$ where f_i is differentiable and g_i is a convex private function, and if an exact potential function exists, then the latter function can also be written as the sum of a differentiable function and a convex function in a single (block of) variable. This result is particularly relevant to the convergence analysis using the potential approach because the player objective functions of the game $\mathcal{G}$ and their surrogates are assumed to be of this type.

Proposition 4.6 *Suppose that the family of functions $\{\theta_i\}_{i=1}^n$ has an exact potential function P. The following statements hold.*

(a) If each function θ_i is directionally differentiable, then so is P; moreover, for any pair (x, d) with $d = (0, \ldots, 0, d^i, 0, \ldots, 0)$ for some $i = 1, \ldots, n$, it holds that $P'(x; d) = \theta_i'(x; d)$.

(b) Let each $\theta_i(x) = f_i(x) + g_i(x^i)$. Then $P(x) = f(x) + \sum_{i=1}^n g_i(x^i)$, where

$$f(x) \triangleq P(0) + \sum_{i=1}^n [f_i(x^1, \ldots, x^i, 0, \ldots, 0) - f_i(x^1, \ldots, x^{i-1}, 0, \ldots, 0) - g_i(0)].$$

(c) If each f_i is differentiable then so is f. Moreover, $\nabla_{x^i} f(x) = \nabla_{x^i} f_i(x)$ for all i. □

A generalized potential game covers a wider class of problems than exact potential games; a special case of a generalized potential game is a *weighted potential game* defined in the paper [47] that corresponds to forcing functions being $\xi_i(t) = tw_i$ for some positive scalar w_i. This paper also defines an "ordinal potential game" and gives an example of such a game that is not an exact potential game. The following example gives a generalized potential game that is not an exact potential game. Unlike the latter, there is presently no easily checkable condition to determine if a game is of the generalized potential type. More research on this class of extended games is needed.

Example 4.7 *Consider a two-player game where the two players' optimization problems are respectively*

$$\begin{array}{ll}
\underset{x_1 \in \mathbb{R}}{\text{minimize}}\ \theta_1(x_1, x_2) \triangleq x_1 & \Big|\ \underset{x_2 \in \mathbb{R}}{\text{minimize}}\ \theta_2(x_1, x_2) \triangleq x_1 x_2 + x_2 \\
\text{subject to } -2 \le x_1 \le 2 & \Big|\ \text{subject to } 1 \le x_2 \le 3.
\end{array}$$

It can be shown that this is not an exact potential game, while it is a generalized potential game with potential $P(x_1, x_2) = x_1 x_2 + x_2$. □

The setting

As stated before, our goal in this section is to analyze the convergence of the proposed unified framework for potential games. For this analysis, we assume that each $\theta_i(x) = f_i(x) + g_i(x^i)$ for some differentiable function f_i defined on Ω and convex function g_i defined on Ω^i. To this end, we recall that at each iteration ν, the players in the block σ_k^ν for $k = 1, \ldots, \kappa_\nu$ update their strategies by solving the subgame $\mathcal{G}^{\sigma_k^\nu}(x^\nu)$ consisting of the optimization problems: $\{\min_{x^i \in \mathcal{X}^i} \widehat{\theta}_i^{\sigma_k^\nu}(x^{\sigma_k^\nu}; x^\nu) \mid i \in \sigma_k^\nu\}$, where $\widehat{\theta}_i^{\sigma_k^\nu}(x^{\sigma_k^\nu}; z) = g_i(x^i) + \widehat{f}_i^{\sigma_k^\nu}(x^{\sigma_k^\nu}; z)$, with $\widehat{f}_i^{\sigma_k^\nu}$ being differentiable and g_i being the same as in the original objective θ_i. We assume that each family of functions $\{\widehat{f}_i^{\sigma_k^\nu}(\bullet; x^\nu)\}_{i \in \sigma_k^\nu}$ admits an exact potential function $\widehat{f}_{\sigma_k^\nu}(\bullet; x^\nu)$ satisfying the following assumptions.

- **Strong convexity**: there exists a constant $\eta > 0$ such that

$$\widehat{f}_{\sigma_k^\nu}(\widetilde{x}^{\sigma_k^\nu}; y) \geq \widehat{f}_{\sigma_k^\nu}(x^{\sigma_k^\nu}; y) + \nabla_{x^{\sigma_k^\nu}} \widehat{f}_{\sigma_k^\nu}(x^{\sigma_k^\nu}; y)^T (\widetilde{x}^{\sigma_k^\nu} - x^{\sigma_k^\nu}) + \frac{\eta}{2} \| \widetilde{x}^{\sigma_k^\nu} - x^{\sigma_k^\nu} \|^2$$

  for all $\widetilde{x}^{\sigma_k^\nu}$ and $x^{\sigma_k^\nu}$ in $\mathcal{X}^{\sigma_k^\nu}$, and y in $\boldsymbol{\mathcal{X}}$.
- **Gradient consistency**: $\nabla_{x^i} f_i(x)^T (u^i - x^i) = \left(\nabla_{x^i} \widehat{f}_i^{\sigma_k^\nu}(\bullet, x^{\sigma_k^\nu;-i}; x)|_{x^i} \right)^T (u^i - x^i)$ for all u^i and x^i in $\mathcal{X}^i$, $x^{-i} \in \mathcal{X}^{-i}$ and $i \in \sigma_k^\nu$.

Using a standard derivation, we can deduce from the strong convexity condition the following inequality for any two points $\widetilde{x}^{\sigma_k^\nu}$ and $x^{\sigma_k^\nu}$ in $\mathcal{X}^{\sigma_k^\nu}$ and any $y \in \boldsymbol{\mathcal{X}}$:

$$\left[\nabla_{x^{\sigma_k^\nu}} \widehat{f}_{\sigma_k^\nu}(\widetilde{x}^{\sigma_k^\nu}; y) - \nabla_{x^{\sigma_k^\nu}} \widehat{f}_{\sigma_k^\nu}(x^{\sigma_k^\nu}; y) \right]^T (\widetilde{x}^{\sigma_k^\nu} - x^{\sigma_k^\nu}) \geq \eta \, \| \widetilde{x}^{\sigma_k^\nu} - x^{\sigma_k^\nu} \|^2, \tag{4.10}$$

which simply expresses the strong monotonicity of the gradient map $\nabla_{x^{\sigma_k^\nu}} \widehat{f}_{\sigma_k^\nu}(\bullet; y)$ with modulus η. If P is an exact potential function of the family $\{\theta_i\}$, then by part (a) of Proposition 4.6 and the gradient consistency condition, we have

$$\begin{aligned} P(\bullet; x^{-i})'(x^i; u^i - x^i) &= \theta_i(\bullet, x^{-i})'(x^i; u^i - x^i) \\ &= \nabla_{x^i} f_i(x)^T (u^i - x^i) + g_i'(x^i; u^i - x^i) \\ &= \left(\nabla_{x^i} \widehat{f}_i^{\sigma_k^\nu}(\bullet, x^{\sigma_k^\nu;-i}; x)|_{x^i} \right)^T (u^i - x^i) + g_i'(x^i; u^i - x^i) \\ &= \widehat{\theta}_i^{\sigma_k^\nu}(\bullet, x^{\sigma_k^\nu,-i}; x)'(x^i; u^i - x^i). \end{aligned} \tag{4.11}$$

This string of equations results in the equality version of the directional derivative consistency condition (4.8) in the contraction analysis.

Example 4.8 *Consider a four-player game* $\mathcal{G}$ *with an exact smooth potential* $P(x^1, x^2, x^3, x^4)$. *Assume* $\boldsymbol{\sigma}^\nu = \{\sigma_1^\nu = \{1, 2\}, \sigma_2^\nu = \{3\}\}$ *and we choose* α *large enough so that*

$$\widehat{\theta}_{\sigma_1^\nu}(x^1, x^2; x^\nu) \triangleq P(x^1, x^2, x^{\nu;3}, x^{\nu;4}) + \frac{\alpha}{2} \|x^1 - x^{\nu;1}\|^2 + \frac{\alpha}{2} \|x^2 - x^{\nu;2}\|^2$$

is strongly convex in (x^1, x^2)*. Then the gradient consistency condition holds for the functions*

$$\widehat{f}_1^{\sigma_1^\nu}(x^1, x^2; x^\nu) = \widehat{\theta}_1^{\sigma_1^\nu}(x^1, x^2; x^\nu) \triangleq P(x^1, x^2, x^{\nu,3}, x^{\nu;4}) + \frac{\alpha}{2}\|x^1 - x^{\nu;1}\|^2$$

and

$$\widehat{f}_2^{\sigma_1^\nu}(x^1, x^2; x^\nu) = \widehat{\theta}_2^{\sigma_1^\nu}(x^1, x^2; x^\nu) \triangleq P(x^1, x^2, x^{\nu,3}, x^{\nu;4}) + \frac{\alpha}{2}\|x^2 - x^{\nu;2}\|^2.$$

Moreover, solving the subgame $\{\min_{x^i \in \mathcal{X}^i} \widehat{\theta}_i^{\sigma_1^\nu}(x^{\sigma_1^\nu}; x^\nu) \mid i \in \sigma_1^\nu\}$ *is equivalent to the minimization of the strongly convex function* $\widehat{\theta}_{\sigma_1^\nu}(x^1, x^2; x^\nu)$ *on* $\mathcal{X}^1 \times \mathcal{X}^2$*.* □

Having made the above assumptions, we study the convergence of the unified algorithm for the following two types of player selection rules in the algorithm.

- **Essentially covering** We say the player selection rule in the proposed unified algorithm is *essentially covering* if there exists an integer $T \geq 1$ such that

$$\mathcal{N}_\nu \cup \mathcal{N}_{\nu+1} \cup \ldots \cup \mathcal{N}_{\nu+T-1} = \{1, 2, \ldots, n\}, \quad \forall \nu = 1, 2, \ldots;$$

  note that this definition allows a non-empty intersection of the elements in any two members of the family $\{\mathcal{N}_\nu, \ldots, \mathcal{N}_{\nu+T-1}\}$; the definition does not imply, however, that $\mathcal{N}_\nu = \mathcal{N}_{\nu+T}$. These two features make the player selection rule much less restrictive than that in the contraction approach where the family $\{\mathcal{N}_\nu, \ldots, \mathcal{N}_{\nu+T-1}\}$ is completely determined by an iteration-independent family $\{\widehat{\sigma}^t\}_{t=1}^T$ of non-overlapping player index sets. This broadening is possible because of the existence of an exact potential function of the game.
- **Randomized** The choice of the updates in the algorithm is called *randomized* if the players are chosen randomly, identically, and independently from the previous iterations so that

$$\Pr(j \in \mathcal{N}_\nu) = p_j \geq p_{\min} > 0, \ \forall j = 1, 2, \ldots, n, \ \forall \nu = 1, 2, \ldots.$$

The following theorem establishes the asymptotic convergence of the unified algorithm for exact potential games. We consider only the case of a constant step-size rule, i.e. we assume $\tau_{\sigma_k^\nu} = \tau$ for all σ_k^ν; thus, we have $x^{\nu+1;\sigma_k^\nu} \triangleq x^{\nu;\sigma_k^\nu} + \tau(\widehat{x}^{\nu;\sigma_k^\nu} - x^{\nu;\sigma_k^\nu})$; see [34] for other step-size selection rules and non-asymptotic convergence analysis of the algorithm for solving a nonsmooth, nonconvex optimization problem. The proof of the following theorem can be found in the Appendix.

Theorem 4.9 *Under the following assumptions:*

(a) each function $\theta_i(x) = g_i(x^i) + f_i(x)$ *with* g_i *being convex and* f_i *being* C^1 *with Lipschitz continuous gradients; i.e. there is a scalar* $L > 0$ *such that* $\|\nabla_{x^i} f_i(x) - \nabla_{x^i} f_i(x')\| \leq L\|x - x'\|$ *for all* x *and* $x' \in \mathcal{X}$ *and all* $i = 1, \ldots, n$*;*
(b) the family of objective functions $\{\theta_i\}$ *has an exact potential function* P*;*
(c) $\widehat{\theta}_i^{\sigma_k^\nu}(x^{\sigma_k^\nu}; z) = g_i(x^i) + \widehat{f}_i^{\sigma_k^\nu}(x^{\sigma_k^\nu}; z)$*, with* $\widehat{f}_i^{\sigma_k^\nu}$ *being continuously differentiable;*
(d) the family $\{\widehat{f}_i^{\sigma_k^\nu}(\bullet; x^\nu)\}_{i \in \sigma_k^\nu}$ *satisfies the strong convexity (with modulus* η*) and gradient consistency conditions and admits an exact potential function* $\widehat{f}_{\sigma_k^\nu}(\bullet; x^\nu)$*.*

Let a constant step-size (τ) *rule be employed with* $\tau \in \left(0, \frac{2\eta}{L}\right)$. *Then, for an essentially covering player selection rule, every limit point of the iterates generated by the unified algorithm is a QNE of the game* $\mathcal{G}$. *The same statement holds with probability one for the randomized player selection rule.* □

4.5.1 Generalized potential games

In the previous subsection, we have established the asymptotic convergence of the unified algorithm under the assumption of the existence of an exact potential and having a small enough step size depending on the game potential. In this subsection, we relax these two requirements by considering the generalized potential games and taking a step size that does not depend on the potential of the game. As a price, we are only able to show the convergence under the Gauss–Seidel-type update rule where $\boldsymbol{\sigma}^\nu = \sigma_1^\nu = \{i_\nu\}$ for all ν. More precisely, we focus on the analysis of the case where at each iteration ν, a single player i_ν is selected; then it calculates

$$\widehat{x}^{\nu;i_\nu} = \underset{x^{i_\nu} \in \mathcal{X}^{i_\nu}}{\operatorname{argmin}} \, \widehat{\theta}_{i_\nu}(x^{i_\nu}; x^\nu),$$

where $\widehat{\theta}_{i_\nu}$ is a surrogate function for θ_{i_ν}, and updates its variable using the constant step size $\tau \in (0, 1]$, resulting in $x^{\nu+1;i_\nu} = (1 - \tau) x^{\nu;i_\nu} + \tau \widehat{x}^{\nu;i_\nu}$, while $x^{\nu+1;-i_\nu} = x^{\nu;-i_\nu}$. In addition to the gradient consistency assumption, we make the following assumption, which is motivated by [30].

- **Tight upper-bound assumption** For every $i = 1, \ldots, n$,

$$\widehat{\theta}_i(x^i; y) \geq \theta_i(x^i, y^{-i}) \quad \text{and} \quad \widehat{\theta}_i(x^i; x) = \theta_i(x^i, x^{-i}), \qquad \forall x, y \in \boldsymbol{\mathcal{X}}.$$

  Example 4.8 provides an approximation function satisfying both the gradient consistency and the tight upper-bound assumptions; for more examples the interested readers are referred to [30].

With the above assumptions, we are ready to state Theorem 4.10, which establishes the asymptotic convergence of the unified algorithm for generalized potential games; for the proof of the theorem, see Appendix.

Theorem 4.10 *Assume that* $\mathcal{G}$ *is a generalized potential game and the approximation functions* $\widehat{\theta}_i(\bullet; y)$ *are strictly convex for all* $y \in \mathcal{X}$. *Under the tight upper-bound assumption and the essentially covering player selection rule with* $\boldsymbol{\sigma}^\nu$ *being a singleton for every* ν, *it holds that for every* $\tau \in (0, 1]$, *every limit point of the iterates generated by the algorithm with fixed step* τ *is a QNE. The same result holds with probability one for the randomized player selection rule.* □

Concluding remarks

In this chapter, we have introduced and analyzed the convergence of a unified distributed algorithm for computing a quasi-Nash equilibrium of a multi-player game with nonsmooth, nonconvex player objective functions and with decoupled convex constraints.

The algorithm employs a family of surrogate objective functions to deal with the nonconvexity and non-differentiability of the original objective functions and solves subgames in parallel involving deterministic or randomized choice of non-overlapping groups of players. The convergence analysis is based on two approaches: contraction and potential; the former relies on a spectral condition while the latter assumes the existence of a potential function. Extension of the algorithm and analysis to games with coupled constraints can be done by introducing multipliers (or prices) of such constraints that are updated in an outer iteration. Although we have not discussed nonconvex constraints, we expect that they can also be handled by using surrogate functions; for a most recent reference in the context of an optimization problem, see [50].

References

[1] J. v. Neumann, "Zur theorie der gesellschaftsspiele," *Mathematische Annalen*, vol. **100**, no. 1, pp. 295–320, 1928.

[2] J. F. Nash, "Equilibrium points in n-person games," *Proceedings of the National Academy of Sciences*, vol. **36**, no. 1, pp. 48–49, 1950.

[3] J. Nash, "Non-cooperative games," *Annals of Mathematics*, pp. 286–295, 1951.

[4] Z.-Q. Luo and J.-S. Pang, "Analysis of iterative waterfilling algorithm for multiuser power control in digital subscriber lines," *EURASIP Journal on Advances in Signal Processing*, 2006.

[5] G. Scutari and J.-S. Pang, "Joint sensing and power allocation in nonconvex cognitive radio games: quasi-Nash equilibria," in *IEEE Transactions on Signal Processing*, vol. **61**, IEEE, 2013, pp. 2366–2382.

[6] G. Scutari, F. Facchinei, J.-S. Pang, and D. Palomar, "Real and complex monotone communication games," *IEEE Transactions on Information Theory*, pp. 4197–4231, 2014.

[7] G. Scutari, D. P. Palomar, F. Facchinei, and J.-S. Pang, "Convex optimization, game theory, and variational inequality theory," *IEEE Signal Processing Magazine*, vol. **27**, no. 3, pp. 35–49, 2010.

[8] G. Scutari, D. P. Palomar, J.-S. Pang, and F. Facchinei, "Flexible design of cognitive radio wireless systems," *IEEE Signal Processing Magazine*, vol. **26**, no. 5, pp. 107–123, 2009.

[9] G. Scutari and J.-S. Pang, "Joint sensing and power allocation in nonconvex cognitive radio games: Nash equilibria and distributed algorithms," *IEEE Transactions on Information Theory*, vol. **59**, pp. 4626–4661, 2013.

[10] W. Yu, G. Ginis, and J. M. Cioffi, "Distributed multiuser power control for digital subscriber lines," *IEEE Journal on Selected Areas in Communications*, vol. **20**, no. 5, pp. 1105–1115, 2002.

[11] C. J. Day, B. F. Hobbs, and J.-S. Pang, "Oligopolistic competition in power networks: a conjectured supply function approach," *IEEE Transactions on Power Systems*, vol. **17**, no. 3, pp. 597–607, 2002.

[12] B. F. Hobbs and U. Helman, Complementarity-based equilibrium modeling for electric power markets. Chapter 3 in D. W. Bunn, editor, *Modeling Prices in Competitive Electricity Markets*, Citeseer, 2004.

[13] B. Hobbs, C. B. Metzler, and J.-S. Pang, "Nash–Cournot equilibria in power markets on a linearized DC network with arbitrage: formulations and properties," *Networks and Spatial Economics*, vol. **3**, no. 2, pp. 123–150, 2003.

[14] J.-S. Pang and B. F. Hobbs, "Spatial oligopolistic equilibria with arbitrage, shared resources, and price function conjectures," *Mathematical Programming*, vol. **101**, no. 1, pp. 57–94, 2004.
[15] J. Zhao, B. F. Hobbs, and J.-S. Pang, "Long-run equilibrium modeling of alternative emissions allowance allocation systems in electric power markets," *Operations Research*, vol. **58**, pp. 529–548, 2010.
[16] F. Facchinei and C. Kanzow, "Generalized Nash equilibrium problems," *Annals of Operations Research, [This is an updated version of the survey paper that appeared in AOR (2007) 173–210]*, pp. 177–211, 2010.
[17] F. Facchinei and J.-S. Pang, Nash equilibria: the variational approach. In Y. Eldar and D. Palomar, Eds., *Convex Optimization in Signal Processing and Communications*, Cambridge University Press, 2010.
[18] L. M. Bruce, "Game theory applied to big data analytics in geosciences and remote sensing." *IEEE International Conference on Geoscience and Remote Sensing Symposium (IGARSS)*, pp. 4094–4097, 2013.
[19] H. Baligh, M. Hong, W.-C. Liao, Z.-Q. Luo, M. Razaviyayn, M. Sanjabi, and R. Sun, "Cross-layer provision of future cellular networks: A WMMSE-based approach," *IEEE Signal Processing Magazine*, vol. **31**, no. 6, pp. 56–68, 2014.
[20] I. Guyon and A. Elisseeff, "An introduction to variable and feature selection," *The Journal of Machine Learning Research*, vol. **3**, pp. 1157–1182, 2003.
[21] Y. Saeys, I. Inza, and P. Larrañaga, "A review of feature selection techniques in Bioinformatics," *Bioinformatics*, vol. **23**, no. 19, pp. 2507–2517, 2007.
[22] T. Joachims, *Text Categorization With Support Vector Machines: Learning With Many Relevant Features*, Springer, 1998.
[23] F. Facchinei, A. Fischer, and V. Piccialli, "Generalized Nash equilibrium problems and Newton methods," *Mathematical Programming*, vol. **117**, no. 1–2, pp. 163–194, 2009.
[24] F. Facchinei, V. Piccialli, and M. Sciandrone, "Decomposition algorithms for generalized potential games," *Computational Optimization and Applications*, vol. **50**, no. 2, pp. 237–262, 2011.
[25] A. Alvarado, G. Scutari, and J.-S. Pang, "A new decomposition method for multiuser DC-programming and its applications," *IEEE Transactions on Signal Processing*, vol. **62**, no. 11, pp. 2984–2998, 2013.
[26] G. Scutari, F. Facchinei, P. Song, D. P. Palomar, and J.-S. Pang, "Decomposition by partial linearization: parallel optimization of multi-agent systems," *IEEE Transaction on Signal Processing*, pp. 641–656, 2014.
[27] J.-S. Pang and G. Scutari, "Nonconvex games with side constraints," *SIAM Journal on Optimization*, vol. **21**, no. 4, pp. 1491–1522, 2011.
[28] B. R. Marks and G. P. Wright, "Technical note – a general inner approximation algorithm for nonconvex mathematical programs," *Operations Research*, vol. **26**, no. 4, pp. 681–683, 1978.
[29] M. Razaviyayn, "Successive convex approximation: analysis and applications," Ph.D. dissertation, University of Minnesota, 2014.
[30] M. Razaviyayn, M. Hong, and Z.-Q. Luo, "A unified convergence analysis of block successive minimization methods for nonsmooth optimization," *SIAM Journal on Optimization*, vol. **23**, no. 2, pp. 1126–1153, 2013.
[31] J. Bolte, S. Shoham, and M. Teboulle, "Proximal alternating linearized minimization for nonconvex and nonsmooth problems," *Mathematical Programming*, pp. 1–36, 2013.

[32] J. Mairal, "Optimization with first-order surrogate functions," arXiv preprint arXiv: 1305.3120, 2013.

[33] M. Hong, X. Wang, M. Razaviyayn, and Z.-Q. Luo, "Iteration complexity analysis of block coordinate descent methods," arXiv preprint arXiv:1310.6957, 2013.

[34] M. Razaviyayn, M. Hong, Z.-Q. Luo, and J.-S. Pang, "Parallel successive convex approximation for nonsmooth nonconvex optimization," arXiv preprint arXiv:1406.3665, 2014.

[35] F. Facchinei, S. Sagratella, and G. Scutari, "Flexible parallel algorithms for big data optimization," arXiv preprint arXiv:1311.2444, 2013.

[36] M. Hong, T. H. Chang, X. Wang, M. Razaviyayn, S. Ma, and Z.-Q. Luo, "A block successive upper bound minimization method of multipliers for linearly constrained convex optimization," arXiv preprint arXiv:1401.7079, 2014.

[37] M. Razaviyayn, M. Sanjabi, and Z.-Q. Luo, "A stochastic successive minimization method for nonsmooth nonconvex optimization with applications to transceiver design in wireless communication networks," arXiv preprint arXiv:1307.4457, 2013.

[38] J. M. Ortega and W. C. Rheinboldt, *Iterative Solution of Nonlinear Equations in Several Variables*, SIAM Classics in Applied Mathematics, 2000, vol. **30**.

[39] R. W. Cottle, J.-S. Pang, and R. E. Stone, *The Linear Complementarity Problem*. SIAM, 2009, vol. **60**.

[40] J.-S. Pang and D. Chan, "Iterative methods for variational and complementarity problems," *Mathematical Programming*, vol. **24**, no. 1, pp. 284–313, 1982.

[41] J.-S. Pang, "Asymmetric variational inequality problems over product sets: applications and iterative methods," *Mathematical Programming*, vol. **31**, no. 2, pp. 206–219, 1985.

[42] D. P. Bertsekas and J. N. Tsitsiklis, *Parallel and Distributed Computation: Numerical Methods*, Athena Scientific, 1989.

[43] J.-S. Pang, G. Scutari, F. Facchinei, and C. Wang, "Distributed power allocation with rate constraints in Gaussian parallel interference channels," *IEEE Transactions on Information Theory*, vol. **54**, no. 8, pp. 3471–3489, 2008.

[44] J.-S. Pang, G. Scutari, D. P. Palomar, and F. Facchinei, "Design of cognitive radio systems under temperature-interference constraints: a variational inequality approach," *IEEE Transactions on Signal Processing*, vol. **58**, no. 6, pp. 3251–3271, 2010.

[45] F. Facchinei and J.-S. Pang, *Finite-Dimensional Variational Inequalities and Complementarity Problems*, Springer, 2003, vols. I and II.

[46] J.-S. Pang, M. Razaviyayn, and A. Alvarado, "Computing b-stationary points of nonsmooth dc programs," Manuscript. Department of Industrial and Systems Engineering, University of Southern California, 2014.

[47] D. Monderer and L. S. Shapley, "Potential games," *Games and Economic Behavior*, vol. **14**, no. 1, pp. 124–143, 1996.

[48] R. T. Rockafellar, *Convex Analysis*, Princeton University Press, 1970.

[49] R. T. Rockafellar and R. J. B. Wets, *Variational Analysis*, Springer, 1998, vol. **317**.

[50] G. Scutari, F. Facchinei, L. Lampariello, and P. Song, "Parallel and distributed methods for nonconvex optimization," In *Proceedings of IEEE International Conference on Acoustics, Speech, and Signal Processing (ICASSP)*, 2014.

[51] B. E. Fristedt and L. F. Gray, *A Modern Approach to Probability Theory*, Springer, 1997.

[52] D. P. Bertsekas, *Nonlinear Programming*, Athena Scientific, 1999.

[53] D. P. Bertsekas and J. N. Tsitsiklis, *Neuro-dynamic Programming*, Athena Scientific, 1996.

Appendix

Proof of Theorem 4.3

We begin the convergence analysis by invoking the variational principle applied to the optimization problem (4.3), obtaining at each iteration $\nu \equiv t$ modulo T and for each $i \in \sigma_k^t$ and $k = 1, \ldots, \kappa_t$,

$$g_i(x^i) - g_i(x^{\nu+1;i}) + \left(x^i - x^{\nu+1;i}\right)^T \nabla_{x^i} \widehat{f}_i\left(x^{\nu+1;\sigma_k^t}; x^\nu\right) \geq 0, \quad \forall x^i \in \mathcal{X}^i.$$

Substituting $x^i = x^{\nu;i}$, we deduce

$$g_i\left(x^{\nu;i}\right) - g_i\left(x^{\nu+1;i}\right) + \left(x^{\nu;i} - x^{\nu+1;i}\right)^T \nabla_{x^i} \widehat{f}_i\left(x^{\nu+1;\sigma_k^\nu}; x^\nu\right) \geq 0.$$

Consider two consecutive cycles, $\widehat{\omega} = \omega - 1$ and ω for some integer $\omega \geq 2$, of T successive iterations each; see Figure 4.1. During these two cycles, the components indexed by $\left\{\sigma_k^t\right\}_{k=1}^{\kappa_t}$ are sequentially updated in iterations $\widehat{\omega}_t = (\widehat{\omega} - 1)T + t$ and $\omega_t = (\omega - 1)T + t$ for $t = 1, \ldots, T$, respectively. Starting with the first iteration within these two cycles, we have, for all $i \in \sigma_k^1$,

$$g_i\left(x^{\omega_1 - 1;i}\right) - g_i\left(x^{\omega_1;i}\right) + \left(x^{\omega_1 - 1;i} - x^{\omega_1;i}\right)^T \nabla_{x^i} \widehat{f}_i\left(x^{\omega_1;\sigma_k^1}; x^{\omega_1 - 1}\right) \geq 0,$$

and $g_i(x^{\omega_1;i}) - g_i(x^{\widehat{\omega}_1;i}) + (x^{\omega_1;i} - x^{\widehat{\omega}_1;i})^T \nabla_{x^i} \widehat{f}_i(x^{\widehat{\omega}_1;\sigma_k^1}; x^{\widehat{\omega}_1 - 1}) \geq 0$. Noting that $x^{\omega_1 - 1;i} = x^{\widehat{\omega}_1;i} = \mathbf{x}^{(\omega-1);i}$ and $x^{\omega_1;i} = \mathbf{x}^{(\omega);i}$ by (4.6), and adding the above two inequalities, we obtain, by the mean-value theorem for multivariate functions, for every $i \in \sigma_k^1$,

$$\begin{aligned}
0 &\leq \left(x^{\omega_1 - 1;i} - x^{\omega_1;i}\right)^T \left[\nabla_{x^i} \widehat{f}_i\left(x^{\omega_1;\sigma_k^1}; x^{\omega_1 - 1}\right) - \nabla_{x^i} \widehat{f}_i\left(x^{\widehat{\omega}_1;\sigma_k^1}; x^{\widehat{\omega}_1 - 1}\right)\right] \\
&= \left(\mathbf{x}^{(\omega-1);i} - \mathbf{x}^{(\omega);i}\right)^T \sum_{j \in \sigma_k^1} \nabla^2_{x^j x^i} \widehat{f}_i\left(u^{\omega_1;\sigma_k^1}; z^{\omega_1}\right)\left(x^{\omega_1;j} - x^{\omega_1 - 1;j}\right) \\
&\quad + \left(\mathbf{x}^{(\omega-1);i} - \mathbf{x}^{(\omega);i}\right)^T \nabla^2_{zx^i} \widehat{f}_i\left(u^{\omega_1;\sigma_k^1}; z^{\omega_1}\right)\left(x^{\omega_1 - 1} - x^{\widehat{\omega}_1 - 1}\right) \\
&= \left(\mathbf{x}^{(\omega-1);i} - \mathbf{x}^{(\omega);i}\right)^T \sum_{j \in \sigma_k^1} \nabla^2_{x^j x^i} \widehat{f}_i\left(u^{\omega_1;\sigma_k^1}; z^{\omega_1}\right)\left(\mathbf{x}^{(\omega);j} - \mathbf{x}^{(\omega-1);j}\right) \\
&\quad + \left(\mathbf{x}^{(\omega-1);i} - \mathbf{x}^{(\omega);i}\right)^T \sum_{\ell=1}^{n} \nabla^2_{z^\ell x^i} \widehat{f}_i\left(u^{\omega_1;\sigma_k^1}; z^{\omega_1}\right)\left(\mathbf{x}^{(\widehat{\omega});\ell} - \mathbf{x}^{(\widehat{\omega}-1);\ell}\right),
\end{aligned}$$

where $(u^{\omega_1;\sigma_k^1}; z^{\omega_1}) = \tau_k^1(x^{\omega_1;\sigma_k^1}; x^{\omega_1 - 1}) + (1 - \tau_k^1)(x^{\widehat{\omega}_1;\sigma_k^1}; x^{\widehat{\omega}_1 - 1})$ for some scalar $\tau_k^1 \in [0, 1]$. For each cycle count ω and player label i, let $e^{(\omega);i} \triangleq \|\mathbf{x}^{(\omega);i} - \mathbf{x}^{(\omega-1);i}\|$. By the Cauchy–Schwartz inequality, the above inequality yields

$$\gamma_{ii} e^{(\omega);i} - \sum_{j \in \sigma_k^1,\ j \neq i} \gamma_{ij} e^{(\omega);j} \leq \sum_{\ell=1}^{n} \widetilde{\gamma}_{i\ell} e^{(\omega-1);\ell}, \quad \forall i \in \sigma_k^1. \tag{4.12}$$

Similarly in the second iteration within the cycles $\widehat{\omega} = \omega - 1$ and ω, we have, for $k = 1, \ldots, \kappa_2$ and for all $i \in \sigma_k^2$,

$$g_i\left(x^{\omega_2-1;i}\right) - g_i\left(x^{\omega_2;i}\right) + \left(x^{\omega_2-1;i} - x^{\omega_2;i}\right)^T \nabla_{x^i} \widehat{f}_i\left(x^{\omega_2;\sigma_k^2}; x^{\omega_2-1}\right) \geq 0$$

and $g_i(x^{\omega_2;i}) - g_i(x^{\widehat{\omega}_2;i}) + (x^{\omega_2;i} - x^{\widehat{\omega}_2;i})^T \nabla_{x^i} \widehat{f}_i(x^{\widehat{\omega}_2;\sigma_k^2}; x^{\widehat{\omega}_2-1}) \geq 0$. For some $(u^{\omega_2;\sigma_k^2}; z^{\omega_2}) = \tau_k^2(x^{\omega_2;\sigma_k^2}; x^{\omega_2-1}) + (1 - \tau_k^2)(x^{\widehat{\omega}_1;\sigma_k^2}; x^{\widehat{\omega}_2-1})$ with $\tau_k^2 \in [0, 1]$, we deduce

$$\begin{aligned}
0 &\leq \left(x^{\omega_2-1;i} - x^{\omega_2;i}\right)^T \left[\nabla_{x^i} \widehat{f}_i\left(x^{\omega_2;\sigma_k^2}; x^{\omega_2-1}\right) - \nabla_{x^i} \widehat{f}_i\left(x^{\widehat{\omega}_2;\sigma_k^2}; x^{\widehat{\omega}_2-1}\right)\right] \\
&= \left(\mathbf{x}^{(\omega-1);i} - \mathbf{x}^{(\omega);i}\right)^T \sum_{j \in \sigma_k^2} \nabla^2_{x^j x^i} \widehat{f}_i\left(u^{\omega_2;\sigma_k^2}; z^{\omega_2}\right)\left(x^{\omega_2;j} - x^{\omega_2-1;j}\right) \\
&\quad + \left(\mathbf{x}^{(\omega-1);i} - \mathbf{x}^{(\omega);i}\right)^T \sum_{\ell=1}^{n} \nabla^2_{z^\ell x^i} \widehat{f}_i\left(u^{\omega_1;\sigma_k^1}; z^{\omega_1}\right)\left(x^{\omega_2-1;\ell} - x^{\widehat{\omega}_2-1;\ell}\right) \\
&= \left(\mathbf{x}^{(\omega-1);i} - \mathbf{x}^{(\omega);i}\right)^T \sum_{j \in \sigma_k^2} \nabla^2_{x^j x^i} \widehat{f}_i\left(u^{\omega_2;\sigma_k^2}; z^{\omega_2}\right)\left(\mathbf{x}^{(\omega);j} - \mathbf{x}^{(\omega-1);j}\right) \\
&\quad + \left(\mathbf{x}^{(\omega-1);i} - \mathbf{x}^{(\omega);i}\right)^T \sum_{k'=1}^{\kappa_1} \sum_{\ell \in \sigma_{k'}^1} \nabla^2_{z^\ell x^i} \widehat{f}_i\left(u^{\omega_1;\sigma_k^1}; z^{\omega_1}\right)\left(x^{\omega_2-1;\ell} - x^{\widehat{\omega}_2-1;\ell}\right) \\
&\quad + \left(\mathbf{x}^{(\omega-1);i} - \mathbf{x}^{(\omega);i}\right)^T \sum_{t=2}^{T} \sum_{k'=1}^{\kappa_t} \sum_{\ell \in \sigma_{k'}^t} \nabla^2_{z^\ell x^i} \widehat{f}_i\left(u^{\omega_1;\sigma_k^1}; z^{\omega_1}\right)\left(x^{\omega_2-1;\ell} - x^{\widehat{\omega}_2-1;\ell}\right).
\end{aligned}$$

Hence, similar to the inequalities in (4.12), we have, $\forall i \in \sigma_k^2$,

$$-\sum_{k'=1}^{\kappa_1} \sum_{\ell \in \sigma_{k'}^1} \widetilde{\gamma}_{i\ell} e^{(\omega);\ell} + \gamma_{k;ii}^2 e^{(\omega);i} - \sum_{i \neq j \in \sigma_k^2} \gamma_{ij} e^{(\omega);j} \leq \sum_{t=2}^{T} \sum_{k'=1}^{\kappa_t} \sum_{\ell \in \sigma_{k'}^t} \widetilde{\gamma}_{i\ell} e^{(\omega-1);\ell}. \tag{4.13}$$

We may continue the above derivation with the third till last iterations within the two cycles ω and $\widehat{\omega}$ and obtain inequalities that extend (4.12) and (4.13). Let $\mathbf{E}^{(\omega)} \triangleq (\mathbf{e}^{(\omega);t})_{t=1}^T$, where $\mathbf{e}^{(\omega);t} \triangleq \{(e^{\omega;i})_{i \in \sigma_k^t}\}_{k=1}^{\kappa_t}$. The inequalities (4.12) and (4.13) and their extensions can then be written in the simple matrix inequalities:

$$(\overline{\mathbf{\Gamma}} - \widetilde{\mathbf{L}})\mathbf{E}^{(\omega)} \leq (\widetilde{\mathbf{D}} + \widetilde{\mathbf{U}})\mathbf{E}^{(\omega-1)}.$$

Assumption (A_f^4) implies that the matrix $\overline{\mathbf{\Gamma}} - \widetilde{\mathbf{L}}$ has a nonnegative inverse, thus recalling the definition of the matrix $\widehat{\mathbf{M}} = [\overline{\mathbf{\Gamma}} - \widetilde{\mathbf{L}}]^{-1}(\widetilde{\mathbf{D}} + \widetilde{\mathbf{U}})$, we deduce

$$\mathbf{E}^{(\omega)} \leq \widehat{\mathbf{M}}\mathbf{E}^{(\omega-1)}.$$

Since $\widehat{\mathbf{M}}$ has spectral radius less than unity, it follows that the sequence $\{\mathbf{x}^{(\omega)}\}$ is a contraction and therefore converges to a limit $x^\infty \triangleq (x^{\infty;i})_{i=1}^n$, which must necessarily satisfy the following optimality conditions: for all $t = 1, \ldots, T$, $k = 1, \ldots, \kappa_t$, and

$i \in \sigma_k^t$, $x^{\infty;i} \in \underset{x^i \in \mathcal{X}^i}{\operatorname{argmin}}\, \widehat{\theta}_i(x^i, x^{\infty;\sigma_k^t;-i}; x^\infty)$. Thus,

$$\widehat{\theta}_i(\bullet, x^{\infty,\sigma_k^t;-i}; x^\infty)'(x^{\infty,i}; x^i - x^{\infty,i}) \geq 0, \qquad \forall x^i \in \mathcal{X}^i.$$

Consequently, if (4.8) holds, then x^∞ is a QNE of the original game $\mathcal{G}$. □

Proof of Theorem 4.4

Without referring to the randomized algorithm, we first define a (deterministic) self-map Φ from $\boldsymbol{\mathcal{X}}$ into itself as follows. For a subset σ_k for $k = 1, \ldots, K$, and for a given point $y \in \boldsymbol{\mathcal{X}}$, let $\{\widehat{x}^i(y)\}_{i\in\sigma_k}$ be the solution of the subgame

$$\left\{ \min_{x^i \in \mathcal{X}^i} \widehat{\theta}_i \left(x^i, x^{\sigma_k;-i}; y\right) \right\}_{i\in\sigma_k}.$$

Let $\Phi(y)$ be the tuple of these solutions; i.e. $\Phi(y) \triangleq \{\{\widehat{x}^i(y)\}_{i\in\sigma_k}\}_{k=1}^K$. Similar to the analysis for the deterministic case, exploiting the variational principle, one can deduce, for all $i \in \{1, \ldots, n\}$, and for all $y, w \in \boldsymbol{\mathcal{X}}$:

$$g_i(\widehat{x}^i(y)) - g_i(\widehat{x}^i(w)) + \left(\widehat{x}^i(y) - \widehat{x}^i(w)\right)^T \nabla_{x^i} \widehat{f}_i(\widehat{x}^{\sigma(i)}(w); w) \geq 0,$$

and $g_i(\widehat{x}^i(w)) - g_i(\widehat{x}^i(y)) + (\widehat{x}^i(w) - \widehat{x}^i(y))^T \nabla_{x^i} \widehat{f}_i(\widehat{x}^{\sigma(i)}(y); y) \geq 0$. Adding these two inequalities leads to

$$\left(\widehat{x}^i(y) - \widehat{x}^i(w)\right)^T \left[\nabla_{x^i} \widehat{f}_i\left(\widehat{x}^{\sigma(i)}(w); w\right) - \nabla_{x^i} \widehat{f}_i\left(\widehat{x}^{\sigma(i)}(y); y\right)\right] \geq 0, \tag{4.14}$$

for any $y, w \in \boldsymbol{\mathcal{X}}$ and for all $i = 1, 2, \ldots, n$. Again, similar to the deterministic case, we may deduce, by a simple application of the mean-value theorem,

$$\gamma_{ii}\|\widehat{x}^i(y) - \widehat{x}^i(w)\| - \sum_{j\neq i} \gamma_{ij}\|\widehat{x}^j(y) - \widehat{x}^j(w)\| \leq \sum_{\ell=1}^{n} \widetilde{\gamma}_{i\ell}\|y^\ell - w^\ell\|,$$

for any player i and any two points $y, w \in \boldsymbol{\mathcal{X}}$. Concatenating these inequalities, we obtain,

$$\overline{\mathbf{\Gamma}} \begin{bmatrix} \|\widehat{x}^1(y) - \widehat{x}^1(w)\| \\ \|\widehat{x}^2(y) - \widehat{x}^2(w)\| \\ \vdots \\ \|\widehat{x}^n(y) - \widehat{x}^n(w)\| \end{bmatrix} \leq \left(\widetilde{\mathbf{L}} + \widetilde{\mathbf{D}} + \widetilde{\mathbf{U}}\right) \begin{bmatrix} \|y^1 - w^1\| \\ \|y^2 - w^2\| \\ \vdots \\ \|y^n - w^n\| \end{bmatrix}.$$

Since the matrix $\widehat{\mathbf{M}} = (\overline{\mathbf{\Gamma}} - \widetilde{\mathbf{L}})^{-1}(\widetilde{\mathbf{D}} + \widetilde{\mathbf{U}})$ has spectral radius less than unity, it follows that the same is true for the matrix $\widetilde{\mathbf{M}} = \overline{\mathbf{\Gamma}}^{-1}(\widetilde{\mathbf{L}} + \widetilde{\mathbf{D}} + \widetilde{\mathbf{U}})$; see Lemma 4.11 below. Thus, the map Φ is a contraction and hence has a unique fixed point x^∞ that must belong to the set $\boldsymbol{\mathcal{X}}$. Thus x^∞ satisfies (4.8) and is a QNE of the game $\mathcal{G}$.

Now, consider the probabilistic algorithm that generates a randomized sequence $\{x^\nu\}$ such that given x^ν at the beginning of iteration ν, for each i, $x^{\nu+1,i} = \widehat{x}^i(x^\nu)$ conditional on the index set σ_k containing i being picked. Therefore, taking expectations, we deduce,

$$\begin{aligned}\mathbb{E}\left[\|x^{\nu+1;i} - x^{\infty;i}\| \mid x^\nu\right] &= p_k \|\widehat{x}^i(x^\nu) - x^{\infty;i}\| + (1-p_k)\|x^{\nu;i} - x^{\infty;i}\| \\ &\le p_k \sum_{j=1}^n \widetilde{\mathbf{M}}_{ij} \|x^{\nu;j} - x^{\infty;j}\| + (1-p_k)\|x^{\nu;i} - x^{\infty;i}\|.\end{aligned}$$

Define $\mathbf{P} \triangleq \operatorname{diag}[p_{ii}]_{i=1}^n$, where $p_{ii} \triangleq p_k$ if $i \in \sigma_k$, and $\Upsilon \triangleq \mathbf{I} - \mathbf{P} + \mathbf{P}\widetilde{\mathbf{M}}$. Taking expectations in the above inequality with respect to the whole iteration trajectory implies $\mathbb{E}[\|x^{\nu+1;i} - x^{\infty;i}\|] \le \sum_{j=1}^n \Upsilon_{ij}\mathbb{E}[\|x^{\nu;j} - x^{\infty;j}\|]$, which by rewriting in the matrix form leads to

$$\begin{bmatrix} \mathbb{E}\left[\|x^{\nu+1;1} - x^{\infty;1}\|\right] \\ \mathbb{E}\left[\|x^{\nu+1;2} - x^{\infty;2}\|\right] \\ \vdots \\ \mathbb{E}\left[\|x^{\nu+1;n} - x^{\infty;n}\|\right] \end{bmatrix} \le \begin{bmatrix} \Upsilon_{11} & \Upsilon_{12} & \dots & \Upsilon_{1n} \\ \Upsilon_{21} & \Upsilon_{22} & \dots & \Upsilon_{2n} \\ \vdots & \vdots & \ddots & \vdots \\ \Upsilon_{n1} & \Upsilon_{n2} & \dots & \Upsilon_{nn} \end{bmatrix} \begin{bmatrix} \mathbb{E}\left[\|x^{\nu;1} - x^{\infty;1}\|\right] \\ \mathbb{E}\left[\|x^{\nu;2} - x^{\infty;2}\|\right] \\ \vdots \\ \mathbb{E}[\|x^{\nu;n} - x^{\infty;n}\|] \end{bmatrix}.$$

Defining $e^\nu \triangleq [\mathbb{E}[\|x^{\nu;1} - x^{\infty;1}\|]]_{i=1}^n$, we can write the above inequality as

$$0 \le e^{\nu+1} \le \Upsilon e^\nu. \tag{4.15}$$

By Lemma 4.12 below, it follows that $\rho(\Upsilon) < 1$. Hence, there exists an induced (monotonic) ℓ_q-norm $\|\cdot\|_q$ for some $q \ge 1$ such that $\|\Upsilon\|_q < 1$. Using (4.15), we obtain, for all i, $\mathbb{E}[\|x^{\nu;i} - x^{\infty;i}\|] = e^{\nu;i} \le \|e^\nu\|_q \le (\|\Upsilon\|_q)^\nu \|e^0\|_q$. Therefore, Markov's inequality implies

$$\Pr\left(\|x^{\nu;i} - x^{\infty;i}\| > \varepsilon\right) \le \frac{\left(\|\Upsilon\|_q\right)^\nu \|e^0\|_q}{\varepsilon}, \quad \forall i$$

for any $\varepsilon > 0$, and hence, a simple application of the Borel–Cantelli lemma [51] yields the almost sure convergence of the sequence $\{x^\nu\}_{\nu=1}^\infty$ to x^∞. □

Lemma 4.11 *Let $\Gamma \in \mathbb{R}^{n\times n}$ be a Minkowski matrix and let A and B be nonnegative matrices of the same order, If $\rho(\Gamma^{-1}(A+B)) < 1$, then $\rho((\Gamma - A)^{-1}B) < 1$.*

Proof Since $\rho(\Gamma^{-1}(A+B)) < 1$, by [39, Lemma 5.3.14], it follows that $\Gamma - (A + B) = (\Gamma - A) - B$ is a Minkowski matrix. Since $\Gamma - A$ is a Z-matrix and $\Gamma - A \ge \Gamma - A - B$, thus $\Gamma - A$ is Minkowski. Hence by the cited lemma again, we deduce $\rho((\Gamma - A)^{-1}B) < 1$ as claimed. □

Lemma 4.12 *For any nonnegative matrix $A \in \mathbb{R}_+^{n\times n}$ and diagonal matrix D with diagonal entries $D_{ii} \in (0, 1)$ for all $i = 1, \dots, n$, $\rho(I - D + DA) < 1$ if and only if $\rho(A) < 1$.*

Proof By [39, Lemma 5.3.14], $\rho(I - D + DA) < 1$ if and only if the matrix

$$D(I - A) = D - DA = I - (I - D + DA)$$

is Minkowski. Since D is a positive diagonal matrix, $D - DA$ is Minkowski if and only if $I - A$ is Minkowski, which in turn holds if and only if $\rho(A) < 1$ by the same lemma in the cited reference. □

Proof of Proposition 4.5

The equivalence of (a) and (b) follows from [49, Theorem 12.5]. Since $\partial\psi(x) \subseteq \prod_{i=1}^{n} \partial_{x^i}\psi(x)$, by [48, Proposition 2.3.15], it follows that if (b) holds, then $\partial_{x^i}\theta_i(x) \subseteq \partial_{x^i}\psi(x)$ and equality must hold by the maximality of the (partial) subdifferential. The existence of the functions $A_i(x^{-i})$ then follows from [48, Theorem 25.9]. We next show that (c) is equivalent to (d). [Convexity is not needed in the remaining proof of the proposition.] Indeed if (c) holds, then it is easy to see that ψ is an exact potential function. Conversely, suppose that P is an exact potential function for the game $\mathcal{G}$. We then have $\theta_i(x) = P(x) + \theta_i(0, x^{-i}) - P(0, x^{-i})$, which shows that P satisfies (c) with $A_i(x^{-i}) \triangleq \theta_i(0, x^{-i}) - P(0, x^{-i})$. □

Proof of Proposition 4.6

We can write

$$
\begin{aligned}
P(x) - P(0) &= P(x^1, \ldots, x^n) - P(0) \\
&= \sum_{i=1}^{n} \left[P(x^1, \ldots, x^i, 0, \ldots, 0) - P(x^1, \ldots, x^{i-1}, 0, \ldots, 0) \right] \\
&= \sum_{i=1}^{n} \left[\theta_i(x^1, \ldots, x^i, 0, \ldots, 0) - \theta_i(x^1, \ldots, x^{i-1}, 0, \ldots, 0) \right] \\
&= \sum_{i=1}^{n} \left[f_i(x^1, \ldots, x^i, 0, \ldots, 0) - f_i(x^1, \ldots, x^{i-1}, 0, \ldots, 0) + g_i(x^i) - g_i(0) \right],
\end{aligned}
$$

which leads to the desired expression of $P(x)$. Finally, to prove the last assertion in part (c) of the proposition, we have, for any direction $d = (0, \ldots, 0, d^i, 0, \ldots, 0)$, $P'(x; d) = \nabla f(x)^T d + g_i'(x^i; d^i) = \theta_i'(x; d) = \nabla f_i(x)^T d + g_i'(x^i; d^i)$. Hence the equality $\nabla_{x^i} f(x) = \nabla_{x^i} f_i(x)$ follows. □

Proof of Theorem 4.9

By Proposition 4.6 and following the notation therein, we have, for any two vectors w and z in $\mathcal{X}$,

$$
\begin{aligned}
\|\nabla f(w) - \nabla f(z)\|^2 &= \sum_{i=1}^{n} \|\nabla_{x^i} f(w) - \nabla_{x^i} f(z)\|^2 = \sum_{i=1}^{n} \|\nabla_{x^i} f_i(w) - \nabla_{x^i} f_i(z)\|^2 \\
&\leq \sum_{i=1}^{n} L^2 \|w^i - z^i\|^2 = L^2 \|w - z\|^2.
\end{aligned}
$$

Thus ∇f is Lipschitz continuous with constant L. Consider the essentially covering player selection rule. Using the descent lemma [52] and defining the scalar

$\beta \triangleq \frac{\tau}{2}(2\eta - L\tau)$ which is positive due to the restriction on τ, we can write

$$
\begin{aligned}
P(x^{\nu+1}) - P(x^{\nu}) &= f(x^{\nu+1}) - f(x^{\nu}) + \sum_{i=1}^{n} \left(g_i(x^{\nu+1;i}) - g_i(x^{\nu;i})\right) \\
&\leq \nabla f(x^{\nu})^T(x^{\nu+1} - x^{\nu}) + \frac{L}{2}\|x^{\nu+1} - x^{\nu}\|^2 + \sum_{i=1}^{n} \left(g_i(x^{\nu+1;i}) - g_i(x^{\nu;i})\right) \\
&= \sum_{\sigma_k^{\nu} \in \sigma^{\nu}} \sum_{i \in \sigma_k^{\nu}} \left[\nabla_{x^i} f(x^{\nu})^T(x^{\nu+1;i} - x^{\nu;i}) + \frac{L}{2}\|x^{\nu+1;i} - x^{\nu;i}\|^2 + g_i(x^{\nu+1;i}) - g_i(x^{\nu;i})\right] \\
&= \sum_{\sigma_k^{\nu} \in \sigma^{\nu}} \sum_{i \in \sigma_k^{\nu}} \left[\tau \nabla_{x^i} f_i(x^{\nu})^T(\widehat{x}^{\nu;i} - x^{\nu;i}) + \frac{L\tau^2}{2}\|\widehat{x}^{\nu;i} - x^{\nu;i}\|^2 + g_i(x^{\nu+1;i}) - g_i(x^{\nu;i})\right] \\
&\leq \sum_{\sigma_k^{\nu} \in \sigma^{\nu}} \sum_{i \in \sigma_k^{\nu}} \Big[\tau\left\{\left(\nabla_{x^i} \widehat{f}_i^{\sigma_k^{\nu}}(\bullet, x^{\nu;\sigma_k^{\nu};-i}; x^{\nu})|_{x^i}\right)^T(\widehat{x}^{\nu;i} - x^{\nu;i}) + g_i(\widehat{x}^{\nu;i}) - g_i(x^{\nu;i})\right\} \\
&\qquad + \frac{L\tau^2}{2}\|\widehat{x}^{\nu;i} - x^{\nu;i}\|^2\Big] \qquad \text{by convexity of } g_i \text{ and gradient consistency of } \widehat{f}_i \\
&\leq \sum_{\sigma_k^{\nu} \in \sigma^{\nu}} \sum_{i \in \sigma_k^{\nu}} \Big[\tau\left\{\nabla_{x^i} \widehat{f}_i^{\sigma_k^{\nu}}(\widehat{x}^{\nu;\sigma_k^{\nu}}; x^{\nu})^T(\widehat{x}^{\nu;i} - x^{\nu;i}) + g_i(\widehat{x}^{\nu;i}) - g_i(x^{\nu;i})\right\} \\
&\qquad - \beta\|\widehat{x}^{\nu;i} - x^{\nu;i}\|^2\Big] \qquad \text{by (4.10) and the definition of } \beta.
\end{aligned}
$$

Since $\widehat{x}^{\nu;\sigma_k^{\nu}}$ is a Nash equilibrium of the subgame $\mathcal{G}^{\sigma_k^{\nu}}(x^{\nu})$, by the variational principle of the optimization problem (4.3), we have

$$
0 \leq \nabla \widehat{f}_i(\widehat{x}^{\nu;\sigma_k^{\nu}}; x^{\nu})^T(x^{\nu;i} - \widehat{x}^{\nu;i}) + g_i(x^{\nu;i}) - g_i(\widehat{x}^{\nu;i}), \qquad \forall i \in \sigma_k^{\nu}. \tag{4.16}
$$

Hence, we obtain

$$
P(x^{\nu+1}) - P(x^{\nu}) \leq -\beta\|\widehat{x}^{\nu;\mathcal{N}_{\nu}} - x^{\nu;\mathcal{N}_{\nu}}\|^2. \tag{4.17}
$$

Thus, the sequence of potential values $\{P(x^{\nu})\}$ is non-increasing. Therefore, assuming the existence of a limit point $\bar{x}$ with a subsequence $\{x^{\nu_\ell}\}_{\ell=1}^{\infty}$ converging to $\bar{x}$, we must have

$$
\lim_{\nu \to \infty} P(x^{\nu}) = P(\bar{x}). \tag{4.18}
$$

Moreover, since the player selection rule is essentially covering, there exist update sets $\bar{\sigma}^0, \ldots, \bar{\sigma}^{T-1}$ with $\bar{\mathcal{N}}^m \triangleq \bigcup_{\sigma \in \bar{\sigma}^m} \sigma$ such that by restricting to a subsequence, we have $\sigma^{\nu_\ell + m} = \bar{\sigma}^m$ for all $m = 0, \ldots, T-1$ and all ℓ and $\bar{\mathcal{N}}^0 \cup \cdots \cup \bar{\mathcal{N}}^{T-1} = \{1, \ldots, n\}$. By rewriting (4.17) for the restricted subsequence, we have

$$
P(x^{\nu_\ell+m+1}) - P(x^{\nu_\ell+m}) \leq -\beta\|\widehat{x}^{\nu_\ell+m;\bar{\mathcal{N}}^m} - x^{\nu_\ell+m;\bar{\mathcal{N}}^m}\|^2, \qquad \forall m = 0, \ldots, T-1.
$$

Setting $m = 0$ in the above inequality and combining it with (4.18) will lead to

$$
\lim_{\ell \to \infty} \|\widehat{x}^{\nu_\ell;\bar{\mathcal{N}}^0} - x^{\nu_\ell;\bar{\mathcal{N}}^0}\| = 0. \tag{4.19}
$$

On the other hand, based on the definition of $\widehat{x}^{\nu_\ell;\mathcal{N}^0}$, we have that

$$\widehat{\theta}_i^{\sigma}(\widehat{x}^{\nu_\ell;i}, \widehat{x}^{\nu_\ell;\sigma;-i}; x^{\nu_\ell}) \leq \widehat{\theta}_i^{\sigma}(x^i, \widehat{x}^{\nu_\ell;\sigma;-i}; x^{\nu_\ell}), \quad \forall x^i \in \mathcal{X}^i, \ \forall i \in \sigma, \ \forall \sigma \in \bar{\boldsymbol{\sigma}}^0,$$

which combined with (4.19) yields

$$\widehat{\theta}_i^{\sigma}(\bar{x}^i, \bar{x}^{\sigma;-i}; \bar{x}) \leq \widehat{\theta}_i^{\sigma}(x^i, \bar{x}^{\sigma;-i}; \bar{x}), \quad \forall x^i \in \mathcal{X}^i, \ \forall i \in \sigma, \ \forall \sigma \in \bar{\boldsymbol{\sigma}}^0.$$

The variational principle (first-order optimality condition) of the above inequality leads to $\widehat{\theta}_i^{\sigma}(\bullet, \bar{x}^{\sigma;-i}; \bar{x})'(\bar{x}^i; u^i - x^i) \geq 0, \quad \forall u^i \in \mathcal{X}^i, \ \forall i \in \sigma$, and $\forall \sigma \in \bar{\boldsymbol{\sigma}}^0$. Therefore, by (4.11) we deduce

$$\theta_i(\bullet, \bar{x}^{-i})'(\bar{x}^i; u^i - x^i) \geq 0, \quad \forall u^i \in \mathcal{X}^i, \ \forall i \in \mathcal{N}^0.$$

Repeating the above argument for $m = 1, \ldots, T-1$ leads to

$$\theta_i(\bullet, \bar{x}^{-i})'(\bar{x}^i; u^i - x^i) \geq 0, \quad \forall u^i \in \mathcal{X}^i, \ \forall i = 1, 2, \ldots, n.$$

Hence, $\bar{x}$ is a QNE of the game $\mathcal{G}$.

Next consider the randomized player selection rule. For any player i, define $\widetilde{\sigma}(i)$ to be the most probable set containing index i, i.e. $i \in \widetilde{\sigma}(i)$ and

$$\Pr(\widetilde{\sigma}(i) \in \boldsymbol{\sigma}^\nu) \geq \Pr(\sigma \in \boldsymbol{\sigma}^\nu), \quad \forall \sigma \subseteq \{1, 2, \ldots, n\} \text{ with } i \in \sigma.$$

Since $p_{\min} > 0$, there must exist a constant $\widetilde{p}_{\min} > 0$ so that $\Pr(\widetilde{\sigma}(i) \in \boldsymbol{\sigma}^\nu) \geq \widetilde{p}_{\min} > 0, \forall i$. Define $\widetilde{x}^{\nu;i}$ to be the strategy of player i after solving the subgame $\mathcal{G}^{\widetilde{\sigma}(i)}$ with the optimization problems $\{\min_{x^j \in \mathcal{X}^j} \widehat{\theta}_j^{\widetilde{\sigma}(i)}(x^{\widetilde{\sigma}(i)}; x^\nu)\}_{j \in \widetilde{\sigma}(i)}$ at iteration ν. As an immediate consequence of (4.17), we obtain

$$\mathbb{E}\left[P(x^{\nu+1}) - P(x^\nu) \mid x^\nu\right] \leq -\beta \widetilde{p}_{\min} \|\widetilde{x}^{\nu;i} - x^{\nu;i}\|^2, \quad \forall i.$$

Therefore, the process $\{P(x^\nu)\}_{\nu=1}^\infty$ is a supermartingale and by the supermartingale convergence theorem [53, Proposition 4.2], we deduce $\sum_{\nu=1}^\infty \|\widetilde{x}^{\nu;i} - x^{\nu;i}\|^2 < \infty, \forall i$, with probability one, which in turn implies that

$$\lim_{\nu \to \infty} \|\widetilde{x}^\nu - x^\nu\|^2 = 0, \quad \text{almost surely.}$$

Similar to the proof of the essentially covering case, we can easily use the above fact and the variational principle combined with (4.11) to show that any limit point of the sequence is a QNE of the game (with probability one). □

Proof of Theorem 4.10

We first prove the result for the the essentially covering player selection rule. Consider a limit point $\bar{x}$ that is the limit of the convergent subsequence $\{x^{\nu_\ell}\}_{\ell=1}^\infty$. Owing to the tight upper-bound assumption, one can write

$$\theta_{i_\nu}(x^{\nu+1}) \leq \widehat{\theta}_{i_\nu}(x^{\nu+1;i_\nu}; x^\nu) \leq \widehat{\theta}_{i_\nu}(x^{\nu;i_\nu}; x^\nu) = \theta_{i_\nu}(x^\nu), \ \forall \nu, \tag{4.20}$$

where the second inequality is because of the update rule of the algorithm and also the convexity of $\widehat{\theta}_{i_\nu}(\bullet; x^\nu)$. Since the approximation function $\widehat{\theta}_{i_\nu}(\bullet, x^\nu)$ is strictly convex, the above inequality implies that either $\theta_{i_\nu}(x^{\nu+1}) < \theta_{i_\nu}(x^\nu)$, or $x^{\nu+1} = x^\nu$. Clearly in both cases we must have

$$P(x^{\nu+1}) \leq P(x^\nu), \quad \forall \nu, \tag{4.21}$$

and therefore $\lim_{\nu\to\infty} P(x^\nu) = P(\bar{x})$, due to the continuity of the potential function $P(\cdot)$. Furthermore, since the essentially covering update rule is employed, by restricting to a subsequence, it can be assumed that there exists a tuple of player indices $(\alpha_1, \ldots, \alpha_T)$ such that

$$i_{\nu_\ell+t} = \alpha_t, \quad \forall t = 1, \ldots, T, \ \forall \ell = 1, 2, \ldots$$

with $\alpha_t \in \{1, \ldots, n\}$ and $\{\alpha_1, \alpha_2, \ldots, \alpha_T\} = \{1, 2, \ldots, n\}$. Next, we will show that

$$\lim_{\ell\to\infty} \theta_{\alpha_1}(x^{\nu_\ell+1}) = \theta_{\alpha_1}(\bar{x}), \tag{4.22}$$

by deriving a contradiction. First, let us rewrite (4.20) for the subsequence of interest:

$$\theta_{\alpha_1}(x^{\nu_\ell+1}) \leq \widehat{\theta}_{\alpha_1}(x^{\nu_\ell+1;\alpha_1}; x^{\nu_\ell}) \leq \widehat{\theta}_{\alpha_1}(x^{\nu_\ell;\alpha_1}; x^{\nu_\ell}) = \theta_{\alpha_1}(x^{\nu_\ell}).$$

Thus, $\limsup_{\ell\to\infty} \theta_{\alpha_1}(x^{\nu_\ell+1}) \leq \theta_{\alpha_1}(\bar{x})$. Combining this fact with the contrary of (4.22) implies that, by restricting to a subsequence, we have

$$\theta_{\alpha_1}(x^{\nu_\ell+1}) \leq \theta_{\alpha_1}(\bar{x}) - \beta, \tag{4.23}$$

for some scalar $\beta > 0$. Therefore,

$$P(x^{\nu_\ell+1}) \leq P(\bar{x}) - \xi(\theta_{\alpha_1}(\bar{x}) - \theta_{\alpha_1}(x^{\nu_\ell+1})). \tag{4.24}$$

Clearly, $\liminf_{\ell\to\infty} \xi(\theta_{\alpha_1}(\bar{x}) - \theta_{\alpha_1}(x^{\nu_\ell+1})) > 0$ due to (4.23). Therefore, letting $\ell \to \infty$ in (4.24) yields $P(\bar{x}) < P(\bar{x})$, which is a contradiction. Therefore (4.22) must hold true. Next, we show that

$$\lim_{\ell\to\infty} x^{\nu_\ell+1} = \bar{x}. \tag{4.25}$$

Suppose the contrary is true. By restricting to a subsequence, there exists $\bar{\gamma} > 0$ such that $\|x^{\nu_\ell+1;\alpha_1} - x^{\nu_\ell;\alpha_1}\| \triangleq \gamma_{\nu_\ell} \geq \bar{\gamma}$, $\forall \ell$. Define $S^{\nu_\ell} \triangleq \dfrac{x^{\nu_\ell+1;\alpha_1} - x^{\nu_\ell;\alpha_1}}{\gamma_{\nu_\ell}}$. Using the properties of the approximation function, one can write

$$\theta_{\alpha_1}(x^{\nu_\ell+1}) \leq \widehat{\theta}_{\alpha_1}(x^{\nu_\ell+1;\alpha_1}; x^{\nu_\ell}) = \widehat{\theta}_{\alpha_1}(x^{\nu_\ell;\alpha_1} + \gamma_{\nu_\ell} S^{\nu_\ell}; x^{\nu_\ell}) \tag{4.26}$$

$$\leq \widehat{\theta}_{\alpha_1}(x^{\nu_\ell;\alpha_1} + \varepsilon\bar{\gamma} S^{\nu_\ell}; x^{\nu_\ell}), \quad \forall \varepsilon \in [0, 1] \tag{4.27}$$

$$\leq \widehat{\theta}_{\alpha_1}(x^{\nu_\ell;\alpha_1}; x^{\nu_\ell}) = \theta_{\alpha_1}(x^{\nu_\ell}), \tag{4.28}$$

where the equations (4.26), (4.27), and (4.28) are due to the update rule of the algorithm and the convexity of the function $\widehat{\theta}_{\alpha_1}(\bullet; x^\nu)$. Since S^{ν_ℓ} lies in a compact ball, it has a limit point $\bar{S}$. By restricting to a subsequence, letting $\ell \to \infty$, and using (4.22), we can rewrite the above inequality as

$$\theta_{\alpha_1}(\bar{x}) \leq \widehat{\theta}_{\alpha_1}(\bar{x}^{\alpha_1} + \varepsilon\bar{\gamma}\bar{S}; \bar{x}) \leq \theta_{\alpha_1}(\bar{x}), \quad \forall \varepsilon \in [0, 1],$$

which contradicts the strict convexity of $\widehat{\theta}(\bullet;\bar{x})$. Therefore, (4.25) holds. Furthermore, $\lim_{\ell\to\infty} \widehat{x}^{\nu_\ell;\alpha_1} = \lim_{\ell\to\infty} \frac{x^{\nu_\ell+1;\alpha_1} - (1-\tau)x^{\nu_\ell;\alpha_1}}{\tau} = \bar{x}^{\alpha_1}$. On the other hand, the update rule of the algorithm simply implies that

$$\widehat{\theta}_{\alpha_1}(\widehat{x}^{\nu_\ell;\alpha_1}; x^{\nu_\ell}) \le \widehat{\theta}(x^{\alpha_1}; x^{\nu_\ell}), \qquad \forall x^{\alpha_1} \in \mathcal{X}^{\alpha_1}.$$

Letting $\ell \to \infty$ leads to $\widehat{\theta}_{\alpha_1}(\bar{x}^{\alpha_1};\bar{x}) \le \widehat{\theta}(x^{\alpha_1};\bar{x})$, $\forall x^{\alpha_1} \in \mathcal{X}^{\alpha_1}$. Therefore, the first-order optimality condition implies that $\widehat{\theta}'_{\alpha_1}(\bullet;\bar{x})(\bar{x}^{\alpha_1}; x^{\alpha_1} - \bar{x}^{\alpha_1}) \ge 0$, $\forall x^{\alpha_1} \in \mathcal{X}^{\alpha_1}$. Using the gradient consistency property of the approximation function, we obtain

$$\theta'_{\alpha_1}(\bar{x};d) \ge 0, \qquad \forall d = (0,\ldots,0,x^{\alpha_1} - \bar{x}^{\alpha_1},0,\ldots,0) \text{ with } x^{\alpha_1} \in \mathcal{X}^{\alpha_1}.$$

Repeating the above argument for the other players $\alpha_2,\ldots,\alpha_T$ will complete the proof for the essentially covering player selection rule.

Next, let us consider the randomized player selection rule. Let $\widetilde{x}^{\nu;i}$ denote the updated point at iteration ν if block i is chosen, i.e. $\widetilde{x}^{\nu;i} = (1-\tau)x^{\nu;i} + \tau\widehat{x}^{\nu;i}$ with $\widehat{x}^{\nu;i} = \operatorname{argmin}_{x^i\in\mathcal{X}^i} \widehat{\theta}_i(x^i;x^\nu)$. First of all, we will show that $\lim_{\nu\to\infty} \theta_i(x^\nu) - \theta_i(\widetilde{x}^{\nu;i}, x^{\nu;-i}) = 0$, almost surely, for all i. To this end, we will first prove, similar to (4.21), the decrease of the potential function at each iteration and therefore the objective value converges for any realization of the random choices. In other words, for any realization, we must have $\lim_{\nu\to\infty} P(x^\nu) - P(x^{\nu+1}) = 0$. Then similar to (4.20), we obtain

$$\theta_i(\widetilde{x}^{\nu;i}, x^{\nu;-i}) \le \widehat{\theta}_i(\widetilde{x}^{\nu;i}; x^\nu) \le \widehat{\theta}_i(x^{\nu;i}; x^\nu) = \theta_i(x^\nu), \ \forall i = 1,\ldots,n, \ \forall \nu. \tag{4.29}$$

Therefore, the existence of the generalized potential for the game yields

$$P(x^\nu) - P(\widetilde{x}^{\nu;i}, x^{\nu;-i}) \ge \xi\left(\theta_i(x^\nu) - \theta_i(\widetilde{x}^{\nu;i}, x^{\nu;-i})\right), \qquad \forall i = 1,\ldots,n, \ \forall \nu,$$

which, combined with the randomized choice of players, implies

$$\begin{aligned}
\mathbb{E}\left[P(x^\nu) - P(x^{\nu+1}) \mid x^\nu\right] &\ge \sum_{i=1}^n p_i \xi\left(\theta_i(x^\nu) - \theta_i(\widetilde{x}^{\nu;i}, x^{\nu;-i})\right) \\
&\ge p_{\min} \sum_{i=1}^n \xi\left(\theta_i(x^\nu) - \theta_i(\widetilde{x}^{\nu;i}, x^{\nu;-i})\right),
\end{aligned}$$

where $p_{\min} \triangleq \min_i p_i$. Rearranging terms, we can write

$$\mathbb{E}\left[P(x^{\nu+1}) \mid x^\nu\right] \le P(x^\nu) - p_{\min} \sum_{i=1}^n \xi\left(\theta_i(x^\nu) - \theta_i(\widetilde{x}^{\nu;i}, x^{\nu;-i})\right).$$

Evidently the random process $\{P(x^\nu)\}_{\nu=1}^\infty$ is a supermartingale and it follows from the supermartingale convergence theorem [53, Proposition 4.2] that

$$p_{\min} \sum_{\nu=1}^\infty \sum_{i=1}^n \xi\left(\theta_i(x^\nu) - \theta_i(\widetilde{x}^{\nu;i}, x^{\nu;-i})\right) < \infty,$$

with probability one, which in turn implies that

$$\lim_{\nu\to\infty} \xi\left(\theta_i(x^\nu) - \theta_i(\widetilde{x}^{\nu;i}, x^{\nu;-i})\right) = 0, \qquad \text{almost surely, } \forall i.$$

Since ξ is a forcing function, we must have

$$\lim_{\nu\to\infty} \theta_i(x^\nu) - \theta_i(\widetilde{x}^{\nu;i}, x^{\nu;-i}) = 0, \qquad \text{almost surely, } \forall i.$$

Therefore, (4.29) implies

$$\lim_{\nu\to\infty} \widehat{\theta}_i(\widetilde{x}^{\nu;i}; x^\nu) - \widehat{\theta}_i(x^{\nu;i}; x^\nu) = 0, \qquad \text{almost surely, } \forall i. \tag{4.30}$$

Furthermore, the convexity of the function $\widehat{\theta}_i(\bullet, x^\nu)$ leads to

$$\widehat{\theta}_i(\widetilde{x}^{\nu;i}; x^\nu) \le \tau\widehat{\theta}_i(\widehat{x}^{\nu;i}; x^\nu) + (1-\tau)\widehat{\theta}_i(x^{\nu;i}; x^\nu), \qquad \text{almost surely, } \forall i.$$

Thus using (4.30), we have

$$\limsup_{\nu\to\infty} \left(\widehat{\theta}_i(x^{\nu;i}; x^\nu) - \widehat{\theta}_i(\widehat{x}^{\nu;i}; x^\nu)\right) \le 0, \quad \text{almost surely, } \forall i,$$

which combined with the definition of $\widehat{x}^{\nu;i}$ implies

$$\lim_{\nu\to\infty} \widehat{\theta}_i(\widehat{x}^{\nu;i}; x^\nu) - \widehat{\theta}_i(x^{\nu;i}; x^\nu) = 0, \quad \text{almost surely, } \forall i.$$

Now consider a subsequence $\{x^{\nu_\ell}\}_{\ell=1}^\infty$ converging to a limit point $\bar{x}$. The above equation and the definition of $\widehat{x}^{\nu;i}$ lead to

$$\widehat{\theta}_i(\bar{x}^i; \bar{x}) \le \widehat{\theta}_i(x^i; \bar{x}), \qquad \forall x^i \in \mathcal{X}^i, \ \forall i,$$

almost surely. The proof can now be completed as in the previous theorems. □

Part II

Big data over cyber networks

5 Big data analytics systems

Ganesh Ananthanarayanan and Ishai Menache

Performing timely analysis on huge datasets is the central promise of big data analytics. To cope with the high volumes of data to be analyzed, computation frameworks have resorted to "scaling out" – parallelization of analytics that allows for seamless execution across large clusters. These frameworks automatically compose analytics jobs into a DAG of small tasks, and then aggregate the intermediate results from the tasks to obtain the final result. Their ability to do so relies on an efficient scheduler and a reliable storage layer that distributes the datasets on different machines.

In this chapter, we survey the above two aspects, *scheduling* and *storage*, which are the foundations of modern big data analytics systems. We describe their key principles, and how these principles are realized in widely deployed systems.

5.1 Introduction

Analyzing large volumes of data has become the major source for innovation behind large Internet services as well as scientific applications. Examples of such "big data analytics" occur in personalized recommendation systems, online social networks, genomic analyses, and legal investigations for fraud detection. A key property of the algorithms employed for such analyses is that they provide better results with increasing amount of data processed. In fact, in certain domains (like search) there is a trend towards using relatively simpler algorithms and instead relying on more data to produce better results.

While the amount of data to be analyzed increases on the one hand, the acceptable time to produce results is shrinking on the other hand. Timely analyses have significant ramifications for revenue as well as productivity. Low latency results in online services leads to improved user satisfaction and revenue. Ability to crunch large datasets in short periods results in faster iterations and progress in scientific theories.

To cope with the dichotomy of ever-growing datasets and shrinking times to analyze them, analytics clusters have resorted to *scaling out*. Data are spread across many different machines, and the computations on them are executed in *parallel*. Such scaling out is crucial for fast analytics and allows coping with the trend of datasets growing faster than Moore's laws increase in processor speeds.

Big Data over Networks, ed. Shuguang Cui, Alfred O. Hero III, Zhi-Quan Luo, and José M. F. Moura. Published by Cambridge University Press.

Table 5.1 Definitions of terms used in data analytics frameworks

Term	Description
Task	Atomic unit of computation with a fixed input
Phase	A collection of tasks that can run in parallel, e.g. map, aggregate
Workflow	A directed acyclic graph denoting how data flows between phases
Job	An execution of the workflow
Block	Atomic unit of storage by the distributed file system
File	Collection of blocks
Slot	Computational resources allotted to a task on a machine

Many *data analytics frameworks* have been built for large scale-out parallel executions. Some of the widely used frameworks are MapReduce [1], Dryad [2] and Apache Yarn [3]. The frameworks share important commonalities, and we will use the following common terminology throughout this chapter. Frameworks compose a computation, referred to as a *job*, into a DAG of *phases*, where each phase consists of many fine grained *tasks*. Tasks of a phase have no dependencies among them and can execute in parallel. The job's input (*file*) is divided into many *blocks* and stored in the cluster using a distributed file system. The input of each task consists of one or more blocks of a file. A centralized scheduler assigns a *compute slot* to every task.[1] Tasks in the input phase produce *intermediate* outputs that are passed to other tasks downstream in the DAG. Table 5.1 summarizes the terminology.

As a concrete example, consider an analysis of web access logs (of many TBs). Each row in the log consists of a URL being accessed (e.g. www.cnn.com) and details about its access (e.g. time of access, user accessing it); the aim of the analysis is to count the number of accesses of each web URL to understand popularities. The distributed file systems splits the access logs into small 256 MB blocks and stores them over the different machines. A job to analyze them would consist of two phases. Every task in the first phase would read a block of data and generate a $\langle url, 1\rangle$ tuple for each row in the log (the "1" indicates a single access). Each task in the second phase is responsible for some fraction of the URLs and collects the corresponding tuples from the first phase's outputs. It then sums up the accesses per URL and produces the output.

The key benefit of the analytics frameworks is their ability to scale out to thousands of commodity machines for computation as well as storage. They do so by automatically and logically dividing data as well as analyses on them into fine-grained units of blocks and tasks, respectively. As commodity machines are susceptible to failures and unavailabilities, the frameworks are resilient to these failures to ensure data availability as well as successful execution of the tasks. In this chapter, we survey the important aspects of systems running big data analytics frameworks. In particular, we focus on two fundamental requirements: (i) scheduling jobs efficiently, and (ii) managing data storage across distributed machines. Scheduling principles and solutions are described

[1] Slot is a virtual token, akin to a quota, for sharing cluster resources among multiple jobs. One task can run per slot at a time.

in Section 5.2, and storage architectures in Section 5.3. Additional related and upcoming topics are briefly outlined in Section 5.4.

5.2 Scheduling

Analytics frameworks typically use a centralized scheduler where all the tasks in the cluster are queued. The scheduler manages the machines in the cluster, where each machine has a *worker* process. The worker processes send periodic *heartbeats* to the scheduler informing it of the status of running tasks as well as the resource usages and general health of the machines. The scheduler aggregates the information from the heartbeats and makes scheduling decisions to allocate tasks to machines. When a machine does not send heartbeats for a certain period of time, the scheduler assumes the machine to be lost and does not schedule any further tasks to it. It also reschedules the unfinished tasks on the machine elsewhere.

Recent designs have advocated (and deployed) a hierarchical two-level scheduling model where the individual jobs talk to a central "resource manager" to register their demands and queue their tasks (e.g. Mesos [4], Apache Yarn [3]). The resource manager, then, allocates compute slots to the different tasks. Nonetheless, the abstraction of a central scheduler allocating slots to tasks holds.

Our focus in this section is on describing the principles and logic behind the central scheduler decisions. Scheduling models have been considered for numerous applications in both academia in industry for at least half a century. The emergence of big data jobs, running on large-scale clusters, has brought novel challenges that cannot be readily solved by traditional scheduling mechanisms. We therefore highlight below the distinctive aspects of scheduling big data jobs. In Section 5.2.1 we outline how fairness policies are adapted to deal with several complexities such as multiple resource types, intra-job dependencies and machine fragmentation. In Section 5.2.2 we discuss how scheduling solutions accommodate *placement constraints*, such as the necessity to place job tasks alongside their input data. In Section 5.2.3, we describe scheduling solutions that incorporate different objective criteria, such as mean completion time and finishing jobs by pre-specified deadlines. Finally, in Section 5.2.4 we present the problem of *stragglers* (tasks running on slow machines), and the main approaches for mitigating their effect.

5.2.1 Fairness

Allocating resources to multiple jobs has been a fundamental problem in shared computer systems. While several design criteria have been considered for scheduling mechanisms, *fairness* is perhaps the most prominent one. There are different definitions for what constitutes a fair allocation; we do not attempt to cover all here, but rather focus on the ones that are used in modern big data clusters. *Max-min fairness* is one such policy – it simply maximizes the minimum allocation across users. If each user has enough "demand" for the resource, the policy boils down to allocating the resource in equal shares among users. A natural generalization of this policy is *weighted* max-min fairness, in which each user receives resources in proportion to its pre-specified weight. Several algorithms have been proposed to implement weighted max-min fairness in different engineering contexts (e.g. deficit round robin [5] and weighted fair queuing [6]). However, the original algorithms do not cover important considerations associated with big data jobs, which require adjustments of both the fairness policies and the algorithms to sustain them. We list below the main considerations and how they have been addressed.

Multiple resources

Big data jobs such as map-reduce jobs utilize different resources, such as CPU, memory and I/O resources. Consider the following numeric example (taken from [7]). Suppose the system consists of 9 CPUs, 18 GB RAM. Two users ought to share the system – user A runs tasks with demand vector (1 CPU, 4 GB), and user B runs tasks with demand vector (3 CPUs, 1 GB) each. Note that each task of user A consumes 1/9 of the total CPU and 2/9 of the total memory. Each task from user B consumes 1/3 of the total CPUs and 1/18 of the total memory. In order to divide the system resources, [7] defines the notion of *dominant resource*, which is the resource that is utilized the most (percentage-wise) by the tenant. In our example, user A's dominant resource is memory, while user B's dominant resource is CPU. The authors in [7] propose a new policy, dominant resource fairness (DRF), which extends max-min fairness to the multiple resource case. DRF simply applies max-min fairness across users' dominant shares. In the example, the DRF allocation would be three tasks for user A and two tasks for user B, which will equalize the dominant resource shares of A and B (2/3 of the RAM for A, and 2/3 of the CPU for B). The DRF solution has some appealing properties such as pareto efficiency, bottleneck fairness, sharing incentive, envy-free, and strategy proofness. Without going into details, the first two properties indicate that the DRF solution is efficient and fair, while the latter properties mean that users have incentives to participate in a system which divides resources according to DRF, and further would not game the system. Reference [7] also proposes an iterative greedy algorithm, which gears the allocation towards the DRF solution. The algorithm was implemented in the Mesos cluster resource manager, and leads to better fairness compared to schemes that divide single-dimension slots.

Intra-job dependencies

As mentioned before, big data jobs often consist of multiple phases with data dependencies among them. A fair but naive scheduler that does not take the inner structure of jobs into account might actually result in an unfair allocation of resources. For simplicity, we focus here on the issues arising in the MapReduce framework, although the problems and solutions can apply to other analytics frameworks. Hadoop launches reduce tasks for a job as soon as some mappers finish, so that reduces can start copying the maps outputs while the remaining maps are still running. Assuming that the cluster is initially not congested, a large job with many mappers would keep getting reduce slots from a naive fair scheduler. With a lack of preemption mechanism, the problem here is that these reduce slots would not be released until all mappers finish, because only then can the reduce function be carried out. This means that small jobs that arrive to the system during the map phase of the big job might be starved. The following solution to this problem is proposed in [8]: split reduce tasks into two logically distinct types of tasks – copy tasks and compute tasks, and have separate admission control mechanisms for both types. In particular, [8] limits the total number of slots for reduce-compute on each machine, and further sets a per-job upper bound on the reduce-copy slots on each machine. The combination of these mechanisms constrains the amount of simultaneous resources given to jobs without needing to use more aggressive preemption mechanisms.

Fragmentation and over-allocation

As described above, the basic fair schedulers divide resources to slots, where each slot represents some amount of memory and CPU; slots are allocated to the job that is furthest from its (weighted) fair-share. Such bundling of resources is obviously inefficient, because it might lead either to wasting resources in some dimensions or to over-allocation if some dimensions are overlooked. DRF partially resolves these issues by considering the allocation problem as a multi-dimensional one. However, DRF might leave some resources idle, since it attempts only to maximize the dominant resource share across all jobs. Further, DRF does not explicitly take into account the available capacity in each machine, but rather considers the total available capacity for each resource.

To overcome fragmentation, [9] proposes a multi-resource scheduler that *packs* tasks to machines based on their vector of requirements. The underlying scheduling mechanism is based on an online (multi-dimensional) *bin-packing* heuristic, which attempts to utilize the available resources in each machine as much as possible. The tasks are the "balls" and the "bins" are machines over time. Intuitively, because the bin-packing heuristic attempts to pack balls to bins while using a minimal number of bins, resources are well utilized and jobs can finish quickly.

We illustrate the advantages of a packing-based scheduler via the following example [9]. Consider a cluster with 18 cores and 36 GB of memory. Three jobs A, B, and C have one phase each consisting of 18, 6, and 6 tasks, respectively. Each task in job A requires 1 core and 2 GB of memory, while job B's and C's tasks require 3 cores and 1 GB of memory. Assume all tasks run for t time units. DRF will schedule six tasks of job A and two tasks each of jobs B and C, at a time, giving each job a dominant resource share of $1/3$. This leads to an average job duration of $3t$. Such an allocation, however, leaves 20 GB of memory in the cluster idle. A scheduler that packs the tasks schedules all tasks of job A initially because they use up all the cores and memory in the cluster, followed by tasks of job B and then job C. This leads to job durations of t, $2t$ and $3t$ for the jobs for an average of $2t$ and a 33% improvement over DRF.

The advantages of packing tasks carry over to jobs with multiple *phases* (e.g. a map-reduce computation consists of a map phase and a reduce phase). Extending the above example, let all three jobs have two phases separated by a barrier, i.e., tasks of their second phase begin only after all tasks of their first phase finish. For simplicity, assume tasks of the second phase require 1Gbps of network resource but no cores or memory. Suppose that the cluster has total network bandwidth of 3Gbps; tasks in the

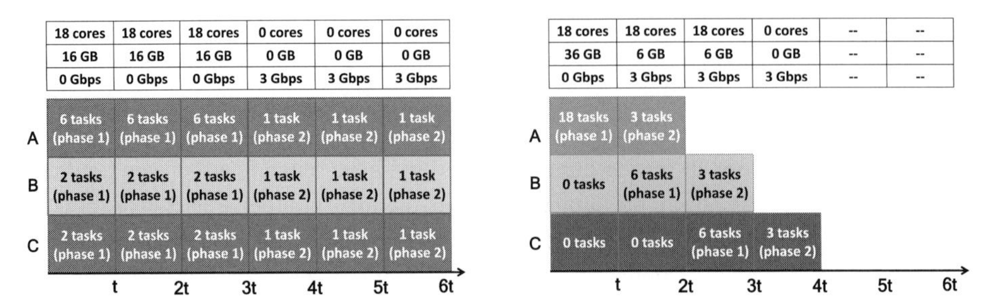

Figure 5.1 Impact of packing on DAGs of tasks compared to fairness based allocation (like DRF). Jobs A and B have two phases with a strict barrier between them. The tables on top show utilization of the resources. Packing results in better utilization of resources using complementarity of task requirements.

first phase have no network usage but need cores and memory as listed above.[2] Let the jobs have 18, six, and six tasks in their first phase, as before, and three tasks each in their second phases (each task, again, of t time units). Figure 5.1 compares DRF allocation with a packing based scheduler. Again, by scheduling tasks of job A initially and not wasting resources, the packing-based scheduler is able to exploit the complementarity in resource requirements between tasks of the first and second phases. This results in all three jobs finishing faster than under a DRF-based allocation. The average job duration is $3t$ versus $6t$, or a speed up of 50% compared to DRF. The makespan improves by 33% from $6t$ to $4t$.

Reference [9] incorporates the packing-based scheduler in a system called Tetris. Tetris achieves substantial *makespan* reductions (over 30%) over existing schedulers. When resources on a machine become available, Tetris chooses a task that can fit on the machine, and whose score is maximal; the score is an inner product of the residual capacity of the machine and the pick resource requirement of the task. This choice maximizes the utilization and diminishes fragmentation. Tetris takes fairness into account by considering only a fraction $(1 - f)$ of the tasks, which are ordered according to a fairness criterion (e.g. max-min), and then picking the highest-score job among this subset.

Placement constraints

Finally, scheduling algorithms have to take into account placement constraints, such as scheduling tasks where their input data are. Since this topic is quite broad, we address it in a separate subsection below.

5.2.2 Placement constraints

Our focus so far has been on scheduling under fairness considerations. In its simplest form, sustaining fairness means that each user gets "enough" cluster resources according

[2] Note that this example is typical of a map-reduce computation; map tasks are indeed CPU and memory intensive while reduce tasks are network-intensive.

to pre-specified fairness criterion (e.g. max-min fairness, or weighted fair sharing). However, in the big data analytics context, it greatly matters exactly *which* resources are given to jobs; jobs might be constrained on the set of resources (e.g. machines) they can run on. Placement constraints can be roughly classified into three classes [10]: (i) hard, (ii) soft, and (iii) combinatorial. Examples of hard constraints include jobs that must run on machine with public IP address, particular kernel version, specific SKU or hardware (e.g. GPU machines or machines with SSDs). The most prominent soft constraint is data locality; while job tasks could be scheduled on machines that do not necessarily hold the required data for the task, a "cost" is incurred, in the form of additional latency and excess network bandwidth usage. Combinatorial constraints specify rules for a collection of machines that are used for the job, such as fault tolerance constraints. We survey below the related work for each class.

Hard constraints

Reference [11] provides a comprehensive study of Google workloads, and in particular the impact of hard placement constraints on task scheduling delays. It turns out that these constraints increase task scheduling delays by a factor of 2–6. Accordingly, that paper develops a methodology that enables predicting the impact of hard constraints on task scheduling delays. Specifically, a metric termed utilization multiplier (UM) is introduced; this metric measures for each relevant resource the utilization ratio between tasks with constraints and the average utilization of the resource. Accordingly, the higher UM is, the higher is the expected scheduling delay. Finally, [11] describes how to incorporate placement constraints into existing performance benchmarks, so that they are properly taken into account when evaluating new cluster architectures or scheduling mechanisms. Reference [10] proposes scheduling algorithms for dealing with hard placement constraints, while sustaining some notion of fairness between users (jobs). In particular, this paper defines a new notion of fairness, termed constrained max-min fairness (CMMF), which extends max-min fairness while taking into account that some jobs cannot run on some machines. A CMMF allocation is a machine assignment in which it is not possible to increase the minimum allocation within any subset of users by reshuffling the machines given to these users. A CMMF allocation has appealing properties such as incentive compatibility with regard to a user reporting his placement constraints. Calculating a CMMF solution, however, might be too costly in practice, as it involves solving a sequence of LPs. Consequently, the authors propose a simple greedy online scheduler called Choosy. Whenever a resource becomes available, it is assigned to the user with the lowest current allocation, which is allowed to get that resource according to its placement constraints. Perhaps surprisingly, Choosy achieves allocations which are very similar to those of a CMMF solution, and as a result, the latencies of jobs are on average at most 2% higher than the CMMF solution.

Soft constraints

We focus here on a particular soft constraint, termed *data locality*. Preserving data locality means placing the computation of the job near its input data. Locality is important because network bandwidth might be a bottleneck, hence transferring input data over the network might lead to substantial task delays [12]. Simply scheduling jobs near their input data would in general violate the fairness requirements across multiple users. Hence, it is crucial to design schedulers which are both *fair* and locality-aware. Reference [12] uses a simple algorithm to address the tension between locality and fairness: when a job that should be scheduled next according to the underlying fairness criterion cannot launch a local task, it waits for a small amount of time, while letting other jobs use the available resource. If after a pre-specified timeout the job still cannot execute tasks locally, it is assigned available resources which are not necessarily local to the task. Accordingly, this basic algorithm is termed *delay scheduling*.

The authors in [12] extend the basic algorithm described above to address several practical considerations, including rack locality, hotspots, long tasks, and more; we omit the details here for brevity. The resulting scheduling system is termed the Hadoop fair scheduler (HFS). One important extension in HFS is a two-layer hierarchical architecture. The higher layer divides resources between organizations in a fair manner, while the lower layer divides resources between the organization's jobs according to a local policy (e.g. FIFO or fair share). Evaluation on Facebook workloads shows that delay scheduling achieves nearly 100% locality, leading to substantial improvements in job response times (especially for small jobs) and throughput (especially for IO-heavy workload).

The underlying engineering principle which makes delay scheduling appealing in practice, is that slots running Hadoop free up in a *predictable rate*. The time-threshold used for delay scheduling is set accordingly – i.e. taking into account typical execution times of tasks, it is expected that some local slots would free up by the time-threshold. We now describe an alternative scheduler which does not rely on such predictions. Reference [13] maps the problem of scheduling with locality and fairness considerations into a *min-flow problem* on a graph. The underlying idea here is that every scheduling decision can be assigned a cost. For example, there is a data transfer cost when task is scheduled far from its input, there is a cost for killing a task that takes too long, etc. These costs are embedded in a graph, where the nodes are the tasks and physical resources such as racks and servers; the graph-based algorithm operates by assigning a unit flow to each task node in the system (either running or waiting for execution). There is a single sink to which all flows are drained. The flows could traverse through either a path consisting of resources (meaning that the respective task is scheduled on the respective resources), or through their job's "unscheduled" node, in which case the task is (still) not executed. Fair sharing constraints are added by setting lower and upper bounds on the edges from the unscheduled node to the sink. The cost of killing a job is modeled by gradually increasing the costs of all edges related to the task, but the edge that "connects" the task node to its executing resource. The authors manage to scale the solution to large instances by reducing the effective number of nodes through cost aggregation techniques. In [13] this graph-based algorithm is implemented in a system called Quincy. Quincy is evaluated against a queue-based algorithm, and shows sizable gains in both amount of data transferred and throughput (up to 40%).

It is of interest to qualitatively compare the delay-scheduling approach to the graph min-flow approach. The graph-based algorithm is more sophisticated, and can incorporate the *global* state of the cluster and multiple cost considerations (e.g. further delaying a waiting task vs. killing a task that has been running for a long time). While fast min-cost procedures can be used for the solution, scale issues might arise when the cluster and number of jobs is very large. Another potential weakness of the graph-based algorithm is that it is greedy by nature, and optimizes based on the current snapshot of the cluster. The delay-scheduling algorithm is arguably simpler and easier to implement in production. The framework does not "code" all cost trade-offs, which could be problematic, especially in more complicated job models which do not obey systematic execution patterns. Another advantage of delay scheduling is that it uses knowledge about task durations which allows it to take into account the future evolvement of the cluster state (rather than act solely as a function of the present state).

Combinatorial constraints

Combinatorial constraints may arise, e.g. due to security and fault tolerance considerations. An example of such constraints could be "not more than x% of each job's tasks should be allocated in the same fault domain". Reference [14] considers the problem of assigning physical machines to applications while taking into account fault tolerance performance, and tackles the difficult combinatorial problem by using a smooth convex cost function, which serves as a proxy for fault tolerance. In particular, this function incentivizes stripping machines belonging to the same application across fault domains. Reference [14] also deals with the fundamental cost trade-off between fault tolerance and bandwidth consumption – a solution that spreads machines across fault domains is usually bad in terms of the bandwidth consumption, as communication might heavily use the network core. Accordingly, [14] introduces an additive penalty to the above cost function, which carefully balances the optimization of the two metrics.

5.2.3 Additional system-wide objectives

Sustaining fairness has been the primary objective of cluster schedulers. However, there are additional objectives that may be as important. For example, Tetris [9] (described above) optimizes the *makespan* by packing tasks to machines. Tetris also has a knob for reducing *average job completion time*. It sorts jobs by estimated remaining work and chooses tasks whose jobs have the least remaining work; this is commensurate with the shortest remaining time (SRTF) scheduling discipline, which is known to be effective in minimizing job completion times. Tetris combines the packing score (described in Section 5.2.1) and the remaining-work score into a single score per-task, with a tuneable parameter that controls how much weight is given to each component.

Another potential system-wide objective is meeting job *deadlines*. Big data jobs are often used for business critical decisions, hence have strict deadlines associated with them. For example, outputs of some jobs are used by business analysts; delaying job completion would significantly lower their productivity. In other cases, a job computes the charges to customers in cloud computing settings, and delays in sending the bill might have serious business consequences. Recently, several works have addressed the resource allocation problem where meeting job deadlines is the primary objective. Jockey [15] dynamically predicts the remaining run time at different resource allocations, and chooses an allocation which is expected to meet the job's deadline without wasting unnecessary resources. Jockey treats each job in isolation with the assumption that allocating resources "economically" would lead to solutions that satisfy multiple jobs deadlines. [16] proposes an interface in which users submit resource requirements along with deadlines, and the scheduler objective is to meet the jobs deadlines. The resource requirements are *malleable* in the sense that the scheduler can often pick alternative allocations that would lead to job completion. As such, the scheduler can choose the specific allocations which will allow multiple jobs to finish without violating the cluster's capacity. In a similar context, references [17–19] develop scheduling algorithms for the model where jobs have different values, and the scheduler objective is to maximize the value of jobs that complete before the deadline. References [17–19] provide constant-factor approximation algorithms for the problem. As may be expected, the approximation quality improves when deadlines are less stringent with respect the available cluster capacity.

5.2.4 Stragglers

Classic scheduling models usually assume that a task has a predictable/deterministic execution time, regardless of the physical server chosen to run it; e.g. a task would complete in 30 s if run on any server at any point in time. However, in the context of current big data analytics framworks, there are plenty of factors that could vary the actual running time of the task. The original map-reduce paper [1] introduces the notion of a *straggler* – "a machine that takes an unusually long time to complete one of the last few map or reduce tasks in the computation". A couple of reasons for stragglers are listed in [1], such as machine with bad disk, other tasks running on the same machine, bugs in machine initialization, etc. The paper proposes a general mechanism to deal with stragglers. When a MapReduce job is close to completion, *backup* executions of the remaining in-progress tasks are scheduled. The task completes whenever either the primary or backup execution completes. This assignment of backup task(s) is often termed *speculative* execution (or speculative task). The simple mechanism described above can lead to substantial improvement in job response times (up to 44% according to [1]). While the principle of speculative execution is natural for dealing with stragglers, it is not a-priori clear under which conditions duplication should take place. Obviously, this knob can lead to negative congestion effects if used injudiciously. We survey below recent works that have addressed this problem for different scenarios.

The Hadoop scheduler uses a simple threshold rule for speculative execution. It calculates the average progress score of each category of tasks (maps and reducers) and defines a threshold accordingly. When the task's progress score is less than the average minus some parameter (say 0.2), it is considered a straggler. The shortcomings of this mechanism are emphasized in [20]. It does not take into account machine heterogeneity, it ignores the hardware overhead in executing stragglers, it does not take into account the *rate* at which the task progresses; further, it assumes that tasks finish in waves and that tasks require the same amount of work, assumptions that are imprecise even in homogenous environment. Accordingly, [20] designs a new algorithm: longest approximate time to end (LATE). As can be inferred from its name, the distinctive feature of LATE compared to previous approaches is that it selects tasks for speculative execution based on the estimated time left, rather than based solely on the progress itself. In particular, it uses (1 − *ProgressScore*)/*ProgressRate* to rank each task, where *ProgressScore* is the progress estimator given by Hadoop. The authors point out that this formula has some drawbacks, yet it is simple and works well in most cases. In addition to the above, LATE also sets a hard constraints on the number of speculative tasks allowed in the system. When a slot becomes available and the number of speculative tasks running is below the threshold, LATE launches a copy of the highest-ranked task whose progress rate is low enough (below some pre-specified threshold). Finally, LATE does not launch copies on machines that are statistically slow. Implementation of LATE on an EC2 virtual cluster with 200 machines shows $2\times$ improvement in response time compared to Hadoop scheduler.

Mantri [21] proposes a refined method for limiting the resource overhead of speculative execution. A speculative task is scheduled if the probability of it finishing before the original task is high. Mantri provides more knobs with regard to the speculative execution itself – it supports duplication (including multiple duplicates), kill-restart and pruning based on re-estimations of progress. In addition, Mantri extends the scope of speculative execution to preventing and mitigating stragglers and failures. It does so in different ways. First, it replicates critical task output to prevent situations where tasks wait for lost data; as before the decision to do so is based on estimating the probability of the bad event and evaluating the trade-off between re-computation and excess resource cost. Second, Mantri executes tasks in descending order of their input size. The idea here is that tasks with heavier input take longer to execute, and thus should be prioritized since what matters is the makespan of the tasks. Finally, Mantri chooses the placement of reduce tasks in a *network-aware* manner, taking into account the data that has to be read across the network and the available bandwidth. Mantri has been deployed on Bing's cluster with thousands of machines, and exhibited median job speedup of 32%.

The aforementioned techniques for handling stragglers generally operate at the task level, e.g. rank tasks based on progress or completion time, limit the number speculative tasks, etc. Instead, [22] proposes to apply different mitigation logic at the *job* level. The main observation here is that for *small* jobs, i.e. jobs that consist of few tasks, a system can tolerate a more aggressive issuing of duplicates. In particular, [22] uses full *cloning* of tasks belonging to small jobs, avoiding waiting and speculating, hence improving the chances that the tasks output is ready on time. Analyzing production traces, [22] shows that the smallest 90% of jobs consume only 6% of the total resources. Hence, cloning can be done to a substantial chunk of the jobs with a rather small resource overhead. On the algorithmic front, cloning introduces the following challenge: efficient cloning means that the system uses the (intermediate) output data of the upstream clone that finishes first; however, this creates contention for the IO bandwidth at that upstream clone, since multiple downstream clones require its data. This problem is solved in [22] through a heuristic reminiscent to delay scheduling, hence named *delay assignment*. Every downstream clone waits for a small amount of time (a parameter ω of the heuristic) to see if it can get an exclusive copy of the intermediate data from a preassigned upstream clone. If the downstream clone does not get its exclusive copy by ω, it reads from the upstream clone that finishes first. Reference [22] implements the cloning framework in a system called Dolly. The paper reports substantial improvements in completion time of small jobs – 34%–46% compared to LATE and Mantri, using less than 5% extra resources.

5.3 Storage

Storage is a key component of big data analytics clusters. Clusters store petabytes of data, distributed over many machines. Key to analytics is *reliable* and *efficient* storage of the data. Data should be accessible even in the face of machine failures (which are common) and should be read/written efficiently (without significant overheads).

Distributed file systems present the abstraction of a single unified storage to applications by abstracting away as many of the details of the individual storage machines as possible; the machines store the data on their local disks. We describe how distributed file systems enable such storage while presenting the unified storage abstraction in Section 5.3.1. An important and upcoming class of storage solutions, driven by falling RAM prices is in-memory caching. Such caching offers much faster access to data compared to disks but requires special handling along certain aspects, which we cover in Section 5.3.2.

5.3.1 Distributed file system

Typically, a *file* in a distributed file system is divided into many smaller *blocks*, which are stored on different machines. Every file has a unique identifier, and every block within a file is also referenced uniquely. The distributed file system is oblivious to the local storage mechanisms used by the disk subsystems of the machines. The machine could employ various disk architectures like just-bunch-of-disks, RAID or simple striping. It could also employ its own error-recovery and caching mechanisms.

Architecture

Distributed file systems have a centralized architecture: a single central master that maintains metadata about the blocks in the cluster. Figure 5.2 presents a simple architectural representation. Metadata information for every block contains the locations storing it, its last access time, size, the file it is part of, and so forth. Each of the machines have a file system worker that is responsible for managing the local data blocks. It interfaces with the local storage subsystem and allows the master to be oblivious. It also periodically informs the master about the available space on the machine, and other performance characteristics.

The widely used distributed file systems provide interfaces similar to the POSIX interface [23] for reading and writing data. File namespaces are hierarchical and can be identified via path names. The operators supported are *create*, *open*, *read*, *write*, *close* and *delete*.

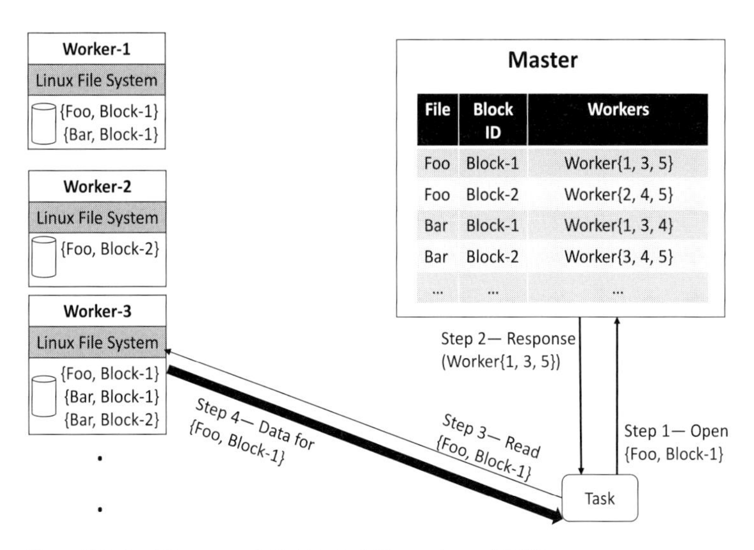

Figure 5.2 Architecture of distributed file systems (DFS).

Writing Writing data to the file system involves the following steps.

(1) Application calls *create* to the master with a file name. The master responds positively with a handle if the filename is not in use currently.
(2) Application uses the handle to *open* a file to the master. The master checks its metadata to pick the machine to write the first block of the file. It returns the details of the corresponding worker process.
(3) Application calls *write* to the worker to store its data until the size is equal to the upper limit for a block. Once the block is full, the worker informs the application.
(4) Application gets another worker to write its next block of data. A file is written only one block at a time.
(5) When all the blocks are written, the application calls *close* on the file to the master.

While the master controls where the data is written and automatically obtains metadata information, the actual writes themselves go directly to the worker. Such a design helps significantly with the scalability of the master. Imagine an alternate design where the application provides its data to the master and lets it send to the worker machine transparently. While such an approach would marginally simplify the application, it places a huge burden on the master.

Append-only

Modern distributed file systems do not support *updates* to the blocks, only *appends*. While applications can add blocks to an existing file, they cannot modify any of the existing blocks. Such an append-only decision is suited for these clusters where data is written once and read many times over, and rarely updated.

While the single master scales well since it only deals with requests as opposed to data, its scalability can be improved by federating its namespace. Recall that the files are organized as a hierarchical namespace. Therefore, the master can be scaled by partitioning the namespace appropriately. Thus each subspace in the hierarchy will be independently handled which reduces the load on any single machine. Partitioning happens based on hierarchical boundaries as well as popularity of data access.

Reading Reading data from the file system involves the following steps.

(1) Application calls *open* to the master with the desired file name. The master returns with the handle.
(2) Application calls *read* to the master and optionally provides block identifiers. The master responds with the set of worker machines on which the desired blocks of data are stored.
(3) Application contacts the worker nodes and requests data blocks.
(4) After reading all the blocks, application calls *close* on the file to the master.

As earlier, the master deals only with providing applications with the locations of the workers, and does not directly pass the data blocks. However, different from above, applications can read from multiple workers in parallel or in series. Not allowing applications to write in parallel simplifies the design of the master as it does not have to deal with load balancing the write speeds across different machines.

Authentication An important piece missing in the above workflows for reading and writing data is authentication. Files should be written to and read by only authenticated users. This is highly important in multi-tenant clusters. For this purpose, the master enforces explicit access controls and requires users to authenticate themselves. Further, when writing data, applications are provided an explicit token that helps authenticate themselves to the workers. The tokens are time-specific and cannot be reused.

Fault tolerance

Fault tolerance is an important concern for distributed file systems. The master is a single point of failure as the loss of its metadata will require hours to rebuild as each worker has to inform the new master of the blocks stored on its machine. To prevent such expensive rebuilding, file systems use two approaches: periodic checkpointing and hot standby. The master periodically checkpoints its metadata to persistent storage. In addition there is also a hot standby that can take over immediately when the primary master fails. The standby picks up from the persisted metadata state.

Data are replicated (typically, three times) to provide fault tolerance. Replication, however, is transparent to the application. The worker to which a block is written contacts the master to obtain another location where data are to be replicated. The second location, in turn, creates another replica of the data.

File systems also support reverting to previous versions via snapshots. Snapshots allow users to revert to previous versions. Traditionally, snapshotting is implemented by time-stamping and storing the old copy of a block whenever it changes. The append-only model greatly simplifies snapshotting by avoiding the need to store any old copies. Whenever blocks are appended to a file, a simple log that timestamps the action is sufficient to roll back to any desired time. Of course, when files are deleted, they still have to be persisted for rollback.

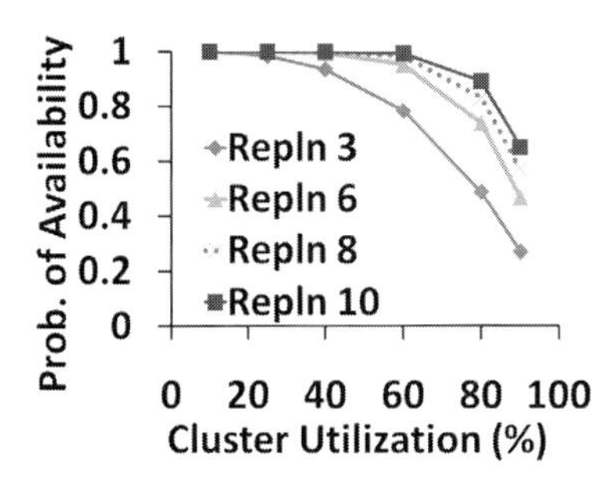

Figure 5.3 The probability of finding a replica on a free machine for different values of file replication factor and cluster utilization.

Variable replication

In time, replication is used for both performance as well as reliability. In these distributed clusters, reading from local storage is often faster than reading from remote machines. Thus, analytics frameworks schedule their tasks on the machines that contain data locally. Achieving locality for concurrently executing tasks, however, is dependent on the number of replicas as well as cluster utilization.

We present a simple analysis that demonstrates the intuition behind how increased replication reduces contention. With m machines in the cluster, k of which are available for a task to run, the probability of finding one of r replicas of a file on the available machines is $1 - (1 - \frac{k}{m})^r$. This probability increases with the replication factor r, and decreases with cluster utilization $(1 - \frac{k}{m})$.

Figure 5.3 plots the results of a numerical analysis to understand how this probability changes with replication factors and cluster utilizations. At a cluster utilization of 80%, for example, with the current replication factor ($r = 3$), we see that the probability of finding a replica among the available machines is less than half. Doubling the replication factor raises the probability to over 75%. Even at higher utilizations of 90%, a file with 10 replicas has a 60% chance of finding a replica on a free machine. By replicating files proportionally to their number of concurrent accesses, the chances of finding a replica on a free machine improves.

Therefore, some file systems perform automatic variable replication of files based on their popularity. Using historical access statistics – total number of accesses as well as number of concurrent accesses – systems create extra replicas of popular data blocks. Such variable replication ensure sufficient number of replicas that help in providing locality for future tasks concurrently acccesing the same data block.

5.3.2 In-memory storage

Hardware trends, driven by falling costs, indicate a steep increase in memory capacities of large clusters. This presents an opportunity to store the input data of the analytics jobs in memory and speed them up. As mentioned earlier, data-intensive jobs have a phase where they process the input data (e.g. *map* in MapReduce [1], *extract* in Dryad [2]). This phase simply reads the raw input and writes out parsed output to be

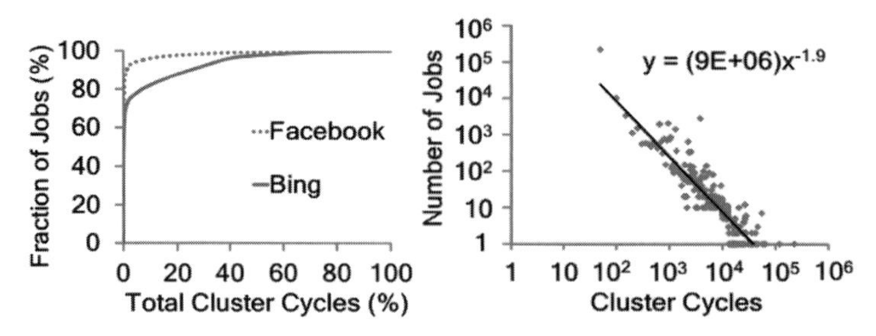

Figure 5.4 Power-law distribution of jobs in the resources consumed by them. Power-law exponents are 1.9 and 1.6 in the two traces, when fitted with least squares regression [22].

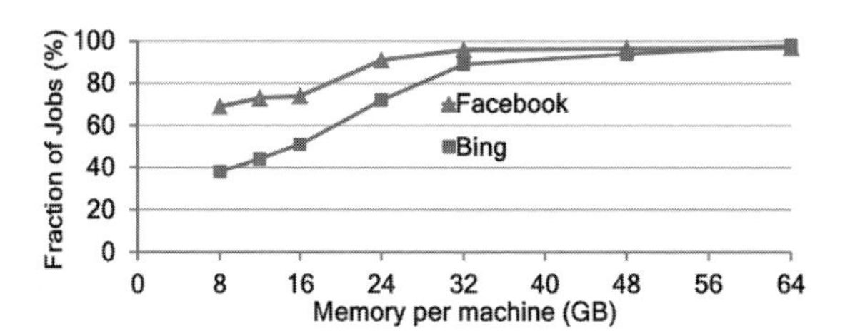

Figure 5.5 Fraction of active jobs whose data fit in the aggregate cluster memory, as the memory per machine varies [24].

consumed during further computations. Naturally, this phase is IO-intensive. Workloads from Facebook and Microsoft Bing datacenters, consisting of thousands of servers, show that this IO-intensive phase constitutes 79% of a job's duration and consumes 69% of its resources.

Storing *all* the data currently present in disks is, however, infeasible because of the three orders of magnitude difference in the available capacities between disk and memory, notwithstanding the growing memory sizes. Datacenter jobs, however, exhibit a heavy-tailed distribution of their input sizes thus offering the potential to cache the inputs of a large fraction of jobs.

Heavy-tailed input sizes

Workloads consist of many small jobs and relatively few large jobs. In fact, 10% of overall data read is accounted by a disproportionate 96% and 90% of the smallest jobs in the Facebook and Bing workloads. As Figure 5.4 shows, job sizes indeed follow a power-law distribution, as the log-log plot shows a linear relationship [22].

The skew in job input sizes is so pronounced that a large fraction of active jobs can simultaneously fit their entire data in memory.[3] Consider a simple simulation that looks at jobs in the order of their arrival time. The simulator assumes the memory and computation slots across all the machines in the cluster to be aggregated. It loads a job's entire input into memory when it starts and deletes it when the job completes. If the available memory is insufficient for a job's entire input, none of it is loaded. Figure 5.5 plots the results of the simulation. For the workloads from Facebook and Bing, 96% and

[3] By active jobs we mean jobs that have at least one task running.

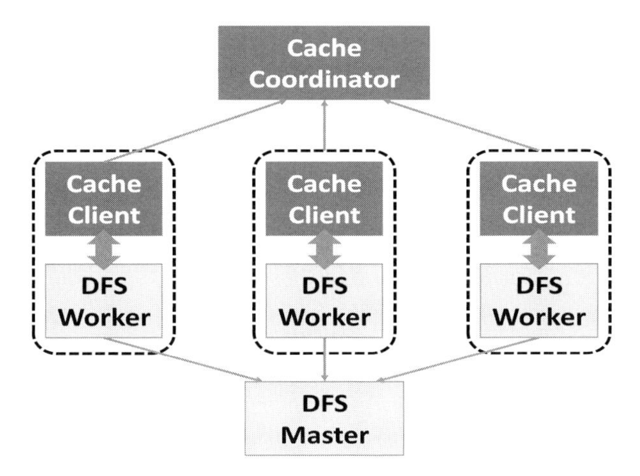

Figure 5.6 Coordinated cache architecture [24]. The central *coordinator* manages the distributed *clients*. Thick arrows represent data flow while thin arrows denote metadata flow.

89% of the active jobs respectively can have their data entirely fit in memory, given an allowance of 32 GB memory per server for caching [24].

Coordination architecture

Data are cached on the different distributed machines, and globally coordinated. Global coordination enables the abstraction of viewing different input blocks in unison to implement cache replacement policies. A coordinated cache infrastructure (a) supports queries for the set of machines where a block is cached, and (b) mediates cache replacement globally across the machines (covered shortly).

The architecture of a caching system consists of a central *coordinator* and a set of *clients* located at the storage nodes of the cluster (see Figure 5.6). Blocks are added to the cache clients. The clients update the coordinator when the state of their cache changes (i.e. when a block is added or removed). The coordinator uses these updates to maintain a mapping between every cached block and the machines that cache it. As part of the map, it also stores the file that a block belongs to.

The client's main role is to serve cached blocks, as well as cache new blocks. Blocks are cached at the *destination*, i.e. the machine where the task executes, as opposed to the *source*, i.e. the machine where the input is stored. This allows an uncapped number of replicas in cache, which in turn increases the chances of achieving memory locality especially when there are hotspots due to popularity skew [25]. Memory local tasks contact the local cache client to check if its input data is present. If not, they fetch it from the distributed file system (DFS). If the task reads data from the DFS, it puts it in cache of the local cache client and the client updates the coordinator about the newly cached block. Data flow is designed to be local in the architecture as remote memory access could be constrained by the network.

Fault tolerance

The coordinator's failure does not hamper the job's execution as data can always be read from disk. However, the architecture includes a secondary coordinator that functions as

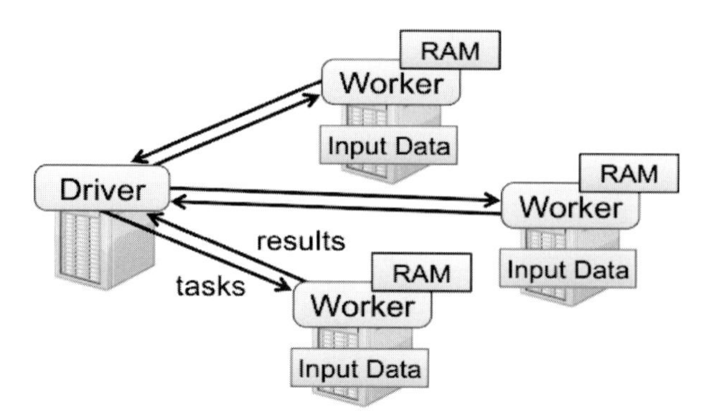

Figure 5.7 Resilient distributed datasets in Spark [26].

a cold standby. Since the secondary coordinator has no cache view when it starts, clients periodically send updates to the coordinator informing it of the state of their cache. The secondary coordinator uses these updates to construct the global cache view. Clients do not update their cache when the coordinator is down.

Resilient distributed datasets

An important class of applications *reuse* data across computations. Examples include iterative machine learning and graph algorithms like K-Means clustering and logistic regression. In-memory storage is an efficient way to support such data reuse. Storing it to the distributed file system incurs substantial overheads due to data replication and serialization, which can dominate execution times. Reuse also occurs in interactive data explorations when the same input is loaded once and used by many queries.

To enable efficient reuse of data, there is an abstraction proposed called *resilient distribtued datasets (RDDs)*. RDDs are fault-tolerant parallel data structures that let users store data in memory, optimize their placement and execute generic queries over them. RDDs can persist input data as well as intermediate results of a query to memory. A simple example to illustrate the programming model is as follows.

```
lines = spark.textFile("hdfs : //")
errors = lines.filter(_.startsWith("ERROR")
errors.persisit()
```

The main challenge in designing RDDs is providing efficient fault tolerance. Traditional options like replication or logging updates across machines are resource-expensive and time-consuming. RDDs provide fault tolerance by storing the *lineage* of the dataset. The lineage represents the set of steps (e.g. map, filter) used to create the RDD. Therefore, when a machine fails, its data can be regenerated by replaying the corresponding lineage steps. Maintaining lineage is a simple and inexpensive way to provide fault tolerant in-memory storage *without* replication.

The Spark system, which supports RDDs, has a simple Scala programming interface for managing RDDs. Spark's execution engine automatically converts the Scala program to a data parallel job of many parallel tasks, and executes them on the cluster. Figure 5.7 illustrates this execution.

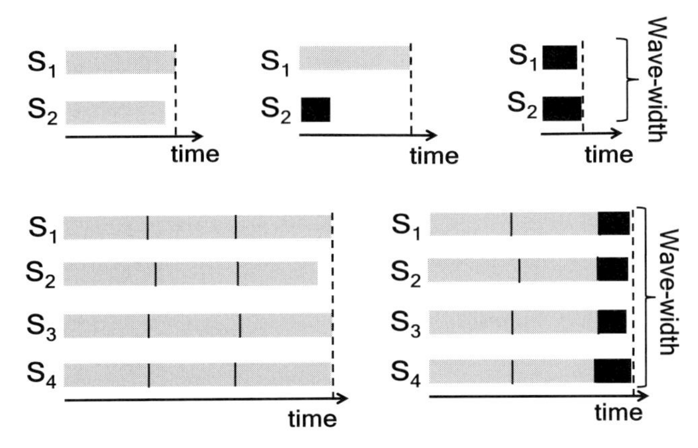

Figure 5.8 Example of a single-wave (two tasks, simultaneously) and multi-wave job (12 tasks, four at a time). Here S_is are slots. Memory local tasks are dark blocks. Completion time (dotted line) reduces when a wave-width of input is cached.

Cache replacement for parallel jobs

Maximizing cache hit-ratio does not minimize the average completion time of parallel jobs. In this section, we explain that using the concept of wave-width of parallel jobs and use it to devise the LIFE cache replacement algorithm.

All-or-nothing property

Achieving memory locality for a task will shorten its completion time. But this need not speed up the job. Jobs speed up when an entire *wave-width* of input is cached (Figure 5.8).[4] The wave-width of a job is defined as the number of simultaneously executing tasks. Therefore, jobs that consist of a single wave need 100% memory locality to reduce their completion time. We refer to this as the *all-or-nothing* property. Jobs consisting of many waves improve as we incrementally cache inputs in multiples of their wave-width. In Figure 5.8, the single-waved job runs both its tasks simultaneously and will speed up only if the inputs of both tasks are cached. The multi-waved job, on the other hand, consists of 12 tasks and can run four of them at a time. Its completion time improves in steps when any four, eight, and 12 tasks run memory locally.

Average completion time

In a cluster with multiple jobs, favoring jobs with the smallest wave-widths minimizes the average completion time of jobs. Assume that all jobs in the cluster are single-waved. Every job j has a wave-width of w and an input size of I. Let us assume the input of a job is equally distributed among its tasks. Each task's input size is $\left(\frac{I}{w}\right)$ and its duration is proportional to its input size. Memory locality reduces its duration by a factor of μ. The factor μ is dictated by the difference between memory and disk bandwidths, but limited by additional overheads such as deserialization and decompression of the data after reading it.

[4] This assumes similar task durations, which turns out to be true in practice. The 95th percentile of the coefficient-of-variance ($\frac{stdev}{mean}$) among tasks in the data-processing phase (e.g. map) of jobs is 0.08.

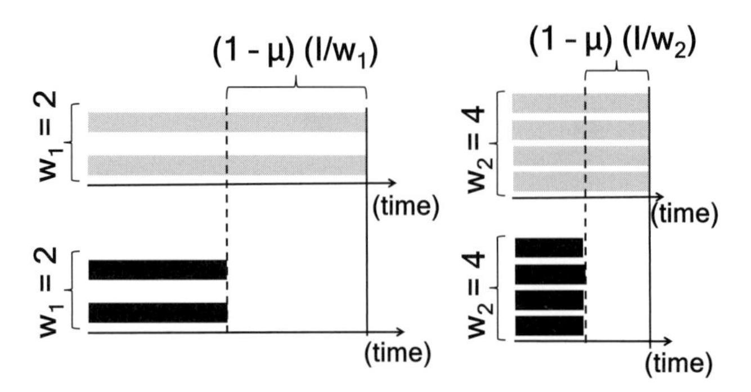

Figure 5.9 Gains in completion time due to caching decreases as wave-width increases. Solid and dotted lines show completion times without and with caching (for two jobs with input of I but wave-widths of 2 and 4). Memory local tasks are dark blocks, sped up by a factor of μ.

To speed up a single-waved job, we need I units of cache space. On spending I units of cache space, tasks would complete in $\mu\left(\frac{I}{w}\right)$ time. Therefore the saving in completion time would be $(1-\mu)\left(\frac{I}{w}\right)$. Counting this savings for every access of the file, it becomes $f(1-\mu)\left(\frac{I}{w}\right)$, where f is the frequency of access of the file. Therefore, the ratio of the job's benefit to its cost is $f(1-\mu)\left(\frac{1}{w}\right)$. In other words, it is directly proportional to the frequency and inversely proportional to the wave-width. The smaller the wave-width, the larger the savings in completion time per unit of cache spent. This is illustrated in Figure 5.9 comparing two jobs with the same input size (and of the same frequency), but wave-widths of 2 and 4. Clearly, it is better to use I units of cache space to store the input of the job with a wave-width of two. This is because its work per task is higher and so the savings are proportionately more. Note that even if the two inputs are unequal (say, I_1 and I_2, and $I_1 > I_2$), caching the input of the job with lower wave-width (I_1) is preferred despite its larger input size. Therefore, in a cluster with multiple jobs, *average completion time is best reduced by favoring the jobs with smallest wave-widths* (LIFE).

This can be easily extended to a multi-waved jobs. Let the job have n waves, c of which have their inputs cached. This uses $cw\left(\frac{I}{nw}\right)$ of cache space. The benefit in completion time is $f(1-\mu)c\left(\frac{I}{nw}\right)$. The ratio of the job's benefit to its cost is $f(1-\mu)\left(\frac{1}{w}\right)$, hence best reduced by still picking the jobs that have the smallest wave-widths.

5.4 Concluding remarks

The concepts surveyed in this chapter enable the execution of big data analytics on commodity clusters. These concepts address the principal aspects of large-scale computation, such as data locality, consistency and fault tolerance, provided by analytics frameworks while abstracting the details away from the user/application. While the concepts were presented in the context of batch analytics, we would like to point out that the template of jobs composed as DAG of tasks is by no means restricted to batch analytics alone. Many *interactive* frameworks [27–29] are using this template for fast interactive analytics, while also employing in-memory caching techniques explained in the chapter. Recently, *stream* processing – continuous processing of data as it streams in – is also employing this template. These systems divide the streaming input data into time buckets, and execute jobs on each of the time buckets. Thus, we expect the solutions presented here to remain the relevant core as analytics frameworks evolve in the future with new models and requirements of big data processing.

Our focus here has been on scheduling and storage, which can be viewed as basic ingredients of any such systems. We conclude this chapter by briefly outlining additional elements that are subject to ongoing research.

Approximate computing

An emerging class of applications is geared towards *approximation*. Approximation jobs are based on the premise that providing a timely result, even if on only part of the dataset is *good enough*. Approximation jobs are growing in interest to cope with the deluge in data since processing the entire dataset can take prohibitively long.

Approximation is explored along two dimensions – time to produce the result (deadline), and error (accuracy) in (of) the result [30, 31].

(1) *Deadline-bound* jobs strive to maximize the accuracy of their result within a specified time limit. Such jobs are common in real-time advertisement systems and web search engines. Generally, the job is spawned on a large dataset and accuracy is proportional to the fraction of data processed.
(2) *Error-bound* jobs strive to minimize the time taken to reach a specified error limit in the result. Again, accuracy is measured in the amount of data processed (or tasks completed). Error-bound jobs are used in scenarios where the value in reducing the error below a limit is marginal, e.g. counting of the number of cars crossing a section of a road to the nearest thousand is sufficient for many purposes.

Approximation jobs raise several challenges that are part of ongoing research. Sample selection is crucial towards meeting the desired approximation bounds. The selected sample impacts the achieved accuracy within a deadline, as well as the ability to meet an error bound. Also, understanding the query patterns and data-access characteristics help provide *confidence intervals* on the produced result. Designing smart samplers that balance the accuracy of the result, sample storage space and function generically is an open problem.

Another important problem is straggler mitigation. Approximation jobs require schedulers to *prioritize the appropriate subset of their tasks* depending on the deadline or error bound. Optimally prioritizing tasks of a job to slots is a classic scheduling problem with known heuristics. Stragglers, on the other hand, are unpredictable and require dynamic modification to the priority ordering according to the approximation bounds. The challenge is to achieve the approximation bound by dynamically weighing the gains due to speculation against the cost of using extra slots.

Energy efficiency

Energy costs are considered to be a major factor in the overall operational costs of datacenters. Therefore, there is strong motivation for utilizing clusters in *energy-efficient* manner. In our context of big data jobs, there is an opportunity of saving on energy costs, as some of the workloads can tolerate delays in execution, and can be scheduled when the energy costs are cheaper. The associated decision process involves several factors, such as time-varying energy prices, the existence of renewable energy sources, cooling overhead, and more; see e.g. [32–34] and references therein.

On top of the original scheduling decision of when to run jobs, there are additional control knobs that may be incorporated. For example, tasks can be executing using different power levels [35], and perhaps some resources can be turned off when overall demand is not high [36]. At a higher level, it may be possible to migrate jobs across datacenters in order to execute the jobs where energy is currently cheap [37].

Pricing

In Section 5.2 we have briefly discussed mechanisms which incentivize users to utilize resources fairly and truthfully. Such mechanisms are mostly relevant for internal enterprize clusters (a.k.a. private clouds). In the public cloud context, however, the main provider objective is maximizing the net profit, consisting of revenues minus operating costs. Current cloud providers rent virtual machines (VMs) on an hourly basis using a variety of pricing schemes (such as on-demand, reserved, or spot pricing, see e.g. [38]); recent research has accordingly been devoted to maximizing the customer's utility in face of the different pricing schemes (see e.g. [39] and reference therein). From the provider perspective, public cloud offerings, such as Amazon EC2, have recently incorporated an additional premium price for using big-data services such as Hadoop. Nevertheless, an interesting research direction is to investigate more specific pricing schemes, e.g. such that take into account job SLAs, deadlines, etc.

References

[1] J. Dean and S. Ghemawat, "Mapreduce: simplified data processing on large clusters," *Communications of the ACM*, vol. **51**, no. 1, pp. 107–113, 2008.

[2] M. Isard, M. Budiu, Y. Yu, A. Birrell, and D. Fetterly, "Dryad: distributed data-parallel programs from sequential building blocks," in *ACM EuroSys*, 2007.

[3] V. Vavilapalli *et al.*, "Apache hadoop yarn: yet another resource negotiator," in *ACM SoCC*, 2013.

[4] B. Hindman, A. Konwinski, M. Zaharia, *et al.*, "Mesos: a platform for fine-grained resource sharing in the data center," in *USENIX NSDI*, 2011.

[5] M. Shreedhar and G. Varghese, "Efficient fair queuing using deficit round-robin," *IEEE/ACM Transactions on Networking*, vol. **4**, no. 3, pp. 375–385, 1996.

[6] A. Demers, S. Keshav, and S. Shenker, "Analysis and simulation of a fair queueing algorithm," in *ACM SIGCOMM Computer Communication Review*, vol. **19**, no. 4. ACM, 1989, pp. 1–12.

[7] A. Ghodsi, M. Zaharia, B. Hindman, *et al.*, "Dominant resource fairness: fair allocation of multiple resource types." in *NSDI*, vol. **11**, 2011, pp. 24–24.

[8] M. Zaharia, D. Borthakur, J. S. Sarma, *et al.*, "Job scheduling for multi-user mapreduce clusters," *EECS Department, University of California, Berkeley, Tech. Rep. UCB/EECS-2009-55*, 2009.
[9] R. Grandl, G. Ananthanarayanan, S. Kandula, S. Rao, and A. Akella, "Multi-resource packing for cluster schedulers," in *ACM SIGCOMM*, 2014, pp. 455–466. [Online]. Available: http://doi.acm.org/10.1145/2619239.2626334.
[10] A. Ghodsi, M. Zaharia, S. Shenker, and I. Stoica, "Choosy: max-min fair sharing for datacenter jobs with constraints," in *Proceedings of the 8th ACM European Conference on Computer Systems*, ACM, 2013, pp. 365–378.
[11] B. Sharma, V. Chudnovsky, J. L. Hellerstein, R. Rifaat, and C. R. Das, "Modeling and synthesizing task placement constraints in google compute clusters," in *Proceedings of the 2nd ACM Symposium on Cloud Computing*, ACM, 2011, p. 3.
[12] M. Zaharia, D. Borthakur, J. Sen Sarma, *et al.*, "Delay scheduling: a simple technique for achieving locality and fairness in cluster scheduling," in *Proceedings of the 5th European Conference on Computer Systems*, ACM, 2010, pp. 265–278.
[13] M. Isard, V. Prabhakaran, J. Currey, U. Wieder, K. Talwar, and A. Goldberg, "Quincy: fair scheduling for distributed computing clusters," in *Proceedings of the ACM SIGOPS 22nd Symposium on Operating Systems Principles*, ACM, 2009, pp. 261–276.
[14] P. Bodík, I. Menache, M. Chowdhury, *et al.*, "Surviving failures in bandwidth-constrained datacenters," in *Proceedings of the ACM SIGCOMM 2012 Conference on Applications, Technologies, Architectures, and Protocols for Computer Communication*, ACM, 2012, pp. 431–442.
[15] A. D. Ferguson, P. Bodik, S. Kandula, E. Boutin, and R. Fonseca, "Jockey: guaranteed job latency in data parallel clusters," in *Proceedings of the 7th ACM european conference on Computer Systems*, ACM, 2012, pp. 99–112.
[16] C. Curino, D. E. Difallah, C. Douglas, *et al.*, "Reservation-based scheduling: If you're late don't blame us!" in *Proceedings of the ACM Symposium on Cloud Computing*, ACM, 2014, pp. 1–14.
[17] N. Jain, I. Menache, J. Naor, and J. Yaniv, "Near-optimal scheduling mechanisms for deadline-sensitive jobs in large computing clusters," in *SPAA*, 2012, pp. 255–266.
[18] B. Lucier, I. Menache, J. Naor, and J. Yaniv, "Efficient online scheduling for deadline-sensitive jobs: extended abstract," in *SPAA*, 2013, pp. 305–314.
[19] P. Bodík, I. Menache, J. S. Naor, and J. Yaniv, "Brief announcement: deadline-aware scheduling of big-data processing jobs," in *Proceedings of the 26th ACM symposium on Parallelism in algorithms and architectures*, ACM, 2014, pp. 211–213.
[20] M. Zaharia, A. Konwinski, A. D. Joseph, R. H. Katz, and I. Stoica, "Improving mapreduce performance in heterogeneous environments," in *OSDI*, vol. **8**, no. 4, 2008, p. 7.
[21] G. Ananthanarayanan, S. Kandula, A. G. Greenberg, I. Stoica, Y. Lu, B. Saha, and E. Harris, "Reining in the outliers in map-reduce clusters using mantri," in *OSDI*, vol. **10**, no. 1, 2010, p. 24.
[22] G. Ananthanarayanan, A. Ghodsi, S. Shenker, and I. Stoica, "Effective straggler mitigation: attack of the clones." in *NSDI*, vol. **13**, 2013, pp. 185–198.
[23] "Posix," http://pubs.opengroup.org/onlinepubs/9699919799/.
[24] G. Ananthanarayanan, A. Ghodsi, A. Wang, *et al.*, "Pacman: coordinated memory caching for parallel jobs," in *USENIX NSDI*, 2012.
[25] G. Ananthanarayanan, S. Agarwal, S. Kandula, *et al.*, "Scarlett: coping with skewed popularity content in mapreduce clusters," in *ACM EuroSys*, 2011.

[26] M. Zaharia, M. Chowdhury, T. Das, *et al.*, "Resilient distributed datasets: a fault-tolerant abstraction for in-memory cluster computing," in *USENIX NSDI*, 2012.
[27] S. Melnik, A. Gubarev, J. J. Long, *et al.*, "Dremel: Interactive analysis of web-scale datasets," in *Proceedings of the 36th International Conf on Very Large Data Bases*, 2010, pp. 330–339.
[28] R. Xin, J. Rosen, M. Zaharia, *et al.*, "Shark: SQL and rich analytics at scale," in *Proceedings of the 2013 ACM SIGMOD International Conference on Management of Data*, 2013.
[29] S. Agarwal, B. Mozafari, A. Panda, *et al.*, "Blinkdb: queries with bounded errors and bounded response times on very large data," in *Proceedings of the 8th European Conference on Computer Systems*, ACM, 2013.
[30] J. Liu, W.-K. Shih, K.-J. Lin, R. Bettati, and J.-Y. Chung, "Imprecise computations." in *IEEE*, 1994.
[31] S. Lohr, "Sampling: design and analysis," in *Thomson*, 2009.
[32] Y. Chen, S. Alspaugh, D. Borthakur, and R. Katz, "Energy efficiency for large-scale mapreduce workloads with significant interactive analysis," in *Proceedings of the 7th ACM European Conference on Computer Systems*, ACM, 2012, pp. 43–56.
[33] Z. Liu, Y. Chen, C. Bash, *et al.*, "Renewable and cooling aware workload management for sustainable data centers," in *ACM SIGMETRICS Performance Evaluation Review*, vol. **40**, no. 1, ACM, 2012, pp. 175–186.
[34] A. Beloglazov, R. Buyya, Y. C. Lee, *et al.*, "A taxonomy and survey of energy-efficient data centers and cloud computing systems," *Advances in Computers*, vol. **82**, no. 2, pp. 47–111, 2011.
[35] A. Gandhi, M. Harchol-Balter, R. Das, and C. Lefurgy, "Optimal power allocation in server farms," in *ACM SIGMETRICS Performance Evaluation Review*, vol. **37**, no. 1, ACM, 2009, pp. 157–168.
[36] A. Gandhi, V. Gupta, M. Harchol-Balter, and M. A. Kozuch, "Optimality analysis of energy-performance trade-off for server farm management," *Performance Evaluation*, vol. **67**, no. 11, pp. 1155–1171, 2010.
[37] N. Buchbinder, N. Jain, and I. Menache, "Online job-migration for reducing the electricity bill in the cloud," in *NETWORKING 2011*, Springer, 2011, pp. 172–185.
[38] "EC2 pricing," http://aws.amazon.com/ec2/pricing/.
[39] I. Menache, O. Shamir, and N. Jain, "On-demand, spot, or both: dynamic resource allocation for executing batch jobs in the cloud," in *11th International Conference on Autonomic Computing (ICAC)*, 2014.

6 Distributed big data storage in optical wireless networks

Chen Gong, Zhengyuan Xu, and Xiaodong Wang

We consider a distributed storage system employing some existing regenerate codes where the storage nodes are scattered in an optical wireless network. The data collector (DC) connects to the storage nodes via orthogonal channels and downloads data symbols from these nodes. In the existing data reconstruction schemes for distributed storage systems, the data collector downloads all symbols from a subset of the storage nodes. Such a full downloading approach becomes inefficient in wireless networks since due to fading, the wireless channels may not offer sufficient bandwidths for full downloading. Moreover, full downloading is also less power efficient than partial downloading. Given a coding scheme employed by the wireless distributed storage system, we propose a partial downloading scheme that allows downloading a portion of the symbols from any storage node. We formulate a cross-layer wireless resource allocation problem for data reconstruction in distributed storage systems employing such partial downloading. To derive the fundamental properties of partial downloading as well as to reduce the complexity of wireless resource allocation, we derive necessary and sufficient conditions for data reconstructability for partial downloading, in terms of the numbers of downloaded symbols from the storage nodes. We also propose channel and power allocation schemes for partial downloading in wireless distributed storage systems.

6.1 Introduction

The purpose of distributed storage is to store a data file in a distributed manner where the individual storage nodes may be unreliable. This has attracted significant research interests in both communication and computer science fields. The original data file is firstly encoded into multiple coded symbols, which are stored into various storage nodes. Note that, encoded by advanced coding schemes, the original data can be reconstructed if the number of collected data symbols is no less than the number of original data symbols.

Recently, two data encoding schemes for distributed data storage have been proposed, based on rateless coding and network coding [1, 2]. A criterion for distributed data storage is the transmission bandwidth for the reconstruction of the original data

Big Data over Networks, ed. Shuguang Cui, Alfred O. Hero III, Zhi-Quan Luo, and José M. F. Moura. Published by Cambridge University Press.

file and the repair of a failed storage node. For data reconstruction, a data collector (DC) downloads the symbols in some storage nodes to reconstruct the data. For node regeneration, assuming that a storage node has failed, a new storage node downloads the symbols from some other storage nodes to regenerate the symbols in the failed node. Although rateless code-based distributed data storage can achieve almost the minimum bandwidth for data reconstruction, it requires large node regeneration bandwidth since a failed node needs to reconstruct the original file first before regenerating its stored data symbols. To decrease the node regeneration bandwidth, for the network coding based distributed data storage, a data coding scheme named regenerated code has been proposed [1, 2], which achieves both efficient original data file reconstruction and failed node regeneration.

Currently, the wake of emerging applications of distributed storage systems with regard to wireless mobile devices like smart phones, tablets, etc., calls for the research and development of wireless distributed storage networks. It can be used for a mobile wireless cloud without infrastructure, including military wireless communication network of vehicles and ships, and some emergency cases, for example earthquake, tsunami, where infrastructures are destroyed. Owing to the fundamental difference between wireline and wireless transmission media, data reconstruction and node regeneration schemes need to be revisited for the wireless distributed storage networks. More specifically, for a wireless distributed storage system, given the regeneration code employed and the channel state information of the wireless links, a data downloading scheme that minimizes the total power consumption is desired. The exponential nature of the power consumption with respect to the data downloading amount implies it is more desirable for more uniform data downloading from all storage nodes. This is different from the wireline network, which aims to minimize the total number of downloaded symbols. As a result, different from the wireline network data downloading that downloads either all symbols from a storage node or nothing, such more-uniform data downloading requirements call for downloading part of the symbols from each storage node for data reconstruction. This chapter will first review a partial downloading scheme for data reconstruction for wireless distributed networks [3].

In the future big data era, the distributed storage of big data in a mobile wireless network will become an interesting topic for both research and application purposes. In such a wireless storage network, a large communication bandwidth is required for low-delay data reconstruction. One option is to adopt an optical wireless communication network as the communication backbone, which provides a large transmission bandwidth especially in the indoor scenario. An typical wireless distributed system for big data based on the optical wireless communication network can be set in an indoor conference room, where the lights in the ceiling are equipped with transmission units and data storage devices. Upon the data downloading request, the lights extract the data symbols in the storage devices and transmit the data symbols to the DC.

The remainder of this chapter is organized as follows. Section 6.2 has an overview of big data storage in a wireless network. Section 6.3 reviews the reconstructability condition for partial downloading. Section 6.4 reviews the wireless resource allocation. Finally, Section 6.5 provides several interesting future research directions.

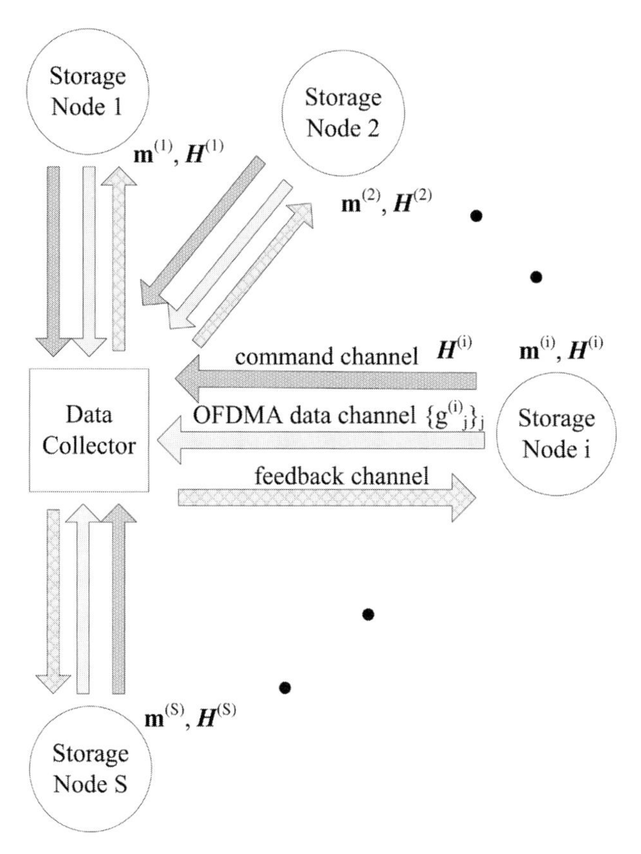

Figure 6.1 Illustration of wireless distributed storage system.

6.2 Big data distributed storage in a wireless network

6.2.1 Wireless distributed storage network framework

Data encoding for distributed storage

The basic structure of the wireless distributed storage system is shown in Figure 6.1, where a DC aims to download coded symbols from the storage nodes and reconstruct the original data file.

Assume that a data block of M symbols $\boldsymbol{s} = [s_1, s_2, \ldots, s_M]$ is stored in a distributed storage system consisting of S storage nodes denoted as $\mathcal{S} = \{1, 2, \ldots, S\}$, where each node stores α symbols. Each symbol is a packet of subsymbols in the field $GF(q)$, and contains B bits. Each storage node i stores a linear combination of the data symbols denoted as $\boldsymbol{m}^{(i)} = [m_1^{(i)}, \ldots, m_\alpha^{(i)}]$, given by

$$m_j^{(i)} = \sum_{k=1}^{M} h_{kj}^{(i)} s_k = \boldsymbol{s}\boldsymbol{h}_j^{(i)}, \quad 1 \leq j \leq \alpha, \tag{6.1}$$

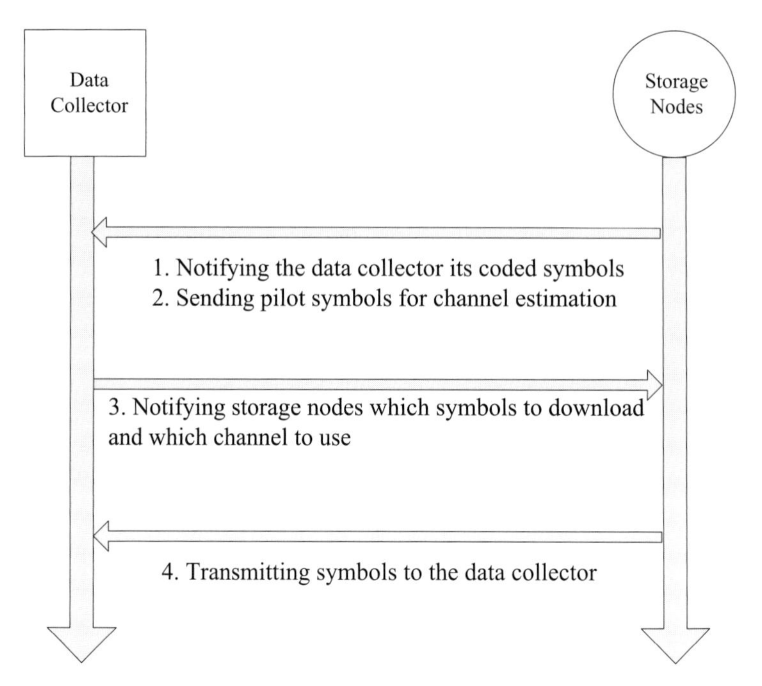

Figure 6.2 Basic data downloading protocol for distributed storage.

where the coefficients $\boldsymbol{h}_j^{(i)} = [h_{1j}^{(i)}, h_{2j}^{(i)}, \ldots, h_{Mj}^{(i)}]^T \in GF(q)^M$ for $1 \leq j \leq \alpha$. Denoting the encoding matrix $\boldsymbol{H}^{(i)} \triangleq [\boldsymbol{h}_1^{(i)}, \boldsymbol{h}_2^{(i)}, \ldots, \boldsymbol{h}_\alpha^{(i)}] \in GF(q)^{M\times\alpha}$, then $\boldsymbol{m}^{(i)} = \boldsymbol{s}\boldsymbol{H}^{(i)}$ for $i \in \mathcal{S}$. The DC downloads symbols from the storage nodes to reconstruct the data.

The advantages of using coded data storage over the uncoded schemes lie in the efficiency of data reconstruction. More specifically, encoded based on a Vandermonde matrix, the original data can be reconstructed if the number of downloaded symbols is larger than the number of original data symbols. However, this cannot be achieved by a simple repetition code.

Wireless storage system framework and basic data downloading protocol

Assume that the DC and the storage nodes are connected via N orthogonal wireless channels $\mathcal{N} = \{1, 2, \ldots, N\}$, a command channel from each storage node to the DC and a feedback channel from the DC to each storage node. Each channel is of bandwidth W and duration T. The N orthogonal channels can be either in time-division multiplexed (TDM) or in frequency-division multiplexed (FDM), usually in FDM. For $i \in \mathcal{S}$ and $j \in \mathcal{N}$, let $g_j^{(i)}$ be the gain of channel j from storage node i to the DC. Assume that each storage node i knows only its own data symbols $\boldsymbol{m}^{(i)}$ and linear combination coefficients $\boldsymbol{H}^{(i)}$.

The basic data downloading protocol is shown in Figure 6.2. The handshake process consists of three steps.

(1) Each storage node i sends its $\boldsymbol{H}^{(i)}$ to the DC through the command channel, and pilot symbols for channel estimation to the DC through the data channel.
(2) The DC then estimates the channel gains $\{g_j^{(i)}\}_{j\in\mathcal{N}}$ of the wireless link that connects itself to each storage node i. It then performs a wireless resource allocation to decide which symbols to download from each storage node, and the corresponding channels and powers for downloading these symbols; the results are fed back to the storage nodes through the feedback channel.
(3) Finally each storage node sends its chosen symbols to the DC through the data channel according to the resource allocation results.

6.2.2 Optical wireless framework

Optical channel transmission

An indoor visible light transceiver pair typically consists of a laser emitter diode (LED) transmitter and an avalanche photodiode (APD) or photodiode (PD) receiver. Owing to the incoherent phase nature of the LED transmission signal, the transmitter employs an intensity modulation (IM) and the receiver employs direct detection (DD) to detect the transmitted signal. Because of this, the baseband signals are real and positive, which is different from the radio-frequency (RF) communication.

Usually, the transmitted signal consists of a direct current component and a bipolar signal component, where the direct current component guarantees the real positive signal and the bipolar signal component is used to carry the information. The upper and lower bounds on the capacity of such a channel are investigated in [4]. Single- and multi-carrier transmission formats have been elaborated in [5–9]. One important feature is that, for orthogonal-frequency division-multiplexing (OFDM) transmission, in order to guarantee the real inverse fast Fourier transform (IFFT) output signal, the input signal should satisfy a certain conjugate constraint.

The advantages of using visible light over RF lie in its potential high transmission bandwidth and security, which are brought by its ultra-wide band optical spectrum and impenetrability into the wall. Moreover, the semi-directional transmission feature of the visible light implies lower interference in a multi-user network.

Indoor lighting system for big data storage

The indoor lighting system for big data storage is shown in Figure 6.3. Typically, each room is equipped with significant number of light bulbs for lighting coverage of the whole space. These densely distributed lighting fixtures are connected with data storage nodes for convenient data storage and transfer. Data files are stored in a distributed manner, and each light is connected to a data storage node. There is a DC that aims to download the data stored in the storage nodes and reconstruct the original file. Each light serves as an optical wireless communication transmitter, which can retrieve the data stored in the storage nodes and transmits the data to the DC.

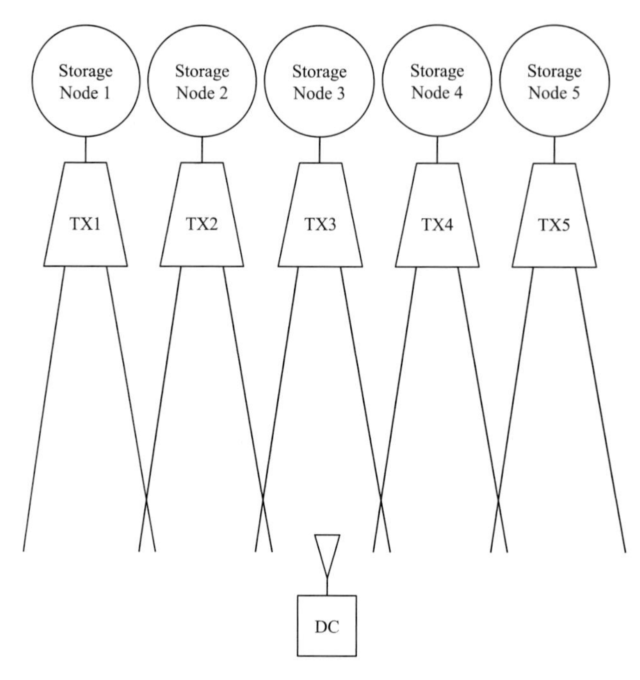

Figure 6.3 Indoor lighting system for big data storage.

Optical wireless communication backbone

Let P_j be the transmission power of channel j. The number of symbols that can be transmitted from storage node i to the DC over channel j is then given by $c(|g_j^{(i)}|^2 P_j)$. Assume that the minimum unit of data transmitted over a channel is *a symbol*, i.e. only *one or several symbols* can be transmitted over each channel. Our objective is to minimize the total transmission power $\sum_{j \in \mathcal{N}} P_j$ while guaranteeing successful data reconstruction.

Note that for the wireless additive Gaussian white noise (AWGN) channel, the number of symbols that can be downloaded from channel j to storage node i is given by

$$c\left(\left|g_j^{(i)}\right|^2 P_j\right) = \frac{WT}{B} \log_2 \left(1 + \kappa \frac{\left|g_j^{(i)}\right|^2 P_j}{\sigma^2}\right), \tag{6.2}$$

where σ^2 is the power of background noise and $\kappa < 1$ accounts for the rate loss due to the practical modulation and coding, compared with the ideal case of Gaussian signaling and infinite-length code. For optical wireless communications, the capacity is different. For more details, the reader may refer to [4].

The remainder of this section will address three data coding and downloading schemes, i.e. the data coding based on rateless codes, the data coding based on network coding with full downloading, and the data coding based on network coding with partial downloading.

6.2.3 Rateless coded distributed data storage

Rateless codes

One type of distributed storage system is to use rateless codes [10, 11] to encode the original file. A rateless code over the N information symbols is characterized by the following two features.

(1) Infinite encoding length: the potential coding length of a rateless code is infinite. For each coded symbol, there is a coding mapping over the N information. symbols, which is usually characterized by random coding.
(2) Rateless code decoding: the receiver can decode the N information symbols upon receiving any $N(1 + \Delta)$ coded symbols, for a small $\Delta > 0$.

Usually encoding a rateless code is performed according to a random coding rule. The rule is characterized by a random profile $\{\rho_j\}_{j \geq 1}$, where ρ_j is probability that a coded symbol uniformly selects j among the N information symbols and adds them in the Galois symbol domain for the encoding. Note that we have $\sum_{j \geq 1} \rho_j = 1$.

Rateless coded distributed storage

Assume that a data file consists of N data symbols, which are encoded by a rateless code [12]. The coded symbols are stored in the storage nodes, where each storage node stores α coded symbols. Since for the rateless code, the receiver can decode the N information symbols upon receiving $N(1 + \Delta)$ coded symbols, a data collector needs to download all symbols from $\lceil \frac{N(1+\Delta)}{\alpha} \rceil$ storage nodes to reconstruct the original data file.

However, one disadvantage of the rateless code-based data storage is that, if the data symbols in a data storage node are destroyed, regenerating those symbols requires first regenerating the original data file and then performing the coded symbol encoding. In other words, the data regenerating requires the failed storage node download $N(1 + \Delta)$ data symbols to reconstruct the original data file, which consumes large transmission bandwidth. To overcome this, the distributed data storage based on network coding is proposed [1, 2].

6.2.4 Network coded system with full downloading

Distributed storage based on regenerates codes [1, 2] is characterized by the parameters $(S, K, d, \alpha, \gamma)$, where $\gamma = d\beta$, which satisfies the following conditions.

(1) There are S storage nodes, where each node stores α symbols.
(2) A DC can reconstruct the data via downloading all the $K\alpha$ symbols from any K storage nodes.
(3) A new storage node can regenerate the α symbols in a failed node by downloading β symbols from each of any d surviving nodes; and after that a DC can still reconstruct the data via downloading all the $K\alpha$ symbols from any K storage nodes.

An optimal trade-off curve between the amount of downloaded symbols $K\alpha$ for data reconstruction and $\gamma = d\beta$ for node regeneration is reported in [1, 2], with two extremes called the minimum-storage regenerating (MSR) point and the minimum-bandwidth regenerating (MBR) point, corresponding to the coding schemes with the best efficiency $K\alpha = M$ and $d\beta = \alpha$ for data reconstruction and node regeneration, respectively. The design of explicit coding schemes for distributed storage systems is addressed in [13–20]. Up to now, the best results are the coding schemes in [13] and [20], which achieve the optimal trade-off at the MSR and MBR points, respectively.

MSR data coding point

The MSR data coding point requires the minimum amount of data transmission for the original data file reconstruction in the DC. For such a coding point, we have $M = K\alpha$. For any size-K subset $\mathcal{R} \subseteq \mathcal{S}$, the data $\boldsymbol{s}$ can be reconstructed from the M downloaded symbols $\boldsymbol{s}[\boldsymbol{H}^{(i)}]_{i\in\mathcal{R}}$, i.e. the *square matrix* $[\boldsymbol{H}^{(i)}]_{i\in\mathcal{R}}$ is of rank M. In other words, for any subset $\mathcal{R} \subseteq \mathcal{S}$ with $|\mathcal{R}| \leq K$, the columns of matrix $[\boldsymbol{H}^{(i)}]_{i\in\mathcal{R}}$ are *linearly independent.*

MBR data coding point

The MBR data coding point requires the minimum amount of data transmission for the repair of a failed storage node. We have that $\alpha = \gamma = \beta d$. For any storage node i and any size-d subset $\mathcal{R} \subseteq \mathcal{S}, i \notin \mathcal{R}$, there exists column-$\beta$ sub-matrix $\bar{\boldsymbol{H}}^{(j)} \subseteq \boldsymbol{H}^{(j)}$ for $j \in \mathcal{R}$, such that there exists an $\alpha \times \alpha$ full-rank matrix $\boldsymbol{T}$ such that $\boldsymbol{H}^{(i)} = \bar{\boldsymbol{H}}^{\mathcal{R}}\boldsymbol{T}$.

6.2.5 Network coded system with partial downloading

Consider the wireless distributed storage systems as shown in Figure 6.1. Owing to the channel fading and the power and bandwidth constraints of the wireless links, the data collector may not be able to download all symbols from a storage node. In other words, when the downloading bandwidth from all storage nodes to the DC is limited, e.g. in a wireless distributed storage system, it could be more efficient to download a small number of symbols from all storage nodes, i.e. partial downloading, than to download all symbols from a few storage nodes, i.e. full downloading, due to the *exponential nature* of the power consumption as a function of the amount of information downloaded. Note that when designing such a partial downloading scheme for data reconstruction, we still adopt the network-coded distributed storage [21], in order to maintain the same efficiency of the node regeneration as that in full downloading.

Consider the reconstructability of the original data if the DC downloads a portion of the symbols stored in each storage node. For $i \in \mathcal{S}$, let μ_i be the number of symbols downloaded from storage node i. Since $\mu_i \leq \alpha$, we assume that the downloaded symbols are linear combinations of the symbols in node i given by $\boldsymbol{s}^T\boldsymbol{H}^{(i)}\boldsymbol{A}^{(i)}$ where $\boldsymbol{A}^{(i)}$ is an $\alpha \times \mu_i$ matrix. From [17], the data $\boldsymbol{s}$ can be reconstructed from the downloaded symbols $\boldsymbol{s}^T[\boldsymbol{H}^{(i)}\boldsymbol{A}^{(i)}]_{i\in\mathcal{S}}$ if and only if

$$\operatorname{rank}\left(\left[\boldsymbol{H}^{(i)}\boldsymbol{A}^{(i)}\right]_{i\in\mathcal{S}}\right) = M. \tag{6.3}$$

For each $i \in \mathcal{S}$ the matrix $\boldsymbol{A}^{(i)}$ is assumed to be of full column rank, since otherwise at least one downloaded symbol can be expressed as a linear combination of other symbols downloaded from the same storage node, which means that this symbol is redundant and should be removed to reduce the downloading bandwidth.

However, the search for the linear combination matrices $\{\boldsymbol{A}^{(i)}\}_{i\in\mathcal{S}}$ that satisfy (6.3) can be computationally prohibitive. A simpler scheme is to directly download the stored symbols from each storage node, without performing linear combination. An immediate question is whether such a simpler approach may lose optimality, i.e. is it possible that for some $\{\mu_i\}_{i\in\mathcal{S}}$, (6.3) can be satisfied by performing linear combination but it can not be satisfied by simply downloading the stored symbols without linear combination. The following result states that in terms of the number of symbols downloaded from the storage nodes, downloading the symbols stored in the storage nodes directly and downloading their linear combinations are equivalent. The proof has been provided in [3] and thus omitted here.

Theorem 6.1 *If there exist $\alpha \times \mu_i$ matrices $\boldsymbol{A}^{(i)}$ for $i \in \mathcal{S}$ such that (6.3) is satisfied, then there exist $M \times \mu_i$ column extracted submatrices $\bar{\boldsymbol{H}}^{(i)}$ of $\boldsymbol{H}^{(i)}$, $i \in \mathcal{S}$, such that $\bar{\boldsymbol{H}}^{\mathcal{S}} = [\bar{\boldsymbol{H}}^{(i)}]_{i\in\mathcal{S}}$ is of rank M.*

Then, based on Theorem 6.1, a natural question is that, for which types of $\{\mu_i\}_{i\in\mathcal{S}}$ matrix $[\bar{\boldsymbol{H}}^{(i)}]_{i\in\mathcal{S}}$ is of rank M such that the original data file can be reconstructed. The following definition of data $\boldsymbol{\mu}$-reconstructability captures this question.

Definition 1 ($\boldsymbol{\mu}$-reconstruction) Given $\boldsymbol{\mu} = [\mu_1, \mu_2, \ldots, \mu_S]$, $\mu_i \in \{0, 1, \ldots, \alpha\}$ for $1 \le i \le S$, the data are *$\boldsymbol{\mu}$-reconstructable* if they can be reconstructed via downloading μ_i symbols from storage node i for $i \in \mathcal{S}$, which is equivalent to saying that there exist $M \times \mu_i$ submatrices $\bar{\boldsymbol{H}}^{(i)} \subseteq \boldsymbol{H}^{(i)}$, $i \in \mathcal{S}$, such that the matrix $\bar{\boldsymbol{H}}^{\mathcal{S}} = [\bar{\boldsymbol{H}}^{(i)}]_{i\in\mathcal{S}}$ is of rank M.

Note that $\boldsymbol{\mu}$-reconstructability implies that $\mu_i \in \{0, 1, \ldots, \alpha\}$ for $1 \le i \le S$, since downloading more than α storage symbols from a storage node is not feasible.

6.3 Reconstructability condition for partial downloading

This part presents the $\boldsymbol{\mu}$-reconstructability for both the MSR point and the MBR point. In particular, for the MSR point, a necessary and sufficient condition for the $\boldsymbol{\mu}$-reconstructability for any MSR coding scheme is provided. On the other hand, for the MBR point, a specific MBR coding scheme is considered and the necessary and sufficient condition for the $\boldsymbol{\mu}$-reconstructability is provided. In the following discussion, for a family of matrices $\{\boldsymbol{H}^{(i)}\}_{i\in\mathcal{A}}$ with the same number of rows, let $\boldsymbol{H}^{\mathcal{A}} \triangleq [\boldsymbol{H}^{(i)}, i \in \mathcal{A}]$ be the matrix obtained by horizontally concatenating $\boldsymbol{H}^{(i)}$ for $i \in \mathcal{A}$.

6.3.1 $\boldsymbol{\mu}$-Reconstructability for MSR point

Consider the $\boldsymbol{\mu}$-reconstructability for all coding schemes satisfying the constraint of the MSR point that $K\alpha = M$. Since for any size-K subset $\mathcal{R} \subseteq \mathcal{S}$, the data $\boldsymbol{s}$ can be reconstructed from the M downloaded symbols $\mathbf{s}^T[\mathbf{H}^{(i)}]_{i\in\mathcal{R}}$, it follows that the *square matrix* $[\mathbf{H}^{(i)}]_{i\in\mathcal{R}}$ is of rank M.

Based on the above observation, we show that the trivial necessary condition $\sum_{i\in\mathcal{S}} \mu_i \geq M$, i.e. that the number of downloaded symbols should be no less than the total number of data symbols to be reconstructed, is also a *sufficient condition* for $\boldsymbol{\mu}$-reconstructability.

Theorem 6.2 *For the MSR point the necessary and sufficient condition for the* $\boldsymbol{\mu}$-*reconstructability is* $\sum_{i\in\mathcal{S}} \mu_i \geq M$.

Given $\{\mu_i\}_{i\in\mathcal{S}}$, a constructive proof is provided in [3], where the DC keeps extracting linearly independent columns from the linear combination matrices $[\mathbf{H}^{(i)}]_{i\in\mathcal{S}}$, until getting stuck. Then, it performs some stuck processing. The detailed proof is omitted here.

6.3.2 $\boldsymbol{\mu}$-Reconstructability for a practical MBR point coding scheme

To the best of our knowledge, up to now the best and also practical coding scheme for the MBR point is proposed in [21]. The $\boldsymbol{\mu}$-reconstructability of this coding scheme is analysed as follows.

MBR data encoding scheme

First the coding scheme given in [21] is outlined as follows. Assume that a data block of

$$M = \frac{K(K+1)}{2} + K(d-K) \tag{6.4}$$

symbols is stored among S storage nodes, where each node stores $\alpha = d$ symbols. The M symbols are represented using the following $d \times d$ symmetric matrix

$$\mathbf{B} = \begin{bmatrix} \mathbf{B}^{(1)} & \mathbf{B}^{(2)} \\ \left(\mathbf{B}^{(2)}\right)^T & \mathbf{0} \end{bmatrix} \tag{6.5}$$

where $\mathbf{B}^{(1)}$ is a $K \times K$ symmetric matrix storing $K(K+1)/2$ symbols and $\mathbf{B}^{(2)}$ is a $K \times (d-K)$ matrix storing $K(d-K)$ symbols. For encoding, the matrix $\mathbf{B}$ is pre-multiplied by an $S \times d$ Vandermonde matrix given by $\boldsymbol{\Psi}$, and each node $i \in \mathcal{S}$ stores the d symbols corresponding to the ith row of $\boldsymbol{\psi}_i\mathbf{B}$, where $\boldsymbol{\psi}_i$ is the ith row of $\boldsymbol{\Psi}$. It is shown in [21] that the data can be reconstructed by downloading all symbols stored in any K storage nodes, and by downloading one symbol from each of any d surviving storage nodes the new storage node can regenerate the same symbols as those in a failed node.

Partial downloading scheme

Next a partial downloading scheme for data reconstruction based on the above coding scheme is outlined, as follows. Since the matrix $\mathbf{B}^{(1)}$ is symmetric, from (6.5) only the

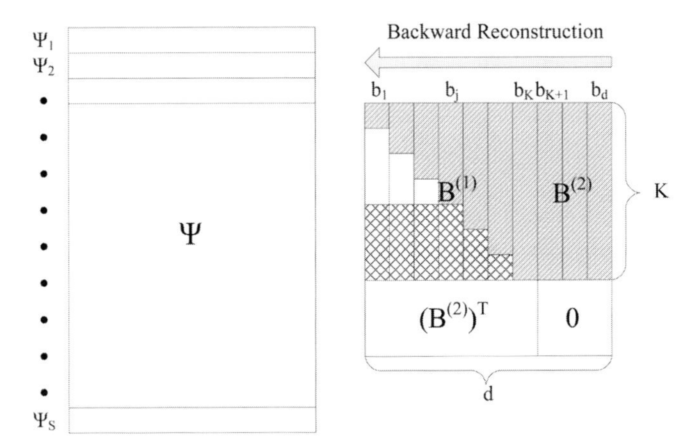

Figure 6.4 Illustration of data reconstruction for the coding scheme at the MBR point.

data symbols in $\boldsymbol{B}^{(2)}$ and in the upper triangular part of $\boldsymbol{B}^{(1)}$ need to be decoded. As shown in Figure 6.4, the data symbols are divided into d columns $\{\boldsymbol{b}_j\}_{1\le j\le d}$, where $\boldsymbol{b}_j$ for $1 \le j \le K$ are in the upper triangular part of $\boldsymbol{B}^{(1)}$, and $\boldsymbol{b}_j$ for $K+1 \le j \le d$ are the columns of $\boldsymbol{B}^{(2)}$. The data symbols are reconstructed in the backward order of $\boldsymbol{b}_d$, $\boldsymbol{b}_{d-1}, \ldots, \boldsymbol{b}_1$. Note that the number of symbols in $\boldsymbol{b}_j$, $1 \le j \le d$, is given by $\min\{j, K\} \triangleq \theta_j$. Let $(\boldsymbol{\psi}_i\boldsymbol{B})_j$ be the jth symbol in storage node i which is the product of $\boldsymbol{\psi}_i$ and the jth column of $\boldsymbol{B}$.

The following partial downloading scheme is proposed to perform the backward reconstruction.

(1) First reconstruct $\boldsymbol{B}^{(2)}$. For $K+1 \le j \le d$, from Figure 6.4 it is seen that to reconstruct $\boldsymbol{b}_j$ the DC needs to download the symbols $(\boldsymbol{\psi}_i\boldsymbol{B})_j$ for i belonging to a subset $\mathcal{R}_j \subseteq \mathcal{S}$, where the size $|\mathcal{R}_j| \ge K = \theta_j$. Since $\boldsymbol{\Psi}$ is a Vandermonde matrix, with K downloaded symbols $\boldsymbol{b}_j$ can be reconstructed.
(2) Then reconstruct $\boldsymbol{B}^{(1)}$ via reconstructing $\boldsymbol{b}_j$ for $1 \le j \le K$ in the order of $\boldsymbol{b}_K$, $\boldsymbol{b}_{K-1}, \ldots, \boldsymbol{b}_1$. As shown in Figure 6.4, when reconstructing $\boldsymbol{b}_j$, since $\boldsymbol{b}_l$ for $j < l \le K$ have been reconstructed and $\boldsymbol{B}^{(1)}$ is symmetric, the part of $\boldsymbol{B}^{(1)}$ in the square shadow is known. Then, as $\boldsymbol{B}^{(2)}$ has been reconstructed and thus is known, reconstructing $\boldsymbol{b}_j$ amounts to downloading the symbols $(\boldsymbol{\psi}_i\boldsymbol{B})_j$ for $i \in \mathcal{R}_j \subseteq \mathcal{S}$, where the size $|\mathcal{R}_j| \ge j = \theta_j$. Again, since $\boldsymbol{\Psi}$ is a Vandermonde matrix, with j downloaded symbols $\boldsymbol{b}_j$ can be reconstructed.

Based on the above partial downloading scheme, let $\eta_j^{(i)} = 1$ if the DC downloads $(\boldsymbol{\psi}_i\boldsymbol{B})_j$ to reconstruct $\boldsymbol{b}_j$, and $\eta_j^{(i)} = 0$ otherwise. The minimum requirement for data reconstruction is that

$$\sum_{i=1}^{S} \eta_j^{(i)} = \theta_j, \quad 1 \le j \le d. \tag{6.6}$$

The data is $\boldsymbol{\mu}$-reconstructable if there exists $\eta_j^{(i)} \in \{0, 1\}$ for $1 \leq i \leq S$ and $1 \leq j \leq d$ such that

$$\mu_i = \sum_{j=1}^{d} \eta_j^{(i)}, \quad i \in \mathcal{S}, \tag{6.7}$$

and (6.6) is satisfied. The following subsection provides a necessary and sufficient condition in terms of $\{\mu_i\}_{i \in \mathcal{S}}$ for (6.6) and (6.7) to hold.

μ-Reconstructability for the MBR point

A straightforward necessary condition for (6.6) and (6.7) is that, for any subset $\mathcal{A} \subseteq \mathcal{S}$,

$$\sum_{i \in \mathcal{A}} \mu_i = \sum_{i \in \mathcal{A}} \sum_{j=1}^{d} \eta_j^{(i)} = \sum_{j=1}^{d} \sum_{i \in \mathcal{A}} \eta_j^{(i)} \leq \sum_{j=1}^{d} \min\{\theta_j, |\mathcal{A}|\}, \tag{6.8}$$

since $\sum_{i \in \mathcal{A}} \eta_j^{(i)} \leq |\mathcal{A}|$ and $\sum_{i \in \mathcal{A}} \eta_j^{(i)} \leq \sum_{i \in \mathcal{S}} \eta_j^{(i)} = \theta_j$ for all $1 \leq j \leq d$. Denote the sorted $\{\mu_i\}_{i \in \mathcal{S}}$ in decreasing order as $\mu^{(1)} \geq \mu^{(2)} \geq \cdots \geq \mu^{(S)}$, then it follows that

$$\sum_{i=1}^{l} \mu^{(i)} \leq \sum_{j=1}^{d} \min\{\theta_j, l\}, \text{ for } 1 \leq l \leq d, \text{ and } \sum_{i=1}^{S} \mu^{(i)} = \sum_{j=1}^{d} \theta_j. \tag{6.9}$$

Since $\theta_j = \min\{j, K\}$, (6.9) becomes

$$\sum_{i=1}^{l} \mu^{(i)} \leq dl - \frac{l(l-1)}{2}, \quad \text{for } 1 \leq l \leq d; \text{ and } \sum_{l=1}^{S} \mu^{(l)} = M. \tag{6.10}$$

The following result shows that (6.9) or (6.10) is also sufficient for data reconstruction.

Theorem 6.3 *Denote the sorted $\{\mu_i\}_{i \in \mathcal{S}}$ by $\mu^{(1)} \geq \mu^{(2)} \geq \cdots \geq \mu^{(S)}$. Then (6.10) is a necessary and sufficient condition for the $\boldsymbol{\mu}$-reconstructability of the proposed partial downloading scheme for the MBR point.*

The sufficiency can be proved by the following symbol selection scheme. The achievability is provided in [3], and is omitted here.

Symbol selection scheme

The sufficiency can be proved via an explicit partial downloading scheme. More specifically, given $\{\mu_i\}_{i \in \mathcal{S}}$ satisfying (6.10), $\{\eta_j^{(i)}\}_{i \in \mathcal{S}, 1 \leq j \leq d}$ can be computed as follows.

- To initialize, rank $\{\mu_i\}_{i \in \mathcal{S}}$ in decreasing order $\mu_{i_1} \geq \mu_{i_2} \geq \cdots \geq \mu_{i_S}$, and let $\theta_j = \min\{j, K\}$, $j = 1, \ldots, K$.
- Then for $k = 1, 2, \ldots$ to S, compute $\{\eta_j^{(i_k)}\}_{1 \leq j \leq d}$ as follows.
  - Rank $\{\theta_j\}_{1 \leq j \leq d}$ in decreasing order $\theta_{j_1} \geq \theta_{j_2} \cdots \geq \theta_{j_d}$;
  - let $\eta_{j_p}^{(i_k)} = 1$ for $1 \leq p \leq \mu_{i_k}$ and $\eta_{j_p}^{(i_k)} = 0$ for $\mu_{i_k} + 1 \leq p \leq d$;
  - subtracting $\{\eta_j^{(i_k)}\}_{1 \leq j \leq d}$ from θ_{j_p}, update $\theta_{j_p} = \theta_{j_p} - 1$ for $1 \leq p \leq \mu_{i_k}$.

Figure 6.5 illustrates the symbol selection procedure. Let $d = 5$, and $[\theta_1, \theta_2, \theta_3, \theta_4, \theta_5] = [4, 5, 3, 3, 3]$, with the ranking $\theta_2 > \theta_1 > \theta_3 = \theta_4 = \theta_5$. Let $\mu_{i_1} = 3$.

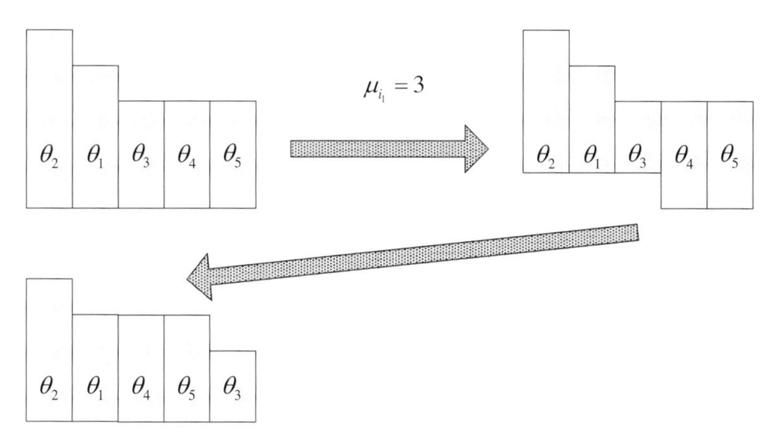

Figure 6.5 Illustration of the symbol selection.

According to the symbol selection procedure, each of the first 3 elements $(\theta_2, \theta_1, \theta_3)$ is subtracted by 1. After the subtraction, we have that $[\theta_1, \theta_2, \theta_3, \theta_4, \theta_5] = [3, 4, 2, 3, 3]$; and then the ranking becomes $\theta_2 > \theta_1 = \theta_4 = \theta_5 > \theta_3$.

Note that due to Theorem 6.1, condition (6.10) cannot be relaxed by allowing downloading the linear combinations of the symbols in the storage nodes. In the full downloading scheme proposed in [21], dK symbols have to be downloaded from K storage nodes for data reconstruction; whereas the above proposed partial downloading scheme downloads a total of M symbols, which is the minimum amount for reconstructing the original M data symbols. Since $M = dK - K(K-1)/2 < dK$, the proposed partial downloading scheme can significantly reduce the bandwidth requirement for symbol downloading. For example, for $d = 6$ and $K = 4$, we have $dK = 24$ but $M = 18$, i.e. the number of downloaded symbols by the proposed partial downloading scheme is 75% of that by the scheme in [21].

Consider further the constraints in (6.10), which read $\mu^{(1)} \leq d$, $\mu^{(1)} + \mu^{(2)} \leq 2d - 1$, and $\mu^{(1)} + \mu^{(2)} + \mu^{(3)} \leq 3d - 3$, etc. That is, downloading all symbols from any two storage nodes, or more than $3d - 3$ symbols from any three storage nodes, is not allowed. Hence these constraints impose that *the DC downloads evenly from all nodes*. Such a requirement matches well with the characteristic of wireless channels. In particular, in general the optimal resource allocation in wireless systems tends to distribute the resource uniformly among the nodes, rather than giving the resources to only a few nodes and starving the other nodes.

6.4 Channel and power allocation for partial downloading

The idea of channel and power allocation in wireless communication systems is to allocate better orthogonal channel to each user, such that each user can better utilize the wireless media. This section provides a wireless resource allocation framework for the partial downloading, as well as a solution.

6.4.1 Wireless resource allocation framework

A common approach for subchannel allocation is to use an indicator to show whether a subchannel has been allocated to a specific user. In this work, for $i \in \mathcal{S}$ and $j \in \mathcal{N}$, let $\beta_j^{(i)} = 1$ if the DC downloads symbols from storage node i using channel j and $\beta_j^{(i)} = 0$ otherwise. Since $\beta_j^{(i)}$ is an indicator and a subchannel cannot be used for more than one user, we have that $\beta_j^{(i)} \in \{0, 1\}$ and $\sum_{i \in \mathcal{S}} \beta_j^{(i)} \leq 1$ for $j \in \mathcal{N}$.

For $j \in \mathcal{N}$, let P_j be the transmission power for channel j. Then the number of symbols downloaded over channel j is given by

$$X_j = c\left(P_j \sum_{i \in S} \beta_j^{(i)} |g_j^{(i)}|^2\right), \tag{6.11}$$

where $c(\cdot)$ is given in (6.2). The number of *symbols* downloaded from storage node i is then

$$\mu_i = \sum_{j=1}^{N} \beta_j^{(i)} X_j, \ i \in \mathcal{S}. \tag{6.12}$$

The wireless resource allocation problem is to minimize the total power across all N channels that guarantees data reconstruction, which can be formulated as follows:

$$\begin{aligned} \min_{\{\beta_j^{(i)}, P_j\}_{i \in \mathcal{S}, j \in \mathcal{N}}} \quad & \sum_{j \in \mathcal{N}} P_j \\ \text{s.t.} \quad & \text{the data is } \boldsymbol{\mu}\text{-reconstructable;} \\ & (6.11) \text{ and } (6.12), \ X_j \in \{0, 1, 2, \ldots, \alpha\} \text{ for } j \in \mathcal{N}; \\ & \sum_{i \in \mathcal{S}} \beta_j^{(i)} \leq 1, \ j \in \mathcal{N}; \ \beta_j^{(i)} \in \{0, 1\}. \end{aligned} \tag{6.13}$$

The last constraint dictates that each channel, if used, can only transmit symbols from one node. The originally full-rank condition shown in Definition 1 (in Section 6.2.5) for the $\boldsymbol{\mu}$-reconstructablity significantly complicates the resource allocation problem, and makes the problem seemingly intractable. The reconstructability condition shown in Section 6.3, which transforms such full-rank condition into the constraint in terms of summation of μ_i, significantly simplifies the wireless resource allocation problem.

The network equivalence theory [22] states that the demand of a network with noisy, independent, and memoryless links can be met if and only if it can be met on another network where each noisy link is replaced by another noiseless one with the same capacity. This means that the codes sent over the wireless channel can be independent of the codes used for data storage, which justifies the problem formulation using the noiseless channel capacity given by (6.11).

6.4.2 Optimal channel and power allocation for the relaxed problem

This section provides a solution to the channel and power allocation problem (6.13) for the MSR and MBR points based on the data reconstructability condition obtained in

Theorems 6.2 and 6.3. First drop the constraint $\mu_i \leq \alpha, i \in S$ for the MSR point, and (6.10) for the MBR point, and then minimize the sum power only with the constraint $\sum_{j \in \mathcal{N}} X_j = \sum_{i \in \mathcal{S}} \mu_i = M$. Then check whether the dropped original constraints are violated, and perform local adjustments if so. Otherwise, the solution to the relaxed problem obtained in the first step is also the optimal solution to the original one (i.e. problem (6.13)); and simulation results show that this is indeed the case most of the time.

According to (6.11), letting $p(\cdot)$ be the inverse function of the capacity function $c(\cdot)$ in (6.2), it follows that

$$P_j = \frac{p(X_j)}{\sum_{i \in S} \beta_j^{(i)} |g_j^{(i)}|^2}. \tag{6.14}$$

The relaxed version of problem (6.13) is given by

$$\begin{aligned} &\min_{\{\beta_j^{(i)}, X_j\}_{i \in \mathcal{S}, j \in \mathcal{N}}} \sum_{j \in \mathcal{N}} \frac{p(X_j)}{\sum_{i \in S} \beta_j^{(i)} |g_j^{(i)}|^2} \\ &\text{s.t.} \sum_{j=1}^{N} X_j = M; \; X_j \in \mathbb{Z}^+ \cup \{0\}; \\ &\sum_{i \in \mathcal{S}} \beta_j^{(i)} \leq 1, \; j \in \mathcal{N}; \; \beta_j^{(i)} \in \{0, 1\}. \end{aligned} \tag{6.15}$$

Then the optimal strategy for channel assignment is to assign each channel to the node with the strongest channel, i.e. for each channel $j \in \mathcal{N}$,

$$\beta_j^{(i_j)} = 1, \; \text{for } i_j = \arg\max \left|g_j^{(i)}\right|; \text{ and } \beta_j^{(i_j)} = 0, \; \text{otherwise}. \tag{6.16}$$

Letting $p_j(X_j) = \frac{1}{|g_j^{(i_j)}|^2} p(X_j)$, the power allocation problem then becomes

$$\begin{aligned} &\min_{\{X_i\}_{i \in \mathcal{N}}} \sum_{j=1}^{N} p_j(X_j) \\ &\text{s.t.} \sum_{j=1}^{N} X_j = M; \; X_j \in \mathbb{Z}^+ \cup \{0\}. \end{aligned} \tag{6.17}$$

Note that the above optimization problem for general function $p_j(x)$ is still intractable. However, for a special type of function $p_j(x)$, where $p_j(X+1) - p_j(X)$ increases with X, (6.17) can be solved optimally in a tractable manner. More specifically, the following greedy algorithm optimally solves (6.17).

- **Initialize** $X_j = 0$ for $j \in \mathcal{N}$.
- **While** $\sum_{j \in \mathcal{N}} X_j < M$ **do**
  - For $j \in \mathcal{N}$, let $\Delta P_j = p_j(X_j + 1) - p_j(X_j)$ be the power increment for channel j.
  - Find the channel $j_0 = \arg\min_{j \in \mathcal{N}} \Delta P_j$ with the minimum power increment, and update $X_{j_0} \leftarrow X_{j_0} + 1$.
- **Output** X_j for $j \in \mathcal{N}$.

The following result shows the optimality of the above greedy approach. The proof is provided in [3] and thus omitted in this chapter.

Theorem 6.4 *The above greedy algorithm provides an optimal solution to the power allocation problem (6.17).*

The solution to power allocation problem (6.17) may violate the constraint that $\mu_i \leq \alpha, i \in S$ for the MSR point, or (6.10) for the MBR point. In that case, additional local adjustment needs to be added to guarantee that number of downloaded symbols satisfies the relaxed constraint. More specifically, based on the solution $\{X_j\}_{j\in\mathcal{N}}$ obtained by the *greedy algorithm*, we compute the number of symbols assigned to each storage node, given by $\mu_i = \sum_{j\in\mathcal{N}} \beta_j^{(i)} X_j$ for $i \in \mathcal{S}$. If $\{\mu_i\}_{i\in\mathcal{S}}$ violates the constraint that $\mu_i \leq \alpha, i \in S$ for the MSR point, or (6.10) for the MBR point, we perform the following local adjustment.

Local adjustment for the MSR point

The basic idea is to find a storage node i with $\mu_i > \alpha$, and reassign one of its subchannels to another storage node to decrease μ_i, until $\mu_i \leq \alpha$ for all $i \in \mathcal{S}$. The local adjustment scheme is given as follows.

- **While** $\mu_i > \alpha$ for some $i \in \mathcal{S}$, **do**
  - Find a storage node i with $\mu_i > \alpha$ and the set of assigned subchannels denoted as $\mathcal{N}_i = \{j : \beta_j^{(i)} = 1\}$;
  - Find the storage node $i' \in \mathcal{S} \setminus \{i\} \triangleq \mathcal{S}'$ and the channel $j \in \mathcal{N}_i$ that minimizes the power increment of reassigning the X_j symbols in channel j to node i' such that $\mu_{i'} + X_j \leq \alpha$, i.e.

$$(i_0, j_0) = \underset{(i',j)\in\mathcal{S}'\times\mathcal{N}_i:\ \mu_{i'}+X_j\leq\alpha}{\arg\min} \quad p_j^{(i')}(X_j) - p_j^{(i)}(X_j). \tag{6.18}$$

  Reassign the X_{j_0} symbols in channel j_0 to storage node i_0, by letting $\beta_{j_0}^{(i_0)} = 1$ and $\beta_{j_0}^{(i)} = 0$.

Local adjustment for the MBR point

The local adjustment for the MBR point follows the same procedure as that for the MSR point. The only difference is that when selecting the storage node–channel pair in (6.18), we reassign one symbol from the storage node with the maximum μ_i to the storage node with the minimum μ_i, while minimizing the power increment of reassigning the symbol.

6.5 Open research topics

6.5.1 General research topics for wireless distributed storage networks

Multiple data files and multiple data collectors

The current research work focuses only on data downloading for single data file and single data collector. However, in practical scenarios, there may be multiple data files to be downloaded by multiple data collectors. Then, how to design a data downloading protocol to minimize the total data downloading cost or increase the multiple data collector fairness remains an open research topic.

Data encoding and uploading

The inverse of data downloading and reconstruction is data encoding and uploading. The data encoding and uploading topics involve the encoding rate and uploading node selections, to find a balance between the data encoding efficiency and reliability.

Heterogeneous data storage networks

The idea comes from the heterogeneous wireless communication networks, which places extra base stations in hotpot communication requirement areas. For the hotpot data request areas, we can also place extra secondary data storage nodes and even networks. Then, how to design data uploading/downloading protocols in such networks remains another open research topic.

6.5.2 Research topics for data storage in optical wireless networks

Distributed optical wireless channel interference

In a typical lighting and optical communication system, different lights are usually distributed on the ceiling to cover a service area. Their beams may cover the same spot where a DC is located due to certain angle of each beam. In such a case, multiple subchannels may be strongly coupled and correlated, causing co-channel interference. Subsequently, the channel interference condition affects the data reconstructability conditions to satisfy. Thus it is valuable to investigate inter-play between optical subchannels and data reconstruction performance.

Storage node relaying

Although optical wireless networks offer potentially high per-link channel capacity, one disadvantage of the optical wireless communication link is that it can be easily blocked by obstacles. If this happens for the communication link between one storage node and the data collector, one solution is to let another data storage node serve as the relay for the data of that storage node. The relay selection and the wireless resource allocation under the relaying scenarios remain as future research topics.

6.5.3 Research topics for data storage in named data networks

Current named data network (NDN) has attracted extensive research interests. A natural idea is to introduce it to the wireless scenario. It essentially adds a cache buffer between the communication backbone network and the user terminals, to locally store the frequently used data such that there is no further need to transmit such data through the communication backbone network each time upon the user request. The user terminals need only to download the frequently used data from the storage nodes. The proposed distributed storage network could provide a roadmap for the user–cache interface of the NDN.

There are two differences between the wireless distributed storage network and the wireless NDN. The NDN adds an interface consisting of cache between the downlink transmission and the communication backbone network. In other words, for NDN the distributed storage nodes are connected to the communication backbone network, and can either transmit data to or receive data from the communication backbone network. Another difference is that the latter does not necessarily use the storage code to store the data in a distributed manner. However, whether to code and store the frequently used data in a distributed manner needs to be investigated, especially to guarantee robust data storage and reconstruction.

References

[1] A. G. Dimakis, P. B. Godfrey, Y. Wu, M. J. Wainwright, and K. Ramchandran, "Network coding for distributed storage systems," *IEEE Trans. Info. Theory*, vol. **56**, no. 9, pp. 4539–4551, September 2010.

[2] Y. Wu, "Existence and construction of capacity-achieving network codes for distributed storage," *IEEE J. Sel. Areas Commun.*, vol. **28**, no. 2, pp. 277–288, February 2010.

[3] C. Gong and X. Wang, "On partial downloading for wireless distributed storage networks," *IEEE Trans. Signal Process.*, vol. **60**, no. 6, pp. 3278–3288, June 2012.

[4] S. Hranilovic and F. R. Kschischang, "Capacity bounds for power-and band-limited optical intensity channels corrupted by Gaussian noise," *IEEE Trans. Info. Theory*, vol. **50**, no. 5, pp. 784–795, May 2004.

[5] J. Armstrong and A. Lowery, "Power efficient optical OFDM," *Electron. Lett.*, vol. **42**, no. 6, pp. 370–372, 2006.

[6] S. Dissanayake, K. Panta, and J. Armstrong, "A novel technique to simultaneously transmit ACO-OFDM and DCO-OFDM in IM/DD systems," in *2011 IEEE GLOBECOM OWC Workshops*, IEEE, 2011, pp. 782–786.

[7] S. C. J. Lee, S. Randel, F. Breyer, and A. M. Koonen, "PAM-DMT for intensity-modulated and direct-detection optical communication systems," *IEEE Photonics Technol. Lett.*, vol. **21**, no. 23, pp. 1749–1751, 2009.

[8] D. Tsonev, S. Sinanovic, and H. Haas, "Novel unipolar orthogonal frequency division multiplexing (U-OFDM) for optical wireless," in *2012 IEEE 75th Vehicular Technology Conference (VTC Spring)*, IEEE, 2012, pp. 1–5.

[9] N. Fernando, Y. Hong, and E. Viterbo, "Flip-OFDM for optical wireless communications," in *2011 IEEE Information Theory Workshop (ITW)*, IEEE, 2011, pp. 5–9.

[10] A. Shokrollahi, "Raptor codes," *IEEE Trans. Inform. Theory*, vol. **62**, no. 6, pp. 2551–2567, June 2006.

[11] M. Luby, "LT codes," in *Proceedings 43rd Annual IEEE Symposium on Foundations of Computer Science*, November 2002, pp. 271–282.

[12] Z. Kong, S. A. Aly, and E. Soljanin, "Decentralized coding algorithms for distributed storage in wireless sensor networks," *IEEE J. Sel. Areas Commun.*, vol. **28**, no. 2, pp. 261–267, February 2010.

[13] C. Suh and K. Ramchandran, "Exact-repair MDS code construction using interference alignment," *IEEE Trans. Info. Theory*, vol. **57**, no. 3, pp. 1425–1442, March 2011.

[14] Y. Wu and A. G. Dimakis, "Reducing repair traffic for erasure coding-based storage via interference alignment," in *2009 IEEE International Symposium Information Theory (ISIT)*, Seoul, Korea, June 28–July 3, 2009, pp. 2276–2280.
[15] C. Suh and K. Ramchandran, "Exact-repair MDS codes for distributed storage using interference alignment," in *2010 IEEE International Symposium Information Theory (ISIT)*, Austin, Texas, June 2010, pp. 161–165.
[16] K. V. Rashmi, N. B. Shah, P. V. Kumar, and K. Ramchandran, "Explicit and optimal exact-regenerating codes for the minimum-bandwidth point in distributed storage," in *2010 IEEE International Symposium Information Theory (ISIT)*, Austin, Texas, June 2010, pp. 1938–1942.
[17] N. B. Shah, K. V. Rashmi, and P. V. Kumar, "Flexible class of regenerating codes for distributed storage," in *2010 IEEE International Symposium Information Theory (ISIT)*, Austin, Texas, June 2010, pp. 1943–1947.
[18] V. R. Cadambe, S. A. Jafar, and H. Maleki, "Minimum repair bandwidth for exact regeneration in distributed storage," in *2010 Wireless Network Coding Workshop (WiNC)*, Boston, Massachusetts, June 2010, pp. 1–6.
[19] K. V. Rashmi, N. B. Shah, P. V. Kumar, and K. Ramchandran, "Explicit construction of optimal exact regenerating codes for distributed storage," in *2009 Allerton Communication, Control, and Computing (Allerton)*, October 2009, pp. 1243–1249.
[20] N. B. Shah, K. V. Rashmi, P. V. Kumar, and K. Ramchandran, "Explicit codes minimizing repair bandwidth for distributed storage," arXiv:0908.2984v2 [cs.IT] 5, September 2009.
[21] K. V. Rashmi, N. B. Shah, and P. V. Kumar, "Optimal exact-regenerating codes for distributed storage at the msr and mbr points via a product-matrix construction," arXiv:1005.4178v1 [cs.IT] 23, May. 2010.
[22] R. Koetter, M. Effros, and M. Medard, "On a theory of network equivalence," in *2009 IEEE Information Theory Workshop (ITW)*, Volos, Greece, June 2009, pp. 326–330.

7 Big data aware wireless communication: challenges and opportunities

Suzhi Bi, Rui Zhang, Zhi Ding, and Shuguang Cui

The fast-growing wireless data service is pushing our communication network's processing power to its limit. The ever-increasing data traffic poses imminent challenges to all aspects of the wireless system design, such as spectrum efficiency, computing capabilities, and backhaul link capacity, etc. At the same time, the massive amount of mobile data traffic may also lead to potential system performance gain that is otherwise not achievable with conventional wireless signal processing models. In this chapter, we investigate the challenges and opportunities in the design of scalable wireless systems to embrace such a "big data" era. We review state-of-the-art techniques in wireless big data processing and study the potential implementations of key technologies in the future wireless systems. We show that proper wireless system designs could harness, and in fact take advantages of the mobile big data traffic.

7.1 Introduction

After decades of rapid growth in data services, modern society has entered the so-called *"big data" era*, where the mobile network is a major contributor. As of the year 2013, the global penetration of mobile subscribers had reached 92%, producing staggeringly 6800 PetaBytes (6.8×10^{18}) of mobile data worldwide [1]. The surge of mobile data traffic in recent years is mainly attributed to the popularity of smartphones, mobile tablets, and other smart mobile devices. Mobile broadband applications such as web surfing, social networking, and online videos are now ubiquitously accessible by these mobile devices, without limitations from time and location. The recent survey shows that the number of smartphone users currently accounts for merely 25%–30% of the entire mobile subscribers. However, the figure will double in the next three years and continue to grow given the considerable room in the smartphone market for further uptake. With a compound annual growth rate of 45%, it is expected that the mobile data traffic will increase by ten times from 2013 to 2019.

In addition to the vast amount of wireless data, wireless signal processing often *amplifies* the system big data effect in the pursuit of higher performance gain. To combat the fading channel, diversity schemes, especially the MIMO antenna technologies,

Big Data over Networks, ed. Shuguang Cui, Alfred O. Hero III, Zhi-Quan Luo, and José M. F. Moura.
Published by Cambridge University Press.

are extensively used in both mobile terminals (MTs) and base stations (BSs). Numerous schemes with co-located and distributed antennas have been proposed over the years to increase the data rate or extend the cellular coverage. This, however, also increases the system data traffic to be processed in proportion to the number of antennas at the BSs. It is foreseeable that the future use of massive MIMO technology will further increase the system data traffic by an order of magnitude [2]. On the other hand, the 5G (fifth-generation) wireless network under development is likely to migrate the current BS-centric cellular system design to a cloud-based layered network structure, consisting of a large number of cooperating wireless access points connected by either wireline or wireless backhaul links to a big data capable processing center. New wireless access structures, such as coordinated multipoint (CoMP or networked MIMO), heterogeneous network (HetNet) and cloud-based radio access network (C-RAN), are proposed to achieve multi-standard, interference-aware and energy-friendly (green) wireless communication [3–5]. In general, the cloud-based structure allows multiple relay-like base stations, named remote radio heads (RRHs), to serve mobile users cooperatively by jointly precoding/decoding the transmitted/received signals at a central unit (CU) via high-speed fiber backhauls to achieve joint signal processing gain. Owing to the use of cooperating RRHs, the backhaul traffic generated from a single user signal of MHz bandwidth could be easily scaled up to multiple Gbps [5]. Under high traffic load, the data transmitting to/from the CU may overwhelm the backhaul links or the CU's computing capacity. Besides, the overhead of exchanging control signals between the CU and RRHs is also significant in the communication network. This system traffic, together with the fast growing mobile user traffic, leaves behind both the processing power improvement speed of our current computing capabilities and the backhaul link rate increase pace of our networking systems.

Luckily, the vast mobile data traffic is not completely chaotic and beyond management. Rather, it often exhibits strong *statistical features*, including the user mobility pattern, the geographical, temporal, and social correlation of data contents, etc. For instance, more than 80% of mobile subscribers visit on average fewer than four locations in a day, such as working places, homes and shopping centers, etc [6]. Mobile data generation concentrates at these popular places and over the routes that connect them. Besides, the contents are rather dissimilar over locations and/or times. Users in a theater frequently query the reviews of a play and the backgrounds of actors and directors; home users are prone to browse the latest news in the morning but popular videos at night after work [7]. Moreover, the popularity of mobile social networking introduces a new concept of "virtual community" where people who share similar interests, social relations, or life experiences generate and require similar contents across the barrier of actual social context [8]. The above features of mobile traffic provide the opportunity to harness the bright-side of big data for potential performance gains in various wireless services, if they could be correctly recognized, extracted, and utilized. For instance, caching popular contents at wireless hot spots could effectively reduce the real-time traffic in the backhaul. Besides, the system control decisions, such as routing, instead of being hardware programmed, could be made data-driven to fully capture the interplay between the big data and network structure. Thanks to the recent development of software defined network (SDN) technology [9], we could achieve such flexible system control with the integration of *big data analytical tools* at a relatively low cost.

Besides the mobile user traffic pattern, the data traffic in the backhaul is also highly correlated, and often contains sizable *redundancy*, mainly due to the broadcasting nature of wireless communication and the efforts to combat the harsh wireless channel conditions. This opens up the opportunities of various redundancy reduction methods that balance between the overall system performance and the volume of backhaul data traffic. For instance, often not the entire received radio signal at a BS needs to be forwarded to the CU. Instead, part of the data traffic could be processed locally by the BS, or even discarded without compromising much of the overall decoding performance. This calls for a *hybrid data processing* structure that is able to opportunistically choose from processing the signals at the local BS-level, the centralized CU-level, or at both levels in parallel. This new signal processing paradigm also spurs a number of traffic reduction techniques that exploit the redundancy and signal correlation in the backhaul traffic. Interesting examples include antenna beamforming/selection, subchannel selection, user–BS association, opportunistic BS-level decoding and data compression, etc. In addition, the combinations of these techniques with different application scenarios, e.g. uplink/downlink transmissions, analog or digital fiber-optic backhaul, web caching and data-aware processing, etc., pose many interesting, and in some cases challenging, system design problems in the hybrid structure to support wireless big data.

In the rest of this chapter, we first introduce in Section 7.2 a general network architecture to support big data traffic. Within the proposed framework, we expand our discussion in Section 7.3 on various physical-layer techniques to control the backhaul traffic. In Section 7.4, instead of treating big data traffic as a hazard, we exploit the potential performance gain from the learning over big data traffic pattern to build a big data aware and SDN-based intelligent wireless network. Illustrative examples are provided along with the discussions in each section. Finally, we conclude this chapter in Section 7.5.

7.2 Scalable wireless network architecture for big data

7.2.1 Hybrid processing structure

The big data challenge to the wireless system calls for a *scalable* network structure. Nonetheless, both the current cellular systems and the C-RAN [5] under development fail to provide a satisfactory solution. The current 3G and 4G cellular systems in general follow a BS-centric design, where a BS carries all the responsibilities of radio access, baseband processing, and mobility management execution to serve the mobile users in the vicinity. To meet the fast-growing mobile data traffic, a smaller cell size is commonly used to improve the frequency reuse, which, however, induces severe inter-cell interference. Besides, small-cell is expensive in further extension due to the high deployment cost of the dense BSs. The cloud-centric network proposed for 5G, on the other hand, mitigates the inter-cell interference by centralized signal processing and

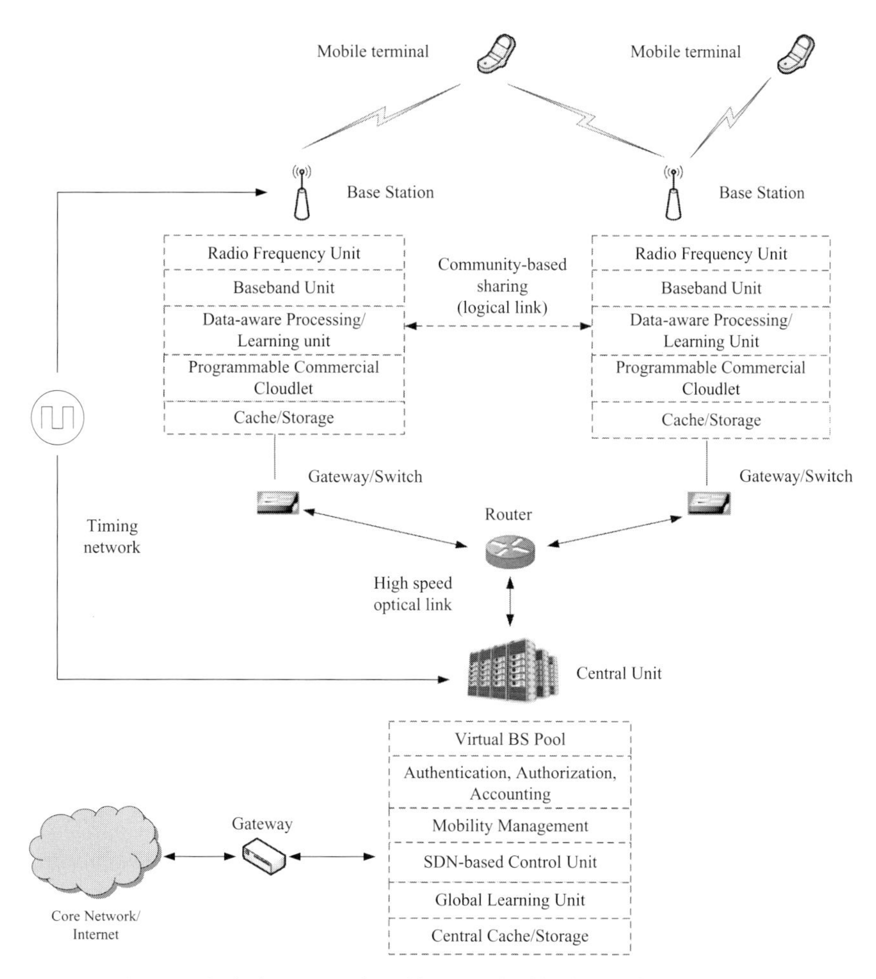

Figure 7.1 A general wireless network architecture for big data traffic.

reduces the deployment cost by moving the computations to the "cloud", while using cheap relay-like RRHs only for radio frequency (RF) level wireless access. However, this fully centralized scheme may suffer from overwhelming data traffic beyond its backhaul link capacities and the CU's computational power. In a word, neither the BS-centric nor the current cloud-centric design provides a scalable solution in the rise of the big data era.

Alternatively, a *hybrid* structure could strike a balance between the two design paradigms; that is, a wireless system that could adaptively choose from local processing at the BS-level, central processing at the CU-level, or at both levels in parallel, based on, for instance, physical channel conditions and correlations in the contents, etc. We thus consider a general system structure shown in Figure 7.1, which mainly inherits the skeleton of C-RAN, but has integrated several programmable modules that enable intelligent signal processing at both the BSs and the CU. Specifically, each BS is equipped with multiple antennas, using spatial multiplexing, frequency division, or other multiplexing methods, to provide multiple access to mobile terminals. Besides, the BSs could serve the MTs cooperatively under the coordination of the CU and with the help of a dedicated timing network for synchronization. The CU has a pool of baseband units, one for each BS, to jointly encode/decode the signals transmitted/received by the BSs. The CU is also connected to the core network via a gateway unit, which has the similar functions as the serving-gateway (S-GW) and packet data network gateway (P-GW) in the long-term evolution (LTE) system, to retrieve the files/services requested by the MTs. In addition, the CU is responsible for all the system administrative and control functions, such as routing, user scheduling, authentication, billing, mobility management, etc. Notice that Figure 7.1 only provides a conceptual framework that allows for some key technologies that we have envisioned, such as joint BS–CU processing and the use of data-analytical tools, etc. The detailed function of each module and the corresponding signal processing methods may vary depending on the applications, and will be specified later in Sections 7.3 and 7.4.

In between, the BSs and the CU are connected via *high-speed fiber backhauls*, where the wireless system designer could select from the optical analog and digital modulation schemes for the uplink/downlink information exchange. The optical analog modulation using radio frequency (RF) signal as the input is commonly referred to as radio-over-fiber (RoF) technology. Alternatively, analog RF input signal could be quantized and encoded into binary codewords for digital fiber-optic communication (DFC) [10]. In both cases, the fiber-optic backhaul has its communication limit, expressed in the forms of bandwidth and data rate, respectively. Transmission rate beyond the capacity limits will result in severe signal distortions, thus poor decoding performance. The communication limit is mainly determined by the fiber material dispersion, and also other factors such as material attenuation and noises, etc. A commercial fiber-optic system normally operates at a bandwidth of tens of GHz (or tens of Gbps for digital communication) over a single optical carrier [10]. To further increase the transmission rate, wavelength-division multiplexing (WDM) technology is commonly used to multiplex a number of optical carrier signals onto a single optical fiber. In practice, RoF and DFC have their respective advantages and limitations. The choice of the fiber-optic communication mode needs to take into consideration several factors, such as the deployment budget, future extension and communication requirements, etc. We will discuss the use of the above backhaul technologies in more detail in Section 7.3.1. Another important issue is the signal delay from the BS- and CU-level signal processing and data transmission in the backhaul, which may significantly degrade the performance of applications with stringent delay requirements. We will introduce various methods to reduce the signal delay in following sections, such as using an analog backhaul, deploying a BS-level cloudlet, and implementing distributed and online signal processing algorithms, etc.

Each BS has a set of RF modules to provide radio access to the mobile users. For the simplicity of illustration, here we mainly consider examples in uplink communication. Similar modules are also applicable to downlink communication and will be detailed in Section 7.3. Upon receiving the analog RF signals from the MTs, a BS could choose to merely amplify and forward (AF) the received radio signals to the CU using the RoF technology. Alternatively, it could also use pulse-amplitude modulation (PCM), or other modulation methods, to sample, quantize and encode the RF signal into binary codewords and transmit digitally to the CU. Other than the simple forwarding methods, joint BS–CU processing could be used to reduce the redundancy in the backhaul traffic. For instance, BSs could further compress the signal using *compression* codebooks optimized by the CU based on the signal correlations across different BSs. Alternatively, antenna beamforming, bandwidth selection and other redundancy reduction methods could be used to discard some of the received signals before forwarding the remaining part to the CU. This could be useful to achieve a balance between the backhaul traffic complexity and the joint decoding performance at the CU.

Each BS also has an optional baseband unit (BU) that is capable of decoding the messages from the mobile users. The baseband unit is optional as it may not be necessary in some system setups, such as the above RoF and PCM schemes. However, the introduction of *BS-level baseband capability* brings in a new set of processing mechanisms that are effective in managing the data traffic in the backhaul. For instance, the BS could now opportunistically decode some of the messages locally, based on the channel condition and the backhaul constraints. This evidently reduces the backhaul data rate, as some related data traffic is no longer necessary to be sent to the CU. Besides, some decoded messages could be selectively sent to the CU as the side information to reduce the complexity of joint decoding. In the downlink case, the BU is able to encode and modulate baseband symbols to RF signals and then transmit to the MTs. Therefore, instead of transmitting the complete signal waveforms (or waveform samples) to the BSs, the CU may save the backhaul bandwidth by transmitting separately the information symbols and the beamforming vector, while leaving RF modulation to the BSs. Another interesting implementation brought by the baseband capability at the BSs is the integration of *web caching* technology in the wireless infrastructure. Intuitively, caching at the BSs and CU prohibits downloading of repeated contents from the core network. Careful design of caching mechanisms could significantly increase the effective system bandwidth, and thus enhancing the wireless system scalability to the soaring data traffic growth, as will be discussed in more detail next.

7.2.2 Web caching in wireless infrastructure

Upon receiving a request for a web object from a mobile user, the CU needs to retrieve the requested content from the core network via the gateway, and responds to the user by transmitting the content to the corresponding BSs. This may lead to severe congestion at the gateway and the CU–BS links in the context of mobile big data. However, despite of its massive volume, mobile traffic also contains significant redundancies, as many requests are indeed serving the same content object, possibly with a number of aliased addresses. For instance, [7] finds out that around 40% of mobile traffic is repeated data, dominated by the popular contents from wideband services such as YouTube videos. Besides, mobile traffic is highly correlated in location and time, whose volume peaks occur in a relatively stable pattern at popular locations and specific time slots. Caching the popular contents in the wireless infrastructure, especially in the base stations serving popular places and at the peak hours, is a *cost-effective* solution to reduce the real-time traffic in the backhaul, considering the yearly decreasing production cost of caching equipments.

The wireless network in Figure 7.1 has caches installed at both the BSs and the CU, which record the popular contents traversed from the core-network. Such a hieratical caching structure is effective in handling user mobility. For instance, as a MT moves to another BS, the CU is aware of the transmission progress of the content to the MT and could send to the new BS only the remaining part of the object. The transition could be made transparent to the MT without interrupting its TCP connection. For cache maintenance, an object in the cache is labeled by its address in the core network (e.g. the URL), the file length, and the hash computed out of its first K bytes. Then, if the URL requested by the user has a match in the BS's (CU's) cache, the BS (CU) does not need to download the content from the CU (core network). In some cases, the same object may have aliased URLs. However, an aliased object could be effectively recognized after downloading the first K bytes from the core network, by comparing its file length and the first-K-byte hash with those in the cache [7]. If a match is found, then the BS/CU could stop downloading and use the cached object instead. The caches are updated regularly in the interval of minutes, to remove the outdated contents and make room for the new popular contents. It is found that the requests for the same popular object often appear in a bursty manner. Caching the popular contents for even only 10 minutes could save comparable bandwidth than that caching for an hour [7]. Some advanced cache organization and maintenance mechanisms are performed by the built-in learning units (LUs), which will be introduced next at Section 7.2.3.

Caching also introduces a new degree of freedom in wireless resource allocation. For instance, caching could now be an important factor in the MT–BS association decision. Intuitively, the achievable downlink data rate of a single CU–BS–MT link is determined by the minimum between the BS–CU wireless channel capacity and the allocated CU–BS backhaul link capacity. The end-to-end data rate of a link may still be very low if the BS–CU backhaul link is congested, despite that the BS–MT wireless channel may be in good condition. In this case, the MT could be better off if associated with a BS that has its requested content in the cache, even if the wireless channel condition is relatively poor compared with the other BSs. A general cache-assisted resource allocation scheme will be discussed in Section 7.3.2. Besides, cache-assisted resource allocation could also exploit the temporal and geographical correlation of mobile traffic. For instance, the CU could pre-feed the popular contents to the BSs in the off-peak hours to reduce the real-time traffic in the peak hour. It could also pre-feed the contents requested by a specific user, in the BSs along the user's route to reduce the connection delay. We notice that in both applications, the CU must be aware of the features of mobile traffic, i.e. the aggregate demand profile and the user mobility patterns. In general, these features are not explicitly informed, but need to be extracted from the vast data traffics. This requires the wireless system to be equipped with data analytical tools, which will be introduced in the next subsection.

7.2.3 Data aware processing units

As mentioned in Section 7.1, the mobile traffic contains strong statistical features. *Data aware processing* units (DPUs) are able to analyze the data traffic, recognize the correlations and extract the features within. In some cases, they could even predict the traffic pattern or mobile user behavior based on the historical data collected. Data analytical tools, such as stochastic modeling, data mining and machine learning, etc., are the key enabling factors. The promising data analytical tools and their applications to wireless data traffic will be detailed in Section 7.4.1. Instead of treating mobile big data as a complete hazard, DPUs leverage the statistical features within to improve the overall wireless service qualities.

One important function of DPUs is cache management. For obvious reasons, contents in the caches cannot be randomly piled up under big data traffic. Instead, they should be carefully categorized, compactly organized, and timely updated. This is achieved by the learning unit (LU), a module of DPU, in both the BSs and the CU. Specifically, the LU at a BS classifies the data traffics into a number of subclasses based on the type of contents, such as music videos and news pictures, etc. As its name suggests, the LU could "learn" from the data traffic to judge the popularity of an object. For instance, an object's popularity could be evaluated based not only on its own access count but also the total access count of its type, which reflects the average frequency of potential future accesses. Popular contents are continuously cached while the unpopular contents are removed regularly. The content classification is also useful to form a *context-aware* cache organization [11]. The global learning unit at the CU could exploit the connection between the contents in the caches and those in the physical world, making use of the context in which the mobile traffic is generated. For instance, the LU could establish the connection between the large volume of film trailer videos in a BS's cache and the actual movie theater in the vicinity of the BS. In practice, the context-aware caching method could optimize the caching strategy in different BSs based on their correlations with the physical world.

Besides, the DPUs could be used to provide *data-driven services* in a *mobile cloud computing* environment. The LUs could look at the data traffic from both the macro and micro perspectives. That is, the LU could learn the aggregate traffic pattern of the general public and also the traffic profile of some specific individuals. The information contained in the mobile traffic may include the aggregate data traffic distribution, the peaks and valleys of data usage in locations and times, the user mobility patterns, and individual demand profiles, etc. Based on the features in the historical data, the wireless cloud could derive the accurate prediction about the data traffic, and provision the resource allocation in the wireless cloud. For instance, based on the predicted traffic volume, the cloud data center could schedule the on and off states of the computing devices to save the power consumptions and reduce the outage probability. Besides, the CU could pre-feed the contents at the popular locations, or along the route of certain users to reduce the backhaul delay and real-time traffic volume. Owing to the recent popularity of smartphones, the data traffic also contains rich sensor information, such as location and peripheral noise level, which could be used to infer the context in the physical world of the mobile user. The context-aware computing blurs the border between the virtual and physical worlds of computing, and is a promising method to improve personalized services. There are also several emerging data-driven mobile cloud computing structures, such as crowd computing (also known as crowd-sourcing), social-aware computing, and mobile cloudlet, etc. We will introduce state-of-the-arts of data-driven mobile cloud computing in Section 7.4.2.

Last but not least, the data-analytical capability enables a *software-defined networking* (SDN) design in the wireless system. Programmable SDN-based modules enable the routing/switching decisions to be adaptive to the wireless data traffic distributions, the wireless channel conditions, the congestion levels of the backhaul links, and the operating conditions of computing units, etc. For instance, the downlink data traffic to a user could either be routed to a BS with a good wireless condition but a congested backhaul link, or some other cooperating BSs with relatively bad wireless conditions but lightly loaded backhaul links, depending on the real-time network data traffic level. Besides, the integration of web-caching technology makes routing a more interesting and challenging problem. In the wireless system of the future, the routing design in the big data era not only stays in the network layer, but becomes a *cross-layer* problem with many joint considerations listed above.

Based on the proposed general wireless network infrastructure in Figure 7.1, we have briefly discussed above the new design opportunities and challenges in the big data era. In the following sections, we will expand our discussions in specific design topics along with illustrative examples.

7.3 Wireless system design in big data era

In the big data era, the conventional metrics of wireless system design, such as data rate and bit-error-rate (BER), must be optimized under the constraints of backhaul link capacity and the processing capability of the CU. In other words, the system design needs to follow a broader *trade-off* between the system performance and the backhaul complexity. In this section, we discuss the wireless system design methods under the hybrid processing structure in Figure 7.1. We first discuss in Section 7.3.1 the choice of analog or digital backhaul technology and its impact on the overall system design. In Section 7.3.2, we focus on using digital backhaul and introduce promising joint BS–CU processing techniques to reduce the backhaul complexity.

7.3.1 Analog vs. digital backhaul

For better illustration, we present in Figure 7.2 an extension of the conceptual wireless architecture given in Figure 7.1. At the wireless access layer, we consider the universal frequency reuse scheme, where the entire radio spectrum is divided into parallel subchannels and shared among all the BSs. BSs could cooperate on each subchannel to serve multiple MTs simultaneously, e.g. BSs 1–3 serve MTs 1–2 using the same frequency subchannel, while BSs 2–4 serve MT3 with another subchannel. In the uplink transmission,

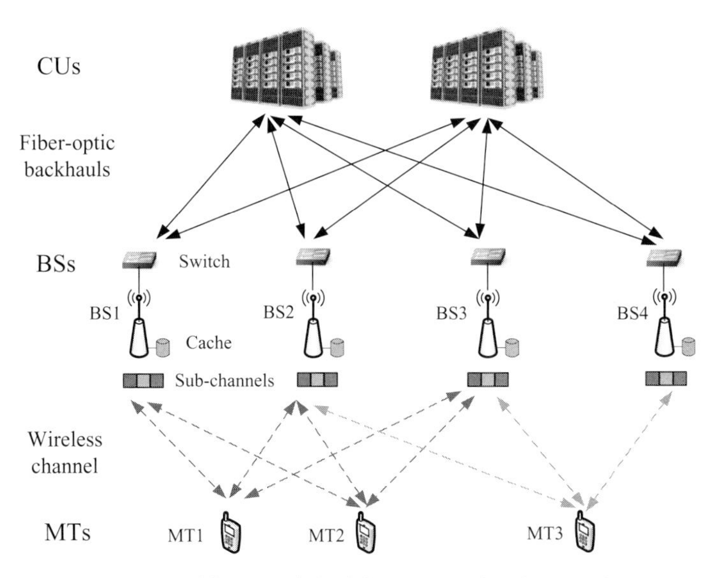

Figure 7.2 A general framework for joint BS–CU signal processing.

the messages transmitted from different MTs on the same subchannel could be resolved at the BS (or CU) using multi-user detection techniques, e.g. MIMO spatial multiplexing. In the downlink case, the co-channel interference could be mitigated using multi-user precoding. In the joint BS–CU processing framework, signal-processing duties could be split among the BSs and the CUs depending on the applications. In general, BSs are connected with multiple CUs by the backhaul infrastructure, which enables flexible data routing based on the link congestion level, occupancy rate of the CUs, and cache availability, etc. To highlight the essence of joint BS–CU signal processing techniques, we consider a simple single-CU case in this section, while the techniques could also be extended to multiple-CU cases. Here, we first discuss the choice of backhaul technology and its impact on the wireless system design. The detailed joint BS–CU processing methods will be discussed in Section 7.3.2.

A wireless system could select between the optical analog and digital modulation schemes for the information exchange in the backhaul. The commonly used analog scheme in wireless systems is the radio-over-fiber (RoF) technology, where the RF signal to/from the MTs directly modulates the optical carrier in the fiber-optic channel, like the conventional amplitude modulation scheme in RF communication. The RoF scheme simplifies the BS to a simple relay used only for wireless access purpose. The simplicity of RoF has several practical advantages. First, this simple scheme reduces the deployment and future extension cost, which accounts for over 40% of the total cost of a carrier operator [5]. In fact, this is the major reason why RoF is highly interesting despite of its many limitations. Second, RoF achieves lower processing delays than its digital counterpart as it does not incorporate any error control or other feedback mechanisms. The reduced delay could benefit many delay-sensitive mobile applications, such as mobile gaming, augmented reality and language processing, etc. Third, RoF is compatible with multiple wireless standards, e.g. 3G, LTE, and WiFi, as it is *oblivious* to the user codebooks and modulation schemes. However, the use of RoF also has its limitations compared with the digital fiber communication (DFC) schemes. First, RoF is susceptible to distortions and noise due to the lack of error control and the analog nature of the information contained. Second, the analog RF signal is difficult to synchronize in the downlink cooperative transmissions of multiple BSs. In contrast, DFC could achieve much easier synchronization using some specific synchronization sequences [12]. Third, the RoF scheme is less flexible in terms of the applicable signal processing methods. For instance, the digital PCM scheme in DFC could be combined with data compression, local decoding and many other advanced digital signal processing techniques to balance the overall decoding performance and backhaul traffic rate, while RoF could not.

Different as they are, both the RoF and DFC schemes may suffer from severe performance degradation under heavy data traffic. RoF carrying large-bandwidth RF signals enhances the noise from high optic transmission power and the intrinsic relative intensity noise from always-on laser source [10]. Besides, some multi-user access schemes, such as orthogonal frequency division multiple access (OFDMA), may produce high peak-to-average power ratios. This could result in some irreversible nonlinear distortion during the intensity modulation when the input signal's amplitude is beyond the modulator's linear range. These combined effects could severely degrade the decoding performance at the CU. As DFC is more noise- and distortion-resistent, it may significantly increase the data rate in the backhaul. For instance, suppose a user signal of 50 MHz bandwidth sampled at a rate of 100 M samples per second, and each sample quantized and represented by 16 bits. When the user is served by five cooperating BSs each with eight antennas, the total data traffic is scaled up to 64 Gbps [5]. This is obviously not a scalable solution for broadband wireless services. From the above discussion, the final choice of backhaul technology needs to take into consideration of several factors, such as the budget, link capacity, and communication quality requirements, etc.

The choice of backhaul technology determines the set of signal processing methods applicable to the joint BS–CU processing framework. We here consider the case when RoF is used and leave the DFC cases to Section 7.3.2. In the uplink, a BS could directly modulate the optical carrier with the received radio signal. In the downlink, the CU computes and generates the entire RF signal waveform for each antenna of BSs. A common technique to support MIMO communication is to set up a dedicated fiber link between each antenna at the BSs and the CU. Evidently, this is costly in deployment and inflexible for future antenna extension. Alternatively, since the RF bandwidth is generally small compared to the bandwidth of a fiber-optic cable, it is viable to use WDM to multiplex a number of non-overlapping bands, each for an antenna, in a single fiber cable. Meanwhile, it is advisable to downconvert the received RF signals to intermediate frequencies (IFs), and then transmits on the optic-fiber cable. For obvious reasons, this saves the backhaul bandwidth to support more concurrent transmissions. At the same time, the reduced bandwidth also help us reduce the noise and distortion effects in the analog optic-fiber channel. An interesting idea to further reduce the signal bandwidth of RoF is to utilize carrier orthogonality [13], i.e. the transmitter multiplies two signals from different antennas respectively with an in-phase and a quadrature IF carrier of the same frequency. At the receiver side, the signals could be recovered using coherent detectors. Ideally, this could save about half of the signal bandwidth.

In the uplink transmission, the traffic reduction methods of RoF are restricted to discarding part of the received RF signals, including *subchannel selection* and *antenna beamforming* (selection). In the case of OFDMA, non-overlapping radio bandwidths are allocated to a dedicated mobile user at all the cooperating BSs. The received signal-to-noise ratio (SNR) of a particular bandwidth is different across the BSs. Hence, the BSs only need to forward a subset of the sub-channels to the CU using a bank of filters, while the joint processing at CU could still guarantee a satisfactory overall SNR performance for each MT. The subchannel selection decision is affected by the wireless channel condition, the allocated backhaul capacity, and the quality-of-service (QoS) requirement of each user, etc. When spatial multiplexing is used, the received user signals are overlapped in the frequency domain, thus cannot be separated by using frequency filtering. In this case, analog beamforming could be used to combine the received signals into lower dimensions. Similar techniques are also applicable to the downlink transmission, where the CU selects only a subset of BSs (bandwidth selection method) or a subset of spatial dimensions (beamforming method) for transmission, possibly with some variations such as per-antenna or sum transmission power constraints.

7.3.2 Joint base station and cloud processing with digital backhaul

The use of digital backhaul technology opens up a new set of joint BS–CU signal processing methods that are effective to control the backhaul traffic. In this subsection, we separate our discussions based on the different techniques used in the BS-level signal processing: (1) data compression; (2) BS-level encoding/decoding; and (3) BS-level caching. In particular, methods (1) and (2) are applicable to both the uplink and downlink transmission, while method (3) is especially useful for exploiting the redundancy in downlink resource allocation.

Data compression

In the uplink transmission, when BS needs to forward to the CU its received analog RF signals from the MTs, it requires infinite backhaul capacity to transmit an analog signal exactly. An analog signal could be more efficiently transmitted through the backhaul if it is quantized and compressed into binary codewords. Conceptually, data compression is able to map the original input sequence to a codeword in a codebook, and send the corresponding index to the decoder. Then, the decoder recovers the codeword from the same codebook at hand based on the index received. From an information theoretical perspective, the effect of data compression could be modeled as a test channel (often Gaussian for simplicity of analysis) with the uncompressed signals as the input and compressed signal as the output. In a Gaussian test channel model, the output signal is generated by the input signal corrupted by an additive Gaussian compression noise. The design of codebook is equivalent to setting the variance of the compression noise. To achieve successful compression, the encoder needs to transmit to the decoder at a rate of at least the mutual information between the input and the output over the

Gaussian test channel (cf. [14] Chapter 3). Since the transmission data rate in a backhaul link is constrained by the link capacity, intuitively, a more stringent backhaul capacity constraint would require a more "coarse compression" with a larger compression noise.

When multiple BSs compress and forward their received signals to the CU, the compression design becomes the setting for the covariance matrix of the compression noises across different BSs. A common objective is to maximize the information rate under the backhaul capacity constraints. In the cloud-based setup, the CU is assumed to have the knowledge of the global channel state information (CSI) at all BSs. Based on the global system knowledge (e.g. CSI, backhaul capacities, and QoS requirements), the CU could compute the optimal compression codebooks and then send to the corresponding BSs. With the received codebooks, each BS then compresses its received RF signals and sends to the CU for joint decompression and decoding.

In particular, from an information theoretical approach, [15] derives an achievable rate region expression for a generic multiple access channel (MAC). In its setting, *distributed Wyner–Ziv lossy compression* is used at the BSs, exploiting the signal correlation across the multiple BSs. However, evaluation of the rate region involves solving an exponential number of minimization problems with regard to the number of MTs in the backhaul, and each minimization has exponentially many constraints over the number of BSs in the network. Evidently, the computational complexity is prohibitively high in a large network and, more importantly, the characterization does not provide much insight into system design. For more tractable analysis, two special scenarios with symmetric geometries are considered, i.e. the circular Wyner model and the "soft-handoff" model, where the achievable per-cell sum rates and the asymptotic characterizations are derived. In particular, it finds that the optimal per-cell multiplexing gain could be achieved (as without backhaul constraints) if the backhaul capacity is allowed to scale at the logarithm of the received SNR at the BSs. Besides, [16] also characterizes the weighted-sum rate and the achievable rate region for a two-user case under an aggregate backhaul capacity constraint. The distributed Wyner–Ziv compression scheme is shown to yield significant capacity gain over *independent quantization* methods especially in the low backhaul capacity region. This is due to the fact that independent quantization is unable to reduce the backhaul redundancy without the exploitation of the signal correlation across BSs. A joint design of compression noise levels at all BSs to maximize the weighted sum rate under an aggregate backhaul capacity constraint is proposed in [17]. In addition, [18] shows that, instead of separating the decompression and decoding operations, the CU could improve the data rate by *jointly decompressing and decoding* the received signals from multiple BSs.

In practice, however, the implementation of distributed Wyner–Ziv compression is difficult mainly because of the high complexity in determining the optimal joint compression codebook and the joint decompressing/decoding at the CU. Alternatively, [19] proposes a reduced-complexity sequential distributed compression method, where at each step a BS compresses its received signal by taking the signals compressed by the BSs at the previous steps as side information. Besides, a similar sequential compression method exploiting the previously decoded user messages from the CU is investigated in [20]. To further reduce the signaling overhead and decoding delay, an independent compression method that uses a simple vector quantizer was also studied in [20], where the quantization codebook of a BS is only determined by its local SNR. Moreover, a per-BS successive interference cancelation (SIC) method is used to decode the user messages in sequence at the CU, which bears a much lower decoding complexity than the joint decoding scheme and leads to an easily computable achievable rate region. The SIC decoding method is proved to achieve the sum capacity of the Wyner soft-handoff model within a constant gap. Again, the backhaul capacity must scale at the logarithm of the signal SNR to approach the infinite-backhaul capacity performance.

Similar compression techniques are also applicable to the downlink. However, unlike the joint processing at the CU in the uplink, the mobile receiver has to cope with the compression noise independently. The conventional uplink/downlink duality in MIMO channel capacity does not hold in general except for some special cases, e.g. the low SNR characterization derived in [21]. Besides, precoding design to mitigate interference is an additional consideration in the downlink transmission design. Reference [21] considers a simple soft-handoff model with single-antenna BSs serving one single-antenna MT in each cell. Dirty paper coding (DPC) is used at the CU, which then compresses the DPC codewords using independent Gaussian quantization codebooks and sends them to each BS. It shows that in the high SNR region, the scheme falls short of achieving the per-cell capacity upper bound by at most one bit. Instead of independent compression, reference [22] takes into consideration the user signal correlation across BSs in the general MIMO downlink channel, and *jointly* designs the compression and precoding across all the CU-to-BS backhaul links. It shows that, by introducing correlation to data compression among the BSs, one could increase the achievable downlink data rate compared to independent compression for individual BSs. Besides, it also finds out that linear precoding, e.g. minimum mean square estimation (MMSE) and zero-forcing precoders, performs very closely to the optimal DPC scheme. In other words, the design of precoding and compression could be made separately without compromising much of the system performance.

BS-level encoding/decoding

In most of the above discussions on compression techniques, we assume that BSs are acting as simple relays to compress/decompress and forward the user signals in the uplink/downlink transmission. In this subsection, we consider BSs with advanced local processing capabilities, such as antenna/bandwidth selection and local encoding/precoding, to further improve the system performance under stringent backhaul constraints.

In the uplink transmission using digital backhauls, the BSs could simply reduce the backhaul traffic by *discarding* a subset of subchannels as discussed in Section 7.3.1. Besides, the antenna selection method could be applied to discard the received data streams from a subset of antennas. For instance, a general optimization framework in the uplink transmission is proposed in [23], where a mobile user's message is allowed to be detected by only a subset of active antennas in the distributed antenna system. Meanwhile, the quantization level of each active antenna is jointly optimized. Base station selection is considered in [19], where a sparsity inducing optimization (l_0-norm optimization with l_1-norm approximation) is proposed to jointly maximize the uplink data rates and the number of inactive BSs in the system.

Other than limiting the cooperating elements in the uplink, *distributed decoding* at the BSs is another effective method to reduce the information exchange in the backhaul. For instance, [15] considers a rate-splitting approach that divides a MT's message into two parts, where one part is decoded locally by the serving BS and the rest is compressed and jointly decoded by the CU. It derives an achievable rate for the rate-splitting approach in the circular Wyner model and demonstrates numerically that a good strategy is time sharing between the pure local decoding and joint processing at the CU. This is essentially a mixed BS-centric and cloud-centric design. Not surprisingly, the flexibility provided by the rate-splitting approach performs better than the joint CU processing scheme when the backhaul capacity is limited. However, the time sharing nature does not fully capture the opportunities of multi-user diversity in the uplink. An alternative method is to let BSs decode some user messages opportunistically, if the SNRs are sufficiently high [24]. In this case, the decoding BS could cancel the signal related to the decoded symbols in the received RF signals, and send to the CU a compressed version of the remaining part for joint processing. The decoded symbols are also selectively sent to the CU as the side information to cancel the interference in the received signals. To avoid the conflict in decoding, the CU should allow a tagged user message to be decoded at no more than one BS. There exists a clear tradeoff in the decision of whether decoding a MT's message at the BS or not. At one extreme, local decoding of all the MTs' messages at BSs is in general unreliable due to high interference levels. At the other extreme, the performance of joint processing over all the user messages at the CU may be impaired by the high compression noise to satisfy the backhaul capacity constraints. In general, the BSs should opportunistically decode the user messages based on the channel condition and allocated backhaul capacity.

In the downlink case, the BS could *encode and modulate* the baseband symbols to RF signals and then transmit them to the MTs. Therefore, instead of transmitting the complete signal waveforms (or waveform samples) to the BSs, the CU may save the backhaul bandwidth by transmitting separately the information symbols and the beam-forming vectors, while leaving RF modulation to the BSs. For instance, an adaptive before/after precoding scheme in a downlink networked MIMO system is proposed in [25]. That is, the CU could choose to precode all the baseband messages and send the data samples to the BSs or send the MTs' baseband symbol vectors and the precoding matrices separately. Since the update of precoding matrix is much slower than that of the information symbols, the separate transmission scheme is in general more favorable. However, the question that is arises, instead of bandwidth-consuming full BS coopera-tion (a user message needs to be sent to all the BSs in a cluster), how do we decide on a subset of cooperating BSs to serve a particular MT under limited backhaul capacity?

To answer the question, a *sparse precoding* method in a simple distributed MIMO system with single-antenna BSs and single-antenna MTs is considered in [26]. With a zero-forcing design criterion, it intends to minimize the total emitted interference in the network over a binary valued routing matrix, which determines the subset of cooperating BSs that each MT's message is sent to. The interference minimization problem is constrained on the number of nonzero entries in the routing matrix, i.e. the total number of symbols sent from the CU to the BSs. To tackle the combinatorial nature of the optimization problem, a sparse approximate inverse method is used. Reference [27] considers a more general sparse linear precoding problem in a multi-cell networked MIMO system, where each cell has a multi-antenna BSs and multiple multi-antenna MTs. The difference from the single-antenna BS case is that, the CU now needs to send a tagged MT's message to a BS if not all the elements of the tagged MT's precoding vector for the BS are zero. It proposes to optimize the beamformer design by jointly maximizing the user utilities (e.g. data rate) and minimizing the total number of data streams in the backhaul. To tackle the nonconvexity in the utility function due to interference, it derives an equivalent convex reformulation in the form of regularized weighted MSE minimization. Besides, a distributed implementation is proposed for each cell to manage its own precoder design with limited information exchange with the other cells. Within a networked MIMO system with multi-antenna BSs and single-antenna MTs, [28] proposes a beamformer design method to jointly minimize the transmit power and the number of data streams in the backhaul, given that each individual MT's SNR requirement is satisfied. A convergent algorithm is proposed to solve the nonconvex optimization based on l_0/l_2 norm minimization and semi-definite relaxation. A similar idea is also investigated in [29], which uses a reweighted l_1-norm relaxation and a heuristic iterative link removal algorithm to minimize the number of data streams in the backhaul.

Except for sparse beamforming, downlink *limited BS-cooperation* methods are also proposed. The general idea is to combine one full BS cooperation scheme (an equivalent MIMO broadcast channel) with another zero/partial cooperation scheme. For instance, [30] considers a two-BS downlink networked MIMO, where a MT's message is separated into a private message and a shared message. In particular, each private message is sent to only one of the BSs to serve the user, while the common message is sent to both BSs for cooperative transmission. In this sense, the system is a mixture of the interference channel and the MIMO broadcast channel. Intuitively, the larger the portion of private messages, the larger the interference caused to the other cells, which induces smaller backhaul traffic as well, and vice versa. It calculates the achievable rate region of the two-user case and numerically shows that the data sharing scheme in general achieves a larger rate region than the conventional interference channel and the network MIMO scenario, especially under stringent backhaul capacity constraints. Reference [31] combines the full BS cooperation with another partial cooperation scheme, i.e. coordinated beamforming. The coordinated beamforming scheme, as its name suggests, requires only CSI, instead of MTs' messages, from the other BSs to jointly mitigate inter-cell interference by controlled beamforming. It considers two downlink transmission modes: soft switch mode, where an MT's message could be separated into two parts, one for coordinated beamforming and the other for full BS cooperation transmission; and the hard switch mode, where a MT must choose from the two cooperation schemes to send its entire message. The favorable operating regions of each transmission scheme are characterized numerically. In addition, [32] proposes another limited BS cooperation scheme for OFDMA systems that allows neighboring BSs to cooperate on a subset of

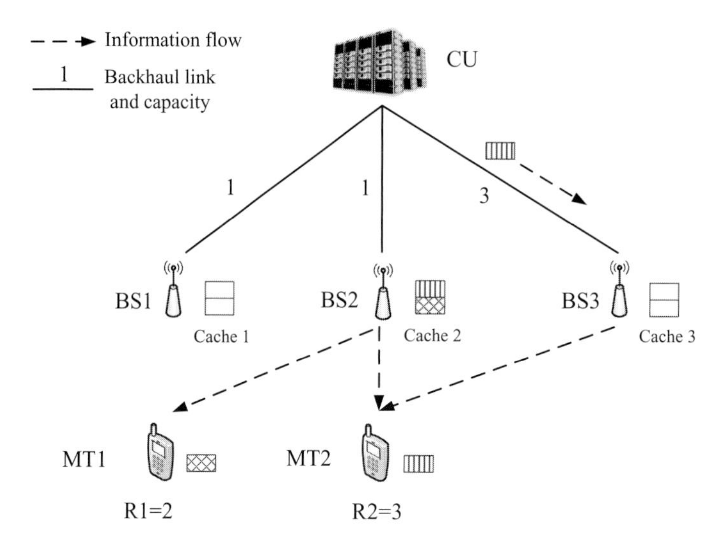

Figure 7.3 An example of cache-assisted and backhaul-constrained rate allocation.

frequency tones for interference mitigation. The idea is that, if a BS (BS1) chooses to cooperate with another BS (BS2) at a certain frequency tones, BS2 could pre-subtract the interference from BS1 using the MT's message and CSI sent by BS1. Then, there is a clear trade-off in the level of interference and the amount of information exchange in the backhaul by tuning the level of cooperations among the BSs.

Cache-assisted processing

In the downlink transmission, a BS could cache the received contents from the CU for future mobile users' quick access. The use of caching at the BSs exploits the spatial correlation of mobile data traffic, and is cost effective to reduce the real-time traffic in the backhaul. Essentially, caching could increase the *effective bandwidth* of the backhaul links, thus enabling significant improvement on the overall cloud-based system performance.

Cache-assisted wireless resource allocation is a cross-layer approach that incorporates the status of application-layer data into wireless physical-layer design. Intuitively, if an object is available in the cache of a BS, any subsequent requests for this object could be responded to by the BS immediately without resorting to the core-network. In the ideal case, a BS could still serve the MTs whose requested files are available in the cache, even if the CU-to-BS link is fully congested with zero residual capacity. In this sense, the selection of the set of BSs transmitting to a MT will be dependent only on the local availability of the files requested by the MT, especially under stringent backhaul constraints that the file cannot be efficiently transmitted to some of the BSs from the CU in real time. In general, this content-aware resource allocation method needs to jointly consider the wireless channel condition, the cache availability, the allocated capacity of backhaul links, and the QoS requirements of different mobile users.

The idea is briefly illustrated in Figure 7.3, where two MTs are requesting two different objects, with each MT having a common minimum data rate requirement of 2 units per second. BS2 has the two objects requested by the two MTs, while the caches of the other two BSs are empty. Although MT1 is close to BS1 with a better wireless channel condition, the maximum downlink data rate is 1 unit per second if selecting BS1 to directly transmit, constrained by the congested link between BS1 and CU. Instead, the CU could select BS2 to send the cached contents to MT1 at a rate of 2 units per second, whose end-to-end data rate is not constrained by the congestion level of the CU-to-BS2 link. On the other hand, MT2 could be served by two cooperating BSs with an improved wireless channel gain from coordinated beamforming. In particular, the CU only needs to transmit the object requested by MT2 to BS3 before the cooperative transmissions of the two BSs commence. Thanks to the cooperation in the wireless channel, MT2 could achieve a higher data rate at 3 units per second.

In a more general setting, web caches could be located not only at the BSs, but also at the routers that interconnect the BSs and potentially multiple CUs. We could foresee that the cache-assisted downlink resource allocation method will achieve significant bandwidth saving as the amount of mobile traffic increases, due to the frequent overlapping of requested objects in the downlink traffic. However, it also becomes a more challenging problem to optimize the system-wide resource allocation due to the interleaving among interference mitigation, routing, and the combinatorial BS selections. A more comprehensive understanding of the design trade-off is still open for future study.

So far we have assumed the availability of the contents in the caches to optimize the downlink resource allocation. Another interesting topic on cache-assisted resource allocation is *cache provisioning* for highly popular contents to reduce the real-time backhaul traffic. In particular, the question concerns what, where, and when to cache in the wireless infrastructure. Reference [33] proposes some initial attempts to study the caching problem in wireless networks, where a number of femtocell BSs, named caching helpers, are introduced in addition to the conventional BSs in cellular network, such that a MT could reduce the connection delay if the requested object is cached in one of the neighboring caching helpers in advance. Based on the knowledge of the aggregate demand profiles, e.g. the access probability of each object, the system operator optimizes its caching strategy, i.e. the subset of objects to be cached in each cache helper, to minimize the expected delay under cache volume constraints. In addition, a recent study by [34] jointly designs the caching strategy with physical layer resource allocation. Specifically, it considers video delivering in a dual-BSs setting, where one BS equipped with cache assists the other's downlink transmission to achieve a beamforming gain when the common message is in its cache. Intuitively, the higher data rate achieved from cooperating transmission helps maintain proper queue (buffer) sizes at the MTs, but also requires extra caching storage and transmission power. Essentially, the design intends to balance the service quality, i.e. lower buffer interruption and overflow probabilities vs. the cost on caching storage and transmit power consumption. Given the user demand profiles and wireless channel distributions, the optimization takes a two-step approach, where the inner optimization uses a Markov decision process (MDP) formulation to optimize the transmit power allocation in the short term, while the outer optimization finds the optimal caching strategy in the long term given the transmit power allocation. In both of the cases above, the use of caching has led to significant increase of the effective system bandwidth in the form of, for instance, larger numbers of supported users and lower probabilities of video buffer interruption. We also notice that the accurate knowledge of the mobile user demand profiles is the key to efficient caching design. The extraction of the demand profiles from the mobile data traffic is one of the topics of wireless big data analytics, which will be discussed in Section 7.4.1.

7.3.3 Section summary

A hybrid data processing structure could tackle the scalability problem in supporting mobile big data, by adaptively distributing the processing complexities among the CU and BSs. Depending on the choice of backhaul technology and the capabilities of BSs, various data compression, encoding/decoding, and caching-assisted processing techniques could be applied to optimize the system performance under some backhaul capacity constraints. However, there are still many challenging problems to be addressed in the future study.

Most of the data compression methods suffer from difficulties in implementing the compression and decompression techniques in practical wireless systems. The real-time calculation of the optimal compression noise covariance matrix is often impeded by the large number of backhaul capacity constraints and the nonconvex nature of many backhaul-constrained problems, letting alone the difficulty in generating practical joint compression codebooks based on the obtained covariance matrix. Therefore, suboptimal but practical compression schemes, such as scalar and vector quantizations, should be given more emphasis in the future study of backhaul-constrained compression scheme design. Similarly, the BS-level encoding/decoding also suffers from high computational complexity on large-scale multi-user detection and the combinatorial nature of many limited cooperation schemes, such as optimal antenna and BS selections. The proposed BS–CU joint processing infrastructure here provides merely a skeleton to build a scalable wireless network. More importantly, it illustrates the needs for key functional units such as the practical complexity-reduction algorithms that are truly scalable to the size of mobile users and the other network components.

BS-level caching is expected to play an important role in future wireless big data processing, due to its simplicity, low cost, and the natural integration with big data analytical tools. However, the research on cache-assisted wireless resource allocation is still in its infancy. Some works have considered distributed caching, which allows mobile users to cache the popular contents and serve the nearby peer users [35, 36]. For cache-assisted cellular networks with BS-level caching, it is currently in lack of both concrete theoretical analysis on the capacity gain of cache-assisted processing and practical optimization frameworks for cache-assisted resource allocation. Besides, how to integrate various big-data characteristics in cache-assisted network design is an interesting problem and is open to future investigations.

7.4 Big data aware wireless networking

The profile-based caching strategies introduced in Section 7.3.2 have shed some light on the potential performance gain from data-aware processing. In this section, we expand

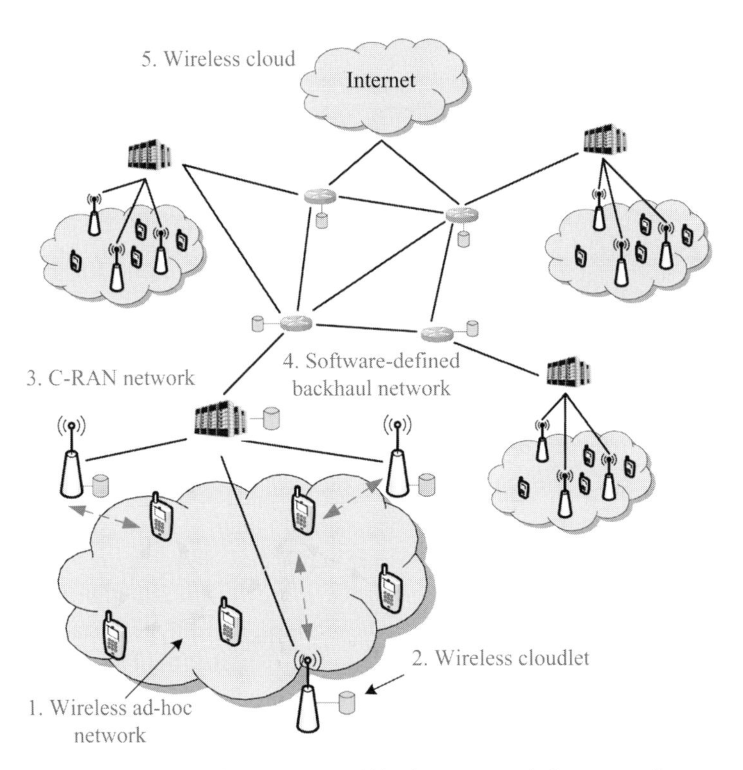

Figure 7.4 An illustrative structure of big data aware wireless network.

our discussions on big data aware processing. Instead of treating mobile big data as a total hazard, we investigate the contained opportunities to build an intelligent big data aware wireless network.

We have envisioned in Figure 7.4 the structure of a big data aware wireless network. In the very bottom layer, MTs could form a peer-to-peer (p2p) wireless ad-hoc network complementary to the conventional cellular connection, in case the cellular connection is unavailable due to link blockage or simply because the user is not willing to pay the high cellular data usage charges. Essentially, the ad-hoc structure enables a crowd computing paradigm among cooperating MTs, with many applications such as p2p file searching or computing delegation. In addition to the wireless ad-hoc network, wireless cloudlet, a simplified version of public wireless cloud, could be installed at the BSs to improve the service quality for delay-sensitive mobile applications, and to facilitate infrastructure-based crowd computing in the local BSs. Above the local cloudlet, the C-RAN cellular network is to provide general wireless coverage and data-aware services considering user mobility. From a macro perspective, the C-RAN could be regarded as a functional unit in the wireless cloud computing framework. Specifically, the remote data centers in the core-network (or internet) oversee the dynamics of all the C-RANs and have the capability to optimize the global resource allocation. All the network components could be unified in a way presented in Section 7.4.2 when we discuss the mobile cloud computing (MCC) structure. In between, the core-network and the C-RANs are interconnected via a SDN backhaul, which enables real-time adaptive data routing and volume control. The details of SDN-based backhaul will be specified in Section 7.4.3.

A well-functioning data-ware wireless network shown in Figure 7.4 relies on the accurate knowledge of the wireless big data traffic characteristics. As most of the characteristics are implicit, we introduce in Section 7.4.1 several effective data-analytical methods to extract these big data features and dynamics. We then discuss in Sections 7.4.2 and 7.4.3 how to leverage these big data characteristics in the designs of MCC and SDN, respectively, to mitigate (and utilize) the big data effects and to enhance the wireless network efficiency.

7.4.1 Wireless big data analytics

As a result of the close connection between wireless service usage and human behavioral patterns in the physical world, wireless data traffic contains strong correlations and statistical features in various dimensions, such as time, location, and the underlying social relations, etc. On one hand, mobile traffic has strong aggregate features. For instance, there exists severe load imbalance over location and time, where 10% of "popular" BSs carry around 50%–60% traffic load, and the peak traffic volume at a given location is much higher than that of regular times [6]. The aggregate features could be used to reduce the real-time backhaul traffic and improve the wireless network efficiency. Example applications include: cell planning according to geographical data usage distribution [37], peak load shifting via load-dependent pricing design [38], and cache provisioning based on aggregate demand profile [34], etc. On the other hand, each individual mobile user also has a unique data usage profile, such as mobility pattern, preference in applications, and service quality requirements. Despite the diversity of travel histories, a mobile user's trajectories often consist of a very limited number of frequent locations and simple repeated patterns. Besides, the recent popularity of mobile social networking interconnects seemingly uncorrelated individual data usages into a unified social profile, creating a new angle to analyze the mobile traffic pattern. These individual and social features are useful for the system operator to improve the personal wireless service quality. Many intelligent data-aware services could be provided based on user profiles, such as resource reservation in handoff using location prediction [39], context-aware personal wireless service adaptation [40], and mobility-based routing and paging control [41], etc.

The ability to comprehend the mobile traffic characteristics is achieved by the data-aware processing units in the proposed system structure shown in Figure 7.1, where the core enabling factors are the embedded *data-analytical algorithms*. The commonly used wireless data analytical methods could be roughly classified into three categories: *stochastic modeling*, *machine learning*, and *data mining*. In a nutshell, stochastic methods use probabilistic models to capture the *explicit* features and dynamics of the data traffic. Data mining, on the other hand, exploits the *implicit* unknown structures hidden in the data set, focusing on accurate representation of the features. Machine learning, which is an assemble of a number of data analytical techniques, focuses on the *generalization* of the structures learnt from data set for accurate predictions. The choice of a data analytical tool depends on the objective of signal processing (e.g. descriptive or predictive) and the properties of data sets, e.g. labeled/unlabeled data sets and explicit/implicit features, etc. In many cases, there is a need for combined applications among the analytical methods to best handle a specific mobile data set.

Stochastic modeling and data mining methods

Using stochastic models to describe the properties of data traffic in general requires expert or some prior knowledge about the structure of the data set. Besides, it often needs some experience-based heuristics to capture the features of the data traffic. Commonly used stochastic models include: order-K Markov model, hidden Markov model, geometric model, time series, linear/nonlinear dynamic systems, etc. The collected data are often used to estimate the parameters in the stochastic models, such as the transition probability of a Markov model. Many of the models are employed to predict the locations of mobile users. For instance, order-K Markov models are widely used to model a mobile subscriber's trajectory across different cells. In particular, [42] combines a geometric model to describe the regular trajectory over a large time scale, and a Markov model to capture the daily random movements. Geometric models are also used in [43, 44] to specify the topology of a cellular system and the constraints of geographical conditions to mobility, such as roads and buildings. Besides, a hidden Markov model (HMM) is often used when the state is not directly observable. For instance, [45] uses HMM to predict a MT's association to a wireless access point from its movement trace data, where the state (AP association) is not observable and follows an order-K Markov precoces. Besides Markov models, user mobility could also be modeled as a linear/nonlinear dynamic system. Accordingly, Kalman and extended Kalman filters could be used to predict the future position in an online manner [43, 46]. Besides mobility prediction, [47] uses Wiener model and time series to predict the resource required by handoff calls in a cell. Reference [48] uses an order-K Markov model to predict the user service requirements, e.g. bandwidth and end-to-end delay.

The applicability of stochastic modeling is often restricted to problems with explicit structures, such as the geometry in mobility pattern. Besides, the heuristics in stochastic models are often in lack of theoretical support, rendering the validity of the results questionable in other application scenarios. To cope with the problems, data mining takes a reverse approach to start from the data sets to exploit the hidden structures within. Common data mining methods for wireless data processing include: pattern matching, text compression, dimension reductions, and clustering, etc. Also taking the mobility prediction problem as an example, individual user's mobility pattern could be mined by finding the most frequent trajectory segments in the mobility log [49]. Prediction could be made accordingly by matching the current trajectory to the mobility profile. An interesting organization of mobility profile is via text compression [50]. Conceptually, it encodes a trajectory into an alphabetic codeword, then parses the trajectories into a set of non-repeating segments, and fianlly organizes them as the nodes in a tree structure, where each node is assigned a weight proportional to the frequency of occurrence. Then, the prediction of the next position could be equivalently modeled as node traversing in a tree. Clustering is another useful technique to identify the different patterns in the data sets. It is widely used in context-aware mobile computing, where a mobile user's context information, such as sleeping and working, is identified from wireless sensing data to provide context-related services [40]. Since clustering and dimension reduction are often recognized as techniques of unsupervised machine learning, we will combine the discussion in the following as a subclass of machine learning methods.

Machine learning methods

While data mining could help us to discover the features in the data traffic, it does not establish a direct mapping (a function) from a specific data to a certain property. For instance, clustering may be able to identify three types of activity patterns from the sensor data, but it cannot decide from a new sensor measurement if a mobile user is running, resting, or walking. Machine learning, on the other hand, focuses on establishing such mapping by learning from the data sets, to achieve the auto-processing capability of the future data. The commonly used machine learning techniques for wireless data traffic could be classified into three genres: supervised learning, unsupervised learning, and reinforcement learning. Supervised learning intends to infer the function from "labeled" data sets with well-defined input and output. For instance, the input sensor measurements are labeled with "running", "walking", and "resting" as the output; or the length of requested files as the input and the probability of such files as the output. A training stage is performed to infer the mapping function, often the parameters of a set of functions. There are many methods to assist the selection of a proper function model, such as the Akaike information criterion (AIC) to balance the performance of representation and prediction (underfitting vs. overfitting). In particular, supervised learning mainly covers two important topics: classification, i.e. determine the type of input data, and regression, i.e. data fitting. For mobile data processing, classification methods are used to identify the context of mobile users [51], predict the traffic level [52], and infer the social relations from mobile data [53], etc. Regression analysis is used to fit the distributions of the length of trajectory and mobile users' locations [54] and the channel holding time [55], etc.

On the other hand, unsupervised learning could be used to exploit the features in the wireless traffic when the data sets are unlabeled, which is very common in practice, e.g. the types of daily routines contained in the sensed data are unknown. Unsupervised learning is mostly used for clustering, which separates the data sets into statistically similar groups. Many classical algorithms are available for clustering analysis, such as K-mean and expectation maximization (EM) algorithms, etc. In mobile data processing, clustering methods have been proposed to exploit the mobile users' daily routines [56], types of user contexts [40] and location clusters in the social graph [57], etc. In many cases, unsupervised learning and supervised learning methods could be combined to deal with semi-labeled data sets. For instance, [40] proposes to use unsupervised learning to identify the types of contexts of mobile users, as it is generally difficult to detect the number of classes in the data. At a higher abstraction level, it applies user reactions to the context-aware services as the label to merge or separate the context classes, i.e. merging two context classes if similar user reactions are detected for separate classes, and vice versa.

Reinforcement learning (RL), unlike supervised or unsupervised learning, is not interested in comprehending the features of the sensed data. Instead, it is interested in making proper actions upon the current state and random event to maximize the rewards to actions. A typical example is the handoff and admission control decision (action) given the current traffic load (state) and incoming new requests (event), and the reward is evaluated based on the lost connections. Essentially, RL constructs a decision look-up table to list its optimal actions upon every combination of state and events. In particular, Q-learning is a widely used RL technique, which maximizes the expected reward by adapting its decision tables to the perceived action-reward pairs. Besides, it could also explore new actions not listed in the decision look-up table. We do not pursue further discussions here. Interested readers could refer to a tutorial paper [58] on the use of RL in wireless networks.

Large-scale data analytics

Wireless big data enforces many challenges to the conventional data-analytical methods due to its high volume and dimension, delay sensitive, uneven data qualities, and the complex features within. To improve the signal processing efficiency, many *complexity reduction* techniques could be combined with the above data-analytical tools for large-scale data processing.

To begin with, distributed computing is an effective design paradigm to relieve both the computational complexity at the CU and the pressures to the backhaul links. Many *distributed optimization algorithms*, such as primal/dual decomposition [59] and alternating direction method of multipliers (ADMM) [60], are very useful to handle large-scale statistical learning problems. In particular, ADMM could decompose a large learning problem into parallel small subproblems, each handling only a separate subset of data locally. By iteratively optimizing the local variables and updating the global variables, the ADMM algorithm is proved to have a strong convergence property towards a globally optimal solution.

On the other hand, *dimension reduction* methods could be used to reduce the data volume and to capture the key features of the big data. Among them, principle component analysis (PCA) may be the most widely used statistical tool for dimension reduction today. Many variant PCA-based algorithms have been proposed to handle different application scenarios. For instance, robust PCA is used to increase the resistance of the conventional PCA to corrupted data samples [61]. Another useful dimension reduction method is *tensor decomposition*, which seeks to approximately represent a high-order multi-way array (tensor) as a linear combination of outer products of low-order tensors. Among the many decomposition techniques, parallel factor analysis (PARAFAC) [62] and Tucker3 decomposition [63] are the most popular. Specifically, PARAFAC uses one-dimensional tensors to represent a high-order tensor. For instance, it could represent a three-dimensional tensor of user mobility data samples, where each entry has user ID, location and time as the coordinates (with cardinality $M \times N \times J$), by a sum of K types of users (e.g. business or home users), and each type has three factors in the form of one-dimensional arrays labeled with user ID ($M \times 1$), location ($N \times 1$) and time ($J \times 1$), respectively. By doing so, the original $M \times N \times J$ entries are reduced to

$K \times (M + N + J)$. Then, data analysis could proceed based on the extracted properties of the types and the associated factors. Besides, tensor compression techniques, such as *compressed sensing*, could be applied to further reduce the size of a sparse tensor's representation [64]. In the context of wireless communications, PARAFAC has extensive applications especially in the area of blind multi-user detection [62]. Tucker3 decomposition is an extension of the singular value decomposition (SVD) from two-dimensional tensors to three-dimensional tensors. Here, we omit the details and refer the interested readers to [63] and Part I of this book.

Another problem often encountered in wireless big data analytics is that many data samples are unlabeled or partially labeled, rendering conventional supervised learning inapplicable. In this case, manual labeling is not feasible simply because of the vast data volume. To tackle this problem, *active learning* is proposed to reduce the manual workload in labeling [65]. As a semi-supervised machine learning technique, active learning starts with a small number of instances in the labeled training set. Based on the learnt knowledge, later it could choose to classify the incoming unlabeled data instances or to selectively query for manual labeling for those very uncertain data instances. Then, the new labeled instance is simply added to the labeled set, and the learner proceeds in a standard supervised way. Another useful semi-supervised learning method is *deep learning*. Conceptually, it is a hieratical *neural network* which models high-level abstractions from multiple lower layers' features [66]. In particular, unsupervised learning could be used within each lower layer to detect the low-level features and supervised learning is used at high-level to adjust the structure of low-level features, e.g. adjusting the weighting parameters. In practice, deep learning may be used to classify the activities of mobile users from the received sensing data. Owing to the rich context in mobile user activity, it is infeasible to classify each context in a standard supervised way. However, many contexts share the similar lower-level features, e.g. both running and walking involve frequent and regular altitude oscillations. Context identification could be more efficient if lower-level feature extraction is combined with high-level reasoning and adjustment.

Last but not the least, *online learning* could be used to respond in real-time to delay-sensitive applications based on the sequentially received data samples. One such application is online cache management in wireless system, where caching decisions need to be made in real-time, even if the data sample size is not big enough to fully characterize the unknown user profiles [67]. Online learning could be combined with convex optimization methods. Given a convex decision set, e.g. caching decision, and a convex loss function of the decision variables, e.g. the real-time backhaul traffic induced by the caching decision, *online convex optimization* intends to minimize the cumulative loss over time [68]. One advantageous property of online convex optimization is that it could guarantee a sublinear increase of the cumulative loss function over time, e.g. square root or logarithm, respect to a best static decision given all the future data samples in a non-causal manner. A related topic is *stochastic learning* [68], where the system operator makes a decision once in a time interval. Intuitively, if the interval is large enough, the system operator could have better estimation of the wireless traffic pattern but may also suffer from a larger loss within the observation period. There is a clearly

Table 7.1 Summary of wireless big data analytic tools

Statistical modeling	order-K Markov model, hidden Markov model, geometric model, time series, dynamic modeling	
Data mining	pattern matching, text compression, clustering, dimension reduction	
Machine learning	*Supervised*	classification, regression, neural network
	Unsupervised	K-mean, EM, PARAFAC, Tucker3
	Semi-supervised	active learning, deep learning
	Reinforced	Q-learning
	Distributed	primal/dual decomposition, ADMM
	Online	online convex optimization, stochastic learning

a design trade-off to determine the optimal decision interval to minimize the expected loss.

For the convenience of reference, we have summarized useful analytical tools for mobile big data in Table 7.1. In the following subsections, we assume the knowledge of big data characteristics and discuss the potential performance gain from data-aware processing to the designs of mobile cloud computing and software defined networking.

7.4.2 Data-driven mobile cloud computing

Mobile cloud computing (MCC) is an emerging computing structure intersecting the paradigms of mobile computing and cloud computing. Owing to the limited processing capability, storage, and battery life of mobile devices, it is a natural attempt to migrate the computation resources to the cloud, which is composed of a cluster of computer hardwares and softwares that offer the services to the general public in a pay per use manner. The cloud computing structure today is mainly designed for enterprise-level applications that have well-organized users, rich local resources and stable connections to the cloud. Mobile cloud computing, on the other hand, is targeted at unorganized mobile users, facing much more critical challenges, for instance, the bursty traffic pattern, scarce radio spectrum for data-intensive applications, intermittent connections to the cloud, large delay, and limited device energy, etc. However, as we discussed in Section 7.4.1, we could exploit the useful features contained in mobile data traffic to improve the performance of MCC.

As illustrated in Figure 7.4, at the very bottom layer of the big data aware wireless network, the ad-hoc nature of mobile terminals fosters a new data-driven computing paradigm, namely *crowd computing*. Crowd computing leverages the resources shared by the peer users in the vicinity, suitable to applications such as location-based query, multimedia search, p2p sharing, etc. Crowd computing could be performed either with or without an infrastructure. When wireless infrastructure is available, the BSs could perform joint computing using the incomplete individual resources, such as photos and videos uploaded by the mobile users. For instance, a BS could piece together the video clips (or photos) shot from different angles to a complete concert live (or 3D street) view. Besides, [69] considers the possible use of crowd computing in emergencies, such as earthquake disaster relief and lost-child rescue. When wireless infrastructure is not available due to intermittent wireless connections, or not preferred due to high cellular mobile data expense, crowd computing could be performed in an ad-hoc manner. In this case, a MT could ask for assistance from its neighboring MTs for files, applications, and relay connection to the cellular network, etc. [70, 71]. Besides, the p2p nature of crowd computing also reduces the necessary backhaul traffic. However, crowd computing also raises questions on the service quality provisioning and its coexistence with the cellular services due to transmission interference.

On top of the ad-hoc crowd computing paradigm, a *wireless cloudlet* installed at the BSs could be used to enhance the local data-ware processing capability. As its name suggests, a cloudlet is a self-organized light cloud with limited storage and computing power at the edge of the public cloud, possibly operated by a simple server, serving only the peripheral users. The concept was first introduced in [72], which envisions that the delay in the conventional MCC is unlikely to improve in the future, due to the extensive use of firewalls, overlay structure, and battery saving techniques. This fundamentally impedes the development of many delay-sensitive mobile applications, such as mobile gaming, augmented reality, and language processing, etc. A cloudlet in the vicinity could effectively reduce the delay by a magnitude, saving the packet traversing time back and forth from the cloud edge to the deep cloud. Here, we consider a commercial cloudlet owned by the operator but leased to the commercial clients to improve their local service quality, as an example. Besides providing local storage and computing power for commercial applications, the cloudlet also allows the commercial applications to access the data stored at the local cache. This helps the commercial clients leverage the local resources to provide better location-based services. For instance, an advertising company could send to its subscribers the latest deals in the vicinity based on the messages posted by the local mobile users; social networking applications could then recommend nearby friends based on the searching histories. The centralized management of the cloudlet by the operator, instead of being self-organized, provides better infrastructure maintenance, platform compatibility, pool of resource, and billing mechanism. Besides, the cloudlet could effectively reduce the real-time traffic in the backhaul links as now many computations could be performed locally instead of resorting to the public cloud.

At a higher level across multiple BSs, a C-RAN could optimize its wireless services based on its knowledge of the mobile traffic pattern, especially involving user mobility within the cell. For instance, [73] introduces a "terminal-centric network" concept to tackle the interruption of user mobility to wireless transmissions in a cloud-access environment. Based on the mobility pattern of an MT, the CU could reserve the resource and pre-feed the contents at the BSs along the MT's route. Then, chunks of contents could be sent from different BSs to achieve seamless handoffs. Similarly, the aggregate behavior of data traffic could also be used to allocate the bandwidth and caching space to some popular locations ahead of the real-time event [33, 34]. Evidently, the mobile traffic features help the MCC reduce the real-time backhaul traffic and enhance the personalized wireless service quality at the same time.

At larger geographical and time scales, the public mobile cloud computing structure could learn and leverage long-term individual mobile user and aggregate public data-usage patterns across different C-RAN cells. As most of data-aware mobile services rely on the knowledge of conventional user profiles such as mobility pattern and demand distribution, context-aware computing and social-aware computing are two emerging processing paradigms for exploiting more complex data characteristics. The idea of context-aware computing is to provide personalized wireless services adaptive to the MT's real-time "context", such as at meeting, working, and resting, either directly reported by the MT or inferred from the received sensor data. The popularity of smartphones proliferates the types of sensing data available (or to be available) at the cloud center, such as the sound, accelerometer and gyroscope readings, preference setting, the battery consumption, the CPU and memory occupations, etc., rendering more accurate context estimations possible. Many context-aware services have been proposed. For instance, [74] proposes a context-aware middleware which automatically responds to the change of context at a MT by selecting the most cost-effective service from the cloud operator, e.g., choosing a lower data rate service at a lower expense when the residual device energy is low. Reference [75] proposes a context-aware mobile battery life management method, which uses the location information to predict future charging opportunities, and phone usage profiles to predict the consumption rate.

On the other hand, social computing calls for the wireless resource allocation to follow closely the interaction within and among social groups. Conceptually, a social group is an assembly of people who share similar interests, professions, life experience, and social relations, etc. In general, a social group has unique "eigenbehaviors", and thus similar data contents are required and generated within the social group [11]. The knowledge of a social community's composition, activities, and interests could be used to improve the wireless services for the targeted social group members. For instance, the agenda of a social group may reveal the demand for some particular types of data contents, e.g. a comic convention indicates a surge of comic-related content demands at the conference venue on the event day. The cloud center could then prefeed the popular contents ahead of the event at proper venues. Interesting work in [76] considers the use of community-level context awareness to bring together context-aware and social-aware computing. It proposes a dynamic community-context management framework that enables to discover, connect and organize relevant people, resources and things, which are socially related. Such social computing could effectively reduce backhaul redundancy with an improved service quality that is otherwise not achievable by conventional data-analytical schemes. It indeed provides a new and macro perspective to study the design of future wireless networks.

In the following subsection, we introduce the design of an SDN-based backhaul network that connects the above functional components in the MCC structure to provide adaptive data-aware routing and flow control.

7.4.3 Software-defined networking design

The existing computer network integrates and couples the functions of control logic and data forwarding at the node level, e.g. over routers and switches. Although such integration facilitates the construction of small networks, it creates tremendous difficulties in efficient control and management of large-scale networks with thousands of switches and hosts, not to mention the future evolvement of complex and hybrid communication networks. In response, software-defined networking (SDN) provides an abstraction of the control logic, which decouples the data-plane from the control and management plane functions with well defined programmable interfaces [77]. In SDN, the network is managed by a central controller and the underlying devices are only responsible for simple packet forwarding. The decoupling largely simplifies the network management and improves the network evolvability, as now the deployment of new network policies could be performed under centralized control by using software programs on top of the existing physical infrastructure. The packet forwarding decisions could be programmed based on many new factors such as the QoS requirement, the application types, and the payload length, in addition to the conventional distance-based metrics. Towards the future extensive deployment of SDN technologies, there have been a number of tentative implementations of SDN, such as OpenFlow [78] and NOX/POX [79], etc.

The dynamic nature of mobile data traffic motivates the development of a programmable wireless network. While the delegation of the physical layer transmission management to software programs has been well-studied, i.e. software-defined radio technology [80], the research on the upper-layer abstraction of wireless systems using SDN is still in its infancy. The design of SDN for wireless systems needs to consider many unique characteristics of wireless channels, such as spectrum scarcity, user mobility, and time varying link capacity, etc. The network decisions on routing, handoff, or packet forwarding within a macro-cell, should be made data-aware, i.e. adaptive to the wireless traffic dynamics. Specifically, the awareness of wireless data may include: the wireless traffic load distribution, the congestion level in the backhaul, the user QoS requirements and the availability of web cache, etc. The OpenRoads platform proposed in [81] is one of the earliest attempts to implement the SDN technology in a wireless environment, which separates the control plane from the datapath using the OpenFlow protocol that allows the flow-tables of the routing/switching devices to be managed by external controllers. In particular, it proposes a FlowVisor application programming interface (API) to isolate different flows, say destined to different mobile users, by delegating the forwarding control of each flow to a specific virtual controller. In this case, the routing decision for a particular mobile user data could be made based on its application or QoS requirement. Since its forwarding control is independent of the physical layer, it is compatible with multiple wireless standards, such as WiFi and LTE.

However, in a large-scale wireless system like the one shown in Figure 7.4, the centralized SDN-based teletraffic control mechanism may be overwhelmed by the enormous number of real-time data flows. The computation power and the memory/storage allocated to virtual controllers could be under stringent pressure and highly susceptible to the risk of single point of failure. To enhance the scalability and robustness of wireless systems, it is necessary to implement distributed control in SDN to support mobile big data. One such scheme is studied in [82], which proposes a SDN-based radio access network named SoftRan that uses a "division-of-labor" mechanism between the central controller and the BSs. Specifically, the controller optimizes all the system-level decisions, where one BS's change of policy may affect the other cells' performance, such as transmitting power and handoff. The BS is left with some independent intra-cell decisions, such as downlink resource block allocation. A similar SoftCell structure is proposed in [83], which provides scalable handling of network traffic using some local agent controller at the BS-level.

The recent development of distributed optimization algorithms provides the theoretical foundation of distributed control, and may find their potential applications to the SDN-based backhaul design. In particular, primal/dual decomposition methods (especially the dual decomposition method) have been extensively applied in teletraffic engineering for distributed routing, scheduling, power control, and congestion management, etc. The dual decomposition method, for instance, essentially decomposes the original constraint-coupling large problem into parallel small subproblems each with a subset of local variables, related by some global dual variables. The dual ascent method is used to iteratively update the local variables and the dual variables until a converging criterion is met. However, the theoretical convergence of the dual decomposition method is based on many assumptions, such as strict convexity (not an affine function) and fitness (e.g. cannot take the value of $+\infty$) of the objective function, rendering its convergence questionable in many practical scenarios that violate these assumptions. The recently emerged ADMM algorithm enhances the robustness of the algorithm by introducing the methods of augmented Lagrangian and multipliers. Moreover, it maintains the decomposition capability for many types of problems with favorable structures. One such implementation for SDN is proposed by [84], which jointly optimizes the routing and beamforming design in a cloud-based wireless network. Specifically, the optimal routing and beamforming solution maximizes the global minimum rate over a set of flows under backhaul link capacity constraints and user QoS requirements. In addition, ADMM decomposition algorithms are used to achieve parallel computing at different network devices, e.g. routers and BSs. Such data-aware schemes show significant performance gain over other traffic-oblivious heuristics.

7.4.4 Section summary

Mobile big data brings in unprecedent opportunities to re-design the wireless communication services and networking structures. Effective methods to extract the big-data characteristics and the applications of data-aware processing in MCC and SDN have been introduced. The design of future wireless networking must be able to fully capture the performance gain from data-aware processing. Because of the complex composition of wireless big data characteristics in diverse temporal and geographical scales, the future data-aware wireless network structure is expected to be heterogenous, consisting of both small distributed ad-hoc wireless networks, and large-scale centralized cloud-computing paradigm as depicted in Figure 7.4.

The heterogenous network structure naturally raises the problem of centralized vs. distributed control. Although both MCC and SDN are inherently centralized schemes upon their introductions, in many cases, however, distributed control/computing algorithms, such as ADMM, could be integrated to alleviate the computational complexity of the CU, reduce the backhaul traffic volume and mitigate the risk of single points of failure without compromising the overall system performance. Thanks to the programmability of SDN-enabled system infrastructure, distributed control mechanisms could be implemented with much more flexibility and lower cost. However, distributed algorithms are often limited by the problem structure, e.g. the coupling constraints in the backhaul and the partial knowledge of data traffic, etc. General distributed control, or a mixed centralized and decentralized control framework, is a promising working direction towards a future wireless networking design supporting mobile big data. Besides, the SDN-based design may also incorporate distributed caching (at BSs and routers) to enhance the efficiency of the routing decision, measured by the redundancy contained in the real-time backhaul traffic.

7.5 Conclusions

Wireless big data brings both big challenges and big opportunities to the design of our future wireless network. In this chapter, we have discussed effective signal processing methods to mitigate the detrimental big data effects and to exploit the big data characteristics for system optimization. Specifically, a hybrid processing structure is proposed to address the network scalability issues of supporting mobile big data, using various data compression, encoding/decoding, and caching techniques at both the CU and BS levels. We have also noticed that the performance of these techniques in the hybrid processing structure, such as caching-assisted resource allocation, is closely related to the degree of data awareness in the wireless system. In light of this, we have introduced effective big data analytical tools to extract useful wireless traffic characteristics, and have discussed the applications of these characteristics in building an intelligent data-aware wireless network. In particular, two emerging data-aware structures have been investigated, i.e. mobile cloud computing and software-defined networking. We propose that, to capture the diverse and complex characteristics of wireless big data traffic, the future wireless system should be heterogenous, consisting of both centralized and distributed network structures and control/computing mechanisms.

The difficulties of big data signal processing and network design mainly lie in the scale of problem size and the complex problem structures. For instance, many backhaul-constrained resource allocation optimizations are nonconvex and combinatorial in nature; and the centralized routing control in SDN-based design may become less efficient than the conventional routing method due to the enormous amount of real-time data flows. In this regard, practical complexity-reduced and distributed algorithms are the keys to build a truly scalable wireless system. However, we also notice that rigorous theoretical performance analysis is still missing in many big data applications, such as cache-assisted networks, and many implementation issues are also open to future investigations. Nonetheless, big data in wireless communication and networking will be a popular research area, considering the continuing data volume explosion, which also opens many opportunities that we cannot miss to revolutionize future wireless system design.

Acknowledgement

The authors of this chapter would like to acknowledge the support of the following funding agencies: DoD with grant HDTRA1-13-1-0029, NSF with grants CNS-1343155, ECCS-1305979, CNS-1265227, CNS-1443870, and ECCS-1307820, and NSFC with grant NSFC-61328102.

References

[1] Ericsson Mobility Report, http://www.ericsson.com/res/docs/2013/ericsson-mobility-report-november-2013.pdf, November 2013.

[2] E. G. Larsson, O. Edfors, F. Tufvesson, and T. L. Marzetta, "Massive mimo for next generation wireless systems," *IEEE Communications Magazine*, vol. **52**, no. 2, pp. 186–195, February 2014.

[3] J. G. Andrews, "Seven ways that hetnets are a cellular paradigm shift," *IEEE Communications Magazine*, vol. **51**, no. 3, pp. 136–144, March 2013.

[4] R. Irmer, H. Droste, P. Marsch, M. Grieger, G. Fettweis, S. Brueck, H. P. Mayer, L. Thiele, and V. Jungnickel, "Coordinated multipoint: concepts, performance, and field trial results," *IEEE Communications Magazine*, vol. **49**, no. 2, pp. 102–111, February 2011.

[5] China Mobile, "C-RAN: the road towards green RAN," White Paper, ver 2 (2011).

[6] U. Paul, A. P. Subramanian, M. M. Buddhikot, and S. R. Das, "Understanding traffic dynamics in cellular data networks," in *Proceedings IEEE International Conference on Computer Communications (INFOCOM)*, Orlando, FL, USA, June 2011.

[7] S. Woo, E. Jeong, S. Park, *et al.*, "Comparison of caching strategies in modern cellular backhaul networks," in *Proceedings ACM International Conference on Mobile Systems, Applications, and Services (MobiSys)*, Taipei, Taiwan, June 2013.

[8] N. Yu and Q. Han, "Context-aware communities and their impact on information influence in mobile social networks," in *Proceedings IEEE International Conference on Pervasive Computing and Communications (PERCOM) Workshops*, Lugano, Switzerland, March 2012.

[9] Open Networking Foundation, "Software-defined networking: the new norm for networks," white paper, 2012.

[10] M. C. Teich and B. E. A. Saleh, *Fundamentals of Photonics*, Canada: Wiley Interscience, 1991.

[11] P. Lukowicz, A. Pentland, and A. Frescha, "From context awareness to socially aware computing," *IEEE Pervasive Computing*, vol. **11**, no. 1, pp. 32–14, January–March 2012.

[12] B. Sundararaman, U. Buy, and A. D. Kshemkalyani, "Clock synchronization for wireless sensor networks: a survey," *Ad Hoc Networks*, vol. **3**, no. 3, pp. 281–323, December 2003.

[13] C. P. Liu and A. J. Seeds, "Transmission of wireless mimo-type signals over a single optical fiber without wdm," *IEEE Transactions on Microwave Theory and Techniques*, vol. **58**, no. 11, pp. 3094–3102, November 2010.

[14] A. E. Gamal and Y. H. Kim, *Network Information Theory*, Cambridge University Press, 2011.

[15] A. Sanderovich, O. Somekh, H. V. Poor, and S. Shamai, "Uplink macro diversity of limited backhaul cellular network," *IEEE Trans. Information Theory*, vol. **55**, no. 8, pp. 3457–3478, August 2009.

[16] A. D. Coso and S. Simoens, "Distributed compression for mimo coordinated networks with a backhaul constraint," *IEEE Transactions on Communications*, vol. **8**, no. 9, pp. 4698–4709, September 2009.

[17] Y. Zhou and W. Yu, "Optimized backhaul compression for uplink cloud radio access network," *IEEE Journal on Selected Areas in Communications*, vol. **32**, no. 6, pp. 1295–1307, June 2014.

[18] S. Park, O. Simeone, O. Sahin, and S. Shamai, "Joint decompression and decoding for cloud radio access networks," *IEEE Signal Processing Letters*, vol. **20**, no. 5, pp. 503–506, May 2013.

[19] ——, "Robust and efficient distributed compression for cloud radio access networks," *IEEE Transactions on Vehicular Technology*, vol. **62**, no. 2, pp. 692–703, February 2013.

[20] L. Zhou and W. Yu, "Uplink multicell processing with limited backhaul via per-base-station successive interference cancellation," *IEEE Journal on Selected Areas in Communications*, vol. **30**, no. 10, pp. 1981–1993, October 2013.

[21] O. Simeone, O. Somekh, and S. Shamai, "Downlink multicell processing with limited-backhaul capacity," *EURASIP Journal on Advances in Signal Processing*, pp. 1–10, May 2009.

[22] S. H. Park, O. Simeone, O. Sahin, and S. Shamai, "Joint precoding and multivariate backhaul compression for the downlink of cloud radio access networks," *IEEE Transactions on Signal Processing*, vol. **61**, no. 22, pp. 5646–5658, November 2013.

[23] P. Marsch and G. Fettweis, "A framework for optimizing the uplink performance of distributed antenna systems under a constrained backhaul," in *Proceedings IEEE International Conference on Communications (ICC)*, Glasgow, Scotland, June 2007.

[24] R. Zhang and J. M. Cioffi, "Exploiting opportunistic multiuser detection in decentralized multiuser mimo systems," *IEEE Transactions on Wireless Communications*, vol. **10**, no. 8, pp. 2474–2485, August 2011.

[25] S. Park, C. B. Chae, and S. Bahk, "Before/after precoded massive mimo in cloud radio access networks," in *Proceedings International Conference on Communications (ICC) Workshop*, Budapest, Hungary, June 2013.

[26] P. D. Kerret and D. Gesbert, "Sparse precoding in multicell mimo systems," in *Proceedings IEEE Wireless Communications and Networking Conference (WCNC)*, Paris, France, April 2012.

[27] M. Hong, R. Sun, H. Baligh, and Z. Q. Luo, "Joint base station clustering and beamformer design for partial coordinated transmission in heterogeneous networks," *IEEE Journal on Selected Areas in Communications*, vol. **31**, no. 2, pp. 226–240, February 2013.

[28] F. Zhuang and V. K. N. Lau, "Backhaul limited asymmetric cooperations for mimo cellular networks via semidefinite relaxation," *IEEE Transactions on Signal Processing*, vol. **62**, no. 3, pp. 684–693, February 2014.

[29] J. Zhao, T. Quek, and Z. Lei, "Coordinated multipoint transmission with limited backhaul data transfer," *IEEE Transactions on Wireless Communications*, vol. **12**, no. 6, pp. 2762–2775, June 2013.

[30] R. Zakhour and D. Gesbert, "Optimized data sharing in multicell mimo with finite backhaul capacity," *IEEE Transactions on Signal Processing*, vol. **59**, no. 12, pp. 6102–6111, December 2011.

[31] Q. Zhang, C. Yang, and A. Molisch, "Downlink base station cooperative transmission under limited-capacity backhaul," *IEEE Transactions on Wireless Communications*, vol. **12**, no. 8, pp. 3746–3759, August 2013.

[32] A. Chowdhery, W. Yu, and J. M. Cioffi, "Cooperative wireless multicell ofdma network with backhaul capacity constraints," in *Proceedings IEEE International Conference on Communications (ICC)*, Kyoto, Japan, June 2011.

[33] N. Golrezaei, K. Shanmugam, A. G. Dimakis, A. F. Molisch, and G. Caire, "Femtocaching: wireless video content delivery through distributed caching helpers," in *Proceedings International Conference on Computer Communications (INFOCOM)*, Orlando, FL, USA, March 2012.

[34] A. Liu and V. Lau, "Cache-enabled opportunistic cooperative mimo for video streaming in wireless systems," *IEEE Transactions on Signal Processing*, vol. **62**, no. 2, pp. 390–402, January 2014.

[35] N. Golrezaei, A. G. Dimakis, and A. F. Molisch, "Wireless device-to-device communications with distributed caching," in *IEEE International Symposium on Information Theory Proceedings (ISIT)*, Cambridge, MA, USA, July 2012.

[36] M. Ji, G. Caire, and A. F. Molisch, "Optimal throughput-outage trade-off in wireless one-hop caching networks," in *IEEE International Symposium on Information Theory Proceedings (ISIT)*, Istanbul, Turkey, July 2012.

[37] K. Tutschku and P. Tran-Gia, "Spatial traffic estimation and characterization for mobile communication network design," *IEEE Journal on Selected Areas in Communications*, vol. **16**, no. 5, pp. 804–811, June 1998.

[38] S. Ha, S. Sen, C. Joe-Wong, Y. Im, and M. Chiang, "Tube: time-dependent pricing for mobile data," *ACM SIGCOMM Computer Communication Review*, vol. **42**, no. 4, pp. 247–258, August 2012.

[39] F. Yu and V. Leung, "Mobility-based predictive call admission control and bandwidth reservation in wireless cellular networks," *Computer Networks*, vol. **38**, no. 5, pp. 577–589, October 2002.

[40] A. Krause, A. Smailagic, and D. P. Siewiorek, "Context-aware mobile computing: learning context-dependent personal preferences from a wearable sensor array," *IEEE Transactions on Mobile Computing*, vol. **5**, no. 2, pp. 113–127, February 2006.

[41] H. Zang and J. C. Bolot, "Mining call and mobility data to improve paging efficiency in cellular networks," in *Proceedings ACM Mobile Computing and Networking (MobiCom)*, Montreal, Canada, September 2007.

[42] G. Liu and G. M. Jr, "A class of mobile motion prediction algorithms for wireless mobile computing and communication," *Mobile Networks and Applications*, vol. **1**, no. 2, pp. 113–121, October 1996.

[43] T. Liu, P. Bahl, and I. Chlamtac, "Mobility modeling, location tracking, and trajectory prediction in wireless atm networks," *IEEE Journal on Selected Areas in Communications*, vol. **6**, no. 6, pp. 922–936, August 1998.

[44] M. M. Zonoozi and P. Dassanayake, "User mobility modeling and characterization of mobility patterns," *IEEE Journal on Selected Areas in Communications*, vol. **15**, no. 7, pp. 1239–1252, September 1997.

[45] P. S. Prasad and P. Agrawal, "Movement prediction in wireless networks using mobility traces," in *Proceedings IEEE Consumer Communications and Networking Conference (CCNC)*, Las Vegas, NV, USA, January 2010.

[46] P. N. Pathirana, A. V. Savkin, and S. Jha, "Mobility modelling and trajectory prediction for cellular networks with mobile base stations," in *Proceedings ACM International Symposium on Mobile ad hoc Networking and Computing (MobiHoc)*, Annapolis, MD, USA, June 2003.

[47] T. Zhang, E. van den Berg, J. Chennikara, *et al.*, "User mobility modeling and characterization of mobility patterns," *IEEE Journal on Selected Areas in Communications*, vol. **19**, no. 10, pp. 1931–1941, October 2001.
[48] I. F. Akyildiz and W. Wang, "The predictive user mobility profile framework for wireless multimedia networks," *IEEE/ACM Trans. Networking*, vol. **12**, no. 6, pp. 1021–1035, December 2004.
[49] W. C. Peng and M. S. Chen, "Mining user moving patterns for personal data allocation in a mobile computing system," in *Proceedings IEEE International Conference on Parallel Processing (ICPP)*, Toronto, Canada, August 2000.
[50] F. Yu and V. Leung, "Mobility-based predictive call admission control and bandwidth reservation in wireless cellular networks," *Computer Networks*, vol. **38**, no. 5, pp. 577–589, April 2002.
[51] E. Miluzzo, N. D. Lane, K. Fodor, *et al.*, "Sensing meets mobile social networks: the design, implementation and evaluation of the cenceme application," in *Proceedings ACM conference on Embedded Network Sensor Systems (Sensys)*, Raleigh, NC, USA, November 2008.
[52] L. Gueguen and B. Sayrac, "Traffic prediction from wireless environment sensing," in *Proceedings IEEE Wireless Communications and Networking Conference (WCNC)*, Budapest, Hungary, April 2009.
[53] N. Eagle and A. Pentland, "Reality mining: sensing complex social systems," *Personal and Ubiquitous Computing*, vol. **10**, no. 4, pp. 255–268, May 2006.
[54] A. Sridharan and J. Bolot, "Location patterns of mobile users a large-scale study," in *Proceedings IEEE International Conference on Computer Communications (INFOCOM)*, Turin, Italy, April 2013.
[55] C. Jedrzycki and V. C. M. Leung, "Probability distribution of channel holding time in cellular telephony systems," in *Proceedings IEEE Vehicular Technology Conference (VTC)*, Atlanta, GA, USA, April 1996.
[56] K. Farrahi and D. Gatica-Perez, "What did you do today?: discovering daily routines from large-scale mobile data," in *Proceedings ACM International Conference on Multimedia*, Vancouver, BC, Canada, October 2008.
[57] N. Li and G. Chen, "Analysis of a location-based social network," in *Proceedings IEEE International Conference on Computational Science and Engineering (CSE)*, Vancouver, BC, Canada, August 2009.
[58] K. L. A. Yau, P. Komisarczuk, and P. D. Teal, "Reinforcement learning for context awareness and intelligence in wireless networks: Review, new features and open issues," *Journal of Network and Computer Applications*, vol. **35**, no. 1, pp. 253–267, January 2012.
[59] D. P. Palomar and M. Chiang, "A tutorial on decomposition methods fornetwork utility maximization," *IEEE Journal on Selected Areas in Communications*, vol. **24**, no. 8, pp. 1439–1451, August 2006.
[60] S. Boyd, N. Parikh, E. Chu, B. Peleato, and J. Eckstein, "Distributed optimization and statistical learning via the alternating direction method of multipliers," *Foundations and Trends in Machine Learning*, vol. **3**, no. 1, pp. 1–122, January 2011.
[61] G. Mateos and G. B. Giannakis, "Robust PCA as bilinear decomposition with outlier-sparsity regularization," *IEEE Transactions on Signal Processing*, vol. **60**, no. 10, pp. 5176–5190, October 2012.
[62] N. D. Sidiropoulos, G. B. Giannakis, and R. Bro, "Blind PARAFAC receivers for DS-CDMA systems," *IEEE Transactions on Signal Processing*, vol. **48**, no. 3, pp. 810–823, March 2000.

[63] E. Acar and B. Yener, "Unsupervised multiway data analysis: A literature survey," *IEEE Transactions on Knowledge and Data Engineering*, vol. **21**, no. 1, pp. 6–20, January 2009.
[64] N. D. Sidiropoulos and A. Kyrillidis, "Multi-way compressed sensing for sparse low-rank tensors," *IEEE Signal Processing Letters*, vol. **19**, no. 11, pp. 757–760, October 2012.
[65] B. Settles, "Active learning literature survey," *Technical Report, University of Wisconsin, Madison*, vol. **52**, pp. 55–66, January 2010.
[66] G. E. Hinton, S. Osindero, and Y. W. Teh, "A fast learning algorithm for deep belief nets," *Neural Computation*, vol. **18**, no. 7, pp. 1527–1554, July 2006.
[67] P. Blasco and D. Gunduz, "Learning-based optimization of cache content in a small cell base station," in *IEEE International Conference on Communications (ICC) 2014*, 2014, accepted and available at http://arxiv.org/abs/1402.3247.
[68] S. Shalev-Shwartz, "Online learning and online convex optimization," *Foundations and Trends in Machine Learning*, vol. **4**, no. 2, pp. 107–194, February 2012.
[69] M. Satyanarayanan, "Mobile computing: the next decade," *ACM SIGMOBILE Mobile Computing and Communications Review*, vol. **15**, no. 2, pp. 2–10, January 2011.
[70] E. E. Marinelli, "Hyrax: cloud computing on mobile devices using mapreduce," Masters Thesis, Carnegie Mellon University, 2009.
[71] G. Huerta-Canepa and D. Lee, "A virtual cloud computing provider for mobile devices," in *Proceedings ACM Workshop on Mobile Cloud Computing and Services: Social Networks and Beyond (MCS)*, San Francisco, CA, USA, June 2010.
[72] M. Satyanarayanan, P. Bahl, R. Caceres, and N. Davies, "The case for vm-based cloudlets in mobile computing," *IEEE Pervasive Computing*, vol. **8**, no. 4, pp. 14–23, October–December 2009.
[73] M. Webb, Z. Li, P. Bucknell, T. Moulsley, and S. Vadgama, "Future evolution in wireless network architectures: towards a 'cloud of antennas'," in *Proceedings IEEE Vehicular Technology Conference (VTC)*, Quebec City, Canada, September 2012.
[74] P. Papakos, L. Capra, and D. S. Rosenblum, "Volare: context-aware adaptive cloud service discovery for mobile systems," in *Proceedings ACM International Workshop on Adaptive and Reflective Middleware*, Bangalore, India, November 2010.
[75] N. Ravi, J. Scott, L. Han, and L. Iftode, "Context-aware battery management for mobile phones," in *Pervasive Computing and Communications (PerCom)*, Hong Kong, China, March 2008.
[76] I. Roussaki, N. Kalatzis, N. Liampotis, *et al.*, "Context-awareness in wireless and mobile computing revisited to embrace social networking," *IEEE Communications Magazine*, vol. **50**, no. 6, pp. 74–81, June 2012.
[77] H. Kim and N. Feamster, "Improving network management with software defined networking," *IEEE Communications Magazine*, vol. **51**, no. 2, pp. 114–119, February 2013.
[78] N. McKeown, T. Anderson, H. Balakrishnan, *et al.*, "Openflow: enabling innovation in campus networks," *ACM SIGCOMM Computer Communication Review*, vol. **38**, no. 2, pp. 69–74, March 2008.
[79] N. Gude, T. Koponen, J. Pettit, *et al.*, "Nox: towards an operating system for networks," *ACM SIGCOMM Computer Communication Review*, vol. **38**, no. 3, pp. 105–110, July 2008.
[80] W. Tuttlebee, *Software Defined Radio: Origins, Drivers, and International Perspectives*, West Sussex, England: John Wiley & Sons, 2002.
[81] K. K. Yap, R. Sherwood, M. Kobayashi, *et al.*, "Blueprint for introducing innovation into wireless mobile networks," in *Proceedings ACM SIGCOMM Workshop*, New Delhi, India, August 2010.

[82] A. Gudipati, D. Perry, L. E. Li, and S. Katti, "Softran: software defined radio access network," in *Proceedings ACM SIGCOMM Workshop*, Hong Kong, China, August 2013.
[83] X. Jin, L. E. Li, and L. Vanbevery, "Softcell: scalable and flexible cellular core network architecture," in *Proceedings ACM Conference on Emerging Networking Experiments and Technologies (CoNEXT)*, Santa Barbara, CA, USA, December 2013.
[84] W. C. Liao, M. Hong, H. Farmanbar, *et al.*, "Min flow rate maximization for software defined radio access networks," http://arxiv.org/abs/1312.5345, 2013.

8 Big data processing for smart grid security

Lanchao Liu, Zhu Han, H. Vincent Poor, and Shuguang Cui

The development of the smart grid, impelled by the increasing demand from industrial and residential customers together with the aging power infrastructure, has become an urgent global priority due to its potential economic, environmental, and societal benefits. Smart grid refers to the next-generation electric power system that aims to provide reliable, efficient, secure, and quality energy generation, distribution, and consumption, using modern information, communications, and electronics technology. A distributed and user-centric system will be introduced in smart grid, which will incorporate end consumers into its decision processes to provide a cost-effective and reliable energy supply. In the smart grid, the modern communication infrastructure will play a vital role in managing, controlling, and optimizing different devices and systems. Information and communication technologies will provide the power grid with the capability of supporting two-way energy and information flows, rapid isolation, and restoring of power outages, facilitating the integration of renewable energy sources into the grid and empowering the consumer with tools for optimizing their energy consumption. The introduction of the cyber infrastructure needed to realize the smart grid also brings with it vulnerability to security breaches. Thus security is a major concern in the development of the smart grid. Moreover, the widespread deployment of smart meters and sensors such as phasor measurement units results in the generation of massive amounts of data that can be exploited in optimizing and securing the grid. This chapter addresses these two issues by investigating the applications of big data processing techniques for smart grid security from two perspectives: exploiting the inherent structure of the data, and dealing with the huge size of the data sets. Two specific applications are included in this chapter: sparse optimization for false data injection detection, and a distributed parallel approach for the security constrained optimal power flow problem.

8.1 Preliminaries and motivations

The smart grid is a modernized power system, which enables bidirectional flows of energy and uses two-way communication and control capabilities to improve the efficiency, reliability, economics, and sustainability of the production and distribution of

Big Data over Networks, ed. Shuguang Cui, Alfred O. Hero III, Zhi-Quan Luo, and José M. F. Moura. Published by Cambridge University Press.

Table 8.1 Domains and roles/services in the smart grid conceptual model

	Domain	Roles/services
1	Customer	The end users of electricity. May also generate, store, and manage the use of energy. Traditionally, three customer types are discussed: residential, commercial, and industrial.
2	Markets	The operators and participants in electricity markets.
3	Service provider	The organizations providing services to electrical customers and to utilities.
4	Operations	The managers of the movement of electricity.
5	Generation	The generators of electricity. May also store energy for later distribution.
6	Transmission	The carriers of bulk electricity over long distances. May also store and generate electricity.
7	Distribution	The distributors of electricity to and from customers. May also store and generate electricity.

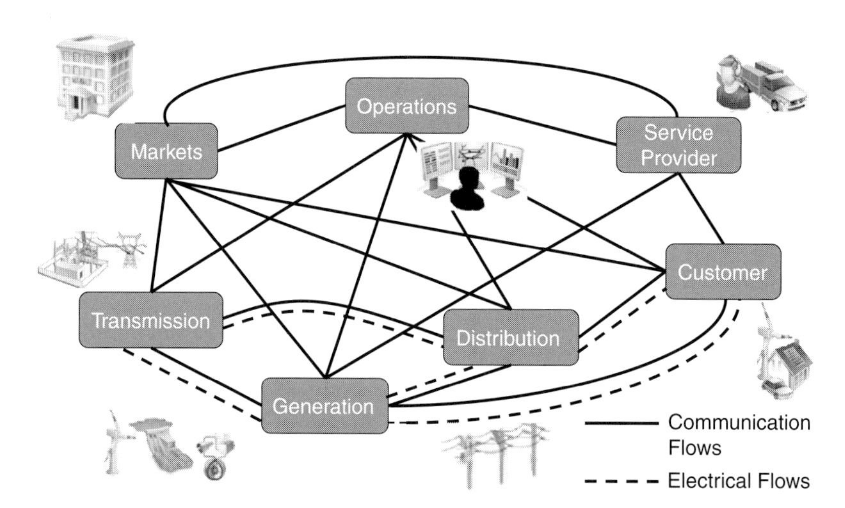

Figure 8.1 An illustration of the updated NIST smart grid framework 3.0 [2].

electricity [1]. In the conceptual model of the smart gird, seven components are introduced as described in Table 8.1 [2], and an illustration of their interactions is given in Figure 8.1. The smart grid is an integration of the electrical and communication infrastructures, such that the inevitable coupling between information/communication technologies and physical operations is expected to present unique challenges as well as opportunities for smart grid research.

On one hand, we are observing increasing integration between cyber operations and physical infrastructures for generation, transmission, and distribution control in the electric power grid. Yet the security and reliability of the power grid are not always guaranteed and some failures can cause significant problems for the grid. For example, in the USA, the 2003 Northeast power blackout showed that even a small failure in a part of the grid has cascading effects causing billions of dollars in economic losses.

Nowadays, the consolidation of physical and cyber components gives rise to security threats in power grids, which can result in power outages and even system blackouts [3], or substantial economic loss due to non-optimal operation of the power grid.

On the other hand, the anticipated smart grid data deluge, generated by the sensing and measurement devices and reinforced by communication and information technologies, provides us with the potential to enhance the security and reliability of the power grid. For example, the deployment of phasor measurement units (PMUs), which provide real-time assessments of power system health to system operators with more accurate information for averting disastrous outages, will generate 4.15 TB of phasor data per day for a full-scale phasor system in the future North American power grid. It is estimated that 61.8 million smart meters have been deployed in the USA by the end of 2013, and the estimated amount of compressed smart meter data for one million users per year is 27.3 TB [4]. Such massive datasets, if effectively managed and translated into actionable insights, have the potential to increase operational efficiency and ensure grid resiliency. The adopted methods should be able to utilize the inherent structure of the huge dataset to extract useful information. Moreover, the data should be processed scalably in a timely fashion. Thus, new computational and mathematical models and methodologies must be explored to effectively operate an ever-complicated power grid and achieve the vision of a smart grid. The rest of this chapter is organized as follows. The sparse optimization for false data injection detection is described in Section 8.2. The distributed parallel approach for the security constrained optimal power flow problem is developed in Section 8.3. Finally, some conclusions are drawn in Section 8.4.

8.2 Sparse optimization for false data injection detection

In this section, we first introduce state estimation and false data injection attacks in power systems. Then we describe two detection methods, nuclear norm minimization and low-rank matrix factorization, which exploit the inherent structure of state estimation data to detect false data injection attacks. Finally we present numerical simulation results for the proposed methods.

8.2.1 State estimation and false data injection attacks

State estimation in power systems

State estimation [5], which estimates the power system operating state based on a real-time electric network model, is a key function of the energy management system (EMS). A linearized measurement model is often used to estimate the states in power systems based on measurements from remote meters on buses or transmission lines. Specifically, every several seconds or minutes, the energy control center (ECC) collects active/reactive power flows and injections from transmission lines and buses across the power grid as measurement data via the supervisory control and data acquisition (SCADA) system. The state estimation results reflect the real-time power grid operation state and are essential for operators to make decisions to maintain security and stability of the system.

In an electric power grid, the control center needs to monitor the magnitude and phase angles of alternate current (AC) voltage at all buses to make real-time decisions on operations. However, it is economically impractical to directly measure all bus voltage phase angles. In this regard, the control center collects the readings from remote electric meters mounted on a subset of buses and branches to estimate the system operation state. Specific measurement data include branch active power flows and bus active power injections, which can be used to estimate bus voltage angles in the system. Let $\boldsymbol{\theta} = (\theta_1, \theta_2, \ldots, \theta_n)^\top$ denote the power system state variables, where θ_i is the phase angle on bus i. The measurement at the control center is expressed as $\mathbf{z} = (z_1, z_2, \ldots, z_m)^\top$ and is related to $\boldsymbol{\theta}$ by

$$\mathbf{z} = \mathbf{h}(\boldsymbol{\theta}) + \mathbf{e}, \tag{8.1}$$

where $\mathbf{h}(\boldsymbol{\theta}) = (h_1(\boldsymbol{\theta}), \ldots, h_m(\boldsymbol{\theta}))^\top$, and $h_i(\boldsymbol{\theta})$ is in general a nonlinear function relating the ith measurement to the state vector $\boldsymbol{\theta}$. The vector $\mathbf{e}$ contains independent Gaussian measurement errors with zero mean and covariance $\mathbf{R}$.

To analyze the efficiency of various state estimation methods considering the measurement configuration in a power system, a simplified direct current (DC) approximation model is usually employed [5]. Assuming that the bus voltage magnitudes are already known and normalized, and neglecting all shunt elements and branch resistances, the active power flow from bus i to bus j can be approximated[1] [6] by the first-order Taylor expansion as

$$P_{ij} = \frac{\theta_i - \theta_j}{X_{ij}} + \omega, \tag{8.2}$$

where X_{ij} is the reactance of the transmission line between bus i and bus j, and ω is the measurement error. Similarly, a power injection measurement at bus i can be expressed as

$$P_i = \sum_j P_{ij} + \nu, \tag{8.3}$$

where ν is the measurement error.

Hence, the DC model for the real power measurement can be written in a linear matrix form as

$$\mathbf{z} = \mathbf{H}\boldsymbol{\theta} + \mathbf{e}, \tag{8.4}$$

where $\mathbf{z}$ is the measurement vector including active power flows and injection measurements, and $\mathbf{H} \in \mathbb{R}^{m \times n}$ is the Jacobian matrix of the power system, which is assumed to be known to the independent system operator (ISO).

Suppose that the measurement errors $\mathbf{e}$ in (8.4) are not correlated, and thus the covariance matrix $\mathbf{R}$ is a diagonal matrix. The weighted least squares estimator of the

[1] In general, one can approximate the impedance of a transmission line with its reactance due to the high reactance over resistance (X/R) ratio.

linearized state vector $\boldsymbol{\theta}$ is given by

$$\hat{\theta} = (\mathbf{H}^\top \mathbf{R}^{-1}\mathbf{H})^{-1}\mathbf{H}^\top \mathbf{R}^{-1}\mathbf{z}. \tag{8.5}$$

Let $\mathbf{K} = (\mathbf{H}^\top \mathbf{R}^{-1}\mathbf{H})^{-1}\mathbf{H}^\top \mathbf{R}^{-1}$, and then the measurement residuals can be expressed as

$$\mathbf{r} = \mathbf{z} - \mathbf{H}\hat{\theta} = (\mathbf{I} - \mathbf{K})(\mathbf{H}\theta + \mathbf{e}) = (\mathbf{I} - \mathbf{K})\mathbf{e}, \tag{8.6}$$

where $\mathbf{I}$ is the identity matrix and the matrix $(\mathbf{I} - \mathbf{K})$ is called the residual sensitivity matrix.

The detection and identification of bad data in measurements can be accomplished by processing the measurement residuals. Specifically, the χ^2-test can be applied on the measurement residuals to detect bad data. Regarding the detection of bad data, two kinds of methods, the largest normalized residual test and the hypothesis testing identification method, can be used to identify the specific measurement that actually contains bad data [5].

False data injection attacks

The accuracy of state estimation can be affected by bad measurements in the grid. Bad data could be due to topology errors in the grid, measurement abnormalities caused by meter failures, or malicious attacks. To detect and identify bad measurements in the power grid state, techniques based on the statistical testing of measurement residuals [5] have been developed and are widely used. However, [7] reveals the fact that false data injection attacks are able to circumvent traditional detection methods based on residual testing, i.e. methods that compare the largest normalized residual with a predefined threshold. By exploiting the configuration of a power system, synchronized data injection attacks on meters can be launched to tamper with their measurements. Moreover, attack vectors can be systematically and efficiently constructed even when the attacker is limited in resources required to compromise meters, which will mislead the state estimation process, and thus affect power grid control algorithms. Hence, attention should be given to the vulnerability of state estimation to false data injection attacks, which may cause catastrophic consequences in the power grid.

Malicious attack vectors are able to circumvent existing statistical tests for bad data detection if they leave the measurement residuals unchanged. One such example is the false data injection attack.

Definition 8.1 (False data injection attack) [7] The malicious attack $\mathbf{a} = (a_1, \ldots, a_m)^\top$ is called a false data injection attack if $\mathbf{a}$ can be expressed as a linear combination of the columns of $\mathbf{H}$; i.e. $\mathbf{a} = \mathbf{Hc}$ for some vector $\mathbf{c}$.

If a false data injection attack is applied to the power system, the collected measurements at the ISO can be expressed as

$$\mathbf{z_a} = \mathbf{z}_0 + \mathbf{a} = \mathbf{H}(\theta + \mathbf{c}) + \mathbf{e}. \tag{8.7}$$

Suppose the state estimate using the malicious measurement $\mathbf{z_a}$ is $\boldsymbol{\theta}_\mathbf{a}$; then the norm of the measurement residual $\|\mathbf{z_a} - \mathbf{H}\boldsymbol{\theta}_\mathbf{a}\|_2$ in this case is

$$\|\mathbf{z_a} - \mathbf{H}\boldsymbol{\theta}_\mathbf{a}\|_2 = \|\mathbf{z}_0 + \mathbf{a} - \mathbf{H}(\boldsymbol{\theta} + \mathbf{c})\|_2 = \|\mathbf{z}_0 - \mathbf{H}\boldsymbol{\theta}\|_2, \tag{8.8}$$

which means that the measurement residuals are unaffected by the injection attack vector $\mathbf{a}$, and the attacker successfully tricks the system into believing that the true state is $\boldsymbol{\theta}_\mathbf{a} = \boldsymbol{\theta} + \mathbf{c}$ instead of $\boldsymbol{\theta}$. Note that $\mathbf{a}$ is the attack vector, which is under the control of the attackers, while $\mathbf{c}$ reflects error induced by $\mathbf{a}$.

Unveiling false data injection attacks is crucial to the security and reliability of power systems. This task is challenging, since attackers may be able to construct false data attacks against a given protection scheme, and inject attack vectors into the power grid that can bypass traditional methods for bad measurement detection. Furthermore, the incomplete measurement data due to intended attacks or meter failures complicate malicious attack detection, and make state estimation even more difficult.

The effects of false data injection attacks have been studied in [7–9]. False data injection attacks against state estimation in electric power grids were presented in [7]. By capitalizing on the configuration of the power system, malicious attacks can be launched to bypass the existing bad measurement detection techniques and manipulate the results of state estimation. References [9] and [8] demonstrated that false data injection attacks are able to circumvent the bad data identification techniques deployed in an EMS, and could lead to congestion of transmission lines as well as profitable financial misconduct in the power market.

On the other hand, various schemes to protect against false data injection attacks are investigated in [10–15]. Reference [10] proposed an efficient method for computing the security index with sparse attack vectors, and describes a protection scheme to strengthen system security by placing encrypted devices in the electric power grid appropriately. A game theoretic model was used in [11] to analyze this situation as a zero-sum game between the attacker and defender. Two kinds of malicious attacks on electric power grids were characterized in [12]: the strong attack regime, in which false data injection attacks exist, and the weak attack regime, in which the generalized likelihood ratio test can be used to detect attacks. In [13], a low-complexity attacking strategy was designed to construct sparse false data injection attack vectors, and strategic protection schemes were also proposed based on greedy approaches. A survey of existing detection methods for false data injection attacks is provided in [14]. Unlike existing works, our methods discussed next utilize the inherent structure of the data to detect false data injection attacks.

Sparse optimization problem formulation

Denote the measurement of the electric power system observed by the ISO at time k as $\mathbf{z}_k$. In the presence of false data injection attacks, the measurement $\mathbf{z_k}$ is contaminated by the attack vector $\mathbf{a_k}$. Denote by $\mathbf{Z}_0 = [\mathbf{z}_1, \mathbf{z}_2, \dots, \mathbf{z_t}] \in \mathbb{R}^{m \times t}$ the measurement of the power state over a time period of t, and by $\mathbf{A} = [\mathbf{a}_1, \mathbf{a}_2, \dots, \mathbf{a_t}] \in \mathbb{R}^{m \times t}$ the false data attack matrix. The obtained temporal observations $\mathbf{Z_a}$ can be expressed as

$$\mathbf{Z_a} = \mathbf{Z_0} + \mathbf{A}. \tag{8.9}$$

Note that gradually changing power system state variables will typically lead to a low-rank measurement matrix $\mathbf{Z_0}$. In addition, owing to the limited capability of the attackers, they are either constrained to have access to only some specific measurement meters or they are unable to compromise measurement meters persistently. Hence, only a small fraction of the observations can be anomalous at a given time instant. This implies that the false data injection matrix $\mathbf{A}$ is sparse across both rows and columns. With a slight abuse of notation, we use $\text{Rank}(\mathbf{Z_0})$ to denote the rank of the matrix $\mathbf{Z_0}$, and $\|\mathbf{A}\|_0$ to represent the number of nonzero entries of the matrix $\mathbf{A}$. Noticing the intrinsic structures of $\mathbf{Z_0}$ and $\mathbf{A}$, the detection and identification of false data injection attacks can be converted to a matrix separation problem as

$$\min_{\mathbf{Z_0},\mathbf{A}} \text{Rank}(\mathbf{Z_0}) + \|\mathbf{A}\|_0, \quad \text{s.t.} \quad \mathbf{Z_a} = \mathbf{Z_0} + \mathbf{A}. \tag{8.10}$$

Solving (8.10) extracts the power state measurement matrix $\mathbf{Z_0}$ and the sparse attack matrix $\mathbf{A}$ from their sum $\mathbf{Z_a}$. Considering the missing measurements due to meter failures or communication link outages in practical applications, (8.10) can be formulated as

$$\min_{\mathbf{Z_0},\mathbf{A}} \text{Rank}(\mathbf{Z_0}) + \|\mathbf{A}\|_0, \quad \text{s.t.} \quad \mathcal{P}_\Omega(\mathbf{Z_a}) = \mathcal{P}_\Omega(\mathbf{Z_0} + \mathbf{A}), \tag{8.11}$$

where Ω is an index subset, and $\mathcal{P}_\Omega(\cdot)$ is the projection operator. Specifically, $\mathcal{P}_\Omega(\mathbf{M})$ is the projection of a matrix $\mathbf{M}$ onto the subspace of matrices whose nonzero entries are restricted to Ω as

$$[\mathcal{P}_\Omega(\mathbf{M})]_{ij} = 0, \quad \forall (i,j) \notin \Omega. \tag{8.12}$$

In the following, we propose two methods to solve this problem.

8.2.2 Nuclear norm minimization

The optimization problem in (8.10) captures the low-rank property of the power state measurement matrix $\mathbf{Z_0}$ as well as the sparseness of the malicious attack major $\mathbf{A}$. However, it is known to be impractical to directly solve (8.10). One possible approach is to replace $\text{Rank}(\mathbf{Z_0})$ and $\|\mathbf{A}\|_0$ with their convex relaxations, $\|\mathbf{Z_0}\|_*$ and $\|\mathbf{A}\|_1$, respectively. Here, $\|\mathbf{Z_0}\|_*$ is the nuclear norm of $\mathbf{Z_0}$, which is the sum of its singular values, and $\|\mathbf{A}\|_1$ is the l_1 norm of $\mathbf{A}$, which is the sum of absolute values of its entries. Hence, (8.10) can be reformulated as the following convex optimization problem:

$$\min_{\mathbf{Z_0},\mathbf{A}} \|\mathbf{Z_0}\|_* + \lambda\|\mathbf{A}\|_1, \quad \text{s.t.} \quad \mathbf{Z_a} = \mathbf{Z_0} + \mathbf{A}, \tag{8.13}$$

where λ is a regularization parameter. Correspondingly, (8.11) can be reformulated as

$$\min_{\mathbf{Z_0},\mathbf{A}} \|\mathbf{Z_0}\|_* + \lambda\|\mathbf{A}\|_1, \quad \text{s.t.} \quad \mathcal{P}_\Omega(\mathbf{Z_a}) = \mathcal{P}_\Omega(\mathbf{Z_0} + \mathbf{A}). \tag{8.14}$$

The optimization problem in (8.14) has been extensively studied in the fields of compressive sensing [16] and matrix completion [17, 18], and can be solved by many off-the-shelf convex optimization algorithms. Motivated by [19], an algorithm that applies the techniques of augmented Lagrange multipliers is utilized here to detect the false data matrix $\mathbf{A}$ as well as to recover the measurement matrix $\mathbf{Z_0}$.

Augmented Lagrange multipliers are used to solve constrained optimization problems as follows:

$$\min f(\mathbf{X}), \quad \text{s.t.} \quad h(\mathbf{X}) = 0, \tag{8.15}$$

where $f : \mathbb{R}^n \to \mathbb{R}$ and $h : \mathbb{R}^n \to \mathbb{R}^m$. The augmented Lagrangian is

$$\mathbf{L}(\mathbf{X}, \mathbf{Y}, \mu) = \mathbf{f}(\mathbf{X}) + \langle \mathbf{Y}, \mathbf{h}(\mathbf{X}) \rangle + \frac{\mu}{2} \|\mathbf{h}(\mathbf{X})\|_2^2, \tag{8.16}$$

where μ is a positive scalar, and $\mathbf{Y}$ contains the Lagrange multipliers. Here $\langle \mathbf{Y}, \mathbf{h}(\mathbf{X}) \rangle$ denotes the inner product. The optimization problem in (8.14) can be solved iteratively via the method of augmented Lagrange multipliers [20]. The Lagrangian for (8.14) is given by

$$\begin{aligned} L(\mathbf{Z_0}, \mathbf{A}, \mathbf{Y}, \mu) = \|\mathbf{Z_0}\|_* + \lambda \|\mathbf{A}\|_1 + \langle \mathbf{Y}, \mathcal{P}_\Omega(\mathbf{Z_a} - \mathbf{Z_0} - \mathbf{A}) \rangle \\ + \frac{\mu}{2} \|\mathcal{P}_\Omega(\mathbf{Z_a} - \mathbf{Z_0} - \mathbf{A})\|_2^2. \end{aligned} \tag{8.17}$$

The value of λ is set to $\frac{1}{\sqrt{\max(m,t)}}$ [19], where $m \times t$ are the dimensions of the measurement matrix $\mathbf{Z_a}$. The optimal $\mathbf{Z_0}$ and $\mathbf{A}$ are found iteratively as

$$(\mathbf{A}_{[k+1]}, \mathbf{Z_0}_{[k+1]}) = \arg\min_{\mathbf{A}} L(\mathbf{Z_0}, \mathbf{A}, u_{[k]}, \mathbf{Y}_{[k]}), \tag{8.18}$$

where $\mathbf{A}_{[k+1]}$ can be explicitly computed from the soft-shrinkage formula, and $\mathbf{Z_0}_{[k+1]}$ can be solved via the singular value shrinkage operator [21]. Specifically, we define this operator as $\mathcal{S}_\tau\{x\} = \text{sgn}(x) \max(|x| - \tau, 0)$ for a real variable x, where sgn is the sign function. This operator can be extended to vectors and matrices by applying it element-wise. The (8.18) is solved in an iterative fashion, where $\mathbf{A}_{[k+1]}$ is updated by

$$\mathbf{A}_{[k+1]}^{[j+1]} = \mathcal{S}_{\lambda u_{[k]}^{-1}} \{\mathbf{Z_a} - \mathbf{Z_0}_{[k+1]}^{[j]} + u_{[k]}^{-1} \mathbf{Y}_{[k]}\}. \tag{8.19}$$

To find $\mathbf{Z_0}_{[k+1]}$, a singular value decomposition (SVD) is first applied to the matrix $\mathbf{Z_a} - \mathbf{A}_{[k+1]}^{[j+1]} + u_{[k]}^{-1} \mathbf{Y}_{[k]}$:

$$\left(\mathbf{Z_a} - \mathbf{A}_{[k+1]}^{[j+1]} + u_{[k]}^{-1} \mathbf{Y}_{[k]}\right) = \mathbf{U}\mathbf{S}\mathbf{V}^\top, \tag{8.20}$$

where $\mathbf{U} \in \mathbb{R}^{m \times m}$ and $\mathbf{V} \in \mathbb{R}^{t \times t}$ are unitary matrices, and $\mathbf{S} \in \mathbb{R}^{m \times t}$ is a diagonal matrix containing the singular values of $\left(\mathbf{Z_a} - \mathbf{A}_{[k+1]}^{[j+1]} + u_{[k]}^{-1} \mathbf{Y}_{[k]}\right)$. The singular values are arranged in decreasing order, and $\mathbf{Z_0}$ is updated via

$$\mathbf{Z_0}_{[k+1]}^{[j+1]} = \mathbf{U} \mathcal{S}_{u_{[k]}^{-1}} \{\mathbf{S}\} \mathbf{V}^\top. \tag{8.21}$$

Equations (8.19)–(8.21) are performed iteratively to solve (8.18). After solving for $(\mathbf{A}_{[k+1]}, \mathbf{Z_0}_{[k+1]})$, both the Lagrange multipliers $\mathbf{Y}$ and μ are updated, which improves the performance of the algorithm:

$$\mathbf{Y}_{[k+1]} = \mathbf{Y}_{[k]} + u_{[k]} \left(\mathbf{Z_a} - \mathbf{Z_0}_{[k+1]} - \mathbf{A}_{[k+1]}\right), \tag{8.22}$$

$$\mu_{[k+1]} = \alpha \mu_{[k]}, \tag{8.23}$$

where α is a positive constant. The algorithm is outlined as Algorithm 1.

Algorithm 1 Nuclear Norm Minimization Approach

Input: $\mathbf{Z_a} \in \mathbb{R}^{m \times t}$; $\lambda = \frac{1}{\sqrt{\max(m,t)}}$;
Initialize: $\mathbf{Y}_{[0]} = 0$; $\mathbf{Z_0}_{[0]} = 0$; $\mathbf{A}_{[0]} = 0$; $\mu_{[0]} > 0$; $\alpha > 0$; $k = 0$;
while not converged **do**
  $\mathbf{Z_0}_{[k+1]} = \mathbf{Z_0}_{[k]}$; $\mathbf{A}_{[k+1]} = \mathbf{A}_{[k]}$; $j = 0$;
  while not converged **do**
    $\mathbf{A}_{[k+1]}^{[j+1]} = \mathcal{S}_{\lambda u_{[k]}^{-1}}\{\mathbf{Z_a} - \mathbf{Z_0}_{[k+1]}^{[j]} + u_{[k]}^{-1}\mathbf{Y}_{[k]}\}$;
    $\left(\mathbf{Z_a} - \mathbf{A}_{[k+1]}^{[j+1]} + u_{[k]}^{-1}\mathbf{Y}_{[k]}\right) = \mathbf{USV}^\top$;
    Obtain $[\mathbf{U}, \mathbf{S}, \mathbf{V}]$;
    $\mathbf{Z_0}_{[k+1]}^{[j+1]} = \mathbf{U}\mathcal{S}_{u_{[k]}^{-1}}\{\mathbf{S}\}\mathbf{V}^\top$;
    $j = j + 1$;
  end while
  $\mathbf{Y}_{[k+1]} = \mathbf{Y}_{[k]} + u_{[k]}\left(\mathbf{Z_a} - \mathbf{Z_0}_{[k+1]} - \mathbf{A}_{[k+1]}\right)$;
  $\mu_{[k+1]} = \alpha\mu_{[k]}$;
  $k = k + 1$;
end while
return $\mathbf{Z_0}_{[k]}$; $\mathbf{A}_{[k]}$;
Output $\mathbf{Z_0}_{[k]}$; $\mathbf{A}_{[k]}$;

8.2.3 Low-rank matrix factorization

The speed and scalability of the nuclear norm minimization approach are limited by the computational complexity of singular value decomposition. When the matrix size and rank increase, the computational operations for singular value decomposition will become quite expensive. To improve the scalability of solving large-scale problems of malicious attack detection in power systems, a low-rank matrix factorization approach is proposed here.

Given the observations $\mathbf{Z_a}$, the measurements $\mathbf{Z_0}$ and false data injection attack matrix $\mathbf{A}$ can be separately obtained by first solving the minimization problem:

$$\min_{\mathbf{U},\mathbf{V},\mathbf{Z_0}} \|\mathbf{Z_a} - \mathbf{Z_0}\|_1, \quad \text{s.t.} \quad \mathbf{UV} - \mathbf{Z_0} = \mathbf{0}, \tag{8.24}$$

where the low-rank matrix $\mathbf{Z_0}$ is expressed as a product of $\mathbf{U} \in \mathbb{R}^{m \times r}$ and $\mathbf{V} \in \mathbb{R}^{r \times n}$ for adjustable rank estimate r. The target matrix $\mathbf{A}$ can be recovered afterwards by $\mathbf{A} = \mathbf{Z_a} - \mathbf{Z_0}$. Correspondingly, (8.14) can be rewritten as

$$\min_{\mathbf{U},\mathbf{V},\mathbf{Z_0}} \|\mathcal{P}_\Omega(\mathbf{Z_a} - \mathbf{Z_0})\|_1, \quad \text{s.t.} \quad \mathbf{UV} - \mathbf{Z_0} = \mathbf{0}. \tag{8.25}$$

Note that a low-rank matrix factorization is explicitly applied to $\mathbf{Z_0}$ instead of minimizing its nuclear norm as in (8.14), which avoids singular value decomposition completely. To solve the minimization problem in (8.25), the augmented Lagrangian can be expressed as

$$L(\mathbf{U}, \mathbf{V}, \mathbf{Z_0}, \mathbf{Y}, \mu) = \|\mathcal{P}_\Omega(\mathbf{Z_a} - \mathbf{Z_0})\|_1 + \langle \mathbf{Y}, \mathbf{UV} - \mathbf{Z_0}\rangle + \frac{\mu}{2}\|\mathbf{UV} - \mathbf{Z_0}\|_2^2, \tag{8.26}$$

Algorithm 2 Low-Rank Matrix Factorization

Input: $\mathbf{Z_a} \in \mathbb{R}^{m \times t}$; Initial rank estimate r.
Initialize: $\mathbf{U} \in \mathbb{R}^{m \times r}$; $\mathbf{V} \in \mathbb{R}^{r \times t}$; $\mathbf{Z_0}_{[0]} = U * V$; $\mathbf{Y}_{[0]} = 0$; $\mu_{[0]} > 0$; $\alpha > 0$; $k = 0$.
while not converged **do**
$\mathbf{U}_{[k+1]} = \left(\mathbf{Z_0} - u_{[k]}^{-1}\mathbf{Y}_{[k]}\right)\mathbf{V}^\top(\mathbf{V}\mathbf{V}^\top)^{-1}$;
$\mathbf{V}_{[k+1]} = (\mathbf{U}^\top\mathbf{U})^{-1}\mathbf{U}^\top\left(\mathbf{Z_0} - u_{[k]}^{-1}\mathbf{Y}_{[k]}\right)$;
$\mathbf{Z_0}_{[k+1]} = \mathcal{S}_{u_{[k]}^{-1}}\left\{\mathbf{U}_{[k+1]}\mathbf{V}_{[k+1]} - \mathbf{Z_a} + u_{[k]}^{-1}\mathbf{Y}_{[k]}\right\}$;
$\mathbf{Y}_{[k+1]} = \mathbf{Y}_{[k]} + u_{[k]}\left(\mathbf{U}_{[k+1]}\mathbf{V}_{[k+1]} - \mathbf{Z_0}_{[k+1]}\right)$;
$\mu_{[k+1]} = \alpha\mu_{[k]}$;
$k = k + 1$;
Check r, possibly re-estimate r and adjust sizes of the iterates;
end while
return $\mathbf{Z_0}_{[k]}$;
Output $\mathbf{Z_0}_{[k]}$; $\mathbf{Z_a} - \mathbf{Z_0}_{[k]}$;

where μ is a penalty parameter and $\mathbf{Y}$ is the vector of Lagrange multipliers corresponding to the constraint $\mathbf{UV} - \mathbf{Z_0} = \mathbf{0}$. Motivated by the idea of the alternating direction method for convex optimization, the augmented Lagrangian can be minimized with respect to the block variables $\mathbf{U}$, $\mathbf{V}$, and $\mathbf{Z_0}$ individually via the following framework at each iteration k [22]:

$$\mathbf{U}_{[k+1]} = \arg\min_{\mathbf{U}} L\left(\mathbf{U}, \mathbf{V}_{[k]}, \mathbf{Z_0}_{[k]}, \mathbf{Y}_{[k]}, \mu_{[k]}\right), \tag{8.27}$$

$$\mathbf{V}_{[k+1]} = \arg\min_{\mathbf{V}} L\left(\mathbf{U}_{[k+1]}, \mathbf{V}, \mathbf{Z_0}_{[k]}, \mathbf{Y}_{[k]}, \mu_{[k]}\right), \tag{8.28}$$

$$\mathbf{Z_0}_{[k+1]} = \arg\min_{\mathbf{Z_0}} L\left(\mathbf{U}_{[k+1]}, \mathbf{V}_{[k+1]}, \mathbf{Z_0}, \mathbf{Y}_{[k]}, \mu_{[k]}\right), \tag{8.29}$$

where (8.27) and (8.28) are least squares problems:

$$\mathbf{U}_{[k+1]} = \left(\mathbf{Z_0}_{[k]} - u_{[k]}^{-1}\mathbf{Y}_{[k]}\right)\mathbf{V}_{[\mathbf{k}]}^\top\left(\mathbf{V}_{[\mathbf{k}]}\mathbf{V}_{[\mathbf{k}]}^\top\right)^{-\mathbf{1}}, \tag{8.30}$$

$$\mathbf{V}_{[k+1]} = \left(\mathbf{U}_{[\mathbf{k+1}]}^\top\mathbf{U}_{[\mathbf{k+1}]}\right)^{-\mathbf{1}}\mathbf{U}_{[\mathbf{k+1}]}^\top\left(\mathbf{Z_0}_{[k]} - u_{[k]}^{-1}\mathbf{Y}_{[k]}\right); \tag{8.31}$$

and (8.29) can be solved by the shrinkage formula:

$$\mathcal{P}_\Omega\left(\mathbf{Z_0}_{[k+1]}\right) = \mathcal{P}_\Omega\left(\mathcal{S}_{u_{[k]}^{-1}}\left\{\mathbf{U}_{[k+1]}\mathbf{V}_{[k+1]} - \mathbf{Z_a} + u_{[k]}^{-1}\mathbf{Y}_{[k]}\right\}\right). \tag{8.32}$$

The Lagrange multipliers $\mathbf{Y}$ and μ are updated during each iteration as

$$\mathbf{Y}_{[k+1]} = \mathbf{Y}_{[k]} + u_{[k]}\left(\mathbf{U}_{[k+1]}\mathbf{V}_{[k+1]} - \mathbf{Z_0}_{[k+1]}\right), \tag{8.33}$$

$$\mu_{[k+1]} = \alpha\mu_{[k]}, \tag{8.34}$$

where α is a positive constant. At the end of each iteration, a rank estimation strategy [23] is applied to update r to ensure the success of the algorithm. The proposed algorithm is illustrated in Algorithm 2.

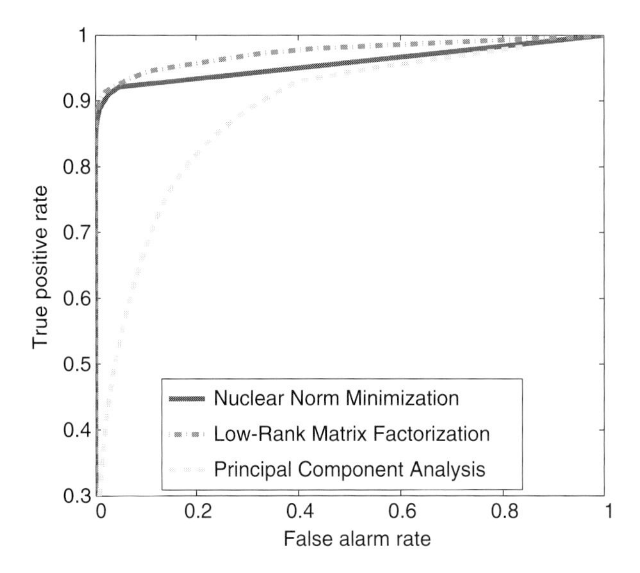

Figure 8.2 ROC performance for the IEEE 57 bus system. SNR = 10 dB.

8.2.4 Numerical results

Numerical simulations are presented here to evaluate the performance of the proposed algorithms. Power flow data for IEEE 57 bus, and IEEE 118 bus test cases, and the Polish system of [24] during the winter time frame in 2007–2008 which has 3012 buses are used to validate the effectiveness of the considered algorithms.

Receiver operating characteristic analysis

Assume the loads on each bus in the power system are uniformly distributed between 50% and 150% of its base load. When the state estimation measurements are collected, a small portion ϵ of the measurement data is compromised by malicious attackers with an arbitrary amount of injection data, and ϵ is defined as the attack ratio in this context. Methods for false data injection attack construction can be found in [12] and [13]. Here, we focus on the protection scheme and suppose that the locations of the attacks are chosen randomly with duration Δt.[2] A total number of T time instance measurements are obtained for analysis. The receiver operating characteristic (ROC) analysis of the proposed algorithms is first given, and then compared with the principal component analysis (PCA)-based detection method.[3] In this analysis, the attack ratio is fixed at $\epsilon = 0.1$ and SNR = 10 dB.

The ROC curves for IEEE 57 bus and IEEE 118 bus cases are shown in Figure 8.2 and Figure 8.3, respectively. It is apparent that the proposed algorithms can detect the false data accurately at a low false alarm rate. For example, in the IEEE 57 bus system, the true

[2] Note that the attack vectors used in this chapter are more general compared to those described in [12] and [13], and will not affect the efficiency of the proposed algorithms.

[3] For PCA, we retain the largest K singular values of the matrix such that $\frac{\sum_1^K s_i}{\sum_1^N s_i} > 95\%$.

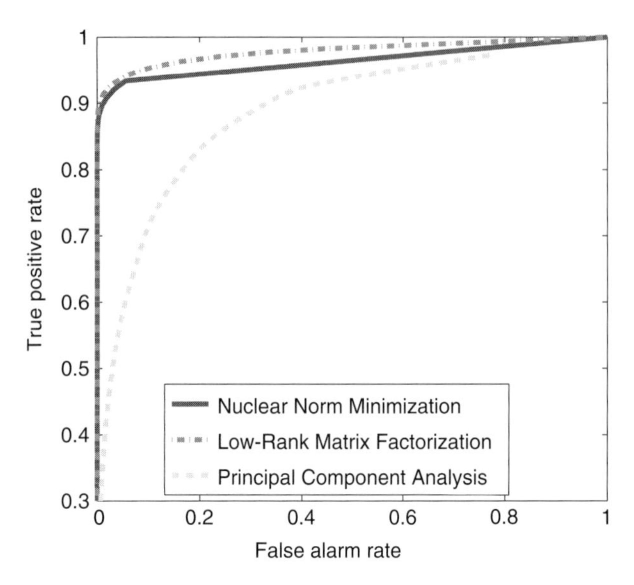

Figure 8.3 ROC performance for the IEEE 118 bus system. SNR = 10 dB.

positive rate of nuclear norm minimization is 93%; and it is 95% with low-rank matrix factorization when the false alarm rate $p_f = 10\%$. The IEEE 118 bus system shows similar results. Moreover, the low-rank matrix factorization approach performs slightly better than the nuclear norm minimization method. The reason for this phenomenon is model-related, and in this case, the sparse attack matrix is not the dominant part in the measurements, which makes the low-rank matrix factorization approach more suitable. Figure 8.2 and Figure 8.3 show that the proposed algorithms outperform the PCA-based approach significantly. The PCA method does not take the corruption of malicious attacks into consideration. Even though the matrix $\mathbf{Z}_0$ is of low rank, the sum of $\mathbf{Z}_0$ and $\mathbf{A}$ will not be of low rank. Thus, directly applying the PCA method will result in poor performance. However, the proposed algorithm exploits the low-rank structure of the anomaly-free measurement matrix, and the fact that malicious attacks are quite sparse, which yields better performance.

Performance vs. measurement missing ratio

Next, we investigate the performance of the proposed algorithms under different measurement missing ratios. In particular, we assume that a portion of the measurements collected at the control center are missing due to meter failures or communication link outages, and evaluate the performance of the proposed algorithms under different measurement missing ratios up to 10% on the IEEE 118 bus system. The attack ratio is fixed at $\epsilon = 0.1$ with SNR = 10 dB.

The ROC curves for the IEEE 118 bus case are depicted in Figure 8.4. From the figure we see that with 10% missing measurements, the proposed algorithms are still able to detect the malicious attacks at acceptable true positive rates, and the low-rank matrix factorization method performs slightly better. By comparing with Figure 8.3, we see

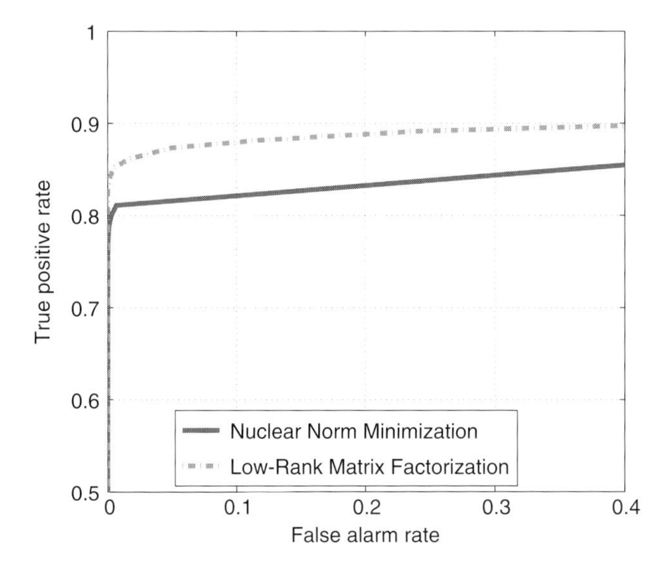

Figure 8.4 ROC curves of the proposed algorithms for the IEEE 118 bus system; 10% measurements are missing with SNR = 10 dB.

that the missing measurements deteriorate the performance of the proposed algorithms as we would expect. Since the PCA-based method is unable to detect the anomalies in this case, we omitted its simulation result. Note that the existence of missing entries will result in an incorrect estimation of the low-dimensional subspace of matrix the $\mathbf{Z}_0$, which leads to the failure of PCA.

To investigate the performance under different missing measurement ratios, the percentage of missing measurements is varied from 0% (no missing) to 10%, and the results are shown in Figure 8.5. The true positive rates are calculated for both algorithms when the false alarm rate equals 10%. It is shown that the performance improves monotonically as more and more measurements are collected. In the worst case when 10% of measurements are missing, the proposed algorithms can still achieve true positive rates of 85% and 90% for the nuclear norm minimization and the low-rank matrix factorization methods, respectively.

Performance vs. attack ratio

Thirdly, we investigate the performance of the proposed algorithms under different attack ratios for the IEEE 118 bus system. In particular, ϵ is varied from 5% to 15%, and SNR = 10 dB.

From Figure 8.6, the true positive rate is quite high at low sparsity ratios for both proposed algorithms. Particularly, when the sparsity ratio is 5%, the true positive rates are 93.6% and 94.3% at $f_a = 10\%$ for the nuclear norm minimization and low-rank matrix factorization methods, respectively. Compared with the PCA-based method, the performance of the proposed algorithms is quite stable as the attack ratio increases. When the attack ratio reaches 15%, the true positive rates for both algorithms are still around

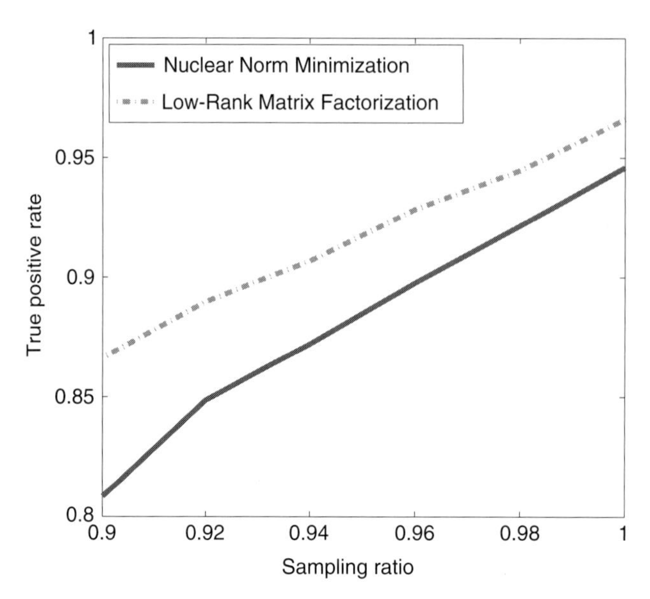

Figure 8.5 Performance of the proposed algorithms under different missing ratios for the IEEE 118 bus system. The false alarm rate is 10% and SNR = 10 dB.

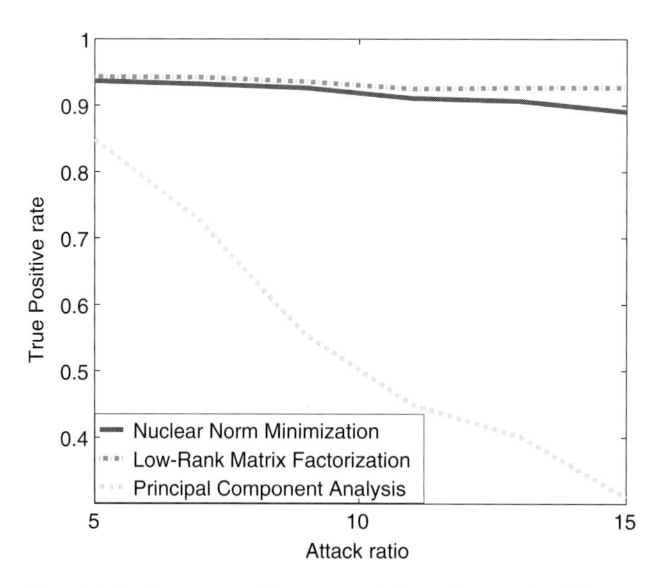

Figure 8.6 Performance of the proposed algorithms under different attack ratios for the IEEE 118 bus system. The false alarm rate is 10% with SNR = 10 dB.

90%. The true positive rates of the proposed algorithms will decrease dramatically when attackers attack the power system massively. This is because, when the attack matrix is not sparse enough, the mixed-norm minimization is not able to separate the low-rank anomaly-free matrix from the attack matrix.

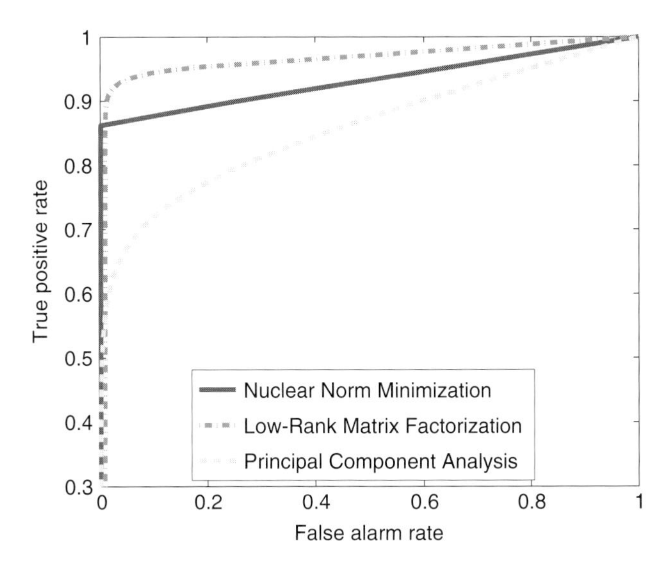

Figure 8.7 Performance on power flow data for the Polish system (3012 buses) during winter peak conditions, 2007–2008; SNR = 10 dB.

Performance on large-scale systems

Finally, we analyze the scalability and computational efficiency of the proposed algorithms on power flow data for the Polish system (3012 buses) during winter peak conditions in 2007–2008. The attack ratio is fixed at $\epsilon = 0.1$ with SNR = 10 dB.

The ROC curves are shown in Figure 8.7. It is seen that the performance of the proposed algorithms is quite stable on the large-scale system compared to the IEEE 57 bus and the IEEE 118 bus systems. A comparison of the computational efficiency of the two proposed algorithms is shown in Figure 8.8. The data matrix row dimension m is varied from 100 to 3400, in steps of 300 rows. The proposed algorithms are applied to a subset of the measurement matrix each time, and the CPU computation time is logged. It is shown in Figure 8.8 that, as the dimension of the measurement matrix increases, the CPU time for computation will increase, and the low-rank matrix factorization approach performs better than the nuclear norm minimization method, which demonstrates a better scalability to large problems, as expected.

The numerical results validate the effectiveness of the proposed algorithms. According to the simulation results, both the low-rank matrix factorization and nuclear norm minimization techniques can solve the matrix separation problem, and the performance of the low-rank matrix factorization is slightly better than that of the nuclear norm minimization technique. From the perspective of recoverability, since the false data attack matrix $\mathbf{A}$ is not the dominant part compared with $\mathbf{Z}_0$ in this setting, the performance of the low-rank matrix factorization technique is better. From the perspective of computation time, the low-rank matrix factorization technique is much faster than the nuclear norm minimization method due to its SVD-free feature.

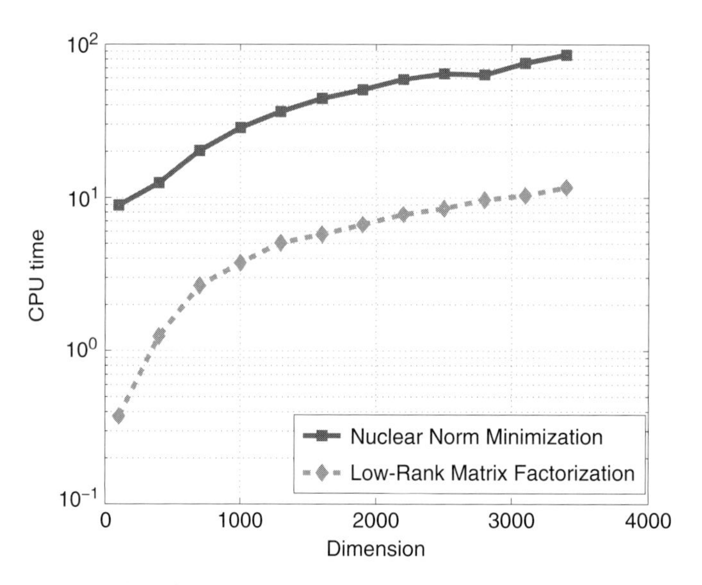

Figure 8.8 CPU time vs. matrix dimension for the proposed nuclear norm minimization and low-rank matrix factorization algorithms.

8.3 Distributed approach for security-constrained optimal power flow

In this section, we first introduce the background of the security-constrained optimal power flow (SCOPF) problem. Then we propose a distributed parallel approach to address it. Finally, numerical simulations are given to validate the effectiveness of the proposed algorithm.

8.3.1 Security-constrained optimal power flow

The deregulation of electric power grids offers the opportunity for electricity market participants to exercise least-cost or profit-based operations [25]. Despite the market-driven tendency of the electric power business, security remains a significant concern of sustainable power system operations, which cannot be compromised. Security-constrained optimal power flow [26, 27] aims at minimizing the cost of system operation while satisfying a set of postulated contingency constraints. It is an important management task allowing optimal control of power systems securely.

The SCOPF problem is an extension of the conventional optimal power flow (OPF) problem [28], whose objective is to determine a generation schedule that minimizes the system operating cost while satisfying the system operation constraints such as hourly load demand, fuel limitations, environmental constraints, and network security requirements. It has been recognized [29] that the optimal control of the normal state (i.e. pre-contingency case) may violate the system operation constraints after the occurrence of some major disturbance events (i.e. post-contingency case), and thus jeopardize the security of power systems. To address this problem, SCOPF is performed by considering both pre-contingency and post-contingency constraints to guarantee sustainable operations of the electric grid. The system security level is improved by taking into account a number of contingencies in a dedicated and selected contingency list. The solution to SCOPF should satisfy the so-called $N-1$ criterion, which requires that the operational limits of the power system should not be violated in case of a single contingency (a line and/or generator outage).

SCOPF can be broadly classified as preventive, where control variables are restricted to their pre-contingency condition settings, and corrective, whose control variables are allowed to be rescheduled once a contingency happens [30]. We will focus on the corrective model in this example. The seminal paper [29] proposed the generalized Benders decomposition method to solve the corrective SCOPF problem. Since then, an extensive literature exists for SCOPF in power systems both for traditional operations and under market environments [27, 31–34]. The nested Benders decomposition method was used in [31] to solve the SCOPF problem for determining the optimal daily generation scheduling in a pool-organized electricity market, and was tested in an actual example of the Spanish power system. SCOPF has been embedded into the security-constrained unit commitment (SCUC) model, and designed an effective corrective/preventive contingency dispatch over a 24-hour period, which balanced the economics and security in the restructured markets [32]. An iterative approach was proposed in [33] to obtain the solution of SCOPF, which aims to efficiently identify an as small as possible superset of the binding contingencies to achieve the SCOPF optimum. The Benders decomposition was applied in [34] to decompose the traditional SCOPF problem, and the underlying computational complexity was analyzed in this approach. The SCOPF problem was solved in [27] by a non-decomposed method based on the compression of the post-contingency networks, which can reduce the size of the security constraints and relieve the computational burden in the problem. Unlike existing works, our methods discussed next solve the SCOPF problem in a distributed way.

Before presenting the distributed parallel approach for this problem, it is useful to recall a general formulation of the conventional SCOPF problem compactly described as follows (further detail on the power flow analysis can be found in [5]):

$$\min_{\mathbf{x}^0,\ldots,\mathbf{x}^C;\mathbf{u}^0,\ldots,\mathbf{u}^C} \quad f^0(\mathbf{x}^0,\mathbf{u}^0) \tag{8.35}$$

$$\text{s.t.} \quad \mathbf{g}^0(\mathbf{x}^0,\mathbf{u}^0) = 0, \tag{8.36}$$

$$\mathbf{h}^0(\mathbf{x}^0,\mathbf{u}^0) \le 0, \tag{8.37}$$

$$\mathbf{g}^c(\mathbf{x}^c,\mathbf{u}^c) = 0, \quad c = 1,\ldots,C, \tag{8.38}$$

$$\mathbf{h}^c(\mathbf{x}^c,\mathbf{u}^c) \le 0, \quad c = 1,\ldots,C, \tag{8.39}$$

$$|\mathbf{u}^0 - \mathbf{u}^c| \le \Delta_c, \quad c = 1,\ldots,C, \tag{8.40}$$

where f^0 is the objective function, through which (8.35) aims to maximize the total social welfare or equivalently minimize the offer-based energy and production cost, $\mathbf{x}^c$ is the vector of state variables, which includes magnitude and voltage angles at all buses, and $\mathbf{u}^c$ is the vector of control variables, which can be generator real powers or terminal voltages. The superscript $c = 0$ corresponds to the pre-contingency configuration, and

$c = 1, \ldots, C$ correspond to different post-contingency configurations. In addition, Δ_c is the maximum allowed adjustment between the normal and contingency states for contingency c. The state and control variables can be concatenated as a single vector, called the operation decisions vector. Here we separate them for presentation simplicity.

In the conventional SCOPF problem, the equality constraints (8.38) on $\mathbf{g}^c, c = 0, \ldots, C$, represent the system nodal power flow balance over the entire grid, and the inequality constraints (8.39) on $\mathbf{h}^c, c = 0, \ldots, C$, represent the physical limits on the equipment, such as the operational limits on the branch currents and bounds on the generator power outputs. Constraints (8.36)–(8.37) capture the economic dispatch and enforce the feasibility of the pre-contingency state. Constraints (8.38)–(8.39) incorporate the security-constrained dispatch and enforce the feasibility of the post-contingency state. Constraint (8.40) introduces the security-constrained dispatch with rescheduling, which couples control variables of the pre-contingency and post-contingency states and prevents unrealistic post-contingency corrective actions. Note that there are some variations on the objective function and constraints of the SCOPF problem, and we focus on the above conventional formulation in this chapter.

Following the standard approach to formulating the SCOPF problem, the objective here is to minimize the cost of generation while safeguarding the power system sustainability. For the sake of simplicity and computational tractability, constraints (8.36)–(8.39) are modeled with the linear DC load flow, and we assume that the list of contingencies is given. Thus, assuming a DC power network modeling and neglecting all shunt elements, the standard SCOPF problem can be simplified to the following optimization problem:

$$\min_{\boldsymbol{\theta}^0,\ldots,\boldsymbol{\theta}^C;\mathbf{P}^{g,0},\ldots,\mathbf{P}^{g,C}} \quad \sum_{i\in\mathcal{G}} f_i^g\left(\mathbf{P}_i^{g,0}\right) \tag{8.41}$$

$$\text{s.t.} \quad \mathbf{B}_{bus}^0\boldsymbol{\theta}^0 + \mathbf{P}^{d,0} - \mathbf{A}^{g,0}\mathbf{P}^{g,0} = 0, \tag{8.42}$$

$$\mathbf{B}_{bus}^c\boldsymbol{\theta}^c + \mathbf{P}^{d,c} - \mathbf{A}^{g,c}\mathbf{P}^{g,c} = 0, \tag{8.43}$$

$$\left|\mathbf{B}_f^0\boldsymbol{\theta}^0\right| - \mathbf{F}_{max} \le 0, \tag{8.44}$$

$$\left|\mathbf{B}_f^c\boldsymbol{\theta}^c\right| - \mathbf{F}_{max} \le 0, \tag{8.45}$$

$$\underline{\mathbf{P}^{g,0}} \le \mathbf{P}^{g,0} \le \overline{\mathbf{P}^{g,0}}, \tag{8.46}$$

$$\underline{\mathbf{P}^{g,c}} \le \mathbf{P}^{g,c} \le \overline{\mathbf{P}^{g,c}}, \tag{8.47}$$

$$|\mathbf{P}^{g,0} - \mathbf{P}^{g,c}| \le \Delta_c, \tag{8.48}$$

$$i \in \mathcal{G}, \quad c = 1, \ldots, C, \tag{8.49}$$

where the notation is given in Table 8.2.

The solution to (8.41) ensures economical dispatch while guaranteeing power system security, by taking into account a set of postulated contingencies. The major challenge of SCOPF is the size of the problem, especially for large systems with numerous contingency cases to be considered. Directly solving the SCOPF problem by simultaneously imposing all post-contingency constraints will result in prohibitive memory requirements and substantial CPU burdens. To achieve efficient and secure operations of the entire electrical grid, a distributed and parallel optimization algorithm is designed to solve the SCOPF problem in large-scale power systems in the next sections.

Table 8.2 Notation definitions for SCOPF

$\mathcal{G}$	Set of generators
$\mathcal{N}$	Set of buses
$\mathcal{B}$	Set of branches
$\boldsymbol{\theta}^c \in \mathbb{R}^{\vert\mathcal{N}\vert}$	Vector of voltage angles
$\mathbf{P}^{g,c} \in \mathbb{R}^{\vert\mathcal{G}\vert}$	Vector of real power flows
f_i^g	Generation cost function
$\mathbf{P}_i^{g,0}$	Displaceable real power of each individual generation unit i for the pre-contingency configuration
$\mathbf{B}_{bus}^c \in \mathbb{R}^{\vert\mathcal{N}\vert\times\vert\mathcal{N}\vert}$	Power network system admittance matrix
$\mathbf{B}_f^c \in \mathbb{R}^{\vert\mathcal{B}\vert\times\vert\mathcal{N}\vert}$	Branch admittance matrix
$\mathbf{P}^{d,c} \in \mathbb{R}^{\vert\mathcal{N}\vert}$	Real power demand
$\mathbf{A}^{g,c} \in \mathbb{R}^{\vert\mathcal{N}\vert\times\vert\mathcal{G}\vert}$	Sparse generator connection matrix, whose (i, j)th element is 1 if generator j is located at bus i and 0 otherwise
$\mathbf{F}_{max}$	Vector for the maximum power flow
$\overline{\mathbf{P}^{g,c}}$	Upper bound on real power generation
$\underline{\mathbf{P}^{g,c}}$	Lower bound on real power generation
Δ_c	Pre-defined maximum allowed variation of power outputs

8.3.2 ADMM method

In this section, a distributed algorithm is proposed to solve the SCOPF problem by decomposing it into a set of simpler and parallel subproblems corresponding to the base case and each contingency case, respectively. The approach is based on the alternating direction method of multipliers (ADMM) [35, 36], whose general form is briefly described as follows (a more detailed introduction of ADMM is given in Chapter 3):

$$\min_{\mathbf{x},\mathbf{z}} \quad f(\mathbf{x}) + g(\mathbf{z}) \tag{8.50}$$

$$\text{s.t.} \quad \mathbf{Ax} + \mathbf{Bz} = \mathbf{c}, \tag{8.51}$$

where $\mathbf{x} \in \mathbb{R}^n$, $\mathbf{z} \in \mathbb{R}^m$, $\mathbf{c} \in \mathbb{R}^p$, $\mathbf{A} \in \mathbb{R}^{p\times n}$ and $\mathbf{B} \in \mathbb{R}^{p\times m}$. The functions f and g are closed, convex and proper. The scaled augmented Lagrangian can be expressed as

$$\mathcal{L}_\rho(\mathbf{x}, \mathbf{z}, \boldsymbol{\mu}) = f(\mathbf{x}) + g(\mathbf{z}) + \frac{\rho}{2}\|\mathbf{Ax} + \mathbf{Bz} - \mathbf{c} + \boldsymbol{\mu}\|_2^2, \tag{8.52}$$

where $\rho > 0$ is the penalty parameter and $\boldsymbol{\mu}$ is the vector of scaled dual variables. Using the scaled dual variables, $\mathbf{x}$ and $\mathbf{z}$ are updated in a Gauss–Seidel fashion. At each iteration k, the update process can be expressed as

$$\mathbf{x}[k+1] = \arg\min_{\mathbf{x}} f(\mathbf{x}) + \frac{\rho}{2}\|\mathbf{Ax} + \mathbf{Bz}[k] - \mathbf{c} + \boldsymbol{\mu}[k]\|_2^2,$$

$$\mathbf{z}[k+1] = \arg\min_{\mathbf{z}} g(\mathbf{z}) + \frac{\rho}{2}\|\mathbf{Ax}[k+1] + \mathbf{Bz} - \mathbf{c} + \boldsymbol{\mu}[k]\|_2^2.$$

Finally, the scaled dual variable vector is updated via

$$\boldsymbol{\mu}[k+1] = \boldsymbol{\mu}[k] + \mathbf{A}\mathbf{x}[k+1] + \mathbf{B}\mathbf{z}[k+1] - \mathbf{c}.$$

The use of ADMM for optimization in power systems has been considered in [37] and [38]. The ADMM-based methods offer a general framework for distributed optimization, and a corresponding distributed and parallel approach to solve SCOPF is introduced below.

8.3.3 Distributed and parallel approach for SCOPF

The optimization problem (8.41) cannot be readily solved using ADMM, since the constraint (8.48) couples the pre-contingency and post-contingency variables, and the inequalities make the problem even more complicated. To address these challenges, the optimization problem (8.41) can be reformulated by introducing a slack variable $\mathbf{p}^c \in \mathbb{R}^{|\mathcal{G}|}$:

$$\begin{aligned}
&\text{minimize} && (8.41) && (8.53)\\
&\text{subject to} && \text{Constraints (8.42)–(8.47)}, && (8.54)\\
& && \mathbf{P}^{g,0} - \mathbf{P}^{g,c} + \mathbf{p}^c = \Delta_c, && (8.55)\\
& && 0 \le \mathbf{p}^c \le 2\Delta_c, \quad c = 1, \ldots, C. && (8.56)
\end{aligned}$$

The above optimization problem can then be solved distributively using ADMM. The scaled augmented Lagrangian can be calculated as

$$\begin{aligned}
&\mathcal{L}_\rho(\mathbf{P}^{g,0}, \ldots, \mathbf{P}^{g,C}; \mathbf{p}^1, \ldots, \mathbf{p}^C; \boldsymbol{\mu}^1, \ldots, \boldsymbol{\mu}^C)\\
&= \sum_{i \in \mathcal{G}} f_i^g\left(\mathbf{P}_i^{g,0}\right) + \sum_{c=1}^{C} \frac{\rho^c}{2} \|\mathbf{P}^{g,0} - \mathbf{P}^{g,c} + \mathbf{p}^c - \Delta_c + \boldsymbol{\mu}^c\|_2^2. \qquad (8.57)
\end{aligned}$$

The optimization variables $\mathbf{P}^{g,0}$, $\mathbf{P}^{g,c}$, and $\mathbf{p}^c$ are arranged into two groups $\{\mathbf{P}^{g,0}\}$, and $\{\mathbf{P}^{g,c}, \mathbf{p}^c\}$ and updated iteratively. The variables in each group are optimized in parallel on distributed computing nodes, and coordinated by the dual variable vector $\boldsymbol{\mu}^c$ during each iteration.

At the kth iteration, the $\mathbf{P}^{g,0}$-update solves the base scenario with squared regularization terms enforced by the coupling constraints and expressed as

$$\begin{aligned}
\mathbf{P}^{g,0}[k+1] = \arg\min_{\mathbf{P}^{g,0}} &\sum_{i \in \mathcal{G}} f_i^g\left(\mathbf{P}_i^{g,0}\right)\\
&+ \sum_{c=1}^{C} \frac{\rho^c}{2} \|\mathbf{P}^{g,0} - \mathbf{P}^{g,c}[k] + \mathbf{p}^c[k] - \Delta_c + \boldsymbol{\mu}^c[k]\|_2^2\\
&\text{subject to Constraints (8.42), (8.44), and (8.46).} \qquad (8.58)
\end{aligned}$$

The $\mathbf{P}^{g,c}$-update solves a number of independent optimization subproblems corresponding to post-contingency scenarios and can be calculated distributively at the cth

Algorithm 3 Distributed SCOPF

Input: $\mathbf{B}^c_{bus}$, $\mathbf{B}^c_f$, $\mathbf{A}^{g,c}$, $\mathbf{P}^{d,c}$, $\overline{\mathbf{P}^{g,c}}$, $\underline{\mathbf{P}^{g,c}}$, Δ_c;
Initialize: $\boldsymbol{\theta}^c$, $\mathbf{P}^{g,c}$, $\mathbf{p}^c$, $\boldsymbol{\mu}^c$, ρ^c, $k = 0$;
while not converged **do**
  $\mathbf{P}^{g,0}$-update:
  $\mathbf{P}^{g,0}[k+1] = \arg\min_{\mathbf{P}^{g,0}} \sum_{i \in \mathcal{G}} f_i^g\left(\mathbf{P}_i^{g,0}\right)$
  $+ \sum_{c=1}^{C} \frac{\rho^c}{2} \|\mathbf{P}^{g,0} - \mathbf{P}^{g,c}[k] + \mathbf{p}^c[k] - \Delta_c + \boldsymbol{\mu}^c[k]\|_2^2$
  subject to Constraints (8.42), (8.44), and (8.46).

  $\mathbf{P}^{g,c}$-update, distributively at each computing node:
  $\mathbf{P}^{g,c}[k+1] = \arg\min_{\mathbf{P}^{g,c},\mathbf{p}^c} \frac{\rho^c}{2} \|\mathbf{P}^{g,0}[k+1] - \mathbf{P}^{g,c} + \mathbf{p}^c - \Delta_c + \boldsymbol{\mu}^c[k]\|_2^2$
  subject to Constraints (8.43), (8.45), (8.47), and (8.56),
  $\boldsymbol{\mu}^c[k+1] = \boldsymbol{\mu}^c[k] + \mathbf{P}^{g,0}[k+1] - \mathbf{P}^{g,c}[k+1] + \mathbf{p}^c[k+1] - \Delta_c$.
  Adjust the penalty parameter ρ^c if necessary;
  $k = k+1$;
end while
return $\boldsymbol{\theta}^c$, $\mathbf{P}^{g,c}$;
Output $\boldsymbol{\theta}^c$, $\mathbf{P}^{g,c}$;

computing nodes via

$$\mathbf{P}^{g,c}[k+1] = \arg\min_{\mathbf{P}^{g,c},\mathbf{p}^c} \frac{\rho^c}{2} \|\mathbf{P}^{g,0}[k+1] - \mathbf{P}^{g,c} + \mathbf{p}^c - \Delta_c + \boldsymbol{\mu}^c[k]\|_2^2$$
$$\text{subject to Constraints (8.43), (8.45), (8.47), and (8.56),} \tag{8.59}$$

where the scaled dual variable vector is also updated locally at the cth computing utility as

$$\boldsymbol{\mu}^c[k+1] = \boldsymbol{\mu}^c[k] + \mathbf{P}^{g,0}[k+1] - \mathbf{P}^{g,c}[k+1] + \mathbf{p}^c[k+1] - \Delta_c. \tag{8.60}$$

At the kth iteration, the original problem is divided into $C+1$ subproblems of approximately the same size. The computing node handling $\mathbf{P}^{g,0}$ needs to communicate with all computing nodes solving (8.59) during the iterations. The results of the $\mathbf{P}^{g,0}$-update, $\{\mathbf{P}^{g,0}\}$, will be distributed among the computing nodes for the $\mathbf{P}^{g,c}$-update. After the $\mathbf{P}^{g,c}$-update, the computed $\{\mathbf{P}^{g,c}, \mathbf{p}^c, \boldsymbol{\mu}^c\}$ will be collected to calculate the pre-contingency control variables. The subproblem data are iteratively updated such that the block-coupling constraints (8.55) are satisfied at the end. Note that since each of the subproblems is a smaller-scale OPF problem, existing techniques for OPF can be applied with minor modifications. The proposed algorithm is illustrated in Algorithm 3.

The ADMM approach is a primal-dual algorithm in which each computing node c solves its own subproblem (8.59), and variations to constraint (8.55) are systematically penalized at certain prices through the scaled dual variable to each individual subproblem. Note that in the ADMM frameworks for distributed computing of the dual variables, or prices, are not uniformly set up for all nodes, which will require costly

Table 8.3 Characteristics of test cases for SCOPF

Case	$\lvert\mathcal{N}\rvert$	$\lvert\mathcal{G}\rvert$	$\lvert\mathcal{B}\rvert$	Number of contingency cases
IEEE 57 bus	57	7	80	50
IEEE 118 bus	118	54	186	100
IEEE 300 bus	300	69	411	100

synchronization. For convex optimization problems, ADMM converges to the optimum geometrically [39], and the convergence rate can be improved by using the warm start technique [40].

8.3.4 Numerical results

In this section, numerical studies are examined to evaluate the performance of the proposed algorithm. Three classical test systems are used to study the SCOPF problem: the IEEE 57 bus, the IEEE 118 bus, and the IEEE 300 bus system [24], whose structures and characteristics are summarized in Table 8.3.

Two kinds of contingencies are considered in the numerical tests: branch outages and generator failures. The contingencies are artificially generated and the numbers of contingencies considered are listed in Table 8.3. We follow the physical limits on the equipment of the test systems. The numerical tests are implemented via MATLAB 7.10 on a PC with an Intel Q8200 2.33GHz processor and 8GB memory. The basic OPF problem solver is the same for all test systems. The performance of convergence and computing time of the proposed algorithm are investigated in the following. The results are averaged over a total of 500 Monte Carlo implementations.

Convergence performance

We first consider the convergence of the proposed algorithm. Since the number of contingencies and the optimal value for each test system differ, the relative error is used here to present the results. Suppose $r[k]$ is the resulting value of the objective function at the kth iteration, and r^* is the optimal solution. The relative error e is defined as $e = \left|\frac{r[k]-r^*}{r[0]-r^*}\right|$. The convergence performance is shown in Figure 8.9. It can be seen that after a moderate number of iterations, the proposed algorithm converges to the optimal values in the cases considered. From Figure 8.9, we see that the IEEE 57 bus system gives the fastest convergence rate. This is mainly a result of the small scale of the test system as well as the lower number of contingencies in the system. A large system leads to a large-scale optimization problem, and a large number of contingencies considered will make the problem scale even larger. Note that, after very few iterations, the algorithm gets very close to the optimal value, which means that the proposed algorithm is able to yield a good approximation to the optimal value in a short time.

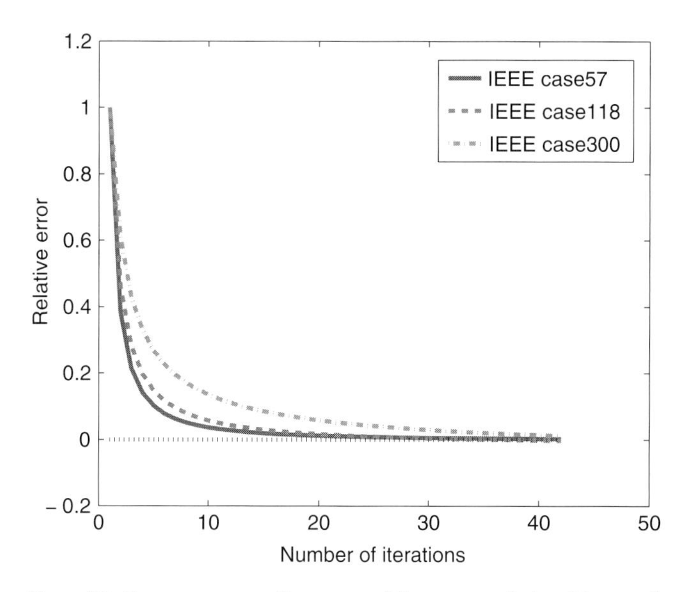

Figure 8.9 Convergence performance of the proposed algorithm on the test systems.

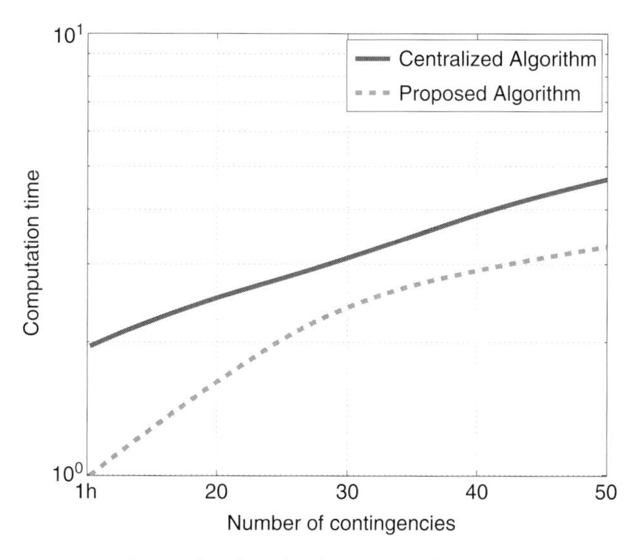

Figure 8.10 Computing time for the IEEE 57 bus system with different numbers of contingency cases.

Computing time performance

The computing time for the test systems with different numbers of contingency cases is investigated and the results are given in Figure 8.10, Figure 8.11, and Figure 8.12. The number of contingencies is increased by 20% each time and the computing time is recorded. It can be seen from these figures that, with an increase in the number of contingency cases for the SCOPF problem, the computing time of the centralized

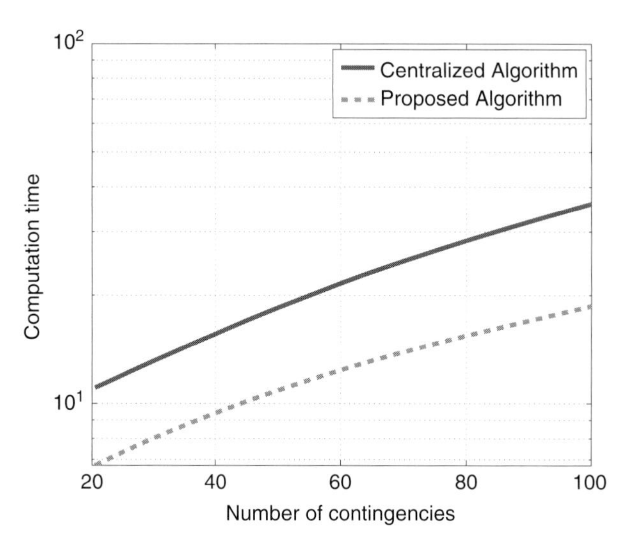

Figure 8.11 Computing time for the IEEE 118 bus system with different numbers of contingency cases.

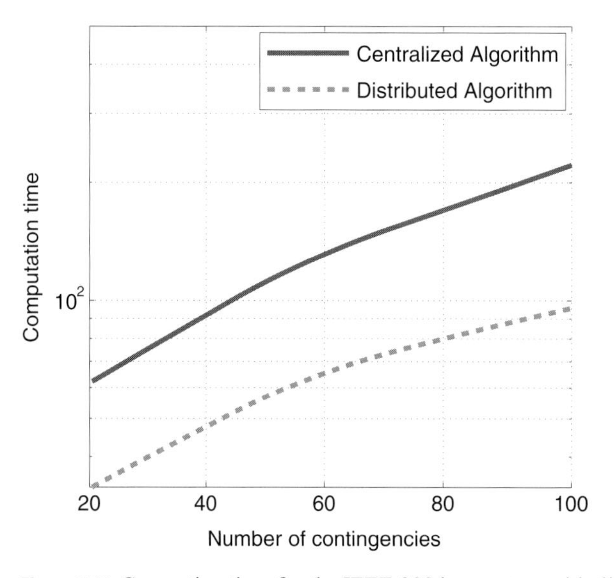

Figure 8.12 Computing time for the IEEE 300 bus system with different numbers of contingency cases.

algorithm increases much faster than that of the proposed algorithm. Thus, the proposed distributed algorithm is more scalable and stable than the centralized approach.

The computing time to achieve an approximate solution with a relative error of $e = 1\%$ is also considered for the distributed case. To better illustrate the numerical results, a speedup factor is defined as $S_p = T_c/T_p$, where T_c is the computing time of the centralized approach, and T_p is the computing time of the distributed approach.

Table 8.4 Computing time of the proposed algorithm on different test systems

	Centralized		Distributed ADMM			
Cases	Cost	Time	Cost	Time	Cost (e = 1%)	Time
IEEE 57 bus	487.53	5.22	487.53	3.55	492.40	1.18
IEEE 118 bus	1606.73	36.03	1606.73	18.92	1622.79	7.93
IEEE 300 bus	9567.12	221.67	9567.12	95.74	9662.80	52.87

The results of the computing time performance are presented in Table 8.4. It is shown in Table 8.4 that the proposed distributed approach obtains the same optimum as the centralized approach, and can achieve a speedup factor S_p of 1.4–2.4. Note that if only an approximate result is needed, the speedup factor can even be improved to S_p of 4.4–4.8 by using the proposed distributed algorithm. The speedup factor for the smallest test system, the IEEE 57 bus system, is the smallest, due to the relatively more-significant communication overhead between different computing nodes during the simulations. A larger S_p can be achieved on a larger-scale test system since the communication overhead is negligible compared with the computing time of the optimization subproblem handled by each computing node.

8.4 Concluding remarks

In this chapter, we have investigated the applications of large-scale data processing techniques for enhancing security in a smart grid. We have introduced two security concerns, false data injection attacks against state estimation and security constrained optimal power flow in power systems. We have explored the possibilities of exploiting the inherent structure of the datasets and effectively processing the large datasets to enhance power system security. We have designed a sparse optimization approach for the false data injection detection problem, and a distributed parallel approach for the security constrained optimal power flow problem. We have performed numerical studies to validate the effectiveness of the proposed approaches. We have shown that effective management and processing of "big data" has the potential to significantly improve smart grid security.

Acknowledgement

The authors of this chapter would like to acknowledge the support of the following funding agencies: the US Department of Defense under Grant HDTRA1-13-1-0029, the US National Science Foundation under Grants CCF-1420575, CCF-1456921, CMMI-1435778, CNS-0905556, CNS-1265227, CNS-1265268, CNS-1343155, CNS-1443917, DMS-1118605, ECCS-1305979, and ECCS-1405121, and the National Science Foundation of China under Grants NSFC-61328102 and NSFC-61428101.

References

[1] E. Hossain, Z. Han, and H. V. Poor, *Smart Grid Communications and Networking*, Cambridge, UK: Cambridge University Press, 2012.

[2] C. Greer, D. A. Wollman, D. E. Prochaska, *et al.*, "NIST framework and roadmap for smart grid interoperability standards, release 3.0," The National Institute of Standards and Technology, Tech. Rep. NIST SP - 1108r3, October 2014.

[3] S. Gorman, "Effect of stealthy bad data injection on network congestion in market based power system," *The Wall Street Journal*, April 2009.

[4] S. Borlase, *Smart Girds: Infrastructure, Technology and Solutions*, Boca Raton, FL: CRC Press, 2012.

[5] A. Abur and A. G. Exposito, *Power System State Estimation: Theory and Implementation*, New York: Marcel Dekker, Inc., 2004.

[6] J. J. Grainger and W. D. Stevenson Jr, *Power System Analysis*, New York: McGraw-Hill, 1994.

[7] Y. Liu, M. K. Reiter, and P. Ning, "False data injection attacks against state estimation in electric power grids," in *Proceedings 16th ACM Conference on Computer and Communications Security*, Chicago, IL, November 2009.

[8] L. Xie, Y. Mo, and B. Sinopoli, "False data injection attacks in electricity markets," in *Proceedings IEEE International Conference on Smart Grid Communications*, Gaithersburg, MD, October 2010.

[9] M. Esmalifalak, Z. Han, and L. Song, "Effect of stealthy bad data injection on network congestion in market based power system," in *Proceedings IEEE Wireless Communications and Networking Conference*, Paris, France, April 2012.

[10] G. Dán and H. Sandberg, "Stealth attacks and protection schemes for state estimators in power systems," in *Proceedings IEEE International Conference on Smart Grid Communications*, Gaithersburg, MD, October 2010.

[11] M. Esmalifalak, G. Shi, Z. Han, and L. Song, "Bad data injection attack and defense in electricity market using game theory study," *IEEE Transactions on Smart Grid*, vol. **4**, no. 1, pp. 160–169, March 2013.

[12] O. Kousut, L. Jia, R. J. Thomas, and L. Tong, "Malicious data attacks on the smart grid," *IEEE Transactions on Smart Grid*, vol. **2**, no. 4, pp. 645–658, December 2011.

[13] T. T. Kim and H. V. Poor, "Strategic protection against data injection attacks on power grids," *IEEE Transactions on Smart Grid*, vol. **2**, no. 2, pp. 326–333, June 2011.

[14] S. Cui, Z. Han, S. Kar, *et al.*, "Coordinated data-injection attack and detection in the smart grid: a detailed look at enriching detection solutions," *IEEE Signal Processing Magazine*, vol. **29**, no. 5, pp. 106–115, September 2012.

[15] Y. Zhao, A. Goldsmith, and H. V. Poor, "Fundamental limits of cyber-physical security in smart power grids," in *Proceedings IEEE 52nd Annual Conference on Decision and Control*, Florence, Italy, December 2013.

[16] Z. Han, H. Li, and W. Yin, *Compressive Sensing for Wireless Communication*, Cambridge, UK: Cambridge University Press, 2012.

[17] E. J. Candès and B. Recht, "Exact matrix completion via convex optimization," *Communications of the ACM*, vol. **55**, no. 6, pp. 111–119, June 2009.

[18] E. J. Candès, X. Li, Y. Ma, and J. Wright, "Robust principal component analysis?" *Journal of the ACM*, vol. **58**, no. 3, pp. 1–37, May 2011.

[19] Z. Lin, M. Chen, L. Wu, and Y. Ma, "The augmented Lagrange multiplier method for exact recovery of corrupted low-rank matrices," UIUC, Tech. Rep. UILU-ENG-09-2215, Urbana, FL, 2009.

[20] D. P. Bertsekas, *Nonlinear Programming*, Belmont, MA: Athena Scientific, 1999.

[21] J. Cai, E. J. Candès, and Z. Shen, "A singular value thresholding algorithm for matrix completion," *SIAM Journal on Optimization*, vol. **20**, no. 4, pp. 1956–1982, January 2010.

[22] Y. Shen, Z. Wen, and Y. Zhang, "Augmented Lagrangian alternating direction method for matrix separation based on low-rank factorization," Rice CAAM, Tech. Rep. TR11-02, Houston, TX, 2011.

[23] Z. Wen, W. Yin, and Y. Zhang, "Solving a low-rank factorization model for matrix completion by a nonlinear successive over-relaxation algorithm," Rice CAAM, Tech. Rep. TR10-07, Houston, TX, 2010.

[24] R. D. Zimmerman, C. E. Murillo-Sánchez, and R. J. Thomas, "MAT-POWER steady-state operations, planning and analysis tools for power systems research and education," *IEEE Transactions on Power Systems*, vol. **26**, no. 1, pp. 12–19, February 2011.

[25] M. Shahidehpour, W. F. Tinney, and Y. Fu, "Impact of security on power system operation," *Proceedings of the IEEE*, vol. **93**, no. 11, pp. 2013–2025, November 2001.

[26] O. Alsac and B. Scott, "Optimal load flow with steady-state security," *IEEE Transaction on Power Apparatus and System*, vol. **93**, no. 3, pp. 745–751, May 1974.

[27] M. V. F. Pereira, A. Monticelli, and L. M. V. G. Pinto, "Security-constrained dispatch with corrective rescheduling," in *Proceedings IFAC Symposium on Planning and Operation of Electric Energy System*, Rio de Janeiro, Brazil, July 1985.

[28] A. J. Wood and B. F. Wollenberg, *Power Generation Operation and Control*, New York: Wiley, 1996.

[29] A. Monticelli, M. V. F. Pereira, and S. Granville, "Security-constrained optimal power flow with post-contingency corrective rescheduling," *IEEE Transactions on Power Systems*, vol. **2**, no. 1, pp. 175–180, February 1987.

[30] F. Capitanescu, J. L. M. Ramos, P. Panciatici, *et al.*, "State-of-the-art, challenges, and future trends in security constrained optimal power flow," *Electric Power System Research*, vol. **81**, no. 8, pp. 1731–1741, August 2011.

[31] J. Martínez-Crespo, J. Usaola, and J. L. Fernández, "Security-constrained optimal generation scheduling in large-scale power systems," *IEEE Transactions on Power Systems*, vol. **21**, no. 1, pp. 321–332, February 2006.

[32] Y. Fu, M. Shahidehpour, and Z. Li, "AC contingency dispatch based on security-constrained unit commitment," *IEEE Transactions on Power Systems*, vol. **21**, no. 2, pp. 897–908, May 2006.

[33] F. Capitanescu and L. Wehenkel, "A new iterative approach to the corrective security-constrained optimal power flow problem," *IEEE Transactions on Power Systems*, vol. **23**, no. 4, pp. 1533–1541, November 2008.

[34] Y. Li and J. D. McCalley, "Decomposed SCOPF for improving efficiency," *IEEE Transactions on Power Systems*, vol. **24**, no. 1, pp. 494–495, February 2009.

[35] S. Boyd, N. Parikh, E. Chu, B. Peleato, and J. Eckstein, "Distributed optimization and statistical learning via the alternating direction method of multipliers," *Foundation and Trends in Machine Learning*, vol. **3**, no. 1, pp. 1–122, November 2010.

[36] D. Bertsekas and J. Tsitsiklis, *Parallel and Distributed Computation: Numerical Methods*, 2nd edn, Belmont, MA: Athena Scientific, 1997.

[37] R. Baldick, B. H. Kim, C. Chase, and Y. Luo, "A fast distributed implementation of optimal power flow," *IEEE Transactions on Power Systems*, vol. **14**, no. 3, pp. 858–864, August 1989.
[38] M. Kraning, E. Chu, J. Lavaei, and S. Boyd, "Dynamic network energy management via proximal message passing," *Foundations and Trends in Optimization*, vol. **1**, no. 2, pp. 1–54, January 2014.
[39] W. Deng and W. Yin, "On the global and linear convergence of the generalized alternating direction method of multipliers," Rice CAAM, Tech. Rep. TR12-14, Houston, TX, 2012.
[40] J. Nocedal and S. J. Wright, *Numerical Optimization*, 2nd edn, New York: Springer, 2006.

Part III

Big data over social networks

9 Big data: a new perspective on cities

Riccardo Gallotti, Thomas Louail, Rémi Louf, and Marc Barthelemy

The recent availability of large amounts of data for urban systems opens the exciting possibility of a new science of cities. These datasets can roughly be divided into three large categories according to their time scale. We will illustrate each category by an example on a particular aspect of cities. At small time scales (of order a day or less), mobility data provided by cell phones and GPS reveal urban mobility patterns but also provide information about the spatial organization of urban systems. At very large scales, the digitalization of historical maps allows us to study the evolution of infrastructure such as road networks, and permits us to distinguish on a quantitative basis self-organized growth from top-down central planning. Finally at intermediate time scales, we will show how socio-economical series provide a nice test for modeling and identifying fundamental mechanisms governing the structure and evolution of urban systems. All these examples illustrate, at various degrees, how the empirical analysis of data can help in constructing a theoretically solid approach to urban systems, and to understand the elementary mechanisms that govern urbanization leaving out specific historical, geographical, social, or cultural factors. At this period of human history that experiences rapid urban expansion, such a scientific approach appears more important than ever in order to understand the impact of current urban planning decisions on the future evolution of cities.

9.1 Big data and urban systems

A common trait shared by all complex systems – including cities – is the existence of a large variety of processes occurring over a wide range of time and spatial scales. The main obstacle to the understanding of these systems therefore resides at least in uncovering the hierarchy of processes and in singling out the few that govern their dynamics. Albeit difficult, the hierarchization of processes is of prime importance. A failure to do so leads either to models which are too complex to give any real insight into the phenomenon or to be validated, or too simple to provide a satisfactory framework which can be built upon. As a matter of fact, despite numerous attempts [1–6], a theoretical understanding of many observed empirical regularities in cities is still missing. This situation is, however,

Big Data over Networks, ed. Shuguang Cui, Alfred O. Hero III, Zhi-Quan Luo, and José M. F. Moura. Published by Cambridge University Press.

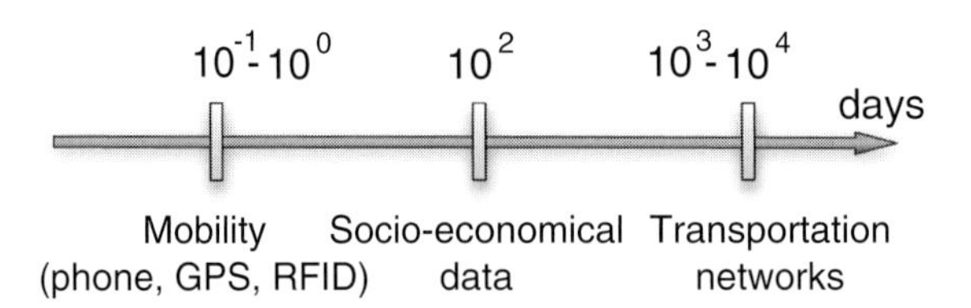

Figure 9.1 Order of magnitude of the time scale of data about cities and their most relevant application.

changing with the recent availability of an unprecedented amount of data about cities and their inhabitants. This fact has made possible new quantitative studies of cities, opening the way to a "New Science of Cities" [7]. The important point is that data can provide stylized facts and quantitative results that allow us to test theoretical models. This feedback between data and theory has been proven to be a successful method in the physical sciences and we can hope that it will lead to new, solid results about a system as complex as a city.

We can roughly classify the data about cities into three large categories according to their time scale (Figure 9.1). At a typical time scale of the order a day, mobility data gathered by mobile phones, GPS, or RFIDs inform us about where and when people move in the city, revealing in depth the spatio-temporal structure of activities in a city and statistical patterns of mobility. The availability of mobility data renewed the interest for understanding the laws governing the trips of individuals, such as the gravity law, questioned its validity, and led to new, more-accurate models. Also, these datasets provide a clear picture of the spatial distribution of activities, and of the existence of multiple activity subcenters, allowing to discuss the possibility of a typology of cities in terms of degree of polycentrism.

At a larger time scale, of a year, socio-economical surveys provide us with relevant information such as the total yearly gasoline consumption, the total yearly number of miles driven, the relation between density and area, etc. The key to understanding how these different quantities scale with population is the mobility spatial pattern, and a simple model inherited from previous studies in spatial economics is able to explain these behaviors. In particular, we can propose theoretical arguments to show that, for large cities, the cost of diseconomies such as congestion can overcome the scale economies for infrastructure. As a result, cities where car traffic is dominant are not sustainable for populations beyond a value typically of order a few millions.

Finally, at very long scales such as decades and centuries, remote sensing and the recent digitization of old maps allows us to study the evolution of urbanized areas, transportation, and road networks. We then observe that large subway networks seem to converge to the same structure, characterized by similar values of morphological indicators, revealing the existence of dominant mechanisms independent from cultural and historical considerations. We also observe the large-scale evolution of road networks, allowing us to characterize quantitatively the natural, self-organized evolution of an urban system. For some other systems, the influence of urban planning was crucial in their evolution and this type of analysis permits us to observe quantitatively the effect of such external, top-down processes.

In this chapter we will discuss these various examples of data and show how we can extract useful information about cities. We will start with long time scale data about the evolution of road networks and show that they reveal some aspects about the elementary processes or urbanization. Next, we discuss mobility data obtained from GPS and mobile phones, and what they teach us. Finally, we will discuss how socio-economical series about cities allow us to identify important mechanisms in the spatial organization of cities, helping in constructing a model with testable predictions.

9.2 Infrastructure networks

Scientists working on complex networks naturally moved to infrastructure networks and their characterization. They quickly realized that these networks were very different from the others due to the fact that they are embedded in space. This led to specific techniques to analyze these spatial networks [8]. In the following, we will discuss studies on infrastructure networks present in cities, such as road and subway networks. Street patterns and subways can both be represented as spatial networks, where nodes – which represent intersections in the case of roads, stations in the case of subways – have a specific location in space, in contrast with purely topological networks, such as the world wide web, where space does not play any role. The network representation allows us to explore the topological and geographical properties of these systems and, in each case, we will introduce the types of data that have been used and discuss what has been understood from their analysis.

9.2.1 Road networks

As mentioned in the introduction, data are central to our approach. So far, the study of road networks has relied more on "little data" than big data, essentially for availability reasons. About ten years ago, when physicists became interested by these systems, data about street networks were difficult to find, and the first studies used small portions of cities that were numerized by the authors themselves [9–16], and consisted mainly of applying various measures to a single portion of road network. Progressively, as more data became accessible, studies on whole cities became possible, allowing useful comparisons between patterns formed in different historical, cultural contexts. More recently, the availability of historical data, allowed by the numerization of old maps by the GIS community, allowed explorations of the growth of streets.

Ultimately, the goal is to build models that are able to account for the similarities that are empirically observed, as well as the processes responsible for the differences. Data are useful at two stages of this process: first, exploratory data analyses – mainly consisting of finding informative measures – give us hints about the relevant mechanisms. Once the model is built, its predictions are constrained by the known empirical regularities. In this section, we will first present empirical studies on street patterns, and classify

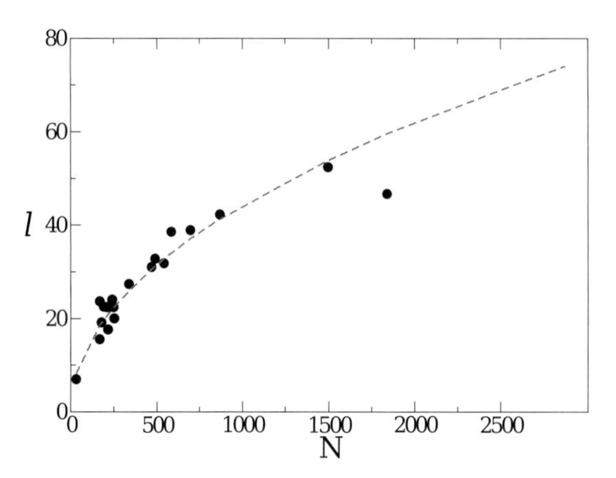

Figure 9.2 Total length of the street network versus the number of nodes. The line is a fit that predicts a $\sqrt{N}$ growth. Data for several cities are from [14], the figure is taken from [17].

their results in three categories: measures on single networks, comparisons between different networks, and study of their time evolution. We will then discuss a model for street growth that accounts for the observed regularities, before discussing briefly perspectives for the near future.

Static properties

The first studies of urban street patterns, driven by the Space Syntax community [9], were made on the dual representation of the street networks, where straight segment of roads become nodes, and where two nodes are connected if the two corresponding streets intersect [13]. However, this dual representation, useful to investigate the navigability properties of the network (for instance the average number of turns needed to reach a given location), does not easily account for the spatial aspect of street networks [12].

The first studies on street neworks concerned their topological properties such as degree distribution, clustering coefficient, and assortativity. These studies, however, left aside the spatial aspects of these graphs. It was later realized that their interesting features were precisely the geometrical patterns formed by these graphs. Indeed, the degree distribution is very peaked around an average value (usually of order $\langle k \rangle \approx 4$), and both the clustering coefficient and the assortativity are large due to neighboring effects [8]. In contrast, all space-related measures contain more information. For example, the total length L_N of the network as a function of the number of nodes N is interesting as it brings information about the structure of the network and is also a crude measure of the total cost associated with the construction of this infrastructure. This quantity was shown in [16] (see Figure 9.2) to follow a power law

$$\frac{L_N}{\sqrt{A}} \sim \sqrt{N}, \tag{9.1}$$

which can be easily explained by considering the network of streets as a network connecting the nodes to their closest neighbor at a typical distance $\ell_R = \sqrt{\frac{A}{N}}$, where A

is the area of the city under consideration. This implicitely implies that the planar graphs which constitute the road network are not far from a perturbed lattice. This preliminary conclusion needs, however, to be reconsidered when studying the distribution of the area a of blocks (i.e. the cells of the planar graph describing the road network). This distribution follows a power law [15, 18]

$$P(a) \sim a^{-\tau}, \tag{9.2}$$

where, remarkably, $\tau \approx 2$ for all the studied cities. This is an example of a result that puts a very strong constraint on any possible model.

The authors of [15, 19, 20] also measured the distribution of the shape of blocks in various cities around the world. The shape of a block can be characterized by the form factor defined by

$$\Phi = \frac{a}{a_C}, \tag{9.3}$$

where a_C is the area of the circumscribed circle to the block. The value of Φ typically varies between 0.2 and 0.6 for all observed cities (see Figure 9.4 for examples of Φ distribution from the Groane study), indicating a great variety of shapes for city blocks. Also, the authors found different Φ distributions for Dresden, Groane, and Paris, indicating that the differences emerging in the distribution of the form factor might arise from local particularities. This diversity of shapes implies that the simple picture of a perturbed lattice is not correct as it would predict a peaked distribution for both the area and Φ distributions. We see here how a simple analysis of data leads to important constraints for possible models of this system.

Another important quantity for complex networks is the betweenness centrality of an edge e, defined as

$$b_e = \sum_{s \neq t} \frac{\sigma_{st}(e)}{\sigma_{st}}, \tag{9.4}$$

where σ_{st} is the number of shortest paths between s and t, and σ_{st} the number of shortest paths between s and t going through e (the betweenness centrality of a node i can be defined in a similar way). If we assume that the traffic between all pairs of nodes is the same, a natural proxy for the traffic is the betweenness centrality. The spatial distribution of this quantity thus gives important information about the coupling between space and the structure of the road network, as well as the flows on this network [8] (see Figure 9.3 for an example of betweenness centrality repartitions).

Time evolving networks

Although studies on street networks at a particular point in time are very useful to demonstrate the similarities and differences between various urban street patterns, historical studies on a given area at different times are essential to uncover the processes governing the growth of the network. Three recent studies have explored the time evolution of street networks. First, the authors of [19] studied the time evolution of the street network in the region of Groane (comprising some 29 urban centers) in Italy with seven points

Figure 9.3 (a) Street network of the Groane region, the gray level corresponds to the time of creation of the streets. (b) Betweenness centrality of the roads in the Groane region in 2007. One can see that the most central streets mainly are the oldest ones. Figure from [19].

in time between 1833 and 2007. The authors of [20] studied the temporal evolution of the street network of Paris (city core) between 1789 and 2010, notoriously reorganized on a large scale by Haussmann in the nineteenth century. Finally, in [21], the authors studied the evolution of the Greater London Area (GLA) street networks between 1786 and 2010, distinguishing the GLA network from the central London network.

In order to be able to compare various networks and to get rid of demographic effects, the number of nodes N is the natural parameter serving as the system's internal clock. Interestingly enough, it has been found [19, 20] that N and the population P of the city are proportional to each other. In other words, the number of intersections per person is constant (different in these studies). This is a first hint at a universal link between population densities and the shape of the street network, which would be interesting to investigate further.

The distribution of the area of cells (or city blocks) behaves as a power law (Eq. (19.2)) with an exponent τ decreasing over time between $\tau \approx 1.2$ in 1833 to $\tau \approx 1.9$ in 2007 (see Figure 9.4). A higher exponent means more homogeneity; the growth of roads seems to correspond to a homogenization of the size of cells.

In addition to the area, it is important to study the shape of the blocks with the factor Φ. The results (see Figure 9.4) show the existence of two regimes: one marked by a single-peak distribution that can be well approximated by a Gaussian distribution, which then evolves to a double-peak distribution, where the new peak hints at the

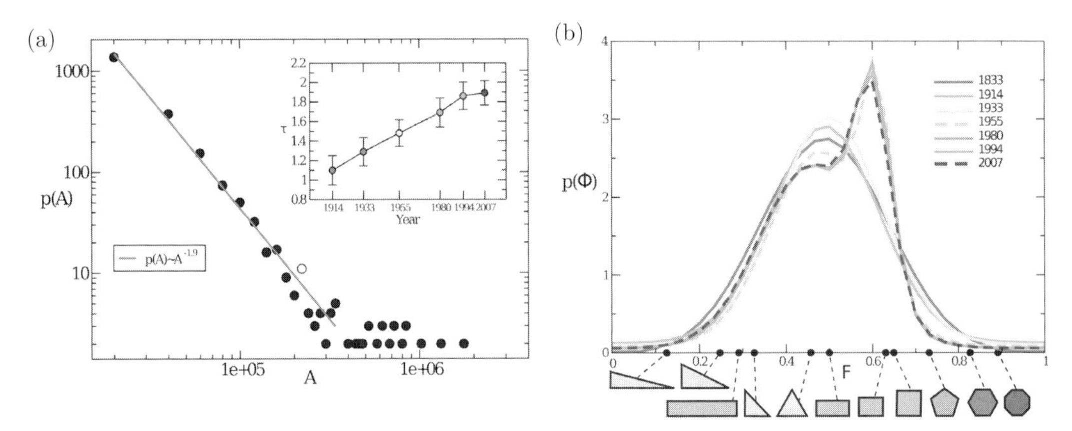

Figure 9.4 (a) The size distribution of cell areas for the street network in the Groane area in 2007. The tail of the distribution can be fitted with a power law $p(A) \sim A^{-\tau}$. The inset shows the evolution of τ with time. (b) Distribution of Φ, quantifying the shape of cells, for the network of the Groane area for all times. Figure from [19].

presence of many rectangles with similar side lengths. This indicates that the pre-urban phase exhibits a large diversity of shapes, while the urban phase is characterized by the appearance of more regular, rectangular faces. This is a particularly nice example where a clear stylized fact will help to identify important mechanisms.

We can also focus on new links and try to characterize them. The authors of [19] showed that the structure of the oldest roads in the region of Groane (Italy) did coincide with the most central (i.e. with the highest betweenness centrality) roads in 2007 (see Figure 9.3), which suggest that the evolution of this system relies on a backbone of older, central streets, on which new streets grow and develop. In order to characterize more precisely new links, the authors of [19] define the "impact" $\delta_b(e*)$ of new edges $e*$ on the average betweenness centrality $\overline{b}$ of the network G_t at a time t, defined as

$$\delta_b(e*) = \frac{\overline{b}(G_t) - \overline{b}(G_t \backslash \{e^*\})}{\overline{b}(G_t)}, \tag{9.5}$$

where $G_t \backslash \{e^*\}$ is the network at time t from which e^* has been removed. The authors found out that the distribution of $\delta_b(e*)$ exhibits at each period two distinct peaks. This means that new links can be classified in two different categories. Remarkably, these two categories correspond to two different growth processes: first, the links that are added between an existing node and a previously unreached place follow a logic of exploration. Second, the links which are built between two already-existing nodes follow a logic of densification. As time passes, the exploration process fades out in favour of the densification process, which suggests that the growth of streets follows a logic of fragmentation of space rather than of exploration.

These results concerned essentially the area of Groane (Italy), which is a region that has not been subjected to large-scale planning efforts and could thus serve as an illustration of a self-organized system. However, the existence of top–bottom central planning is often invoked as a counter-argument to the possibility of understanding

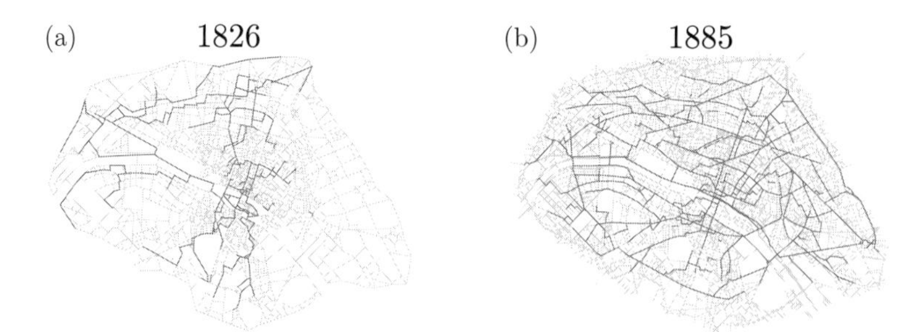

Figure 9.5 Network of the Paris street network in 1826 (a) and 1885 (b), respectively before and after the Hausmann transformations. We represent in dark gray the roads whose betweenness centrality b is greater than $b_{max}/20$. We see that one of the consequences of the Hausmann transformations was a redistribution of betweenness centrality, which can be interpreted in terms of a reorganization of flows (figure courtesy of P. Bordin).

the growth of streets as the result of universal processes. Historical data actually allow us to investigate the impact of such central planning on the urban street patterns. In [20], the authors found on the Paris (France) dataset that all usual network quantities display a smooth behavior over time, even during the Haussmann period (approximately 1850–1870), a short time period during which the street network of Paris was strongly reorganized. The only strong signature of central planning was found in the spatial organization of the most central intersections (see Figure 9.5).

It therefore seems that while the Hausmann modifications act on a global scale and reorganize the spatial distribution of flows, "natural" evolution acts more locally, on the existing substrate. The existence of central planning therefore doesn't seem to be a threat to the possibility of reproducing the statistical properties of urban street patterns with general models.

Building models

Although data are crucial in order to get a good idea of the processes at stake in urban growth, models using simple and realistic ingredients that are able to reproduce observed regularities are important for the identification of essential processes needed to reproduce statistical regularities of street patterns.

The authors of [16] proposed a simple model to account for the properties of urban street patterns. Every $\tau_C > 1$ time steps, new centers (house, business, etc.) are added at random uniformly in the plane. These new settlements are then connected to the vertices of the existing road network in their relative neighborhood (a point P is said to be in the relative neighborhood of S if there are no other points at the same time closer to S and P than P is from S) by addition of small road segments every $\tau_R < \tau_C$ time steps. If a new settlement only "stimulates" one vertex, a straight line is built. When more than one center stimulates the same vertex v, the road that minimizes the total connection length is built. This point of view was also used in a similar model [22].

The results, on top of being visually appealing (see Figure 9.6), are encouraging: the model is able to reproduce the behavior of the total network length $L_N \propto \sqrt{N}$

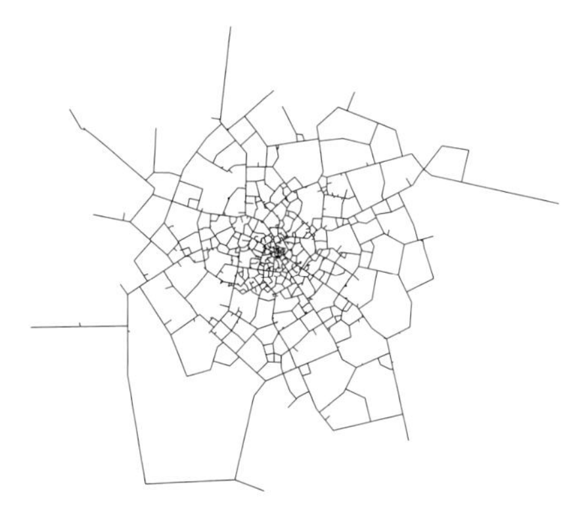

Figure 9.6 One realization of the model presented in [17], locating new settlements at a distance r from the center with probability $P(r) \propto e^{-r/r_c}$.

that is observed on real data. Furthermore, if instead of locating new settlements with uniform probability in the plane they are located at a distance r from the city center with a probability $P(r) \propto e^{-r/r_c}$, the authors find a cell area distribution of the form $P(A) \sim A^{1.9}$ in agreement with previous measures on different cities.

Perspectives

Despite the various studies mentioned here, we still didn't reach a clear understanding of the various street patterns encountered in different cities. However, the availability of data for the whole world made possible by the OpenStreetMap [23] project provides the basis for testing various ideas for constructing a typology of street patterns. Concerning time evolution, various projects of digitalizing historical maps are currently in progress; these also gives hope about the possibly of reaching a good understanding of the time evolution of these spatial systems.

9.2.2 Subway networks

Subways have not been studied as much as the road network, but provide more than 50% of the total mobility within some urban areas. It is thus crucial to understand how they grow if we want to understand the structure of mobility patterns in large cities. Data regarding the structure, growth and use of subway networks exist, but are somewhat tedious to gather. The authors of [24], for instance, gathered data about the opening date and location of every station of the 14 largest metros in the world by hand, using Wikipedia as their main source. Indeed, for most stations, there is a Wikipedia page per station with various pieces of information, such as the exact location, the opening data, and the number of passengers. Other websites maintained by rail enthusiasts (http://cityrailtransit.com/ and http://mic-ro.com/metro/, for instance) also provide useful information, but these have to be gathered by hand into a format suitable for automated statistical analyses. Using these data, the authors of [24] showed that the 14 largest subway systems in the world seem to converge towards the same generic

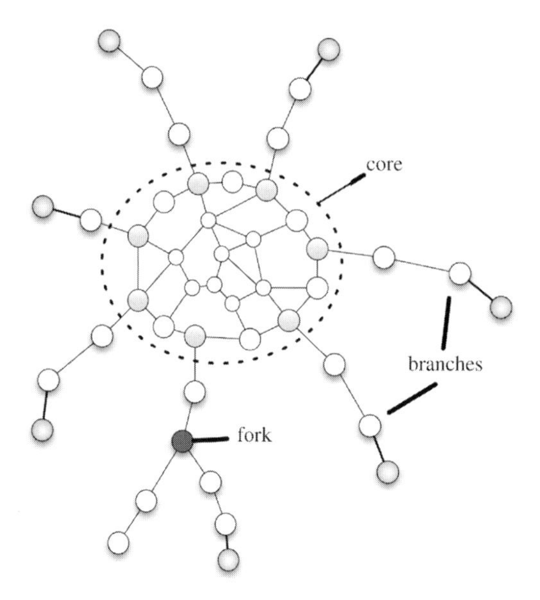

Figure 9.7 Schematic of subway networks. A large “ring” encircles the core of stations. Branches radiate from the core and reach further areas of the urban system. On real network data, the core and branches are extracted algorithmically [24]. Figure taken from [24].

shape, made of a core with branches radiating from it, despite their geographical and economical differences (see Figure 9.7). They also found that the core of the network contains on average half of the total number of stations. This convergence hints at the existence of a universal mechanism driving the growth of these systems.

This picture of a core from which different branches radiate also allows us to understand simply how the number of stations at a given distance r of the barycenter of all stations changes with r, something that was measured in previous studies, but very difficult to explain in the context of fractal geometry [25, 26]. Finally, the authors found that the number of branches $\mathcal{N}_b$ scales with the number of stations N_s as

$$\mathcal{N}_b \sim N_s^{0.60}. \tag{9.6}$$

Assuming a constant distance between branches at the edges of the core, the number of branches should be proportional to the perimeter of the core. If we further assume that the core is a compact arrangement of stations of coverage C, we find that the surface area of the core is of the order $A \propto N_s\, C$. Therefore the perimeter scales as $\sqrt{N_s}$, and we find $\mathcal{N}_b \sim \sqrt{N_s}$, close to the exponent of 0.6 found in the data.

Although the study of the geometry and topology of these networks reveals interesting insights in the network’s structure, it is also important to understand how these networks are used and how their properties depend on the characteristics of the city that is being served. Using historical population data for the city of London between 1871 and 2001, Levinson [27] showed that the growth of the rail network followed a logic of co-development: while high densities seem to encourage the growth of rail infrastructure, the opening of new stations is also positively correlated with population increase.

The city and the subway thus exhibit a sort of symbiotic behavior that is difficult to disentangle and precisely measure. Using contemporary data, the same author tried to relate the topological properties of the subway network to the population size of the cities served [28]. In the same spirit, the authors of [29] try to relate the subway use, as measured by the number of boardings per capita in the city, to various measures of the network. Among other things, they find a good correlation between the total ridership and the total coverage of stations. Finally, although models for the growth of inter-urban networks exist [30, 31], models adapted to the specific growth of subways, and that take into account their relation to the underlying city, are scarce in the literature [32]. The previous data analyses, however, give precious insights into the relevant economical processes at play during the evolution of these networks [33].

9.3 Mobility networks

9.3.1 A renewed interest

Thanks to their increasing availability and precision, GPS measurements probably represent the richest datasource available for the study of human mobility. When the quality of the satellite signal is good the spatial resolution is of order 10 m while the error in the timestamp is negligible. Nevertheless, GPS trajectories are not describing exactly the movements of an individual, but the coordinates of a mobile device and not continuously recorded in time. This sampling problem constitute the real limit of the use of GPS data.

There is a wide range of GPS trace sources, each one having its own sampling issues: (i) small-scale experiments, as in [34], where a small number of volunteers carried a GPS device all day in a limited area and recordings were taken every 10 s; (ii) geo-referenced internet accesses, as in [35], where the derived mobility patterns might be systematically influenced by the complex features of communication habits; (iii) private cars mobility, as in [36], where only the movements of vehicles can be recorded (with a limited spatial sampling of 2 km); and (iv) public transport mobility: trajectories of buses [37] or taxis [38] recorded with a high temporal resolution.

While small-scale/high-resolution data sources are extremely useful for particular cases, the greater and more general development in the use of GPS data comes from taking advantage of the commercial expansion of GPS systems used for vehicle tracking and routing, and for geo-referenced social networks accesses. Both of them are indirect measures for human mobility. Vehicle-related information is limited by nature to the study of the part of the trip with a particular transportation mode. On the other hand, social-media accesses give information only about their users when the communication occurs, at random points of time and space. This problem is shared with mobility patterns derived from cell-phone data, which in addition have spatial resolution imposed on then by antennas. These difficulties can be overcome by imposing appropriate temporal and spatial resolutions. Indeed, when similar spatial and temporal resolutions are used in the definition of mobility pattern from cell-phone traces [39] and vehicular GPS trajectories [40], the observed patterns present remarkable similarities. This similarity on two independent, and rather different, datasets grants that mobility patterns obtained from both communication and private cars' GPS datasets are a good representation of human mobility at a temporal scale of 1 h. Nevertheless, it is important to notice that, using a time frame of one hour, we are leaving out of the analysis a large part of the mobility. For instance, in private cars' mobility, about 40% of parking times are shorter than this value. It has, however, been observed [40] that activities shorter than 2 h tend to cluster around longer ones, forming regularly repeated sequences. These sequences are probably reflecting the spatio-temporal partition of human mobility patterns into habitats [41], groups of locations that tend to be more spatially cohesive than the total mobility. Therefore, it is possible to take advantage of this temporal hierarchy and take into account only activities longer than 2 h, which are playing the leading role in the structure of the mobility patterns.

9.3.2 Individual mobility networks

The raw information obtainable from GPS mobility data is sequences of location, where individuals have gone through or stopped by. In order to understand the dynamics of human mobility, it is necessary to use a more comprehensive approach, which permits us to highlight the behavioral nature of the data. In fact, each person's mobility differs from others' because of her own habits, agenda, and knowledge of the urban environment. To study those differences, we may group the start/end points of each GPS trajectory into several clusters, which become the nodes of the individual mobility networks [42]. We associate to each node a visited location, while each weighted and directed link implies the existence of one or more trips between two locations. Even if GPS data do not give direct information on individual activities, we may assume that each time someone stops at a location, this stop can be associated to a given activity.

Once the individual mobility network is defined, we can study its general features by pointing out the hierarchy between nodes, characterized by their degree. In general, the degree quantifies the total number of connections and corresponds here to the total number of trips that have gone through a node. As a trip starts from the same location where the previous trip stopped, in- and out-degree are equal and we will thus use here the term "degree" without further specifications. A single mobility network, derived from one individual's mobility, is often not enough for an accurate statistical analysis of the degree distribution. For this reason, one may extrapolate the shape of the individual networks' degree distribution by superimposing distributions of the whole ensemble of networks. This aggregated degree distribution is shown in Figure 9.8a (where we indicate as a guide to the eye a power law fit, here with exponent $b \approx 1.6$).

The hierarchy among locations also emerges when considering the duration of stops. But, before showing that, let us consider, for the case of private vehicular data [36], the aggregated distribution of durations for all the parkings of different drivers. This distribution appears to be well described by a power law for $\tau \leq 3$ h ($\simeq 95\%$ of the data), and the fit gives

$$p(\tau) \propto \frac{1}{\tau^{\alpha}}, \tag{9.7}$$

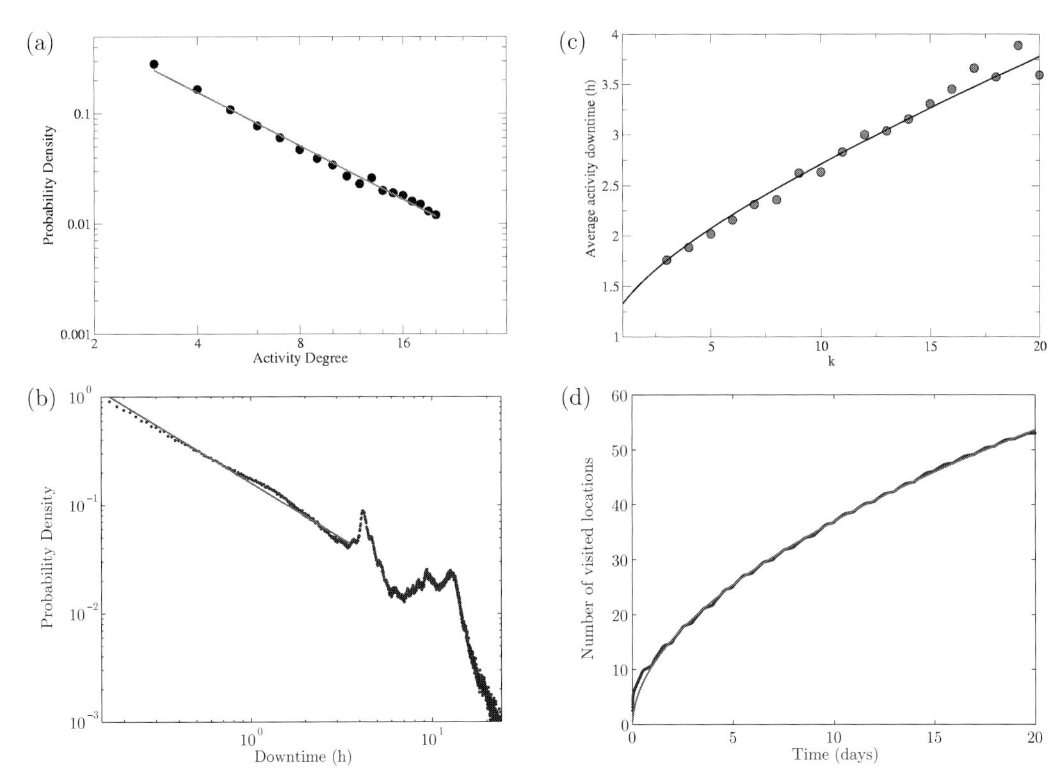

Figure 9.8 (a) Empirical distributions of the node degree (circles) for vehicular mobility in Florence [43]. The continuous curve – serving as a guide to the eye here – is a power law $p(k) = k^{-b}$ with $b \approx 1.6$. (b) Statistical distribution of parking times computed using vehicular GPS data for the Emilia-Romagna region [36] (dots). The straight line suggests the existence of a scaling law regime with $p(t) \propto 1/t$ for $t \leq 3h$. (c) Dependence of the average engine downtime duration τ_k from the node's degree k (same dataset as (a)). The continuous line is a fit using an exponential function. (d) Number of distinct visited locations as a function of time (day unit) in the same dataset as (b); the continuous line is power law fit $n(t) \propto t^{\gamma}$, where $\gamma \approx 0.54$.

with $\alpha = 1.02 \pm 0.02$. This distribution can be obtained in the limit $t \to \infty$ of a multiplicative process of the form $x(t+1) = \xi x(t)$ [44]. As a standard diffusion process leads to a Gaussian distribution, which for $\sigma \to \infty$ becomes a uniform distribution, the multiplicative diffusion leads to a lognormal distribution, which for $\sigma \to \infty$ converges to $1/x$. Also, GPS data suggest that the distribution in Figure 9.8b is robust and does not depend on the spatial scale considered, as the same scaling behavior at short times has been for different cities. Conversely, the peak structure that we observe for stops longer than 3 h is a specific characteristic of each city.

If, instead of taking into account the aggregate distribution $p(\tau)$, we analyze the probability of observing a stop of duration τ in a node of given degree k, experimental observation [43] suggests the existence of a universal probability distribution of the form

$$p(\tau \mid k) = \frac{1}{\langle\tau\rangle_k} f\left(\frac{\tau}{\langle\tau\rangle_k}\right), \tag{9.8}$$

where $\langle\tau\rangle_k$ is the average downtime for the k-degree node. The average value $\langle\tau\rangle_k$ alone thus characterizes the k dependence of the conditional probability $p(\tau \mid k)$. The hierarchy among activities also emerges in this temporal average. Indeed, from the result shown in Figure 9.8c, an almost linearly increasing behavior of $\langle\tau\rangle_k$ appears as the degree k increases. This suggests a tendency for the individual to repeat and to spend time in activities with a relevant added value (individual satisfaction, profit, etc.), which is related to the node degree. Moreover, the exponential fit $\langle\tau\rangle_k = \exp(ck^a)$, suggested in Figure 9.8c, allows us to link the scaling behavior properties of the distribution $p(k)$ to this behavioral feature of time use with the relationship $b = 2 - a$ [43].

The individual mobility network is not static as new locations can be visited every day and new nodes may become part of the network. This "diffusion" process, where the number of visited location grows with time, is limited by the probability of returning to an already visited node. This process can be studied by applying a Markov hypothesis to describe the evolution of the number $n(t)$ of nodes visited in a dataset after a time t. If we denote by $p(r, t)$ the probability that individuals have visited r locations after t days, the key role is played by the location r choice probability. If we assume that this probability at r is proportional to the degree of the node $k(r)$, the growth of n is a power law of the form:

$$n(t) \propto t^{\gamma}, \tag{9.9}$$

with $1/\gamma = (2/(b-1) - 1)$ [36], where b is the scaling coefficient of the degree distribution. In Figure 9.8d, the network growth given by $\gamma \approx 0.54$ is consistent with the value of b for the Emilia-Romagna dataset studied in [36], that is $b \approx 1.7$.

Stop duration is the key quantity associated to nodes of the network. The analogous temporal quantity that could be associated to edges would be travel time. However, it is usually difficult to correctly define travel from social-network accesses data, and to determine the vehicle's speed. The safest approach is then to characterize the edges with a measure of trip length. The distance measure available in all datasets is the Euclidean distance between the origin and destination points. For vehicular GPS data, where the positions along the trajectory are also known, one could use the effective length covered on the street network, but it can be observed that, in an urban environment, this effective distance is statistically equivalent to the Euclidean distance multiplied by a constant [45].

It is natural to consider the daily mobility as being limited for both physiological and economical reasons: any trip has a cost in time, energy, and money. We thus define the total daily length λ for each individual and for each day, the mobility is defined as the sum of trip lengths covered in 24 h. This measure is related to the total daily travel-time expenditure, and while this time displays minor variations among different areas, the total length fluctuates more because of extremely different average speeds in cities. The tail of both the total travel time and the daily length distributions (Figure 9.9b) are exponential [36], also when considering different transportation modes [46]. Similarly, the distribution $p(n)$ of the number of trips performed in a day has an exponential tail [43]. Knowing this, the distribution of the Euclidean trip length in a city can be obtained by randomly distributing at most n destinations within the interval $[0, \lambda]$. The

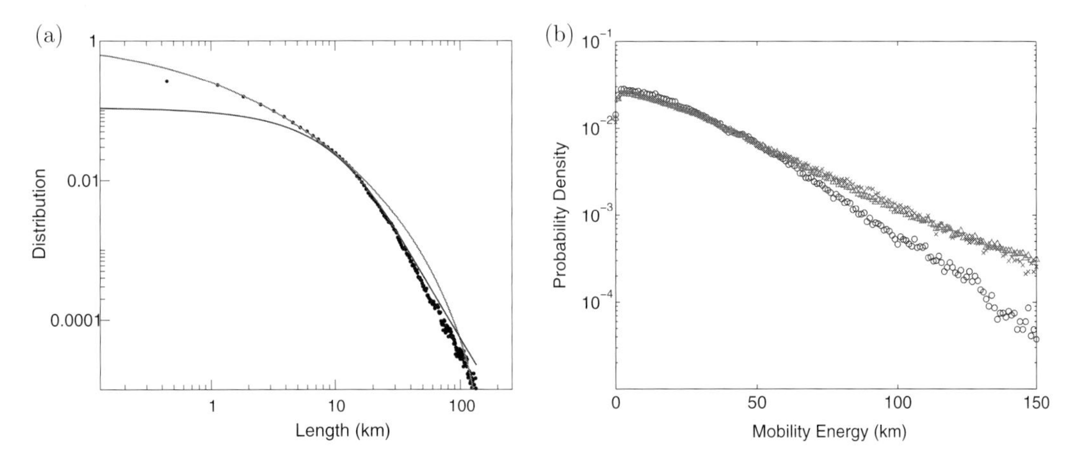

Figure 9.9 (a) Statistical distribution of the trip lengths measured using GPS vehicular data for the Emilia-Romagna region in Italy [36] (dots): the upper curve refers to the distribution computed analytically, whereas the lower curve represents a possible explanation for the long tail of the distribution based on a correlation among λ and n. (b) Daily mobility distributions computed considering individuals performing their mobility inside regions of different size around Bologna: the circles refer to a circle of radius $R \simeq 30$ km, including Bologna province; the crosses refer to a circle that includes the nearby cities ($R \simeq 50$ km); the triangles give the distribution for the whole region.

empirical values for λ_m and n_m, together with an exponential fit of the distributions $p(n)$ and $p(\lambda)$, allows us to compute the trip length distribution analytically. In Figure 9.9a, we see that the analytical result closely fits the experimental data for urban trips $l \leq 15$ km. The discrepancy at very small trips $l < 1$ km is expected, since small trips are overestimated when using the exponential distribution for the total daily length.

9.3.3 From big data to the spatial structure of cities

Pervasive geolocalized data generated by individuals have triggered a renewed interest for the study of numerous dynamics occuring in cities, in the first place those related to the mobility of individuals [47], as has been discussed in the previous section. Various data sources have been used during the past decade, such as car GPS (see previous section), RFIDs for collective transportation [48], or data coming from social applications such as Twitter [49] or Foursquare [35]. A recent, very important, source of data is provided by individual mobile phone data [50, 51]. Mobile phone data have been used to study statistical properties of individuals' mobility [52], but also to revisit some old debates regarding the distribution of daily commute times and distances among individuals [53], the automatic detection of urban land uses [54, 55], or the detection of communities based on human interactions [56].

Morphological properties of cities, such as the form of their density landscape, their space consumption, the number and spatial distribution of their activity clusters, have been studied through a comparative approach for a long time in quantitative geography and spatial economy [57–64]. Until late 2000, these quantitative comparisons of cities were based on census data or remote sensing data, both giving a static estimation of the density of individuals and land use in the city, at a small spatial granularity but with unique or a few points in time. Given the static nature of these studies, they could not allow investigation of some interesting questions related to the dynamical properties of the spatial structure of cities such as the variation of the city shape during the day, the location of hotspots and their changes through the course of the day, etc. In contrast, mobile phone data contain the spatial information about individuals and how it evolves in time. These datasets thus give us the opportunity to give quantitative answers to the aforementioned questions and will hopefully shed a new, quantitative light on many urban planning discussions, as has been pointed out by [65].

Distribution of individuals in the city

Essentially, mobile phone datasets that researchers had access to in the recent years are of two types. The most precise ones are those where the data directly inform on the successive positions in time of individual, anonymized users. Typically, each time a user's mobile phone gives or receives a call, sends or receives a short message, or each time it switches to another network antenna, an activity signal is recorded in the database, containing both the activity timestamp and the spatial coordinates of the closest antenna of the communication network. In the second type of datasets, for privacy reasons the mobile phone company does not give direct access to the individuals successive positions in time, but instead to the number/density of individuals that have transited by a given antenna during a fixed timeframe (typically of length 1 h). In this last case the data give access to the local density $\rho(i, t)$ of users at location i and at time t. The difficulty is then to study this complex object which displays variation in time and space and extract some relevant and interesting features of the city dynamical structure. Two main directions are possible to tackle this problem. The first one is to define global indicators that consider all points in the city and weight them by the users' density. The second approach consists of identifying local maxima of the function $\rho(i, t)$, or in other words, the hotspots[66].

The average weighted distance $D_V(t)$ between individuals in the city and its evolution during the course of a typical weekday provides a first interesting indicator about the organization of a city. Figure 9.10a shows the evolution of this normalized, average, weighted distance during a typical weekday in three Spanish urban areas for which mobile phone data have been collected during a two months' period.

We can essentially distinguish two broad categories according to the spatial organization of residences and activities.

- In the case of the simple picture of a monocentric city with predominant Central Business District (CBD), the city "collapses" in the morning when people living in the suburbs commute to their workplaces, and expands in the evening when they get back home. We then expect in this case a large variation (at the city scale) of the average distance D_V. In this case, activity and residential places are spatially "segregated".

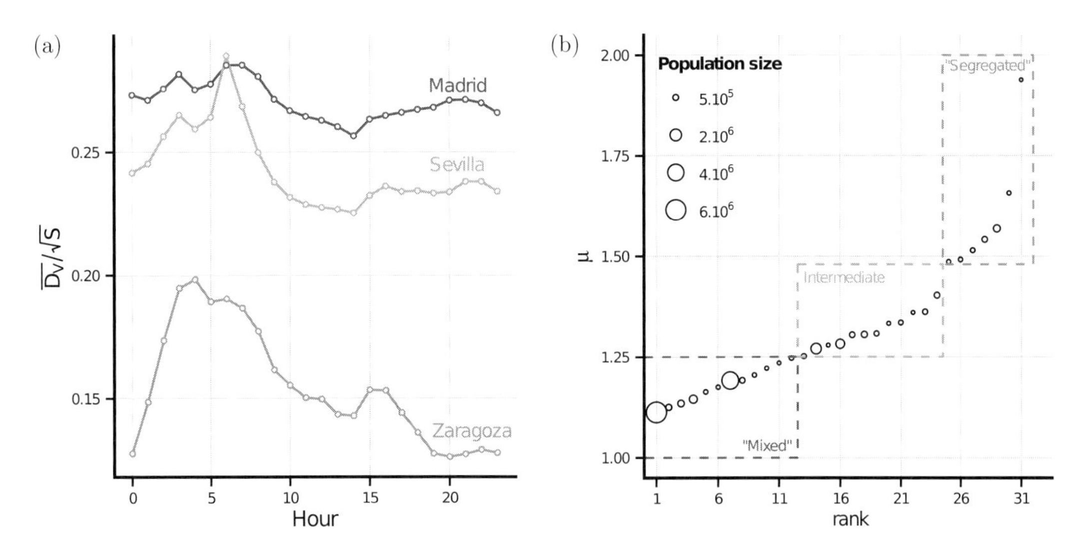

Figure 9.10 Time evolution of the average distance $D_V(t)$ between phone users in the city, and the values of the dilatation index $\mu = \max D_V(t)/\min D_V(t)$ for the 31 Spanish metropolitan areas studied. (a) Illustration of the time evolution of D_V in three urban areas: Madrid, Sevilla, and Zaragoza. This distance D_V is equal to the average of the distances between each pair of cells weighted by the density of each of the cells. The resulting distance is then divided by the typical spatial size of the city (given by $\sqrt{A}$, the square root of the city's area) in order to compare the curves across cities. (b) Rank-size distribution of the dilatation index μ in the 31 metropolitan areas.

- For more polycentric cities, where residential and work places are spatially less separated, we expect a smaller variation of D_V than the one observed for monocentric cities. Here activity places and residential areas are more "mixed".

When analyzing the curve of $D_V(t)$ for the 31 Spanish cities of more than 200 000 inhabitants (in 2011), it is remarkable that for all cities we observe the same typical pattern: we see two peaks, one around 7 a.m., when people switch on their mobile phones, probably at home or when they are in the transportation system's entry points (see Figure 9.10a). We then see a decrease of the distance (the city "collapses"), displaying spatial concentration of individuals during the middle of the day, mainly corresponding to the activity period for most individuals (workers/students). During the afternoon we see a second, smaller peak dispersed over 4–5 p.m., when people start going back home. This afternoon peak is less pronounced, suggesting a higher variety of mobility behaviors at the end of the day, as has also been noticed on other mobile phone datasets [53]. The interesting feature of theses curves is the amplitude variation that informs us about the importance of this collapse phenomenon. Despite the fact that several behavioral factors such as phone use affect these variations, we observe a common pattern: a pronounced peak at the beginning of the day and a minimum usually observed at the middle of the day. From these curves it is natural to calculate for each city a "dilatation coefficient" defined as

$$\mu = \frac{\max_t(D_V(t))}{\min_t(D_V(t))}. \tag{9.10}$$

We show in Figure 9.10b the rank plot of this dilatation index obtained for the 31 biggest Spanish cities. We can distinguish roughly three groups of cities. For the first group the value of μ is around one and the average distance D_V between individuals stays approximately constant throughout the day. This means that, whatever the hour of the day, the spatial spread of high-density locations does not change significantly. High-density locations correspond to different activities depending on the hour of the day, and a small value of the dilatation coefficient implies that daytime activity places (work places, schools, leisure places) are not more spatially concentrated than residences. Homes and activity places are more entangled, supporting the picture of more "mixed" cities, such as Madrid, for example. In the opposite case of large values, the spatial organization of the different high-density locations changes importantly along the day. A typical example would be a monocentric city where individuals are localized in the CBD during the day and where residences are spread all over the city space. In the set of Spanish cities, Zaragoza for example is representative of this type of city.

Activity hotspots

Another approach to understanding the city's structure with spatio-temporal data consists of identifying local maxima of the user's density $\rho(i, t)$ or, in other words, finding the city's hotspots. Looking at hotspots is convenient since it provides a clear picture of the important locations and allows us to concentrate on the "heart" of the city. Identifying hotspots corresponds to identifying local maxima on the surface of density of users. A simple method [66] amounts to choosing a threshold δ and to considering that every point i with a density larger than this threshold $\rho(i, t) > \delta$ is a hotspot at time t. Most of the methods developed in the spatial economy literature have relied so far on this simple argument, but there is obviously some arbitrariness in the choice of δ. A more robust approach consists of determining two extreme choices for the threshold value. The lower threshold δ_{min} corresponds to the average value of the density, which is indeed a reasonable, minimal requirement to be a local maxima. Based on considerations on the curvature of the Lorenz curve of the density, we are also able to determine another value, δ_{max}, which can be considered as the maximal, reasonable value for δ, as illustrated by Figure 9.11. Once the lower and upper bounds for the threshold are determined and allow for the identification of hotspots, all the results related to the hotspots should be robust with respect to the choice of δ. In other words, if a given result is qualitatively the same when considering the lower and upper bounds for δ, the result can safely be considered as an intrinsic feature of the city.

Using both thresholds identified by the method discussed above, we can count the number of hotspots at each hour of the day, compute the average over the day, and see how this average number scales with the population size of the city. This measure has been motivated by theoretical and empirical work [67] that has highlighted a clear sublinear relation between the population size of cities and their number of activity centers. For

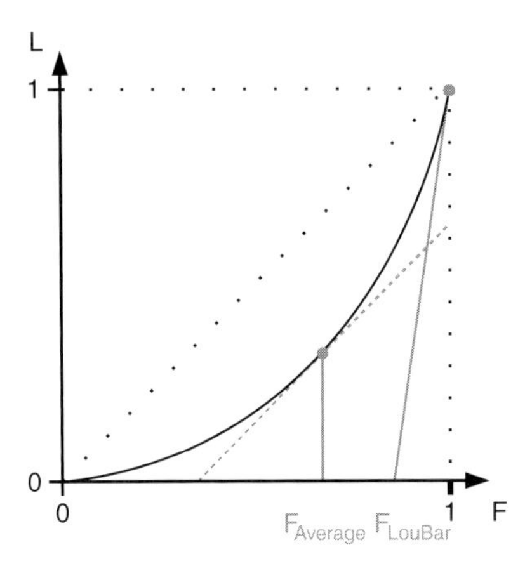

Figure 9.11 Illustration of the two thresholds selection on the Lorenz curve. The Lorenz curve is a graphical representation of the cumulative distribution function of an empirical probability distribution. For a given hour, we have the distribution of densities $\rho(i, t)$ of individuals across the city. We sort them in increasing rank and denote them by $\rho(1, t) < \rho(2, t) < \cdots < \rho(n, t)$ where n is the number of spatial units in the city. The Lorenz curve is constructed by plotting on the x-axis the proportion of cells $F = i/n$ and on the y-axis the corresponding proportion of users density L with $L(i, t) = \frac{\sum_{j=1}^{i} \rho(j,t)}{\sum_{j=1}^{n} \rho(j,t)}$. If all the densities were of the same order the Lorenz curve would be the diagonal from $(0, 0)$ to $(1, 1)$. The stronger the curvature the stronger the inequality and, intuitively, the smaller the number of hotspots. This remark led us to construct a new criterion by relating the number of dominant hotspots (those that have a very high value compared to the other cells) to the slope of the curve at point $F = 1$: the larger the slope, the smaller the number of dominant individuals in the statistical distribution (see [66] for details).

the USA, it has been shown that the number of activity centers N_a (determined from employment data) scales as

$$N_a \sim P^{\beta}, \tag{9.11}$$

with $\beta \sim 0.64$. In Figure 9.12 we display the number H of hotspots versus the population for the set of the 31 largest Spanish cities for which we analyzed mobile phone data. The power law fit confirms the result obtained in [67] that there is a sublinear relation and, remarkably enough, that the value of the exponent is of the same order. We also showed that this result is robust against the thresholding criteria used to define hotspots and the spatial scale at which the density information is aggregated.

Urban rhythms and detection of land-uses

Regarding the time evolution of the number of users along the day, it is interesting to see if it follows the same pattern in every city. Figure 9.13 shows the average number of mobile phone users per hour for different Spanish cities. Globally, the number of phone users is significantly higher during the weekdays than during the weekends, except at night time. From 11 p.m. to 8 a.m., the number of users is relatively low, it reaches a minimum at 5 a.m. during weekdays and at 7 a.m. during the weekend. For all cities we

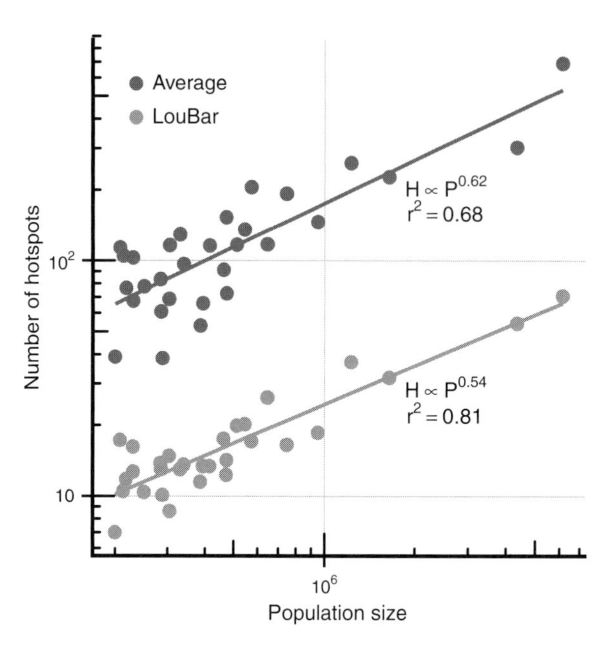

Figure 9.12 Scatter plot and fit of the number of hotspots H vs. the population size P for the 31 biggest Spanish cities. Each point in the scatterplot corresponds to the average number of hotspots determined for each 1 h time bin of a weekday (for five weekdays considered here). The power law fit is consistent, for both hotspots identification thresholds, with a sublinear behavior characterized by an exponent of order 0.6.

observe two activity peaks, one at 12 a.m. during weekdays (1 p.m. during the weekend) and another one at 6 p.m. during weekdays (and at 8 p.m. during the weekend).

In order to compare the curves of different cities, we rescale the values by the total number of users for an average weekday, as shown on Figure 9.13. The rescaled plot suggests the existence of a single "urban rhythm" common to all cities. The data collapse is very good in the morning, while in the afternoon we observe a little more variability from one city to another.

When observing the temporal patterns in the mobile phone data, it is also interesting to study their distribution in the city's space. From these time profiles it is possible to identify clusters of places that share a common daily/weekly rhythm in terms of mobile phone activity. It must be noticed that several authors who have run clustering algorithms on pervasive data have found the same clusters (namely business, residential, leisure and nightlife) in different cities, includig New York (Tweets in Manhattan) [68], Barcelona and Madrid (mobile phone data in urban agglomerations) [69].

Crosschecking different mobility sources

We end this section with a discussion of the comparison between different datasets. A recent and interesting result obtained by a crosscheck of new sources of mobility information (Twitter and mobile phone data) and the classic one (census) in Barcelona and Madrid has shown that when aggregated at the spatial scale of 1 km^2 cells grid and

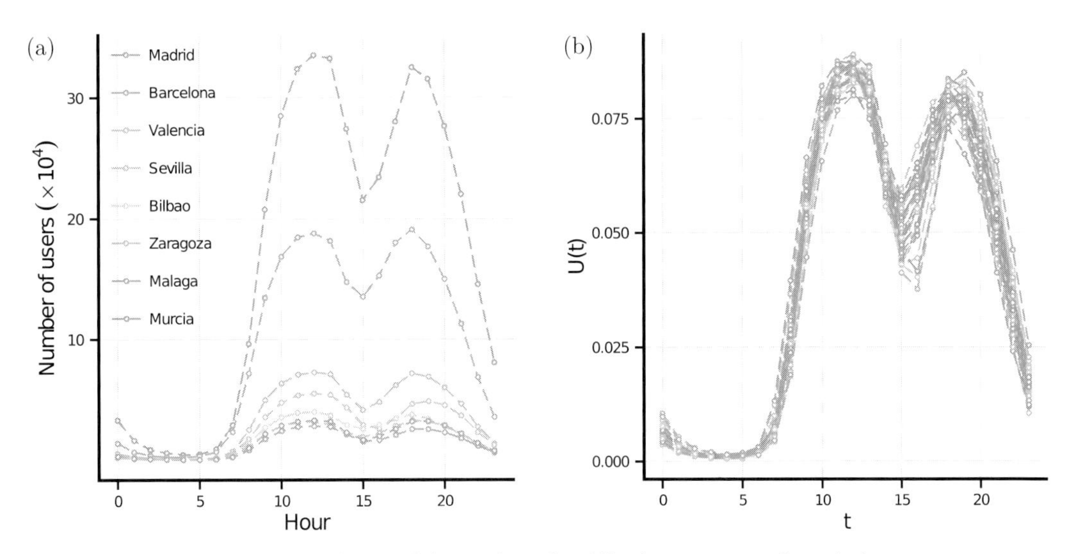

Figure 9.13 Time evolution of the number of mobile phone users per hour during an average weekday. (a) Total number of unique mobile phone users per hour (shown here for the eight biggest Spanish cities). (b) Rescaled numbers of unique users per hour for 31 Spanish cities. Each value $U_i(t)$ is equal to the number of phone users in city i at time t, $N_i(t)$, divided by the total number of phone users in i during the entire day : $U_i(t) = N_i(t)/\sum_{t=1}^{t=24} N_i(t)$. The good collapse suggests the existence of an urban rhythm common to all cities.

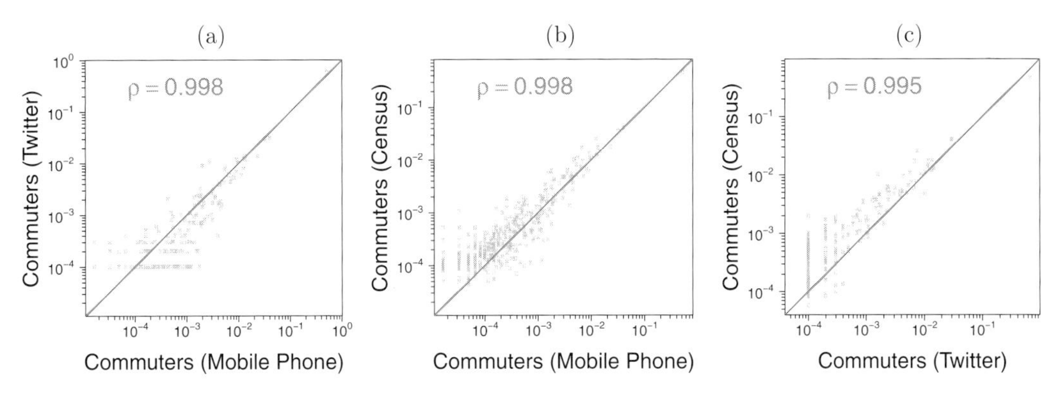

Figure 9.14 Comparison between the nonzero flows from municipality to municipality obtained with three different mobility datasets in Barcelona: Twitter, mobile phone, and census. (a) Twitter and mobile phone. (b) Census and mobile phone. (c) Census and Twitter (from [69]).

at a time resolution of 1 h, these three different sources provide the same quantitative picture of the mobility in cities, as illustrated by Figure 9.14 [69].

In this study, the authors compared the nonzero flows from municipality to municipality obtained with three different mobility datasets in Barcelona. The first set comprises Tweets from 27 707 users recorded during more than a year between September 2012 and December 2013 through the Twitter public API. The second set, of mobile phone data, represents approximately 5% of the mobile phone users living in the Barcelona metropolitan area, and was recorded during a period of two months in 2009. The third set corresponds to origin–destination matrices extracted from the 2009 census led by the urban agglomeration of Barcelona (the values have been normalized by the total number of commuters for both OD tables). Points in the scatter plots represent the number of commuters between each pair of municipalities and line represents the $x = y$ line. We can see here that these different sources give consistent information about mobility. These results are a first step that opens the door to study spatio-temporal properties of cities from spatial information encapsulated in big data generated by social applications.

9.4 Scaling in cities

The discovery of allometric scaling relationships in cities has arguably driven the quantitative research on urban systems in the past years. It has indeed been discovered that different socio-economic indicators in cities, such as the GDP, crime rates, the number of patents, as well as different structural indicators such as the total length of the road network, the urbanized land area, etc., exhibit robust scaling relationships with respect to population [2, 70–76]. The existence of these simple scaling relationships points to the existence of universal processes across several urban systems, and thus the possibility of modeling cities.

In particular, data about intra-urban mobilities, by giving us the structure of mobility patterns, help us understanding why and how people move within cities. However, very few models in the literature try to explain the dominant mechanisms governing the formation and evolution of mobility patterns, and here we show how socio-economical data can provide us with a useful guide for modelling various aspects of cities. We will focus on commuting patterns, from home to workplaces. Mobility appears as an intricate issue: many factors, such as geographical constraints, the location of amenities and availability of transportation – among other things – are susceptible to impact it. Nevertheless, the regularities observed in the behavior of some quantities related to mobility seem to indicate the existence of general behaviors that are common to every city.

We illustrate this discussion on the example of the total driven distance L_{tot}. The authors of [73] have found in US data that L_{tot}, rescaled by the typical size of the city $\sqrt{A}$ (where A is the area of the city), depends on the city's population P (see Figure 9.15) as

$$\frac{L_{tot}}{\sqrt{A}} \sim P^{0.60}. \tag{9.12}$$

This behavior is, however, difficult to explain by using naive arguments. Indeed, if we assume that the city has a single activity center – a "monocentric" city – everyone commutes to the same center, and the average commuting distance is controlled by the typical size of the city $L = \sqrt{A}$. This argument leads to a total commuting distance

$$\frac{L_{tot}}{\sqrt{A}} \sim P, \tag{9.13}$$

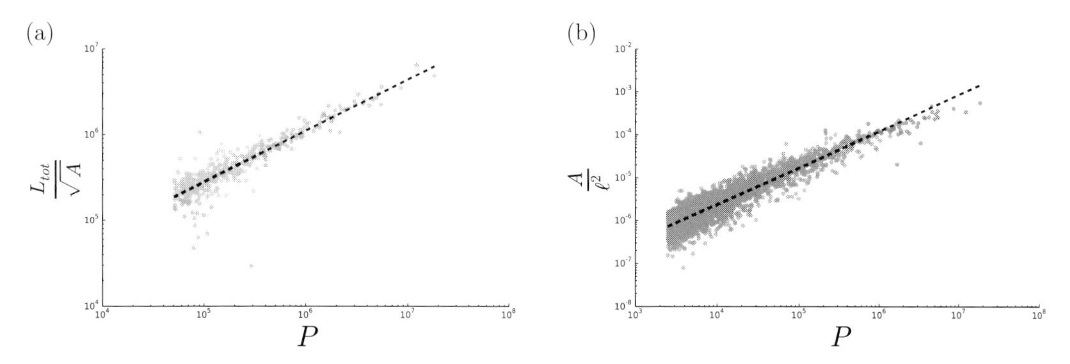

Figure 9.15 (a) Total daily driven distance L_{tot} rescaled by the surface area A as a function of population P. A power law fit gives $\frac{L_{tot}}{\sqrt{A}} = P^{0.60}$ ($R^2 = 0.90$). (b) Total surface area A of urban areas as a function of population. A power law fit gives $\frac{A}{\ell} \sim P^{0.85}$ ($R^2 = 0.93$). Figures taken from [76].

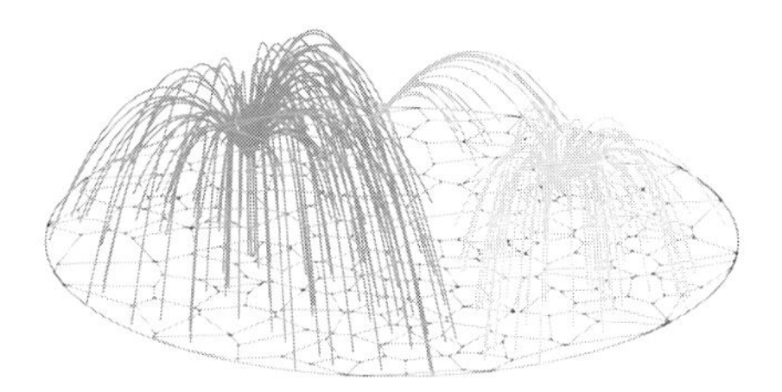

Figure 9.16 The structure of the mobility patterns as given by the model in [67]. Each line corresponds to a commuting trip.

which is at odds with the empirical results. If we assume instead that the city is totally decentralized, the average commuting distance is controlled by the typical distance between two individuals $\sqrt{\frac{A}{P}}$ so that we get

$$\frac{L_{tot}}{\sqrt{A}} \sim \sqrt{P}, \tag{9.14}$$

which, again, does not agree with the empirical results. Reality must therefore lie between the two previous extremes. In fact, it has been known for a long time that cities are neither totally decentralized nor monocentric, whatever the population size. Indeed, as they grow and expand, cities tend do adopt a more distributed organization: activities are spread between several distinct centers. So, if we now assume that there are $k(P)$ centers in the city, and that most people commute to the closest activity center (Figure 9.16), the average commuting distance is given by $\sqrt{Ak}$, leading to

$$\frac{L_{tot}}{\sqrt{A}} \sim \frac{P}{\sqrt{k(P)}}. \tag{9.15}$$

We can thus understand the empirical exponent observed for L_{tot} if we understand the behavior of the number of centers k with P. In [67], the authors tackle this problem, and propose an out-of-equilibrium model in the spirit of statistical physics in order to explain the structure of the observed patterns. The model revisits an older one proposed by economists [77] and assumes random locations for household and activity centers, and focuses on the choice of job location. This choice is determined by a trade-off between the expected income and the total commuting time by an individual car, congestion taken into account. As a result of these assumptions, the authors are able to account for the monocentric to polycentric transition of cities by predicting the existence of a limit value of population P^* over which the city is polycentric (that is, people choose to commute to different centers). Furthermore, they are able to predict the number of centers of activity k as a function of population P:

$$k \sim \left(\frac{P}{P*} \right)^{\frac{\mu}{\mu+1}}, \tag{9.16}$$

where μ characterizes the sensitivity of the road network to congestion. This prediction of a sublinear behavior of k with the population is confirmed using US data giving the employment density per zip code, where a value of 0.64 is measured for the exponent. As explained above, another study on Spanish phone data found a similar value of the exponent using mobile phone data, and a different method for measuring the number of centers [66]. Using the expression for $k(P)$ (Eqn (9.16)) in Eqn (9.15) it is then easy to show that

$$\frac{L_{tot}}{\sqrt{A}} \sim P^{1-\delta} \tag{9.17}$$

with $\delta = \frac{\mu}{2(\mu+1)}$. Using the value of $\mu = 0.64$ measured on the US data, one finds $1 - \delta \approx 0.61$, which agrees strikingly well with the empirical value of 0.60.

A naive interpretation of the previous results would be that the typical commuting length of people is imposed by city size and the number of centers. In fact, it is more plausible that people choose their household and work locations distributed around an average distance ℓ. In other words

$$\frac{L_{tot}}{P} \sim \ell, \tag{9.18}$$

a simple argument that agrees with empirical results (see Figure 9.17). This argument can be consistent with the previous one only if the surface area A depends on population as

$$\frac{A}{\ell^2} \sim P^{2\delta}, \tag{9.19}$$

with $2\delta = 0.64$, consistent with empirical findings (see Figure 9.15). Therefore as cities grow the average commuting distance does not change, while the number of centers increases less than linearly with population size, implying an increase in population density.

Interestingly, a simple model of the structure of the mobility patterns is able to give us useful insights on the behavior of some structural indicators of cities with their population. An important consequence of using cars is, for example, the existence of

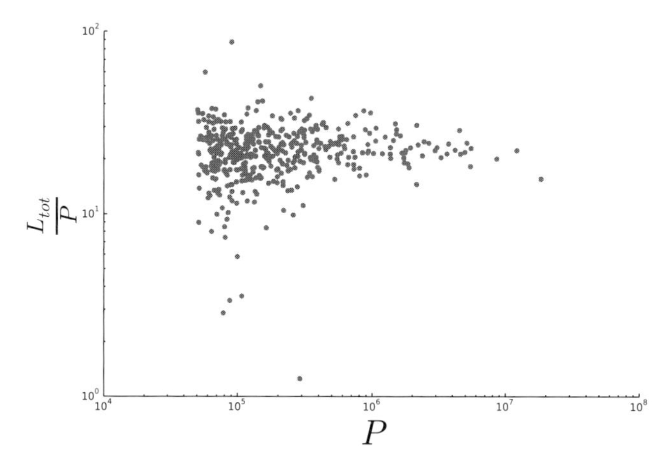

Figure 9.17 The total daily driven distance per capita. The plot is compatible with the hypothesis of a population-independent behavior. We find an average daily driven distance per capita of 23 miles and a standard deviation of 7 miles. The obtained data are from the FHWA for 441 urbanized area in the USA in 2010. Figure taken from [76].

important congestion in large cities and the gas emissions associated with them. In order to estimate these effects, one can write the total commuting time

$$T = \tau_0 + \delta\tau, \tag{9.20}$$

where τ_0 is the free flow (in absence of congestion) commuting time. One can show within the framework of the model that the total commuting time is in fact dominated in large urban areas by the total delay due to congestion $\delta\tau$

$$\delta\tau \sim P^{1+\delta}, \tag{9.21}$$

where $\delta + 1 = 1.39$. In other words, the time spent in congestion per capita increases with population size, and the larger the city, the bigger the impact of congestion on individuals. This behavior agrees with empirical data obtained from the US 2010 Mobility Report for 100 metropolitan areas (see Figure 9.18). A direct consequence of this extra time spent in congestion is the quantity of CO_2 that is emitted due to transportation:

$$Q_{CO_2} \sim P^{1+\delta}. \tag{9.22}$$

The quantity of CO_2 emitted *per person* increases with population size. A result that is confirmed by empirical data obtained from the OECD and the Urban Mobility Report in the USA (see [78] for a more detailed discussion). Cities where individual cars are the main mode of transportation therefore act as catalysts for the production of CO_2. The outlook given by these results on congestion in cities is alarming. Data indeed show that cities are fundamentally unsustainable, as they put more pressure on individuals while accelerating the production of greenhouse gases responsible for climate change.

Further studies are needed in order to investigate some of the assumptions of this simplified model. Among other things, more-specific data are needed in order to

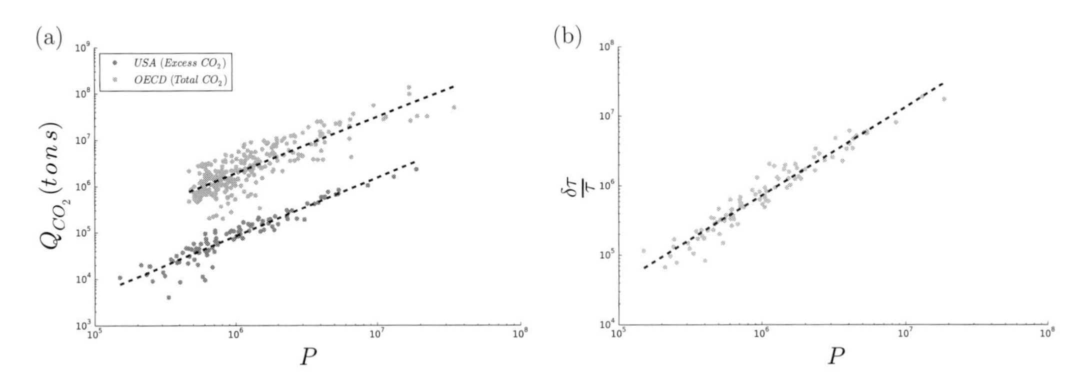

Figure 9.18 (a) Excess CO_2 emitted due to tranportation for 100 metrotpolitan areas in the USA and total quantity of CO_2 emitted due to transportation for 238 metropolitan area accross 28 countries member of the OECD as a function of the population. A power law fit gives $Q_{CO_2}^{US} \sim P^{1.26}$ ($R^2 = 94$), and $Q_{CO_2}^{OECD} \sim P^{1.21}$ ($R^2 = 83$). (b) Total delay due to congestion τ as a function of population P for 100 metropolitan areas in the USA. A power law fit gives $\frac{\delta_\tau}{\tau} \sim P^{1.27}$ $R^2 = 0.97$. Figures taken from [76].

check whether the shape of the mobility patterns is actually close to those depicted on Figure 9.16, and to what extent possible deviations would impact our results. Also, large cities usually have public transportation systems, and it will be necessary to estimate the influence on the model of these alternative transportation modes. Nevertheless, thanks to the simplifying assumptions made, this model is able to capture some of the dominant processes behind the structure of mobility, and lead to a better understanding of how they evolve with population size. From the previous results, we can indeed draw the conclusion that congestion seems to be the dominant process responsible for the polycentric transition of cities, and the structure of mobility patterns. As such, we believe that it provides the building blocks for a future more comprehensive, and hopefully predictive, theory of cities.

9.5 Discussion: towards a new science of cities

More data about cities will continue to become available, and in this chapter we illustrated different uses of different types of data. GPS and mobile phone data, in addition to informing us about the mobility patterns in cities, provide us with valuable information about the spatial organization of cities. Long time scale data inform us about the evolution of critical infrastructures and slow processes acting in cities. Finally, socioeconomic data allow us to test models of mobility, and residence/work location choices. All these different examples bring our focus to an essential ingredient: the availability of data now allow us to test models and their predictions. Spatial economics proposed a wealth of models that are the basis of many urban planning decisions, and we have now the opportunity to test them.

Acknowledgments

T. L. and M. B. acknowledge funding from the European Commission through project EUNOIA (FP7-DG.Connect- 318367). The authors thank EUNOIA consortium members for fruitful scientific discussions. R. G. and M. B. are supported by the European Commission FET-Proactive project PLEXMATH (Grant No. 317614).

References

[1] M. Fujita, P. R. Krugman, and A. J. Venables, *The Spatial Economy: Cities, Regions, and International Trade*, MIT Press, 2001.

[2] H. A. Makse, S. Havlin, and H. E. Stanley, "Modelling urban growth patterns," *Nature*, vol. **377**, no. 6550, pp. 608–612, 1995.

[3] M. Batty, "The size, scale, and shape of cities," *Science*, vol. **319**, no. 5864, pp. 769–771, 2008.

[4] G. F. Frasco, J. Sun, H. D. Rozenfeld, and D. Ben-Avraham, "Spatially distributed social complex networks," *Physical Review X*, vol. **4**, no. 1, p. 011008, 2014.

[5] L. Bettencourt and G. West, "A unified theory of urban living," *Nature*, vol. **467**, no. 7318, pp. 912–913, 2010.

[6] L. M. Bettencourt, "The origins of scaling in cities," *Science*, vol. **340**, no. 6139, pp. 1438–1441, 2013.

[7] M. Batty, *The New Science of Cities*, MIT Press, 2013.

[8] M. Barthelemy, "Spatial networks," *Physics Reports*, vol. **499**, no. 1, pp. 1–101, 2011.

[9] B. Hillier, A. Leaman, P. Stansall, and M. Bedford, "Space syntax," *Environment and Planning B: Planning and Design*, vol. **3**, no. 2, pp. 147–185, 1976.

[10] B. Jiang and C. Claramunt, "Topological analysis of urban street networks," *Environment and Planning B*, vol. **31**, no. 1, pp. 151–162, 2004.

[11] M. Rosvall, A. Trusina, P. Minnhagen, and K. Sneppen, "Networks and cities: an information perspective," *Physical Review Letters*, vol. **94**, no. 2, p. 028701, 2005.

[12] S. Porta, P. Crucitti, and V. Latora, "The network analysis of urban streets: a primal approach," arXiv preprint physics/0506009, 2005.

[13] ——, "The network analysis of urban streets: a dual approach," *Physica A: Statistical Mechanics and its Applications*, vol. **369**, no. 2, pp. 853–866, 2006.

[14] A. Cardillo, S. Scellato, V. Latora, and S. Porta, "Structural properties of planar graphs of urban street patterns," *Physical Review E*, vol. **73**, no. 6, p. 066107, 2006.

[15] S. Lämmer, B. Gehlsen, and D. Helbing, "Scaling laws in the spatial structure of urban road networks," *Physica A: Statistical Mechanics and its Applications*, vol. **363**, no. 1, pp. 89–95, 2006.

[16] M. Barthelemy and A. Flammini, "Modeling urban street patterns," *Physical Review Letters*, vol. **100**, no. 13, p. 138702, 2008.

[17] M. Barthélemy and A. Flammini, "Co-evolution of density and topology in a simple model of city formation," *Networks and Spatial Economics*, vol. **9**, no. 3, pp. 401–425, 2009.

[18] M. Fialkowski and A. Bitner, "Universal rules for fragmentation of land by humans," *Landscape Ecology*, vol. **23**, no. 9, pp. 1013–1022, 2008.

[19] E. Strano, V. Nicosia, V. Latora, S. Porta, and M. Barthélemy, "Elementary processes governing the evolution of road networks," *Scientific Reports*, vol. **2**, 2012.

[20] M. Barthelemy, P. Bordin, H. Berestycki, and M. Gribaudi, "Self-organization versus top-down planning in the evolution of a city," *Scientific Reports*, vol. **3**, 2013.

[21] A. P. Masucci, K. Stanilov, and M. Batty, "Limited urban growth: London's street network dynamics since the 18th century," *PLoS One*, vol. **8**, no. 8, p. e69469, 2013.

[22] T. Courtat, C. Gloaguen, and S. Douady, "Mathematics and morphogenesis of cities: A geometrical approach," *Physical Review E*, vol. **83**, no. 3, p. 036106, 2011.

[23] "OpenStreetMap (OSM) is a collaborative project to create a free editable map of the world," http://www.openstreetmap.org.

[24] C. Roth, S. M. Kang, M. Batty, and M. Barthelemy, "A long-time limit for world subway networks," *Journal of The Royal Society Interface*, vol. **9**, pp. 2540–2550, 2012.

[25] L. Benguigui, "The fractal dimension of some railway networks," *Journal de Physique I*, vol. **2**, no. 4, pp. 385–388, 1992.

[26] K. S. Kim, L. Benguigui, and M. Marinov, "The fractal structure of Seoul's public transportation system," *Cities*, vol. **20**, no. 1, pp. 31–39, 2003.

[27] D. Levinson, "Density and dispersion: the co-development of land use and rail in london," *Journal of Economic Geography*, vol. **8**, no. 1, pp. 55–77, 2008.

[28] ——, "Network structure and city size," *PloS one*, vol. **7**, no. 1, p. e29721, 2012.

[29] S. Derrible and C. Kennedy, "Network analysis of world subway systems using updated graph theory," *Transportation Research Record: Journal of the Transportation Research Board*, vol. **2112**, no. 1, pp. 17–25, 2009.

[30] W. R. Black, "An iterative model for generating transportation networks," *Geographical Analysis*, vol. **3**, no. 3, pp. 283–288, 1971.

[31] R. Louf, P. Jensen, and M. Barthelemy, "Emergence of hierarchy in cost-driven growth of spatial networks," *Proceedings of the National Academy of Sciences*, vol. **110**, no. 22, pp. 8824–8829, 2013.

[32] F. Xie and D. Levinson, "Topological evolution of surface transportation networks," *Computers, Environment and Urban Systems*, vol. **33**, no. 3, pp. 211–223, 2009.

[33] R. Louf, C. Roth, and M. Barthelemy, "Scaling in transportation networks," *PloS one*, vol. **9**, no. 7, p. e102007, 2014.

[34] I. Rhee, M. Shin, S. Hong, *et al.*, "On the levy-walk nature of human mobility," *IEEE/ACM Transactions on Networking (TON)*, vol. **19**, no. 3, pp. 630–643, 2011.

[35] A. Noulas, S. Scellato, R. Lambiotte, M. Pontil, and C. Mascolo, "A tale of many cities: universal patterns in human urban mobility," *PLos One*, vol. **7**, no. 5, p. e37027, 2012.

[36] R. Gallotti, A. Bazzani, and S. Rambaldi, "Toward a statistical physics of human mobility," *Int. J. Mod. Phys. C*, vol. **23**, no. 9, 2012.

[37] C. Coffey, A. Pozdnoukhov, and F. Calabrese, "Time of arrival predictability horizons for public bus routes," in *Proceedings of the 4th ACM SIGSPATIAL International Workshop on Computational Transportation Science*, ACM, 2011, pp. 1–5.

[38] X. Liang, X. Zheng, W. Lv, T. Zhu, and K. Xu, "The scaling of human mobility by taxis is exponential," *Physica A: Statistical Mechanics and its Applications*, vol. **391**, no. 5, pp. 2135–2144, 2012.

[39] C. Song, Z. Qu, N. Blumm, and A.-L. Barabási, "Limits of predictability in human mobility," *Science*, vol. **327**, no. 5968, pp. 1018–1021, 2010.

[40] R. Gallotti, A. Bazzani, M. Degli Esposti, and S. Rambaldi, "Entropic measures of individual mobility patterns," *Journal of Statistical Mechanics: Theory and Experiment*, vol. **2013**, no. 10, p. P10022, 2013.
[41] J. P. Bagrow and Y.-R. Lin, "Mesoscopic structure and social aspects of human mobility," *PLos One*, vol. **7**, no. 5, p. e37676, 2012.
[42] Y. Zheng, Q. Li, Y. Chen, X. Xie, and W.-Y. Ma, "Understanding mobility based on gps data," in *Proceedings of the 10th international conference on Ubiquitous computing*, ACM, 2008, pp. 312–321.
[43] A. Bazzani, B. Giorgini, S. Rambaldi, R. Gallotti, and L. Giovannini, "Statistical laws in urban mobility from microscopic GPS data in the area of florence," *Journal of Statistical Mechanics: Theory and Experiment*, vol. **2010**, no. 05, p. P05001, 2010.
[44] L. Pietronero, E. Tosatti, V. Tosatti, and A. Vespignani, "Explaining the uneven distribution of numbers in nature: the laws of Benford and zipf," *Physica A: Statistical Mechanics and its Applications*, vol. **293**, no. 1, pp. 297–304, 2001.
[45] R. Gallotti, "Statistical physics and modeling of human mobility," Ph.D. dissertation, University of Bologna, 2013.
[46] R. Kölbl and D. Helbing, "Energy laws in human travel behaviour," *New Journal of Physics*, vol. **5**, no. 1, p. 48, 2003.
[47] F. Asgari, V. Gauthier, and M. Becker, "A survey on human mobility and its applications," arXiv preprint arXiv:1307.0814, 2013.
[48] C. Roth, S. M. Kang, M. Batty, and M. Barthelemy, "Structure of urban movements: polycentric activity and entangled hierarchical flows," *PLos One*, vol. **6**, no. 1, p. e15923, 2011.
[49] B. Hawelka, I. Sitko, E. Beinat, *et al.*, "Geo-located twitter as proxy for global mobility patterns," *Cartography and Geographic Information Science*, vol. **41**, no. 3, pp. 260–271, 2014.
[50] J.-P. Onnela, J. Saramäki, J. Hyvönen, *et al.*, "Structure and tie strengths in mobile communication networks," *Proceedings of the National Academy of Sciences*, vol. **104**, no. 18, pp. 7332–7336, 2007.
[51] R. Lambiotte, V. D. Blondel, C. de Kerchove, *et al.*, "Geographical dispersal of mobile communication networks," *Physica A: Statistical Mechanics and its Applications*, vol. **387**, no. 21, pp. 5317–5325, 2008.
[52] M. C. Gonzalez, C. A. Hidalgo, and A.-L. Barabasi, "Understanding individual human mobility patterns," *Nature*, vol. **453**, no. 7196, pp. 779–782, 2008.
[53] K. S. Kung, K. Greco, S. Sobolevsky, and C. Ratti, "Exploring universal patterns in human home-work commuting from mobile phone data," *PLos One*, vol. **9**, no. 6, p. e96180, 2014.
[54] V. Soto and E. Frias-Martinez, "Robust land use characterization of urban landscapes using cell phone data," in *Proceedings of the 1st Workshop on Pervasive Urban Applications, in conjunction with 9th Int. Conf. Pervasive Computing*, 2011.
[55] T. Pei, S. Sobolevsky, C. Ratti, *et al.*, "A new insight into land use classification based on aggregated mobile phone data," *International Journal of Geographical Information Science*, vol. **28**, no. 9, pp. 1988–2007, 2014.
[56] S. Sobolevsky, M. Szell, R. Campari, *et al.*, "Delineating geographical regions with networks of human interactions in an extensive set of countries," *PLos One*, vol. **8**, no. 12, p. e81707, 2013.
[57] A. Anas, R. Arnott, and K. A. Small, "Urban spatial structure," *Journal of Economic Literature*, pp. 1426–1464, 1998.

[58] A. Bertaud and S. Malpezzi, "The spatial distribution of population in 48 world cities: Implications for economies in transition," Center for Urban Land Economics Research, University of Wisconsin, 2003.

[59] Y.-H. Tsai, "Quantifying urban form: compactness versus' sprawl'," *Urban Studies*, vol. **42**, no. 1, pp. 141–161, 2005.

[60] M. Guérois and D. Pumain, "Built-up encroachment and the urban field: a comparison of forty european cities," *Environment and planning. A*, vol. **40**, no. 9, p. 2186, 2008.

[61] N. Schwarz, "Urban form revisited – selecting indicators for characterising european cities," *Landscape and Urban Planning*, vol. **96**, no. 1, pp. 29–47, 2010.

[62] S. Berroir, H. Mathian, T. Saint-Julien, and L. Sanders, "The role of mobility in the building of metropolitan polycentrism," in *Modelling Urban Dynamics*, ISTE-Wiley, 2011, pp. 1–25.

[63] F. Le Néchet, "Urban spatial structure, daily mobility and energy consumption: a study of 34 european cities," *Cybergeo: European Journal of Geography*, 2012.

[64] R. H. M. Pereira, V. Nadalin, L. Monasterio, and P. H. Albuquerque, "Urban centrality: a simple index," *Geographical Analysis*, vol. **45**, no. 1, pp. 77–89, 2013.

[65] C. Ratti, S. Williams, D. Frenchman, and R. Pulselli, "Mobile landscapes: using location data from cell phones for urban analysis," *Environment and Planning B: Planning and Design*, vol. **33**, no. 5, pp. 727–748, 2006.

[66] T. Louail, M. Lenormand, O. G. C. Ros, *et al.*, "From mobile phone data to the spatial structure of cities," *Scientific Reports*, vol. **4**, 2014.

[67] R. Louf and M. Barthelemy, "Modeling the polycentric transition of cities," *Physical Review Letters*, vol. **111**, no. 19, p. 198702, 2013.

[68] V. Frias-Martinez, V. Soto, H. Hohwald, and E. Frias-Martinez, "Characterizing urban landscapes using geolocated tweets," in *Privacy, Security, Risk and Trust (PASSAT), 2012 International Conference on and 2012 International Confernece on Social Computing (SocialCom)*, IEEE, 2012, pp. 239–248.

[69] M. Lenormand, M. Picornell, O. G. Cantú-Ros, *et al.*, "Cross-checking different sources of mobility information," *PLos One*, vol. **9**, no. 8, p. e105184, 2014.

[70] P. W. Newman and J. R. Kenworthy, "Gasoline consumption and cities: a comparison of us cities with a global survey," *Journal of the American Planning Association*, vol. **55**, no. 1, pp. 24–37, 1989.

[71] D. Pumain, F. Paulus, C. Vacchiani-Marcuzzo, and J. Lobo, "An evolutionary theory for interpreting urban scaling laws," *Cybergeo: European Journal of Geography*, 2006.

[72] L. M. Bettencourt, J. Lobo, D. Helbing, C. Kühnert, and G. B. West, "Growth, innovation, scaling, and the pace of life in cities," *Proceedings of the National Academy of Sciences*, vol. **104**, no. 17, pp. 7301–7306, 2007.

[73] H. Samaniego and M. E. Moses, "Cities as organisms: allometric scaling of urban road networks," *Journal of Transport and Land Use*, vol. **1**, no. 1, pp. 21–39, 2008.

[74] H. D. Rozenfeld, D. Rybski, J. S. Andrade, *et al.*, "Laws of population growth," *Proceedings of the National Academy of Sciences*, vol. **105**, no. 48, pp. 18 702–18 707, 2008.

[75] W. Pan, G. Ghoshal, C. Krumme, M. Cebrian, and A. Pentland, "Urban characteristics attributable to density-driven tie formation," *Nature Communications*, vol. **4**, 2013.

[76] R. Louf and M. Barthelemy, "How congestion shapes cities: from mobility patterns to scaling," *Scientific Reports*, vol. **4**, 2014.

[77] M. Fujita and H. Ogawa, "Multiple equilibria and structural transition of non-monocentric urban configurations," *Regional Science and Urban Economics*, vol. **12**, no. 2, pp. 161–196, 1982.

[78] R. Louf and M. Barthelemy, "Scaling: lost in the smog," arXiv preprint arXiv:1410.4964, 2014.

10 High-dimensional network analytics: mapping topic networks in Twitter data during the Arab Spring

Kathleen M. Carley, Wei Wei, and Kenneth Joseph

Social change is often reflected in social talk. The capability to track who is talking about what, where, and with whom, as well as changes in the topics of concern by region, may provide insight into emerging crises and guidance on how to mitigate other crises. Network analytics have been proven successful at analyzing such data. However, such talk is increasingly carried out in social media at dramatically higher volumes than previously analyzed. A high-dimensional network approach for assessing this talk and identifying not just what is being talked about, but the locality and change in that talk and the associated groups, as well as their structure, is presented. This approach is applied to data captured with respect to the Arab Spring. The results provide insight into the co-evolution of topics and groups across the region during this period of dramatic social change.

10.1 Introduction

The wave of revolutions in the Arab world, commonly referred to as the Arab Spring, was a period of major social change. As protests and demonstrations broke out in country after country, questions arose as to what mechanisms supported the diffusion of ideas and actions, promoting or inhibiting violence, and thus enabling successful regime change. New communication technologies and social media were touted as critical to these revolutions. The belief in the power of the Internet was such that in some cases embattled leaders turned off access, e.g. Egypt and Syria [1]. In all cases, as these countries moved from a pre-revolutionary state to a revolutionary state the "talk" changed. Where Wikileaks and sports were topics of interest prior to the onset of the protests, discussion moved towards issues such as liberation, government overthrow, and insurgency once the revolution began. At the same time, groups formed and disbanded, and alliances among diverse actors altered the way they went about their activities.

Throughout the Arab Spring, discussion of the transition and issues potentially related to the transition, such as economic conditions, injustices, and civil rights were discussed

Big Data over Networks, ed. Shuguang Cui, Alfred O. Hero III, Zhi-Quan Luo, and José M. F. Moura. Published by Cambridge University Press.

in the traditional and social media. Various actors, purportedly, used these media to engage discussions to foment or counter rebellion. These media contain information about the set of actors, the set of topics, and the connections among actors and topics. Herein we use the term "topic" to refer to a general idea or issue around which a set of diverse words and sentiments coalesce. Sometimes, such a topic can be meaningfully characterized by a single word – such as Wikileaks; however, in general a longer phrase will be needed to characterize the topic such as "the overthrow of Mubarak and his resignation from power".

A geo-temporal assessment of this information, actors, and topics, should provide insight into the ways in which actors and topics coalesce and disperse during periods of social change. Our key concern is to understand the geo-temporal distribution of topics and groups, and the extent to which these are global or state specific, temporally invariant, or transient. Social media data from the Arab Spring, specifically Twitter data, provide a corpus of interest ideal for studying the geo-temporal dynamics of social and topic networks.

Social network analysis (SNA) supports the understanding of groups using graph theoretic and statistical approaches for assessing the connections among actors. SNA has historically been used to understand how the structure of society, the patterns of connection among actors, influences behavior. The traditional social network analytic approach, however, is limited vis-à-vis its utility for understanding massive social change, particularly when the data source is media based. There are several critical limitations: (1) many of the metrics do not scale well to massive data such as those based on shortest path calculations; (2) social media sites often alter the network structure of the data, e.g. Twitter does not provide the true retweet network but rather connects all retweets only to the original tweet; (3) geo-temporal factors are not accounted for or easily assessed; and (4) typical approaches use only one type of network such as the actor-to-actor network rather than the high-dimensional network data available.

In contrast, dynamic network analysis (DNA) overcomes these limitations [2]. Herein, a DNA assessment of actors and topic networks through the Arab world over the course of the Arab Spring is conducted by using Twitter data. Using a high-dimensional network representation, referred to as the meta-network, complex systems can be represented. We employ this representation to look at two specific questions. First, we study a meta-network of actors, topics, and the sub-networks of actor–actor, topic–topic, and actor–topic in a geo-temporal context. We then consider only the actor–actor network and study the evolution of groups within this network over time. We employ a combination of methods based on techniques from machine learning and statistical network analytics to understand results.

The networks of interest are derived from Twitter data collected for multiple countries over the course of the Arab Spring. Using this corpus, actors (the users) and topics (the critical concepts/hashtags discussed), and the networks connecting these, are extracted per tweet. Temporal information and, if possible, geospatial information are also captured. The result is a set of large high-dimensional geo-temporal networks. These network data are "big" because of their high dimensionality, and the large number of time periods.

10.2 Arab Spring

Beginning in December 2010, a large number of protests, riots, and demonstrations began to occur in country after country in the Middle East. These events are generally referred to as "the Arab Spring". In some cases, e.g. Libya, these protests turned into an insurgency and civil war. In many cases, e.g. Egypt, the incumbent leader of the country was overthrown. When the Arab Spring ended, or whether it has ended, is a point of contention, although many are now arguing that the desired reforms have not occurred.

One of the key elements of the Arab Spring is that it occurred in a region fraught with conflict, revolution [3], and change [4], where the political dialogue since at least the 1920s has been one of identity [5]. Prior to the onset of the revolutionary protests, there was a rise in the number of young educated people with low job prospects, increased urbanization, changing economic basis, and changes in the presence of, and the integration of, terror groups into the local communities. Numerous topics were emerging as points of dissension; some, such as polygamy, were associated with Sharia law, while others, such as soccer, were associated with general past-times.

Social media played a critical role in the Arab Spring [6–8]; debates about freedom, civil liberty, and democracy raged. While not everyone in these countries used Twitter, it nevertheless is thought to provide a good window into the digital conversation. However, the data need to be used with caution as the users are both within and outside the affected countries and the dialogue is carried in English and in Arabic [9] with some, albeit limited, overlap. Moreover, Twitter appears to be used differently by, and different memes appear to be preferred by, protesters at the site and by remote observers [10].

The Arab Spring and the Twitter usage associated with this event present an ideal opportunity for studying at scale the geo-temporal distribution of, and co-evolution of, topics and groups from a network perspective. These prior studies suggest that some topics will be more local and others global, and that there may be greater locality in the topics expressed in Arabic. These prior studies also suggest that the topics will change over the course of the events. We further ask to what extent these changes in topics are geographically local, i.e. specific to a single or small set of countries rather than broadly occurring across the entire Middle East, and/or temporally local, i.e. specific to a single point in time or small window of time rather than occurring continuously, or at least with high frequency through time? Topics that are broader geographically or temporally represent topics of more general concern, whereas those that are very local in time or space may be of less importance.

10.3 General background

The number of studies of Twitter data has exploded in recent years. Key reasons include the relative ease of collection, the fact that the data are held under creative common license, and the interest in large-scale networks. These studies demonstrate that such data can provide early indications of change. Twitter ties are generally predicted by being within the same metropolitan region, being nearby, sharing a common border, sharing the same language, and the frequency of airline flights between the sites [11]. While strong social ties thus exist on Twitter and can be algorithmically uncovered with reasonable accuracy [12], Twitter networks are not necessarily reflective of actual social networks [13].

Despite the breadth of study, the movement of ideas and groups as reflected in Twitter is still poorly understood. To understand groups, retweets, mentions, and reply networks are often extracted from the metadata and then assessed. In many cases these networks are fairly sparse. Understanding the movement of ideas is less straightforward. Media studies, whether using traditional or social media, often turn to sentiment analysis to interpret the flow of information. Recent studies have shown that it is possible to use the information garnered from social media to make "predictions" about future human activity, albeit retrospectively. For example, Leetaru [14], using simple sentiment (positive/negative) and geo-location, was able to show that the level of sentiment expressed in traditional media in Egypt went to an all-time low (considering the past 30 years) prior to Mubarack's resignation. More detailed content analytics that look at the key concepts have also been used to provide general predictions of revolution and violence using diffusion modeling [15]. Further, the comparison of content on Twitter and in the news was used to determine within a few hours that the Benghazi consulate attack was not a ground-up protest [16].

From the perspective of this study, the point here is that most studies of media focus only on identifying sentiment or identifying the most frequently used concepts. In contrast, our concern is with topics where a topic, can be thought of as a general idea or issue around which a number of diverse words and sentiments might coalesce. Text-mining algorithms are generally used to extract topics from texts – the most popular examples are latent Dirichlet allocation [17] (LDA) and latent semantic analysis [18] (LSA). Such algorithms generate a set of "latent" topics for a given text corpora, where each word is associated probabilistically with each topic. Traditionally, the application of these algorithms to very small documents – such as a single tweet – is problematic. To overcome this we combine tweets into documents as will be described in the methodology section.

10.4 Data

For the purposes of the present work, we have extracted information on the time, textual content, geo-coordinates, and social interactions (retweets, mentions, and replies) enclosed within a corpora of tweets related to the Arab Spring. The data collected consists of approximately 95 million tweets gathered from two sources from April 2009 to November 2013. The first source was collected by tracking a manually curated set of keywords, users, and geo-boxes related to the Arab Spring using the Twitter Streaming API, which returns a maximum of around 1% of the full set of tweets at any given time.[1]

[1] For more details on the collection of part of these data, we refer the reader to [19]. For details on the Twitter Streaming API, see [20]. These data were collected by our Minerva research team.

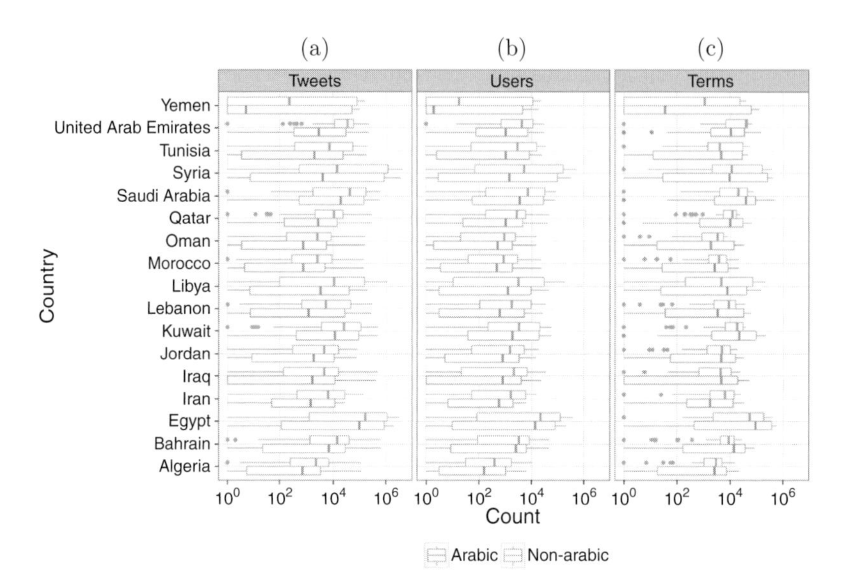

Figure 10.1 Mean and standard deviations of tweets, users and terms from different countries.

Parameters used to search the Streaming API focused mostly on events surrounding Egypt, Libya, Syria, Tunisia, and Yemen, though certain parameters did apply to the entire region associated with the Arab Spring. The second way in which data about tweets were obtained was from an outside researcher who provided us with geo-tagged tweets from a 10% sample of the full set of tweets during this same time period.[2] Information about geo-tagged tweets was obtained only for the set of countries studied by our prior work for the Arab Spring [15].

A high-level statistical overview of these data by country is shown in Figure 10.1, including information on tweets (which includes retweets), users (which are those who tweeted), and terms (which are the isolated sets of three or more characters including hashtags in the textual content of the tweet). The data include information about both Arabic and non-Arabic tweets. In many countries, the statistical profile of the Arabic and non-Arabic tweets are similar.

Figure 10.1 presents box plots for three statistics for each country. In Figure 10.1 and in the analyses below, a tweet was considered to be associated with a country if it (1) was geo-tagged and sent from within that nation's borders; (2) contained the name of the country in English or in Arabic; or (3) contained the name of any of the three largest cities within that country in English. We included the three largest cities in determining the countries associated with a given tweet after noting that discussions around certain important events, such as the Tahrir Square protests, mentioned only a city (Cairo) as opposed to the encompassing nation. Note that, via this methodology, a tweet could be associated with more than one country and thus no straightforward

[2] These tweets were collected through the Language Technology Institute at Carnegie Mellon University under the direction of Brendan O'Connor under an agreement that allowed all CMU researchers to make use of these data.

statistical comparisons can be made across countries comparing the values in Figure 10.1. Also note that all plots in Figure 10.1 are log-scaled.

Figure 10.1a presents of the number of tweets for each country studied for both the set of all tweets and the set of tweets that contained Arabic terms. We see that, on average, countries saw in the thousands to tens of thousands of tweets per month. However, our dataset contains months where Egypt, Syria, and Libya each saw several hundred thousand tweets in a single month. We also see evidence that tweets containing Arabic accounted for a non-negligible portion, and in several cases the majority, of all tweets within a particular country.

Figure 10.1b shows the distribution of the number of unique users for each country for both all tweets and Arabic tweets, where a user is included in the Arabic count if that user ever used an Arabic term. Again, we see that the number of users who tweeted in Arabic in our dataset approached the number of non-Arabic speakers in each nation. This result furthers the point that the Arabic-speaking population played a prominent role in the discussion of the Arab Spring on Twitter, and thus, as implied in [20], that English-only analyses of the events that transpired may be a biased representation of the discussions occurring online.

Figure 10.1c displays the number of non-Arabic and the number of Arabic *terms* in the data. A term was any string in any tweet with greater than three characters. A general stop-word list was applied to remove common terms from the topic list, and tokenization was performed using the widely accepted (e.g. [21]) tokenizer from [22]. The tokenization of Arabic-script and Latin-script texts are unique in a variety of ways, so much so that tokenizers specifically for the Arabic language have been developed [23]. We use the approach described by Taghva *et al.*, as implemented in the Python NLTK library [24]. An Arabic term is a string written in Arabic script. A non-Arabic term is a string containing only Latin letters and numerals. Thus, there may be Arabic words in the non-Arabic terms, they are just written in Latin letters – such as "Insha'Allah." Figure 10.1c shows that, in some cases, the number of Arabic terms was greater than the number of English terms used. There are several reasons for this. First, there may have been more Arabic-speaking/writing users who chose to tweet in Arabic script. Second, this may be in part due to the fact that tokenization of Arabic terminology is unique from English and thus may provide slightly inflated values due to differences in the way in which spaces are used within Arabic script (in English we might use a hyphen instead of a space where in Arabic script a space is used). Nonetheless, this finding serves as validation that both Arabic and non-Arabic tweets and discussions were prominent through the events of the Middle Spring. We also find that most of the tweets are not just showing "allegiance" or interest by inserting an Arabic term into a predominantly non-Arabic tweet; such as when adding the term Insha'Allah in either a Latin or Arabic alphabet in an otherwise English tweet. Moreover, the additional finding that few of the tweets in Arabic script mentioned the non-Arabic script tweets, and that few of the same user-handles tweeted in both non-Arabic and Arabic script, serves in conjunction with the distribution of terms by script to suggest that there were two distinct on-going discussions. It may well be that much of the non-Arabic script discussion was either intended to inform those outside the region, or by

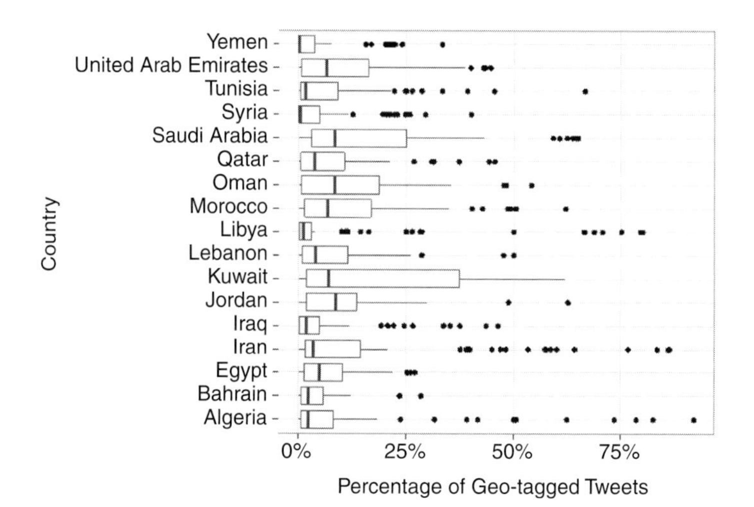

Figure 10.2 Number of geo-tagged tweets in each country.

those outside the region serving as on-lookers. Later work will need to explore these possibilities.

In addition to considering these statistics, it is also interesting to examine the proportion of tweets about each country that are geo-tagged. This is particularly important in the present work, as our geo-temporal analysis of topics is run only on geo-tagged tweets. Figure 10.2 shows box plots of the percentage of tweets that were geo-tagged for each country in each month. While there are several outliers, these tended to be months where data were sparse and thus percentage estimates were highly variable. In sum, the figure shows that, on average, geo-tagged tweets were somewhere between 5% and 15% of the tweets in any given month across all countries in the corpus used.

While our dataset represents a large portion of tweets related to the Arab Spring, it is important to note that this overview provides only a description of our dataset and thus should not be taken as a definitive overview of what the full collection of tweets relating to the Arab Spring looked like. As noted in [25], such considerations are important in cases where samples of tweets are already biased by the search criterion used. In particular, while we believe that results on our dataset may in many cases generalize to the overall sentiment that surrounded the Arab Spring, our emphasis on geo-tagged data from Arabic countries may suggest our results over-represent the general level of discussion of the Arab Spring that occurred globally in Arabic over the allotted time period.

10.5 The social pulse: geo-temporal trends in Twitter topics and users

10.5.1 Methodology

To garner a better understanding of the interrelationships between actors, the topics they discussed, and spatial location, we utilize LDA. In order to do this, we first aggregate all the tweet text by user and treat this aggregated text as a single document, an approach has been adopted by several recent works on Twitter [26–28]. This overcomes the standard problem of using LDA on short texts. Moreover, it makes the "texts" – the tweets of one user – comparable in size to news articles. Given a set of users and the terms associated with them, LDA will extract a number of "latent" topics based on a Bayesian probabilistic model, which assumes that each user discusses a subset of all possible topics. In the model, each latent topic is described by some subset of all terms that tend to be used frequently by the same user. Users are then evaluated using the mined topics, giving an indication of the relevance between each topics and the given user. Note that the number of topics in LDA is specified by the researcher. In the present work, we estimate the model with 100 topics, noting that larger numbers of topics tend to fare better in recovering important latent information [29]. LDA produces two standard outputs – terms by topics, and texts by topics. Each "document" in our topic model is the aggregation of all tweets from a given user, an approach that has been shown to improve the reliability of results [28]. Since we created one "text" per tweeter by combining the tweets by that user into a single document, a result of our LDA analysis is that the texts by topic is a user by topic matrix.

While LDA allows us to associate users with topics, we are also interested in two additional pieces of information, both of which can be inferred by using the posterior distribution given by the LDA model derived by applying LDA to the user-level documents formed of all tweets by that user and not to the individual tweets. First, we are interested in associating particular tweets (and not just users) with each topic. To do this, we make the (reasonable) assumption that each tweet is concerned with only one topic. Given this assumption, we determine the topic of a given tweet by selecting the topic that the terms in the tweet are best associated with. Second, we are interested in linking users by the similarity of the set of topics that they tweet about. This leads to the formation of a "co-topic" network, which is formed by comparing the topic scores between two users. Each user in the data is associated with a topic score vector. The mod score (explained below) between an arbitrary pair of users in the data set is evaluated. If that similarity is larger than a preset threshold, a link will be generated between those two users with the mod score as the tie strength. In our network, we have eliminated self-links so that no node is pointing to itself.

After running LDA on our data, we first explored how the topics clustered in different geographical locations. Here, we analyze in more detail the five top topics uncovered by the model, considering the terms that best represented these topics and the spatial distribution of tweets relevant to the topic. The top five topics were determined by selecting the five topics that had the highest likelihood of occurring across all users. Experimental results showed that topics present high locality and differed significantly from country to country.

10.5.2 Topic overview

Figure 10.3 presents information on the locality of five topics. For each topic we present the five most representative terms for each of the top five topics uncovered in our data. These representative terms are those that ranked high on that topic given the LDA

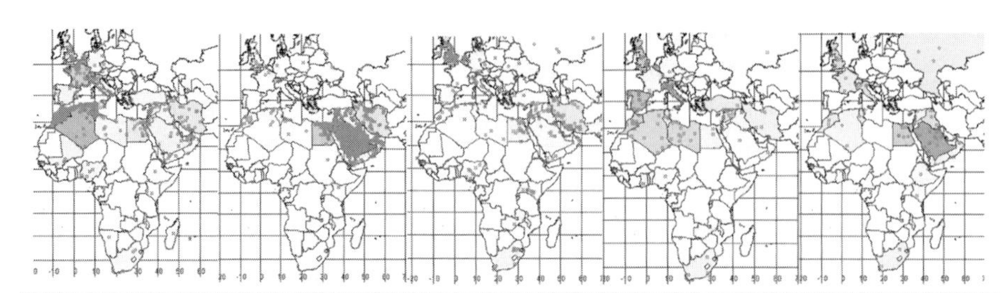

Topic 40 (English)	Topic 46 (English)	Topic 91 (Arabic)	Topic 92 (Arabic)	Topic 98 (English)
social	American	Arab	good	follow
happening	east	Spring	peace	sporting
african	Beirut	Gulf	roses	mosh
families	Jordan	Ali	special	coast
meridien	middle	tyrant	possible	liam
Morocco Algeria Tunisia	Saudi Arabia Egypt	Iran Jordan	Algeria Libya Morocco Syria	Saudi Arabia Egypt

Figure 10.3 Geo-visualizations of top five topics, language, and key associated terms.

scores, and that were interpretable words (note in Twitter there is a tendency to use a high proportion of non-real words or words with exaggerated letters such as soooooooooo) or in the case of Arabic script were translatable terms. Additional details follow. For each topic, the terms that had the highest LDA score for that topic were either entirely in a Latin script – and appeared to be English words – or were entirely in Arabic script; thus we give the "language" of the topic next to its title. For non-Arabic terms, this meant the five terms that had the highest likelihood in the posterior distribution of the topic. For Arabic terms (only the English translation is provided for clarity) we show the five terms with the highest likelihood in the posterior that we could also satisfactorily translate using Google Translate. Figure 10.3 shows that, at least amongst the top five terms for these topics, the foci of discussion were unique. For example, topic 46 talks about Beirut (the capital of Lebanon), Jordan, and America, which were all nations only tangentially involved in the events of the Arab Spring; and topic 98 is focused on sports and possible sports medicine. In contrast, Topic 91 consists of Arabic words directly related to the Arab Spring region and includes the name Ali and the term tyrant, most likely referring to former Tunisian President Zine El Abidine Ben Ali, ousted during the Arab Spring. Quite interestingly, in contrast to the negative sentiment in topic 91, we instead find a set of positive words such as good, peace, rose, and possible characterizing in Topic 92. Thus, our results suggest that notions of peace and tyranny tended to come from distinct segments of the Twittersphere, a claim that would be interesting to substantiate further in future work.

Figure 10.3 also presents a geo-visualization of all the tweets related to each specific topic. Here we see that certain topics present strong localities: topic 40, which talks about African families, is generally concentrated in Morocco, Algeria, Tunisia, and south west Europe. Topic 46, which talks about American activity in Beirut and Jordan

Table 10.1 Temporal dynamics of topics

Time period	10/2010 1/2011	2/2011 2/2011	3/2011 4/2011	5/2011 9/2012	10/2012 11/2013
Topic	94 (Arabic)	74 (Arabic)	41 (Arabic)	46 (English)	62 (Arabic)
Term	people	UaC	Arabs	american	Egypt
Term	god	Elly	people	east	Morsi
Term	life	Quaoui	country	information	Head
Term	solutions	Pak	beloved	Beirut	people
Term	even	OiYai	what is happening	Jordan	President

in particular, and the Middle East more generally, is concentrated in Saudi Arabia and Egypt. The rest of the topics have a greater span across the entirety of the Arab world. The examination of these plots suggests that, in accord with what we would expect, general concepts like “peace” and “tyranny” spread throughout the Arab world, while local topics (e.g. those mentioning a specific location) tended to stay within the confines of certain spatial regions.

10.5.3 Over time analysis

Apart from the geo-spatial distributions of topics, the temporal distribution is also important. In this analysis, we aggregated the topic scores of each tweet assigned by the LDA algorithm and picked up only the topic with highest aggregated score over the whole dataset, and generated a global top topic. Table 10.1 shows the top topic and associated terms in our dataset calculated by month. Note that the top topic moved from a cry looking for solutions to wonderment over the revolution, to more specific discussions of key issues – the role of the Americans and the concern with Morsi (who was elected, took office on June 30, 2012 and then was removed from office on July 3, 2013). It is not clear what topic 74 refers to, although one possibility is that it the associated tweets may contain excerpts from a song.

We can see that, over time, the topics changes from 2010 to 2013, generally in a way that is related to the political movement in the areas where the tweets are being sent out. For example, in 2010, the most prevalent topic is prayers for solutions. This corresponds to the beginning of the Arab Spring movement, which spreads over the whole Arab world and involves a revolutionary wave of demonstrations and protests. This topic is not localized. At the beginning of 2011, topic 41 came to the fore – and within a cry asking what was happening to their countries. This is followed by a year-long debate on the role of Americans in the Middle East – see the associated keywords directly related to key locations such as Beirut and Jordan of the Arab Spring. Interestingly, this is the only time in which an English topic dominated the discussion At the end of 2012, tweet topics moved to political events related to the reign and overthrow of the fifth president of Egypt, Morsi. This trend lasts until the end of the data set.

Several important themes underlie the analysis over time. First, the dominant discussion topic moved over time from general topics – to specific topics. Second, the dominant discussion topics moved from being geographically broad to geographically narrow. Third, the dominant discussion topics moved from being apolitical to political. Thus the topics moved from those to which there was general universal accord and that while specific to the Arab Spring were geographically relevant to the entire region to those that were politically charged and most relevant to a small set of countries. This indicates the qualitatively assumed but, as to our knowledge not quantitatively shown, promise that the Arab Spring region moved over the course of the past few years to a more political focus that surrounded the events of the Arab Spring. Second, we note that the dominant form of discussion was in English for only the period around the most intense actions surrounding the Arab Spring. Naturally, this suggests that the English-speaking world was interested during the height of the conflict but rapidly moved to other topics, while the Arabic-speaking world was (and still may be) predominantly focused on the political events transpiring in the region.

10.5.4 Characterization of user–topic similarity network

Next we consider the relations of users to the geo-temporal distribution of topics. This required constructing networks of users based on whether or not they tweeted on a topic. These networks are based on the user by topic network, where the topics were those previously identified and the links were the number of tweets by that user associated with that topic. Although most tweets tend to be associated with a single topic, most users are associated with most topics. However, many of the associations between users and topics are quite weak, indicating that the user is tweeting using some terms associated with the topic but may not be really interested in the topic. This is, in effect, a limitation of LDA.

The resultant user-by-topic network, which is very dense, is then used to define links between users based on shared topics or similarity in topic usage. In creating a similarity score we needed both to resolve the big data issue and to find the meaningful similarity issue. Since most users are associated with most topics, the simple number of topic shared tends to result in a very dense and non-discriminating network that under-represents the focal interest of the user. Indeed, most approaches to generating a user-to-user network based on tweeting about the same topics generates networks that are problematic for two reasons. First, the resultant matrices are too dense for most network algorithms to run metrics on in a reasonable amount of time, particularly for big data. Second, the algorithms tend to inflate the similarity between users based using the same terms and not based on the extent to which the users are actually associated with that topic. The result is that all users tend to be related and there is little room to discriminate between weak and strong similarity. What is desired is an algorithm that puts a link between two users if those users are strongly interested in a topic and strongly interested in the same topics. By focusing on both strong association with a topic and strong similarity in what topics they are associated with, the resultant similarity network should be both analyzable and meaningful. Therefore, instead of shared topic counts we use a similarity index that weights the topics by focus. This helps ensure both strong association of the user with a topic and strong usage of the same topics by the two users.

Table 10.2 Network statistics of the user–topic similarity network by country in the original data set

	All tweets			Non-Arabic tweets		
Country	Non-isolates	Edges	Density No isolates	Non-isolates	Edges	Density No isolates
Bahrain	4559	206 642	0.012	8698	149 612	0.004
Qatar	6948	378 262	0.016	10 721	230 981	0.004
Iraq	1852	42 257	0.024	3295	23 008	0.004
Iran	975	6304	0.013	1344	4998	0.006
Libya	4394	110 910	0.011	5259	88 827	0.006
Algeria	780	5913	0.019	955	5134	0.011
Egypt	42 060	9 490 034	0.011	62 653	7 964 548	0.004
Kuwait	19 713	6 087 116	0.031	45 955	4 273 476	0.004
Lebanon	6687	226 560	0.010	7722	171 573	0.006
Morocco	5612	258 507	0.016	6689	157 733	0.007
Jordan	3711	79 486	0.012	4887	61 438	0.005
Saudi Arabia	33 663	35 921 282	0.063	136 543	46 843 301	0.005
Oman	2193	45 820	0.019	4491	71 297	0.007
Syria	40 625	8 603 652	0.010	53 350	7 042 616	0.005
Yemen	1109	84 280	0.137	6000	131 767	0.007
United Arab Emirates	24 417	3 542 578	0.012	33 448	3 155 592	0.006
Tunisia	3692	63 728	0.009	4253	49 105	0.005

It is important to note that cosine similarity is the generally accepted solution to this problem; however, it is too slow for the size of our data and this makes it unrealistic for generating the user–topic similarity network in an acceptable running time. We used an alternative more-efficient method to calculate the similarity between the topic usage vectors based on mod scores. This helped to resolve the big data issue. Also, the mod scores are interpretable, generating meaningful links between the users. To create these scores we first define a vector as the real-valued score for that user on all topics. Given two vectors $v_1 \in N^K$ and $v_2 \in N^K$ that are both real-valued vectors in the k-dimensional space, the *mod score* between those two vectors is defined as

$$Sim_{i,j} = \frac{\min(|v_1|, |v_2|)}{\max(|v_1|, |v_2|)}.$$

Since both $|v_1|$ and $|v_2|$ can be calculated in advance before the generation of the network, the magnitude similarity can be calculated fairly efficiently. We then define a link between two users to be 1 if the magnitude of the similarity of the two users is larger than or equal to 0.99, else 0. As a result, we obtain an undirected, binary user–topic similarity. This network can be interpreted as showing those users who have a strong focus on the same topics. For each country, we generate a separate user–topic similarity based on all the tweets associated with that country.

For each country, for each use–topic network, standard network level statistics are calculated – see Table 10.2. First, we consider the number of non-isolated nodes in the network, which is the number of unique users that that have strong topical similarity to at least one other user. Note that this is only a very small subset of the data set since the high threshold filtered out the majority of the users. Second, we consider the number of edges in the user–topic similarity network, which is the number of dyads that have strongly similar topic foci. Finally, we look at the density of the user–topic similarity network after the isolates (those users who were not strongly tied to any other user) are removed. This provides insight into the overall structure of connectivity among the users. For contrast we also show these same statistics for just those tweets that are non-Arabic.

We can see that the majority of the countries have a network density of roughly 0.01, which indicates that only 1 out of 100 users in this strong similarity network share similar topical distributions, and that, on average, each user has a high degree of topic similarity with about 1% of the other users in the network. Some countries, such as Tunisia, have significantly lower densities, indicating the potential for a less homogeneity of topics in the discussion in these areas. Of all countries, Yemen has the highest network density, which indicates that more of the users in that country tend to discuss similar topics on Twitter. The number of nodes in the networks indicated that there are far more active Twitter users talking about dominant topics in Egypt, Saudi Arabia, Syria, and UAE than other countries. Among these countries, Saudi Arabia has an especially high number of links in the network because of the number of Twitter users talking about the same topics.

On the right in Table 10.2, the network statistics of for the non-Arabic user–topic similarity network is shown. More users have a high topic similarity with at least one other user in this non-Arabic discussion network; however, these users are on average connected to fewer other users. That is, in general, the density of the non-Arabic user–topic similarity networks is lower than the overall user–topic similarity network. This is because people tweeting not in Arabic tend to focus on a wider variety of different topics. Most countries that have high node count in the overall data set also have a high node count in the non-Arabic only data set, which is not surprising since those countries have a large number of Twitter users in general.

The differences in densities for the overall data and just the non-Arabic data have some interesting implications. Consider Bahrain. Overall, the density implies that among those users who are strongly tied to at least one other, the average user is strongly tied to about 55 others, but in just the non-Arabic realm to only about 35 others. Whereas, in Saudi Arabia, the values are 2121 users overall and 683 in the non-Arabic realm. This also implies that in the Arabic tweeting part of this network there is substantially more homogeneity in shared topics and more of the Arabic tweeting actors have higher similarity to each other in their topical focus. This could indicate some transference of topics between Arabic and non-Arabic speakers. However, part of the difference is due to the fact that the 100 topics, when assessed overall, are much broader and less discriminatory then the 100 topics for just the non-Arabic tweet content.

10.5.5 Social interaction overview: the reply network

We now consider the social relations among the actors in our dataset. While social relationships in Twitter data require a degree of nuance in interpretation due to the

Table 10.3 Network statistics of the reply network by country in the original data set

Country	Mean nodes	Mean edges	Mean density	Mean clustering coefficient	Mean characteristic path length
Bahrain	804.212	690.115	0.010	0.017	5.749
Qatar	1120.818	959.891	0.010	0.008	5.173
Iraq	306.852	214.111	0.034	0.001	2.305
Iran	253.945	204.182	0.029	0.003	2.924
Libya	1114.686	1158.686	0.053	0.002	2.719
Algeria	132.714	100.653	0.020	0.006	2.015
Egypt	6120.964	6712.636	0.008	0.004	5.209
Kuwait	4745.909	4824.418	0.013	0.014	5.577
Lebanon	693.731	568.923	0.019	0.005	3.982
Morocco	380.472	324.679	0.034	0.003	2.927
Jordan	456.824	388.569	0.016	0.010	3.553
Saudi Arabia	9025.873	9460.709	0.004	0.007	5.305
Oman	316.462	270.346	0.030	0.009	3.348
Syria	3712.906	3717.962	0.016	0.003	3.246
Yemen	502.628	444.581	0.049	0.00	2.493
United Arab Emirates	3785.018	3677.291	0.002	0.014	5.888
Tunisia	395.224	322.061	0.008	0.005	3.531

technological affordances of the media [25, 30], if one is careful insights can nonetheless be gained. In general, most network analytics focus on either the retweet, mentions, or reply network. We focus here on the reply network. The reply network can identify whether the sender hits reply when sending the tweet.

This network changes dramatically over time as new users join Twitter, and as users move between topics and so between groups. In Table 10.3 summary statistics describing the reply network are shown. As there is substantial variation by month, the results shown are the averages across the months. In other words, the reply network was constructed for each month for each country and then the months were averaged by country. By examining this information we see that the sheer volume of users replying to others tweets, and the density of the tweet network, does not correlate with revolutionary activity. There is high country variability. For example, Iran shows a small dense community with very fast information flows (low characteristic path length). It is possible this network is dominated by expatriates. Whereas for Saudi Arabia there is a larger, sparse, community with more distinct clusters with users often needing 5 to 6 steps to move information.

10.5.6 Characterization of group structure

The reply network is not a uniform or random network of connections. Rather, it has a very sparse multi-component structure that changes over time. Figure 10.4 displays various metrics calculated on the reply networks over time, where each point on the line

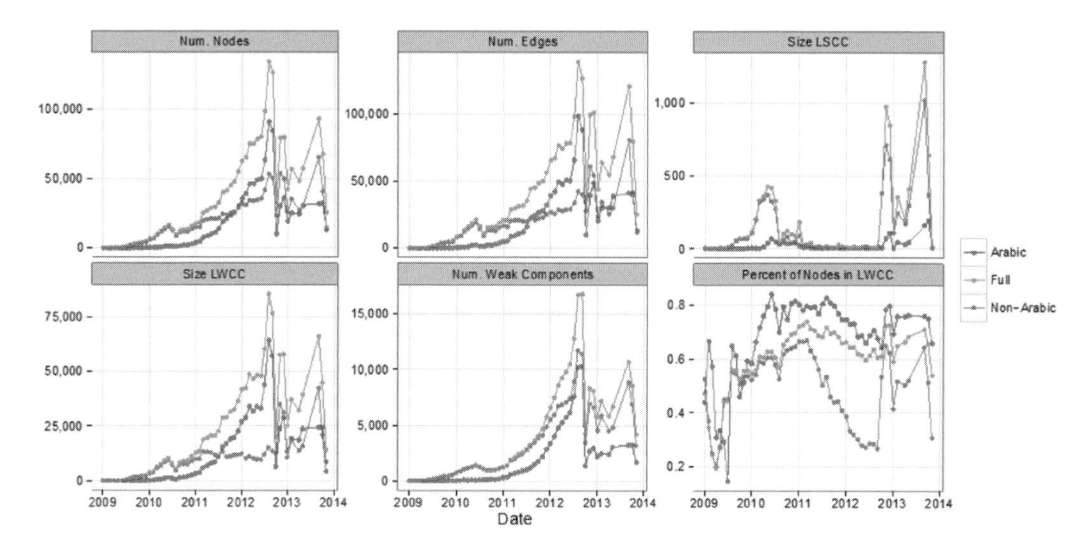

Figure 10.4 Temporal change in the reply network.

represents the network for a given month. In each subplot, there are three lines – a line that represents results for tweets containing one or more Arabic terms, a line for tweets that did not contain any Arabic, and a line for the full dataset. The top row of plots, from the left, displays the number of nodes in the network, the number of edges, and the size of the largest strongly connected component (LSCC), defined as the number of nodes in the largest portion of the directed reply graph where each node is reachable (via following directed lines) from each other node in that portion. The bottom row displays (from the left) the size of the largest *weakly* connected component (LWCC), which considers connectivity assuming that the reply network is an *un*directed network, the number of weak components, and the percentage of actors in the LWCC. The differences between the strong (LSCC) and weak (LWCC) versions is link directionality. Imagine this network A replies to B and B replies to C and C replies to D and F replies to G and D. In the LSCC F and G would not be part of it as you cannot get to F, G from A, B, C, or D. In the LWCC, all six nodes would be included. LSCCs will be smaller than LWCCs. In LWCCs and not LSCCs nodes may be connected because there is one node that connects two disparate groups. Whereas, in LSCCs, all nodes are connected by a single conversation. We choose to consider the LWCCs in the present work, as Twitter replies are often in response to an initial comment towards an individual. Thus, if actor A mentions actor B, and actor B then replies to actor A, use of only the LSCC would only assume a path from B to A, while in the LWCC we relax this assumption and assume there may as well be a social relationship between A and B. The LWCC is thus an overestimate, of sorts, of the actual underlying social ties represented in the data, whereas, the LSCC is an underestimate.

From Figure 10.4, several points of interest can be ascertained. First, as we would expect, as the number of nodes increases the size of the LWCC and LSCC, as well as the number of weak components, steadily increases until July 2012. The peak of

June–July 2012 was a period of great unrest. In June 2012, the former Egyptian president, Hosni Muburak, was sentenced to life imprisonment. Then, in July 2012, the Syrian rebels seized the town of Aleppo. Intense fighting in Aleppo and Damascus at that time provided the backdrop for another attempt in the UN to create a UN Security Resolution, which was vetoed by Russia and China. The second upsurge occurs about a year later with the military coup in which the fifth Egyptian president, Mohamed Morsi, was deposed.

Interestingly, however, the percentage of nodes in the LWCC is highest in early 2011 (full), when things were just starting to flare up in Egypt and Libya, and immediately after Twitter became re-available in February 2011 in Egypt. This suggests that actors may have been more invested in obtaining new information from Twitter during this time as opposed to from traditional media. This is not just due to the "shutting off" and "on" of the internet as: (a) most of our data are not from Egypt, and (b) later suppressions in other country did not lead to similar increases. So while resumption of access to the Internet is part of the story, most of the increase is due to other factors.

Second, we observe that the size of the LSCC is much, much smaller than the size of the LWCC. Because of the one-way directionality of interaction on Twitter, this is to be somewhat expected. However, it also suggests that there may have been little reciprocity in the core of the network, where certain actors were being replied to but were not replying to others who directed communication at them. This may simply be a result of the way in which the reply network is constructed (recall that a reply implies a response and we do not have information on who was mentioned in the initial tweet), but it also suggests that the use of gatekeeping [8] on Twitter strongly structures the resulting network.

Finally, and perhaps most interestingly, we see that while once the Arabic component of the reply network reached a high proportion of nodes in the LWCC this proportion stayed above approximately 60%, the number of nodes in the LWCC of the all-non-Arabic script tweets dipped significantly through 2012–2013. This observation matches the intuition proposed in the sections above that Arabic users were much more invested in the events of the Arab Spring throughout the past four years, while English-speaking (or at least non-Arabic speaking) users tended to be drawn in to the discussion only at dramatic turns in the events. This intuition can be further qualified by suggesting that not only were Arabic speakers more focused on the discussion, they also were more engaged in networked discussions with each other. This point is strengthened by noting that, in the user–topic similarity network, the non-Arabic tweeters show less strong connectivity to each other than do the users overall and those tweeting in Arabic.

Of course, just being together in the same component does not necessarily mean that these actors were interacting with each other. Indeed, within a single network component, there are still groups of actors who discuss particular topics and interact almost exclusively by themselves. To determine the extent to which this is true, ideally one would correlate the various networks. This will be done in future research. For now, we take a faster approach to assess the extent to which there are these clumps of users and topics linked together, by taking the LWCC and running a network grouping algorithm on it. Figure 10.5 presents various metrics from the network grouping that results from taking the LWCC for each month and using the Louvain clustering method on it [31].

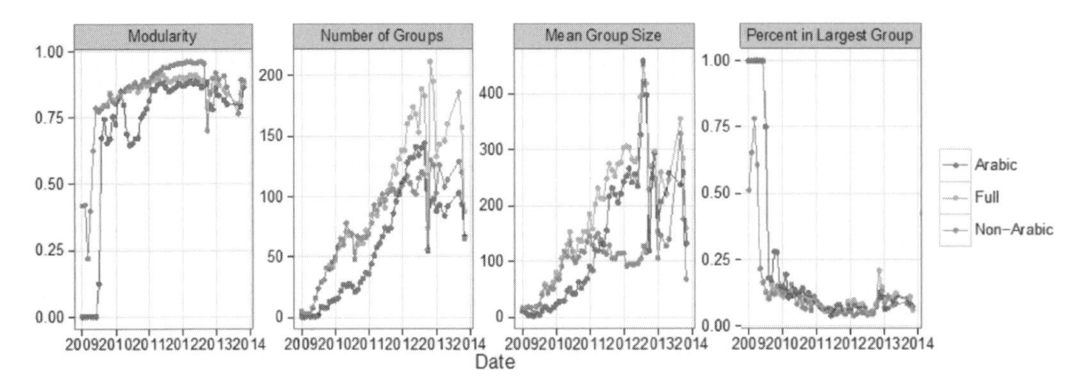

Figure 10.5 Temporal change in groups in the Twitter reply network.

From left to right, the subplots depict the modularity [32] of the graph, which often is used as a measure of the quality of the clustering (i.e. the degree to which the graph is separable), the number of groups into which the clustering split the graph, the mean size of those groups, and, finally, the percentage of actors in the largest found group.

As we can see from Figure 10.5, modularity was very high across all months, thus the network was reasonably separable into groups even within the LWCC. Second, the number of groups and the average size of each group increased steadily over time, indicating that more communities were being added to the network and that existing communities were growing. However, the size of the largest connected component stayed relatively stable, suggesting that while communities increased in size and number, there never became a global social community that infiltrated across the Arab Spring. Thus, one would imagine that brokers of information across communities existed and thus that there existed a select number of individuals that may have shaped interactions across groups.

10.5.7 Key actors

The final question we address is the typical network question – who are the key actors? For the reply network, the five users with the highest total degree centrality are shown in Table 10.4. These are users who reply, or are replied to, the most. These users are distinct by country, suggesting that the country networks may have little interconnection. For this analysis we did not pre-remove news agencies, bots, or any other class of users. This is quite interesting that these users are not news agencies, suggesting that though news agencies are a dominant presence in the Twittersphere, they are not central to the subgroups trying to build consensus. Rather, the interaction is done at the grass-roots level among general users. Some of these users, however, representing the extant government; for example, in Syria one of the most central users in the replay network is SyriaParliament or are freelance journalists such as DubaiWriter. In understanding this result it is important to keep in mind that this is the reply and not the follower network.

Table 10.4 Top user in reply network

Country	1	2	3	4	5
Bahrain	sasbahrain	YasiQannati	Farahfraidoon	bucheeri	SaroooLi
Qatar	Gadgod_	septboog	HEYitsSAL	JamesBryanBG	salkhulaifi
Iraq	yarab14	erdgnhsn	GeorgetteA	cerenationnext	gabitamatos
Iran	matuzalem	Mehrdad	tiagodvaz	dsantamaria	strasboorg
Libya	MaybeLaser	bulltas	Ben_Mussa	FairuzJumain	Sniggah
Algeria	Boubled	Hakim_3i	AmanIrh	Falqallaf	HuskyDaz
Egypt	Betsy_Mo	Hazem_Azim	Monasosh	OFree_zyIV	ZOGHBYZO
Kuwait	BuBarrak	DCiawy	iL3uBiD	HassanALSherazi	m7amdalnasser
Lebanon	iJoePopSlap	myrrnzz	sam_lb	AbirGhattas	Arabear
Morocco	HakimKhadija	neishatorres	yucefj	ravfm	rajk971
Jordan	pminttt	samihtoukan	IbrahimmbI	h_alkhafaji	OmarBiltaji
Saudi Arabia	faisalmeshari	indiesaudi	MohamadAlarefe	Sara_wolf	battalalgoos
Oman	bijoyjacobk	raideraid	MusaabK	Degoat82	samiasmi
Syria	nasermaya	SyriaParliament	monsternotfan	SubhanAksa	resifahma
Yemen	bimbie07	Cyndaquillian	mussoO_	renytacitra	Bilal_ALhamzee
United Arab Emirates	wildpeeta	AskAli	ylootah	binmugahid	DubaiWriter
Tunisia	archcindymonica	iheb911	kcyam5	NaymaMC	Raahma_

This suggests, and future work should confirm this, that while tweeters follow the news agencies, they correspond with non-news agencies.

In some sense, this examination of key actors raises more questions than it addresses. Most of these users are connected to most of the topics and the influence of these users vis-à-vis the topics is unclear. Future work should consider other social interaction relations such as the retweet and mentions network, and consider the relative standing of and influence of verified, news, and government users. The relation of these users to the topics and the relevant opinion leader for each topic will be identified in the future. On another point, many of these actors had accounts that were suspended. Future work needs to examine how different the results would be with and without these suspended accounts.

10.6 Discussion

Online media, and social media in particular, generates a wealth of geo-temporal data that can be used to gain insight into the pulse of a population. Extracting meaning from these data, and using these data to answer research and policy questions can be a daunting task. Data collection, cleaning, and translation, present challenges over and above analysis and visualization. We focused on what can be learned from social media, after the collection, cleaning, and translation using a network approach. The data were segmented by country and time period, and networks connecting users and topics were developed for each country and time period. We used a meta-network strategy in which users were connected to each other in a retweet network, a replies to network, and a similarity in topic usage network. In addition we have a user-by-topic network and a co-topic (i.e. topic-by-topic) network. Many additional analyses can be done with these encoded data – such as determining the extent to which the replies network predicts the similarity network, and determining the paths by which topics change over time and how that relates to the changes in opinion leaders.

On the surface, the data analyzed are big, over 90 million tweets. The data cover a large range, 15 countries, and they cover a multi-year time span. However, there are limitations to these data. First, they are not a comprehensive account of all twitter activity in these countries during this time frame. In some countries, where there is low Twitter usage, it may be close. Second, the data that are not geo-tagged may be coming from outside of the Middle East. Thus, it is possible that the topics identified are mixing internal and external concern. These issues should be examined more in the future.

We recognize that big data increase statistical confidence, not accuracy. Accurate prediction with big data requires triangulation and the identification of patterns; thus, future work should use multiple types of data and triangulation techniques to generate predictions. For social media, a fair amount of triangulation can be afforded by doing broader international studies such as we did for the Arab Spring by comparing countries and comparing Arabic versus non-Arabic data. In general, such comparative work will need the support of data publishers like Twitter who control access to the wealth of data. An alternative form of triangulation is possible by comparing and contrasting results from multiple media. This future work should compare, at scale, distinct sources such as social media and traditional media.

The analyses that were run were all relatively scalable for big data. This means that data processing short cuts that may have impacted the results were sometimes used. For example, topic identification was done at the term level, which means that n-grams were segmented. A good example here is "middle east" which appears as separate terms "middle" and "east" in topics, rather than as a single concept. This speeded analysis, increased the number of terms, and decreased interpretability. Future work should add common n-gram detectors. Another simplification was that translation was done after analysis and at the term level. A good example here is the set of top terms such as Elly associated with topic 74. This approach speeded analysis, but decreased interpretability. Future work should consider alternative translation approaches, used Arabic script thesauri, or analysis without translation but using Arabic n-grams.

The analyses described increased in complexity and scalability. As we moved from one level to the next, the analytics took longer, but the number of results and the operational utility and research insight afforded by those results increased. The first analyses focused on counts. Counts are relatively simple and fast with big data. Unfortunately, counts provide little insight. The most one could learn here was where and when Arabic was more dominant, and changes in the sheer volume of communication. The second level of analysis focused on the clusters in the data. This was done through topic identification. This led to additional insight concerning what was being talked about and how it changed geo-temporally. The third level of analysis focused on the networks themselves. This led to additional insight about major sources of influence and geo-temporal change in that influence. As we move through these levels of analysis new insights are possible, but the scalability of the process somewhat decreases. The issue is not that the network metrics don't scale well – indeed many of them scale as N. Rather, the issue is that, as you move from level one to three, additional data processing is needed to create the right data structures. Creating the data structures and storing them is, in and of itself, at this point, a time-consuming process. Tools that facilitate network construction and the associated data cleaning, and automated workflows would streamline this process.

As we move to the future there are a number of additional challenges that must be addressed when one is interested in network analytics and big data. A general discussion of these challenges has appeared in National Research Council Report (2013) and a more detailed review with specific relevance to crisis management and social media is the subject of a forthcoming publication from the present authors. For data such as the Arab Spring Twitter data, some of the major challenges we have encountered include:

- understanding how the data-collection filters bias the results;
- inferring location for non-geo-tagged data;
- improving the scalability of statistical network tools, such as MRQAP for regression on networks;
- incremental and approximation techniques for path-based network metrics;
- automated techniques for network extraction.

10.7 Conclusion

Social media is increasingly becoming a major source of information for populations. However, the grass-roots nature of social media is changing. The majority of news agencies, e.g. BBC, CNN, and al-Jazeera use Twitter and Facebook to spread breaking news. Social media is also a major outlet for citizens to express their concerns. For example, in the recent Benghazi consulate attack, while the majority of tweets were from individuals, the top "tweeters" were news agencies and the Libyan Youth group. Within the tweet network, individuals and news agencies play different roles and have different geo-temporal tags. Who follows whom, the retweet network, the cyber-norms, the use of hashtags, and incorporated videos or images appear to be different for corporate, group, and individual users. Although social media is a major source of information, so too is traditional media. The information carried via social media is not completely distinct from traditional media. Moreover, the information in social media is not always more timely than that in traditional media. As more organizations and news agencies turn to the use of social media, the relative impacts of social media and traditional media on social change become more complex, as does their role in governance. We have found that news agencies are among some of the most frequent tweeters, and are often re-tweeted within these data; for example, there are approximately 10^5 tweets by BBC World in our data. Future work should consider the relative role of news agencies and other users relative to the change in topics over the course of crisis events.

Throughout the Arab Spring, social media had a presence. Twitter in particular was used both to provide and get information. In assessing data extracted for 15 countries over three years we identified key trends in topics and users. Against a backdrop of increasing Twitter usage, and country level shut downs in access, topics and groups emerge and fade. We found high levels of Arabic and non-Arabic content, but with relatively little overlap. In general the Arabic Twitter network and non-Arabic Twitter network seem to have little connection. Topics identified tend to be predominantly either Arabic or non-Arabic. We found geo-temporal trends in topics. Specifically, temporally topics moved from expressions of concern to detailed political discussions. Geographically, highly localized topics tended to be narrower such as focusing on specific leaders, whereas geographically dispersed topics tended to be more general. This suggests that, in general, the more generic a topic the broader its geographic and temporal footprint. Our results also suggest that the progress to revolution is one involving the incitement of concern and the transition to political specificity. We found that the user community and its connectivity increased over the course of the Arab Spring. Yet this community remained fairly fragmented, held together largely by local opinion leaders.

The strength of these results is due, in part, to the fact that they span a wide geo-temporal swath and are not dependent on the vagaries of specific Twitter users. The strength of the analysis is also due to the co-examination of both topics and users. The strength, however, points to a significant limitation in our ability to assess such large networks, and that is the ability to identify "topic groups", i.e. those sets of users and topics that are tightly linked such as the set of users who only talk about particular terror activity or a specific soccer game. Advances are needed to support the rapid assessment of users and topics together to determine how these communities are evolving. Even without such methodological tools, the foregoing analysis does demonstrate that considering both users and topics from a network perspective, and applying scalable network techniques, results in critical insight into social change. The combination of complex analytical techniques and high-dimensional network data provides the analyst with the tools necessary to go beyond simple trend and sentiment analysis to an improved understanding of the way in which different subgroups are interacting in the Twittersphere.

Acknowledgements

The authors would like to thank Dr. Huan Liu for comments on an earlier draft of this work, and Fred Morstatter and Brendan O'Connor for insights into the data format. This work was supported in part by the Office of Naval Research (ONR) through a MURI N00014081186 on adversarial reasoning and through MINERVA N000141310835 on State Stability. The views and conclusions contained in this document are those of the authors and should not be interpreted as representing the official policies, either expressed or implied, of the Office of Naval Research or the US government.

References

[1] "Internet 'cut off' across Syria,'" *BBC News*. [Online]. Available: http://www.bbc.co.uk/news/technology-20546302. [Accessed: 21-Apr-2014].

[2] K. M. Carley, "Dynamic network analysis," in *Dynamic Social Network Modeling and Analysis: Workshop Summary and Papers*, 2003, pp. 133–145.

[3] W. R. Louis and R. Owen, *A Revolutionary Year: the Middle East in 1958*, IB Tauris, 2002.

[4] K. Selvik and S. Stenslie, *Stability and Change in the Modern Middle East*, IB Tauris, 2011.

[5] M. N. Barnett and E. Goldberg, "Dialogues in Arab politics," *Comp. Polit. Stud.*, vol. **33**, no. 2, pp. 271–272, 2000.

[6] P. N. Howard and M. R. Parks, "Social media and political change: capacity, constraint, and consequence," *J. Commun.*, vol. **62**, no. 2, pp. 359–362, 2012.

[7] G. Lotan, E. Graeff, M. Ananny, *et al.*, "The revolutions were tweeted: information flows during the 2011 Tunisian and Egyptian revolutions," *Int. J. Commun.*, vol. **5**, pp. 1375–1405, 2011.

[8] S. Meraz and Z. Papacharissi, "Networked gatekeeping and networked framing on #Egypt," *Int. J. Press.*, vol. **18**, no. 2, pp. 138–166, April 2013.

[9] A. Bruns, T. Highfield, and J. Burgess, "The Arab Spring and social media: audiences English and Arabic Twitter users and their networks," *Am. Behav. Sci.*, vol. **57**, no. 7, pp. 871–898, 2013.

[10] K. Starbird and L. Palen, "(How) will the revolution be retweeted?: information diffusion and the 2011 Egyptian uprising," in *Proceedings of the ACM 2012 Conference on Computer Supported Cooperative Work*, 2012, pp. 7–16.

[11] Y. Takhteyev, A. Gruzd, and B. Wellman, "Geography of Twitter networks," *Soc. Netw.*, vol. **34**, no. 1, pp. 73–81, Jan. 2012.

[12] E. Gilbert, "Predicting tie strength in a new medium," in *Proceedings of the ACM 2012 conference on Computer Supported Cooperative Work*, New York, NY, USA, 2012, pp. 1047–1056.

[13] H. Kwak, C. Lee, H. Park, and S. Moon, "What is Twitter, a social network or a news media?," in *Proceedings of the 19th International Conference on World Wide Web*, New York, NY, USA, 2010, pp. 591–600.

[14] K. Leetaru, "Culturomics 2.0: Forecasting large-scale human behavior using global news media tone in time and space," *First Monday*, vol. **16**, no. 9, 2011.

[15] K. Joseph, K. M. Carley, D. Filonuk, G. P. Morgan, and J. Pfeffer, "Arab Spring: from newspaper data to forecasting," *Soc. Netw. Anal. Min.*, vol. **4**, no. 1, pp. 1–17, Dec. 2014.

[16] Kathleen M. Carley, Jürgen Pfeffer, Fred Morstatter, and Huan Liu, "Embassies burning: toward a near real time assessment of social media using geo-temporal dynamic network analytics, social network analysis and mining," in the press, 2014.

[17] D. M. Blei, A. Y. Ng, and M. I. Jordan, "Latent dirichlet allocation," *J. Mach. Learn. Res.*, vol. **3**, pp. 993–1022, Mar. 2003.

[18] S. C. Deerwester, S. T. Dumais, T. K. Landauer, G. W. Furnas, and R. A. Harshman, "Indexing by latent semantic analysis," *JASIS*, vol. **41**, no. 6, pp. 391–407, 1990.

[19] F. Morstatter, J. Pfeffer, H. Liu, and K. M. Carley, "Is the sample good enough? Comparing Data from Twitter's Streaming API with Twitter's Firehose," in *The 7th International Conference on Weblogs and Social Media (ICWSM-13), Boston, MA.* Retrieved from http://www.public. asu. edu/˜fmorstat/paperpdfs/icwsm2013. pdf, 2013.

[20] K. Joseph, P. M. Landwehr, and K. M. Carley, "Two 1%s don't make a whole: comparing simultaneous samples from Twitter's Streaming API," in *Social Computing, Behavioral-Cultural Modeling and Prediction*, W. G. Kennedy, N. Agarwal, and S. J. Yang, Eds., Springer International Publishing, 2014, pp. 75–83.
[21] A. Ritter, S. Clark, and O. Etzioni, "Named entity recognition in tweets: an experimental study," in *Proceedings of the Conference on Empirical Methods in Natural Language Processing*, 2011, pp. 1524–1534.
[22] J. Eisenstein, B. O'Connor, N. A. Smith, and E. P. Xing, "A latent variable model for geographic lexical variation," in *Proceedings of the 2010 Conference on Empirical Methods in Natural Language Processing*, Stroudsburg, PA, USA, 2010, pp. 1277–1287.
[23] K. Taghva, R. Elkoury, and J. Coombs, J. "Arabic stemming without a root dictionary," Information Science Research Institute, University of Nevada, Las Vegas, USA. 2005.
[24] S. Bird, "NLTK: the natural language toolkit," in *Proceedings of the COLING/ACL on Interactive Presentation Sessions*, Association for Computational Linguistics, 2006.
[25] Z. Tufekci, "Big questions for social media big data: representativeness, validity and other methodological pitfalls," in *ICWSM '14: Proceedings of the 8th International AAAI Conference on Weblogs and Social Media,* 2014.
[26] L. Hong and B. D. Davison, "Empirical study of topic modeling in twitter," in *Proceedings of the First Workshop on Social Media Analytics*, 2010, pp. 80–88.
[27] D. Ramage, S. Dumais, and D. Liebling, "Characterizing microblogs with topic models," in *ICWSM*, 2010.
[28] L. Hong, and B.D. Davison, "Empirical study of topic modeling in twitter," in *Proceedings of the First Workshop on Social Media Analytics*, ACM, 2010, pp. 80–88.
[29] H. M. Wallach, I. Murray, R. Salakhutdinov, and D. Mimno, "Evaluation methods for topic models," in *Proceedings of the 26th Annual International Conference on Machine Learning*, New York, NY, USA, 2009, pp. 1105–1112.
[30] D. Boyd and K. Crawford, "Critical questions for big data: Provocations for a cultural, technological, and scholarly phenomenon," *Inf. Commun. Soc.*, vol. **15**, no. 5, pp. 662–679, 2012.
[31] V. D. Blondel, J.-L. Guillaume, R. Lambiotte, and E. Lefebvre, "Fast unfolding of communities in large networks," *J. Stat. Mech. Theory Exp.*, vol. **2008**, no. 10, p. P10008, 2008.
[32] M. E. J. Newman, "Modularity and community structure in networks," *Proc. Natl. Acad. Sci.*, vol. **103**, no. 23, pp. 8577–8582, Jun. 2006.

11 Social influence analysis in the big data era: a review

Jianping Cao, Dongliang Duan, Liuqing Yang, Qingpeng Zhang,
Senzhang Wang, and Feiyue Wang

Social influence is a widely accepted phenomenon in social networks, and it has been studied by researchers from various perspectives, including social psychology, sociology, marketing, and computer science, just to name a few. During the past decade, the emergence and fast growth of social media sites (such as Facebook and Twitter) have enabled the measurement, quantitative analysis, and modeling of social influence at a large scale. Therefore, it is essential to re-evaluate these developed algorithms and models in the new era of big data. In this chapter, we review research on social influence analysis in the big data era, with a focus on the computational perspective. We first present the statistical measurements of social influence. Then, we introduce the algorithms and models to characterize the propagation of social influence. Next, we present the issues related to the optimization of the propagation of social influence. In addition, we review research on the diffusion of network influence, which is closely related to the studies of the forecasting and influencing/contagion of information. Towards the end of this chapter, we also discuss the envisioned opportunities and challenges.

11.1 Introduction

Social influence analysis is an intuitive and well-accepted phenomenon by researchers for decades [1, 2]. Since social influence plays a key role in social life and decision making, as discovered by Katz and Lazarsf in the 1950s [3], theories and models have been developed from various perspectives by researchers in many different areas, including sociology, computer science, and management science, etc. With the popularity of social network services, increasing computer science researchers are paying more attention to this field. Social influence has extensive qualitative and quantitative applications, which have been well studied in sociology and computer science. For example, public opinion leaders affect numerous fans, and their opinions are quickly spread to a large population. Since they play an essential role in information dissemination, many studies focused on the identification of those users [4–6]. Social influence analysis has also been applied to other fields, such as recommendation systems [7], information propagation in social networks [1, 8–11], link prediction [12–14], viral marketing [15–21], public health

Big Data over Networks, ed. Shuguang Cui, Alfred O. Hero III, Zhi-Quan Luo, and José M. F. Moura.
Published by Cambridge University Press.

Table 11.1 The definitions of influence in social networks

Source	Definitions	Key words
Rashotte [28]	the phenomenon which is caused by the interaction among adults, and they changed their thoughts, sentiments, manners, and behavior	user behavior, phenomenon change
Watts and Dodds [29]	an influential as an individual in the top $q\%$ of the influence distribution $p(n)$	top $q\%$ Poisson distribution
Cha *et al.* [9]	three types of influence separately based on the following actions, retweet actions, and dialog actions	actions
Bakshy *et al.* [8]	quantified the influence of a given post, then fitted a model that predicted influence using an individual's attributes and average size of past cascades	reposting
Tang *et al.* [25]	the behavioral change of a person, because of the perceived relationship with other people, organizations, and society in general	behavioral change, perceived relationships

[22, 23], expert discovery [24, 25], detection of emergent events [26], and advertising [27], just to name a few. In this chapter, we focus on the "social influence analysis" based on social networks such as Twitter, Facebook, and Weibo.

The definitions of influence in social networks are diverse. Table 11.1 lists several representative ones. (1) Rashotte [28] defined social influence according to a user's behaviors and their effects. They defined social influence as the phenomenon caused by the interactions among adults, changing thoughts, sentiments, manners, and behaviors. (2) Watts and Dodds [29] followed the former studies [6, 30] and defined an influential as being individual in the top $q\%$ of the influence distribution $p(n)$, where $p(n)$ is necessarily Poisson. (3) Using a large amount of data collected from Twitter, Cha *et al.* [9] defined three measures of influence: in-degree, retweets, and mentions, and they investigated the dynamics of user influence over topics and time based on these measures. (4) Bakshy *et al.* [8] adopted the number of repostings as an indicator of influence. They quantified the influence of a given post, and then fitted a model that predicts influence using an individual's attributes and the average size of past cascades. (5) Tang *et al.* [25] argued that social influence is the behavioral change of a person because of the perceived relationship with other people, organizations, and societies in general.

Early studies discovered a few basic phenomenon on social influence, and some interesting conclusions were drawn in sociology. However, social influence analysis largely depends on the data sample space. As most existing conclusions were based on small sample space, they may be unreliable in new scenarios. Therefore, since a large amount of data can be accessed now, we should validate the conclusions based on the large volume of data available. For example, the "word of mouth" can be verified through social networks and the viral marketing [21] used for disseminating the advertisements of new products. Moreover, the underlying mechanism of increasing social phenomena in social media related to influence also needs to be further studied, such as how the information diffusion is facilitated in "cyber-enabled social movement organizations (CeSMOs)" [31].

Now, we are entering the era of "big data". Big data has been characterized in many different ways, from Doug Laney's original "3*V*s" model [32] to the various recent extended "4*V*s" descriptions [33]. Laney's three "*V*s" refer to volume, velocity, and variety; the fourth "*V*" could be variability, virtual, or value. In social network analysis, big data means large-scale networks, various types of data, requirement of high-speed tackling, and the heterogenous value of data. Big data brings great opportunities for social influence analysis. Besides the availability of voluminous, diversified, and real-time data, researchers can get the following benefits. (1) Validate the conclusions based on samples to see whether existing concepts, models and algorithms are still useful. For example, to study the traditional concept of "homophyly", Damon Centola [34] performed a controlled experimental study on the spread of a health innovation through fixed social networks. (2) Conduct social experiments on large-scale social networks with millions of people, which is unimaginable before, and draw more reliable conclusions on social networks. For example, Aral and Walker [35] presented a method that used *in vivo* randomized experiments to identify influence and susceptibility in networks while avoiding the biases inherent in traditional estimates of social contagion. (3) Furthermore, launch new applications based on the new discoveries about social influence through big data mining. For example, recommendation algorithms of "Amazon" or "eBay" can be designed based on the influence of related commodities.

Meanwhile, big data brings big challenges. The first is that the computational scale will grow greatly. This will have significant impact, especially for the algorithms or models with high time or space complexity. The second is that the dilemma of social influence research will be increasingly prominent. On one hand, social influence depends on many factors, and the more factors are considered, the more accurate the model is. On the other hand, more factors bring the "curse of dimensionality" and result in heavy demand on computational resources. It will be much more difficult to construct practical models or algorithms with reasonable complexity and wide acceptance.

The research of social influence is coming to a new start with the backdrop of big data. In this chapter, we survey the topic of social influence analysis with a focus on the computational perspective and, moreover, discuss the advantages and challenges of these models and algorithms. Section 11.2 will present statistical measurements on social influence. It covers several classical measurement methods in social network analysis. In Section 11.3, we describe the models on the propagation of social influence and the optimization algorithms of social influence propagation, as well as practical applications such as marketing and advertisement. In Section 11.4, we will provide discussions on the social influence development in the context of big data, and discuss the advantages and disadvantages of these models and algorithms. Finally, we summarize this chapter and describe some of the future works in this area.

11.2 Social influence measurement

The measurement of social influence is to analyze and predict the social influence strength and evolution rule. The measurement can provide corresponding techniques and theory for the study of social influence. There are roughly four types of social influence measures: network topology-based, user behavior-based, information interaction-based, and topic-based measures. In this section we will introduce popular measures and discuss their advantages and disadvantages in the context of big data.

11.2.1 Network-based measures

Node measures

Node-based centrality is defined in order to measure the importance of a node in a network. Many approaches have already been proposed in the complex network science to measure the influence of nodes [36, 37]. The intuition is the usage of centrality. A node with higher centrality score is usually considered more influential than other nodes in the network. Many centrality measures have been proposed based on the precise definition of influence. The main principle to categorize centrality measures is the type of random walk involved. In particular, centrality measures can be classified into two categories: radial and medial measures [36]. Radial measures assess random walks that start from or end at a given node, while medial measures assess random walks that pass through a given node. The radical measures can be further divided into volume and length measures by the strategy of random walks, where the volume measures fix the length of walks and find the volume (or number) of the walks limited by the length, and the length measures fix the target nodes and find the length of walks to reach them. In the following, we introduce some popular centrality measures.

Degree

Out-degree and in-degree can be applied to measure the influence of nodes in a network. In practice, they usually denote the count of friends, recommendations, or posts. The directions (in- or out-) of a node are also the propagation directions.

The most popular and simple measure in this category is degree centrality, which measures the average influence of a node to its neighbors [37]. Let A be the adjacency matrix of a network, and $deg(i)$ be the degree of node i. The degree centrality is defined as the degree of the node:

$$c_i^{DEG} = deg(i). \tag{11.1}$$

An interpretation of degree centrality is that it counts the number of paths of length 1 starting from a node.

Another category of measures are based on the diffusion behavior in the network. All these measures, e.g. Eigenvector centrality, Katz centrality, and PageRank, can be categorized into random walk models.

Eigenvector centrality computes the weight of a node by its centrality, and takes the sum of the centrality of a node's access nodes as the node's centrality. This method is based on the idea that the influence of a node depends on its neighbors. If its neighbors' influence is large, then the host's influence is correspondingly large, and vice versa.

The Katz centrality [3] counts the number of walks starting from a node and at the same time penalizes longer walks. More formally, the Katz centrality c_i^{KAT} of node i is defined as follows:

$$c_i^{KAT} = \mathbf{e}_\mathbf{i}^\mathbf{T} \left(\sum_{j=1}^{\infty} (\beta \mathbf{A}) \right) \mathbf{1}. \tag{11.2}$$

Here, $\mathbf{e}_\mathbf{i}^\mathbf{T}$ is a column vector where the ith element is 1, and all other elements are 0; β is a positive penalty constant between 0 and 1. A slight variation of the Katz measure is the Bonacich centrality [38], which allows for negative values of β. The eigenvector centrality [39], the principal eigenvector of matrix $\mathbf{A}$, is related to the Katz centrality: the eigenvector centrality is the limit of Katz centrality as β approaches $1/2$ from below [36].

It is well known that a node's degree captures the relation between the node and its neighbors in essence. The advantage of degree measures is that it is simple and intuitive, and the computational load is relatively light. Therefore, in the context of big data, these approaches can also be widely applied in the measurement of social influence. The disadvantage is that these approaches can only reflect the relationship between a user and its neighbors, and they fail to measure the influence of a node to the entire social network.

Closeness

Closeness centrality and betweenness centrality are based on the shortest path of social networks [37, 40]. Closeness centrality can be used to measure a node's indirect influence to other nodes, or the distance from a node to others. It can also be used to measure the strength of a user's social ties. The higher the closeness centrality of a user, the shorter the distance between it and other users, and the faster its influence will be spread to other users. The most popular centrality measure in this group is the Freeman's closeness centrality [37]. It measures the centrality by computing the average of the shortest distances to all other nodes. Then, the closeness centrality c_i^{CLO} of node i is defined as follows:

$$c_i^{CLO} = \mathbf{e}_\mathbf{i}^\mathbf{T} \mathbf{S} \mathbf{y}. \tag{11.3}$$

Here, $\mathbf{S}$ is a matrix whose (i, j)th element contains the length of the shortest path from node i to j, and $\mathbf{y}$ is the all-one vector.

Betweenness

Betweenness is useful as an index of the potential of a point in the control of communications [37]. Betweenness centrality measures the importance of a node in the network. For edges of high betweenness, nodes of high betweenness occupy critical positions in the network structure and are therefore able to play critical roles. Suppose that a large

Table 11.2 Node measures of influence

Type	Methods	Formula
	In-degree	$deg^{in}(v_i) = \sum_j a_{j,i}$
Degree-based	Out-degree	$deg^{out}(v_i) = \sum_j a_{i,j}$
	Degree centrality	$C^{DEG}(v_i) = \frac{\deg(v_i)}{n-1}$
Shortest-path-based	Closeness centrality	$C^{CLO}(v_i) = \frac{1}{\sum_{v_j \in V \setminus v_i} g'_{i,j}}$
	Betweenness centrality	$C^{BET}(v_i) = \frac{\sum_{s<t} \lvert\{g^i_{st}\}\rvert / \lvert\{g_{st}\}\rvert}{n(n-1)/2}$
	Eigenvector centrality	$\lambda x_i = \sum_{j=1}^{n} a_{i,j} x_j, i = 1, 2, \ldots, n$
Random-walk-based	Betweenness centrality	$C^{BET}(v_i) = \frac{\sum_{s<t} \lvert\{g^i_{st}\}\rvert / \lvert\{g_{st}\}\rvert}{n(n-1)/2}$
	Katz centrality	$C^{KAT}(v_i) = \sum_{k=1}^{\infty} \sum_{j=1}^{n} a^k (\mathbf{A}^k)_{ij}$

amount of flows are carried by a certain number of nodes. If the nodes occupy a position at the interface of tightly knit groups, they are considered to have high betweenness. Usually, this measurement represents the amount of information passing through a specific node. If the value is higher, it means that in this course, the node's influence to the information propagation is larger.

Table 11.2 lists the methods of node measures [41] and categorizes these methods into three classes. The first one is the degree-based methods, which can only measure the local influence of a node. However, it is easy to compute against the backdrop of big data. The second one is the shortest-path-based methods. If we assume that the shortest path between nodes plays the key role on influence propagation, the shortest path can be used to measure a node's influence indirectly. Compared with the methods based on degree, shortest-path-based methods can measure a user's influence in the entire network. However, their computational complexity is higher than that of the former ones. In addition, the influence propagating through the shortest path is ideal, and difficult to achieve in reality. The third category includes methods based on the random walk. They can measure the influence of a node by its path nodes. That is the influence exerted by a user associated with the influence of other nodes on the diffusion path. The greater influence the latter has, the bigger influence the original node is, and vice versa.

Edge measures

Edge measures relate the influence-based concepts and measures to a pair of nodes. In early studies, to simplify models or reduce computational complexity, numerous models assign an experimental value to the links between node pairs to represent the influence between them. The value is either a constant or a random value subject to a specific distribution [42–44]. Evidently, the evaluation of edge measures could not capture the essence of social influence.

Tie strength

According to Granovetter's seminal work [45], the tie strength between two nodes depends on the overlap of their neighborhoods. Generally, if the overlap of neighborhoods between v_i and v_j is large, we consider v_i and v_j to have a strong tie. Otherwise, they are considered to have a weak tie. We formally define the strength $S(i, j)$ in terms of their Jaccard coefficient:

$$S(i, j) = \frac{|Ng_i \cap Ng_j|}{|Ng_i \cup Ng_j|}. \tag{11.4}$$

Here, Ng_i and Ng_j denotes the sets of neighbors of v_i and v_j, respectively. Similar to Jaccard coefficient, we can use overlapping similarity and cosine distance to calculate the social influence in the network [14].

Sometimes, tie strength is defined under a different name called "embeddedness". The term was further developed by economic sociologist Mark Granovetter, who argued that even in market societies, economic activity is not as disembedded from the society as economic models would suggest. He applied the concept of embeddedness to market societies, and demonstrated that economic exchanges are not carried out among strangers but rather by individuals involved in long-term continuing relationships [46]. The embeddedness in a social network was measured based on Granovetter's concept. If two nodes on the edge have a high overlap of neighborhoods, the edge's embeddedness is high. When two individuals are connected by an embedded edge, it is easier for them to trust each other, because it is easier to identify dishonest behaviors through mutual friends [46]. On the other hand, when embeddedness is zero, two end nodes have no mutual friends. Therefore, it is riskier for them to trust each other because there is no mutual friend for behavioral verification.

A corollary from tie strength is the hypothesis of triadic closure. This is related to the nature of the ties between a set of three actors A, B, and C. If strong ties connect A to B and A to C, B and C are probably connected by a strong tie as well. Conversely, if A–B and A–C are weak ties, B and C are less likely to have a strong tie. Triadic closure is measured by the clustering coefficient of the network [47, 48]. The clustering coefficient of node A is defined as the probability that two randomly selected friends of A are friends with each other. In other words, it is the fraction of pairs of friends of A that are linked to each other. This is naturally related to the problem of triangle counting in a network. Let n_Δ be the number of triangles in the network and let $|E|$ be the number of edges. The clustering coefficient is formally defined as follows:

$$C = \frac{6n_\Delta}{|E|}. \tag{11.5}$$

The naive way of counting the number of triangles n_Δ is expensive, especially for large-scale networks. An interesting connection between n_Δ and the eigenvalues of the network was discovered by Tsourakakis [49]. This work shows that n_Δ is approximately equal to the third moment of the eigenvalues (or λ_i^3, where λ_i is the ith eigenvalue). Given the skewed distribution of eigenvalues, the triangle counts can be approximated by computing the third moment of only a small number of the top eigenvalues. This also provides an efficient way to compute the clustering coefficient.

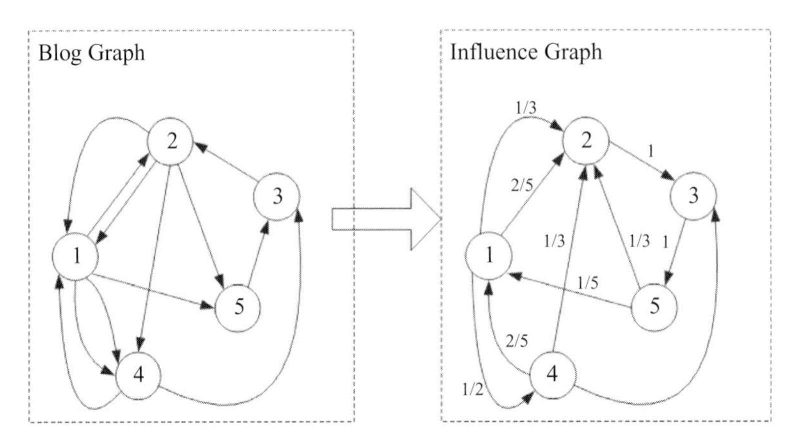

Figure 11.1 This diagram shows the conversion of a blog graph into an influence graph. The graph influence is a weighted graph: the source of the arc denotes the influence actioner, and the weight represents the influence strength.

Edge betweenness

Edge betweenness is an important measure that is generalized from Freeman's betweenness centrality [37, 50] and proposed by Girvan and Newman [51]. Edge betweenness is used to measure the importance of an edge in a network. It is supposed that the information flows between node A and B are evenly distributed on the shortest paths. That is, if there is only one shortest path between them, such a path is given a weight of unity. If there are more than one shortest path between them, each path is given an equal weight such that the total weight of all the paths is unity. The betweenness of an edge is calculated by summarizing the total "weights" of all shortest paths going through it.

One application of edge betweenness is graph partitioning or community discovery. If a network contains communities or groups that are only loosely connected by a few inter-group edges, then all shortest paths between different communities must go along one of these few edges. Therefore, the edges connecting communities will have high edge betweenness. By removing these edges, we separate groups from one another and therefore reveal the underlying community structure of the graph. More detailed studies on clustering methods are presented in the work by Girvan and Newman [51].

Java *et al.* [52] analyzed the propagation of the influence of blogs. In addition, they modeled the influence graph as a weighted, directed graph with edge weights indicating how much influence a particular source node had on its destination. Starting with the influence graph, they aimed to identify a set of nodes for a targeted piece of information such that these nodes caused a large number of bloggers to be influenced by the idea. A link from u to v indicates that u is influenced by v. The edges in the influence graph are the reverse of those in the blog graph to indicate this influence. Multiple edges indicate stronger influence and are weighed higher (Figure 11.1). The formula is given as where $c_{u,v}$ indicates number of arcs from u to v:

$$w_{u,v} = \frac{c_{u,v}}{deg^{in}(v)}. \tag{11.6}$$

To sum up, the measures based on the network topology are simplest for social influence. These measures can extract the network topology between users, but they could miss huge a amount of interaction information. Definitely, these models are computationally simple and easy to understand. As for big data, these measures have natural advantages in applications due to their simplicity. Meanwhile, they have at least two disadvantages. First, the topology network gained by researchers is often a static snapshot of the fast-varying social network. Thus the data collections may miss critical data that contain the influence information between users. For example, the links formed ten years ago are the same as the links formed in the latest second. The friends rarely interacting with each other are the same as the friends who interact with each other frequently, just because of the same actions online. Second, in the network topology, user links have the same influence value or are subject to the same distribution. Clearly, it is not true in reality. This motivates researchers to propose other measures.

11.2.2 Behavior-based measures

User behaviors (actions) in online social networks include posting, purchasing, commenting, retweet, and building friendships, etc. By analyzing these behaviors, we can evaluate a user's influence in a new way. Furthermore, different from network-based measures, behavior based measures are often evaluated by behavior prediction [9, 53–56].

Goyal *et al.* [54] studied the problem of learning influence degrees (called probabilities) from a historic log of user actions. They presented the concept of user influential probability and action influential probability. The assumption is that if user v_i performs an action y at time t and his friend v_j also performs the same action later at t', there is an influence of v_i on v_j. The goal of learning influence probabilities [54] is to find a (static or dynamic) model to capture a user's as well as its actions' influence on the network. They give a general user influential probability and action influential probability definitions as follows.

User influence probability

$$infl(v_i) = \frac{|\{y|\exists v, \Delta t : prop(a, v_i, v_j, \Delta t) \wedge 0 \leq \Delta t\}|}{Y_{v_i}}. \tag{11.7}$$

Action influence probability

$$infl(y) = \frac{|\{v_i|\exists v_j, \Delta t : prop(a, v_j, v_i, \Delta t) \wedge 0 \leq \Delta t\}|}{number\ of\ users\ performing\ y}. \tag{11.8}$$

Here $\Delta t = t_j - t_i$ represents the difference between the time when user v_j performs the action and the time when user v_i performs the action, given $e_{ij} = 1$, and $prop(a, v_j, v_i, \Delta t)$ denotes the action propagation score.

Goyal *et al.* [54] proposed three methods in order to approximate the action propagation: (1) static model (based on Bernoulli distribution, Jaccard index, and partial credits); (2) continuous-time (CT) model; and (3) discrete-time (DT) model. These models could all be learned through a two-stage algorithm. Finally, learned influence probabilities were applied to action prediction, and experiments showed that the CT model achieved a better performance than other models on the Flickr social network in terms of the action of "joining a group".

Saito *et al.* [57, 58] discussed the similar problem of social influence in the study of general cascade model. They cast the influence model into a maximum likelihood problem, and solved the problem by expectation maximization (EM). Here, we point out that such a method is not scalable in the context of big data. Because EM requires the calculation of coefficients for every chain, the time complexity could be extensively high.

All the methods above build their models based on the social network topology. Another interesting method proposed by Yang and Leskovec [11] was beyond the network topology. They argued that the propagation of information is affected by the user's influence, which is independent of the social relationship. They constructed a linear influence model (LIM) to measure the influence. The model depicts a user's influence and relationships with others by constructing the relationships between influence functions and information time. The formula is as follows:

$$V(t+1) = \sum_{u \in A(t)} I_u(t - t_u). \tag{11.9}$$

Here, $V(t)$ denotes the number of nodes that mention the information at time t; each node u has a particular non-negative influence function $I_u(l)$. One can simply treat $I_u(l)$ as the number of followup mentions l time units after node u adopts the information. $I_u(l)$ is set to be a certain parameter equation, such as an exponential: $I_u(l) = c_u e^{-\lambda_u l}$, or a power law: $I_u(l) = c_u l^{-\lambda_u}$. To account for the diversity of reality, they proposed a new parameter-free approach, which represents influence as a decreasing function over time. The measuring effect of these models is closely related to $I_u(l)$. However, researchers recently found that there are many related factors of social influence. Moreover, the influence distribution and its changing rules are still to be revealed. Therefore, it is challenging to represent the influence by a unique equation.

Other actions

Besides posting, retweeting, and commenting, other actions can also be used as an evaluation index of influence. For example, signing on social medias [59] could be used to evaluate the influence of a specific user. If a user signs on frequently and the number of users connected to him increases over the action period, this means that such user has an impact on them.

In order to model and track social influence through user actions, Tan *et al.* [60] proposed a noise-tolerant time-varying factor graph model (NTT-FGM), which is based on three factors: the social network, user attribute, and user historic actions data. NTT-FGM predicts a user's behavior based on his friends' influence to him.

Table 11.3 shows the action-based influence measures discussed above. Compared with discrete models, continuous models have the advantage of characterizing the propagation process of influence. They are also better at prediction, but the computational complexity is too high. Discrete models can be applied with heuristic methods to solve

Table 11.3 User behavior-based method of influence measures

Source	Need network (Y/N)	Discrete(D)/ Continuous (C)	Parametric (Y/N)	Weighted (Y/N)
LIM	N	D	N	N
Goyal *et al.*	Y	D/C	Y	Y
Saito *et al.*	Y	D	Y	N
NTT-FGM	Y	D	Y	Y

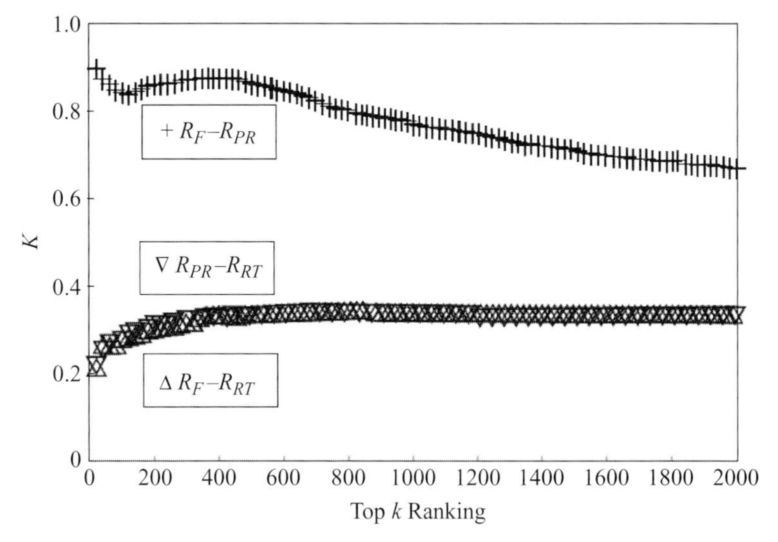

Figure 11.2 Comparison among rankings [61]. Here, R_F, R_{PR}, and R_{RT} denote the number of followers, PageRank and the number of retweets, respectively. [61] plots K for three pairs of rankings with k varying from 20 to 2000. Note that the R_F–R_{PR} pair has high K over 0.6, but both R_F–R_{RT} and R_{PR}–R_{RT} pairs have low K below 0.4. This means that R_F and R_{PR} are similar, but R_{RT} is distinct.

the problem. Non-parametric model is not constrained by social networks, but the description capability is limited, and thus it is appropriate for theoretical analysis rather than applications. Parametric models are excellent in the description of real social networks, but the accuracy of such models depends on that of the parameter estimation.

An interesting comparison was drawn between behavior-based methods and social network-based methods. Kwak *et al.* [61] ranked users by the number of followers, PageRank, and the number of retweets, and then presented a quantitative comparison among them. They found that the former two rankings are similar. Ranking by retweets differs from the previous two rankings, indicating a gap the influence between inferred from the number of followers and that from the popularity of one's tweets (Figure 11.2).

Although the user behavior-based methods are closer to the reality than the network-based ones, their performance is limited by the privacy protection policy, which results in incomplete user behavior data. Therefore, the problem is to improve the prediction accuracy and efficiency under the limitation of user behavior data. As a result, most of researchers integrate the network, user behaviors, and interacting data together to improve the accuracy. In the context of big data, how to combine related factors and how to efficiently reduce computational complexity are still open questions.

11.2.3 Interaction-based measures

In the real world, social networks are formed through locality, activity, family relations, and so on. However, users of online social networks mainly communicate with each other through information diffusion, sharing, comments, retweets, and other interactions. Therefore, the interactions recorded by online social networks are of vital importance to the analysis. However, social networks generate overwhelmingly large amount of interaction data every day. For example, more than 100 million users in 2012 posted 340 million tweets per day. Obtaining and analyzing the data in real time is a task with great challenges. Therefore, researchers take out part of the data as samples and analyze the key interaction information of the users such as topics or key words.

Information content is the carrier of information propagation. It is of great importance to study propagation types and characters. Since a user's social influence can help the propagation of information, the information content and propagation range and time are good reflections of the user's influence.

In a social network, most of popular information is initiated by those who have a large number of followers, and thus the popular information propagation process is a good research sample. Bakshy *et al.* [8] measured a user's influence and made predictions based on the information diffusion tree. They found that many popular topics on Twitter are initiated from users of great influence, either those who were influential in the past or those who have a large number of followers. However, it is very difficult to predict a user's social influence based on the interaction information. Both the spread area and spreading time of information can be used to measure the influence of initial users. Romero *et al.* [62] studied the spread of hashtags on Twitter, and analyzed the way in which tokens known as hashtags spread on a network defined by the interactions among Twitter users. They found significant variations in the ways that widely used hashtags on different topics spread. They also defined an influence curve (or "exposure curve") to describe the character stated above (Figure 11.3). Given a specific hashtag H, $P(k)$ is the fraction of such users who mention H before the $(k+1)$th neighbor does so. Figure 11.3 shows the average exposure curve for the top 500 hashtags. Obviously, most hashtags reach the peak value in [7, 9], and then descend as k increases. The result is consistent with the influence decline phenomena observed by Kempe *et al.* [15]. It also proves that time is of vital importance in the measure of influence. Other researches make it clear that the attribute of users can also be the factors for the propagation of influence. For example, [9] showed that the activity and concentration are such kind of attributes.

The works mentioned above were generally concentrated on the range and time of the interaction information propagation. They mainly focused on qualitative analysis rather than quantitative analysis. Meanwhile, since information is for propagation in nature,

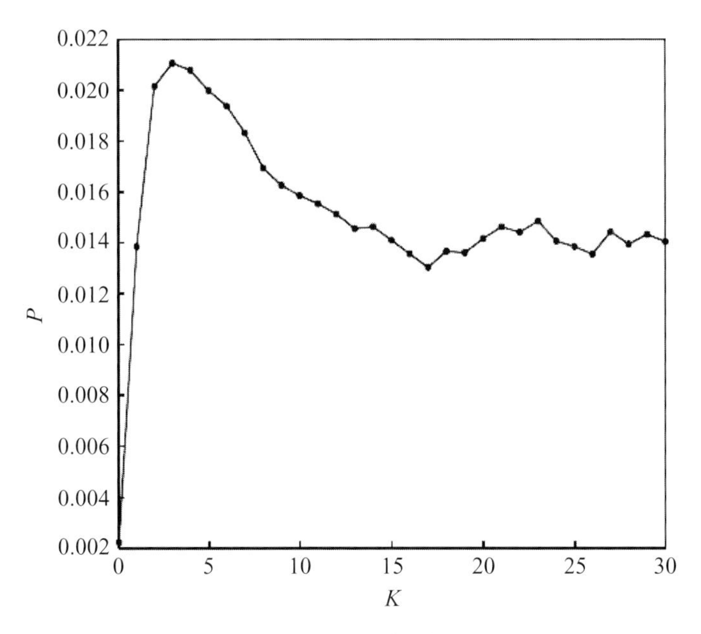

Figure 11.3 The average exposure curve for the top 500 hashtags [62]. $P(K)$ is the fraction of users who adopt the hashtag immediately after their Kth exposure to it, given that they had not yet adopted it.

how do we identify the performance of information itself from a user's influence? Such a question is helpful for us to understand the essence of influence.

11.2.4 Topic-based measures

In reality, a great amount of information is generated and propagated in the form of topics. Researchers found that the influence varies from topic to topic. Moreover, the same topic follows different propagation or influence models in different groups [45, 63, 64]. Therefore, topic, as a basic objective social influence measure, could be applied in the description of users' influence from multiple perspectives. Researchers can build the relationship between topic contents and a user's participation in social influence models. In this way, we can avoid the fans' relationships of social networks. Usually, we call the influence calculated in the former way as "hidden influence" and the latter as "dominant influence". Nevertheless, online entities of social networks include users, texts, and multi-media information, and so on. Constructing heterogeneous networks must be more complex than homogeneous networks.

To differentiate the influence from different topics, Tang *et al.* [25] proposed a topical factor graph (TFG) model to formulate the topic-level social influence analysis into a unified graphical model, and presented topical affinity propagation (TAP) for model learning. In particular, the motivation of the model is to simultaneously capture the user topical distributions (or user interests), similarity between users, and network structure. Figure 11.4 shows the graphical structure of the proposed model. The TFG model has a

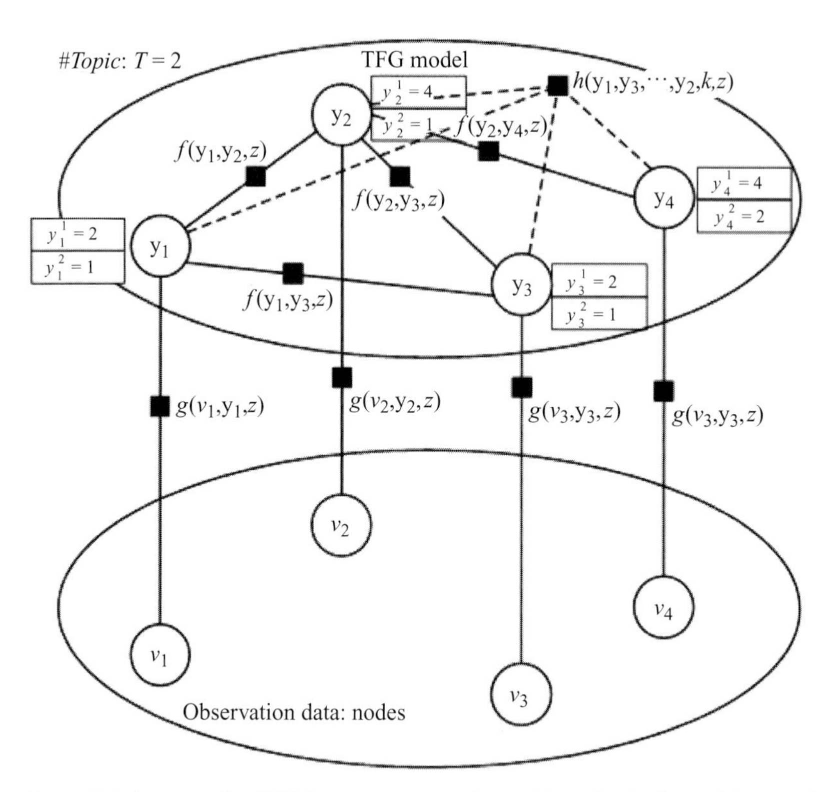

Figure 11.4 An example of TFG: $v_1, \ldots, v_4$ are observable nodes in the social network; $Y = \{y_1, \ldots, y_4\}$ are hidden vectors defined on all nodes, with each element representing the node that has the highest probability to influence the corresponding node; $g(\bullet)$ represents a feature function defined on a node, $f(\bullet)$ represents a feature function defined on an edge; and $h(\bullet)$ represents a global feature function defined for each node with $k \in \{1, \ldots, N\}$.

set of observed variables $\{v_i\}_{i=1}^N$ and a set of hidden vectors $\{y_i\}_{i=1}^N$, which correspond to the N nodes in the input network.

The hidden vector $y_i \in \{1, \ldots, N\}$ models the topic-level influence from other nodes to node v_i. Each element y_i^z takes the value from the set $\{1, \ldots, N\}$, and represents the node that has the highest probability to influence node v_i on topic z. For example, Figure 11.4 shows a simple example of a TFG. The observed data consists of four nodes $v_1, \ldots, v_4$, which have corresponding hidden vectors $y_1, \ldots, y_4$. The edges between the hidden nodes indicate the four social relationships in the original network (or edges in the original network).

Three types of feature functions are defined to capture the network information: node feature function $g(v_i, \mathbf{y_1}, z)$, edge feature function $f(\mathbf{y_k}, \mathbf{y_l}, z)$, $f(\mathbf{y_i}, \mathbf{y_j}, z)$, and global feature function $h(\mathbf{y_1}, \ldots, \mathbf{y_N}, k, z)$.

The node feature function g describes the local information of nodes (e.g. attributes associated with users or topical distribution of users). The edge feature function f describes the correlation between users by the edge on the graph model. The global feature function captures the constraints defined on the network. Based on this formulation, an objective function is defined for likelihood maximization of the observations:

$$P(\mathbf{v}, \mathbf{Y}) = \frac{1}{Z} \prod_{k=1}^{N} \prod_{z=1}^{T} h(\mathbf{y_1}, \ldots, \mathbf{y_N}, k, z) \prod_{i=1}^{N} \prod_{z=1}^{T} g(v_1, \mathbf{y_i}, z) \prod_{e_k l \in E} \prod_{z=1}^{T} f(\mathbf{y_k}, \mathbf{y_l}, z). \tag{11.10}$$

Here, Z is a normalization factor; $\mathbf{v} = [\mathbf{v_1}, \ldots, \mathbf{v_N}]$ and $\mathbf{Y} = [\mathbf{y_1}, \ldots, \mathbf{y_N}]$ correspond to the observed and hidden variables, respectively. The feature function f, g, and h can be defined in different ways. For example, in the work described in [25], f is defined as an indicator function in order to capture the existence of edges between two users; the node feature function g is defined according to the similarity of two users on a topic; and the global feature function h is defined as a constraint.

Based on this formulation, the task of social influence is cast into the identification of the node that has the highest probability to influence another node on a specific topic along with the edge. This is the same as that of maximizing the likelihood function $P(\mathbf{v}, \mathbf{Y})$.

Following the research above, Liu *et al.* [65] took into consideration topic distribution and influence together, and constructed a heterogeneous network with users' information and various texts. They analyzed the influence of topics and constructed a topic model, mining the recessive influence and predicting a user's behavior through text similarity.

Further, Cui *et al.* [66] investigated the item-level social influence to answer the question "Who should share what", which can be extended into information retrieval scenarios. They found various influences of topics, and even the same user's influence is changing with topics. They measured and predicted the recessive social influence through interactions of users, including information items, user profiles, and relationships between users. And they designed the hybrid factor non-negative matrix factorization (HF-NMF) to predict social influence. Such a model was solved through gradient projection matrix factor decomposition.

On the modeling of topic influence, Weng *et al.* [10] designed a two-stage tactic, first extracting the topic of users' interests followed by constructing the relationships within topics, and then applying TwitterRank to analyze users' influence through topic's similarity and network structure. The advantage of this method is the introduction of topics into social networks. Experiments show that this tactic can improve the functionality and prediction accuracy.

Table 11.4 lists the influence analysis methods based on topics. Generally, these methods combine users' interaction information of topics with social networks. By analyzing the relationship between the information and users, such models measures social influence more accurately. Although these methods had great achievements, we believe that there are still some issues to explore: first, there is a great deal of multimedia information in the social network besides texts, and constructing a model including that information is a new challenge. Second, recessive models are better in revealing hidden social relationships, and the problem is how to quantify the relationship between dominant and recessive influences. Third, different topics lead to different results of influence [9, 11, 67], and the propagation model may depend on the topics. However,

Table 11.4 Influence measures based on topics

Model	Need network (Y/N)	Dominant(D)/ Recessive(R)	Homo.(HM)/ Hetero.(HT)	Data source
TFG	Y	D	HM/HT	Wikipedia; ArnetMiner
Liu *et al.* [65]	Y	D/R	HT	Twitter; Digg; Cora
HF-NMF	N	R	HM	Renren
TwitterRank	Y	D	HM	Twitter

the influence of users in reality is relatively static. Therefore, the role of topics and its impact on user influences should be further studied.

11.2.5 Other measures

Online social networks are very complex, and many factors affect the measures of a user's influence. Taking a comprehensive consideration of these factors will improve the precision of the model and provide new aspects for the research of social influence.

Time stamp is of great importance in the generation and propagation of social influence [5], so the influence analysis should combine the social network structure and information time sequence. A novel method has been proposed [68] to find influentials by considering both the link structure and the temporal order of information adoptions in Twitter. This method found distinct influentials who were not discovered by other methods. Therefore, the authors defined the reader who was newly exposed to information as "effective reader", and they found influentials based on the number of effective readers of a user. However, some researchers argued that early adopters are not necessarily influentials [69]. Par *et al.* [70] constructed a model without interactions, and they extended the model to incorporate dynamical parameters and inferred how influence changes over time.

Transfer entropy was applied to measure the causal relationships between nodes, where we have to apply the recessive influence models in the analysis. Such a theoretically grounded measure is based on dynamic information, captures fine-grain notions of influence, and leads to a natural, predictive interpretation [71, 72]. However, these types of methods require large amount of data. Therefore, it is convinced to be more useful in the backdrop of big data.

Social influence acts in the word-of-mouth, improving the marketing performance and enlarging the adoption of innovation. Huang *et al.* [73] found that influence can improve the evaluation index of commodities not only before purchasing, but also after it.

Social influence measurement is critical for the analysis of social influence over networks. Theoretically, it could help us to understand the mechanism of information diffusion, creature adoption, viral marketing, and so on. On the application level, especially in the context of big data, low-complexity measures will still be applied to evaluate the influence of nodes, edges, terms, topics, and events.

11.3 Influence propagation and maximization

The influence propagation and maximization (IP&M) refers to the issue of the dynamics of social influence. From the beginning of social activities, everyone's influence changes with his behaviors and profiles, and propagates by social activities. Therefore, the analysis of social influence processes are beneficial for identifying the nature of social influence evolution. Such a study is critical for us to find the propagation rules and human behavior patterns in social networks. In the classical propagation model proposed by Katz and Lazarsfeld [4], the propagation of information or innovation is initiated from groups of greater influence, then spreads through them to the wider population.

The remarkable applications of influence maximization include viral marketing and online advertising. In this section, we will present the research on influence propagation and maximization, and review recent progresses in this field.

11.3.1 Opinion leader identification

Online advertising

Social influence analysis techniques can also be leveraged for online advertising. For example, the work in [74] proposed a method to identify brand-specific audiences without utilizing users' private information. The proposed method takes advantage of the notion of "seed nodes", which can specifically indicate the users (or browsers) who exhibit brand affinity. Yet another term "brand proximity" is a distance measure between candidate nodes and seed nodes. For each browser b_i, we use $\overrightarrow{\phi_{b_i}} = [\phi_{b_i}^1, \phi_{b_i}^2, \ldots, \phi_{b_i}^P]^T$ to denote the effect of the P proximity measures.

Then we can discover the best audiences for marketing by ranking the candidate nodes b_i with respect to $\overrightarrow{\phi_{b_i}}$ based on the monotonic function: $score(b_i) = f_i(\overrightarrow{\phi_{b_i}} \bullet \overrightarrow{I_q})$. The selection vector $\overrightarrow{I_q} = [0, \ldots, 1, \ldots, 0]^T$ holds a 1 in the qth row. The proximity measure P can be selected from a pool. Finally, the authors showed that the quasi-social network extracted from the data agrees well with a real social network. This means that the modeled "friends" on the virtual network accurately reflect the relationships in the real world.

Another tractable approach for viral marketing is frequent pattern mining, which was studied by Goyal *et al.* in [75]. Their research focused on the actions performed by users, under the assumption that a user could learn his/her friends' actions. The authors formally defined leaders in a social network and introduced an efficient algorithm aimed at discovering the leaders. The essence of the problem is that actions took place in different time steps, and the actions that came up later could be influenced by earlier ones, which is termed as the propagation of influence. The opinion leaders are those who can affect a certain number of people through their actions for a long time. Besides that, those who only affect a small subset of people are called tribe leaders. The algorithms of finding leaders in a social network usually make use of chronological action logs.

Influential blog discovery
In the web 2.0 era, people spend a significant amount of time on user-generated-content websites, where a classical example is blog sites. By visiting others' blogs, blog users form an online social network. Some of the users bring in new information, ideas, and opinions, and disseminate them down to the masses. This affects the opinions and decisions of others by word of mouth. These users are opinion leaders in the blog sphere. Nitin Agarwal *et al.* [76] were the first to define the following properties for each blogger.

Recognition An influential blog post is recognized by many. It can be equated to the case that an influential post p is referenced in many other posts, or its number of inlinks is large.

Activity generation A blog post's capability of generating activity can be indirectly measured by how many comments it receives and the amount of discussions it initiates. In other words, the more comments there are, the more influential the content is.

Novelty Novel ideas exert more influence as suggested in [6]. Normally a novel blog is the one with less outgoing links.

Eloquence Longer articles posted on blog sites tend to be more eloquent, and can thus be more influential.

The work in [76] presented a model which took advantages of the above four properties to describe the influence flow in the influence graph consisting of all the blogger pages. Basically, the influence flow probability is defined as follows:

$$InfluenceFlow(p) = w_{in} \sum_{m=1}^{|I|} I(p_m) - w_{out} \sum_{n=1}^{|\theta|} I(p_n), \tag{11.11}$$

where w_{in} and w_{out} are the weights to capture the contribution of incoming and outgoing links, respectively. Finally, the influence of a blog is defined as:

$$I(p) = w(\lambda) \times (w_{com}\gamma_p + InfluenceFlow(p)), \tag{11.12}$$

where w_{com} denotes the weight to regulate the contribution of the number of comments (γ_p) towards the influence of the blog post p. In another work [77], Song *et al.* associated a hidden node v_e to each node v to represent the source of the novel information in blog v . More specifically, let $Out(v)$ denote the set of blogs that v links to. The information novelty contribution of entry v_e is then calculated as:

$$Nov(v_e|Out(v_e)) = min_{O_e \in Out(v_e)} Nov(v_e|O(v_e)). \tag{11.13}$$

The information novelty provided by the hidden node of blog v is measured as the average of the novelty scores of the entries it contains:

$$Nov(v_e|Out(v)) = \frac{\sum_{v_e \in V}(Nov(v_e|Out(v_e)))}{card(Set(v_e))}, \tag{11.14}$$

where $card(\bullet)$ denotes the total number of entries of interest in blog v. Then, the problem can be formulated as solving the *InfluenceRank* ($\mathbf{IR}^{\mathrm{T}}$):

$$\mathbf{IR}^{\mathrm{T}}(I - (1-\beta)\alpha\mathbf{W} - (1-\beta)\alpha\mathbf{a} \bullet \mathbf{e}^{\mathrm{T}}) = (1-\beta)(1-\alpha)\mathbf{e}^{\mathrm{T}} + \beta Nov^{T} \tag{11.15}$$

with $\mathbf{IR}^{\mathrm{T}} \bullet \mathbf{e} = 1$.

As the *InfluenceRank* can be fitted into a random walk framework, α is the probability that the random walk follows a link; β reflects how significant the novelty is to the opinion leaders we expect to detect; $\mathbf{e}$ is the n-vector of all ones; and $\mathbf{a}$ is the vector with components $a_i = 1$ if ith row $\mathbf{W}$ corresponds to a dangling node and 0 otherwise. That is, $\mathbf{W}$ is the normalized adjacent matrix.

11.3.2 Influence maximization

The problem of influence maximization can be traced back to the research on "word-of-mouth" and "viral marketing" [16–20, 78]. The problem is often motivated by helping companies to determine which are potential customers to market to. If the expected profit from a customer is greater than the cost of marketing, marketing actions for that customer will be executed. Therefore, the goal is to minimize marketing cost and more generally to maximize profit. For instance, a company may wish to market a new product through the natural "word of mouth" effect arising from the interactions in a social network. The goal is to get a small number of influential users to adopt the product, and subsequently trigger a large cascade of further adoptions. In order to achieve this goal, we need a measure to quantify the intrinsic characteristics of the user (e.g. the expected profit from the user) and the user network value (e.g. the expected profit from users that may be influenced by the user). In the past, the problem has mainly been studied in marketing decision or business management. Domingos and Richardson [18] viewed it as a social network and modeled it as a Markov random field. Further, they presented an efficient algorithm to learn the model [20]. However, the method models the marketing decision process in a "black box". Once a set of users have been marketed (selected), how users influence each other was settled. How they would influence their neighbors, and how the diffusion process would continue were problems that were still not fully solved. Kempe *et al.* [79] took the first step to formally define the process in two diffusion models and theoretically proved that the optimization problem of selecting the most influential nodes in these two models is NP-hard. They have developed an algorithm to solve the models and provided the first provable approximation guarantees for efficient algorithms. They showed that a natural greedy strategy obtains a solution that is provably within 63% of the optimum for several classes of models. The efficiency and scalability of the algorithm have been further improved in recent years [80, 81]. We will review the work in marketing and business and focus on problem formulation and model learning.

Diffusion influence model

There are a great number of classical models for this problem. Here, we will review some of them. For the sake of easy explanation, we associate each user with a status: active or inactive. Then, the status of the selected set of users to market (also referred to as "seed nodes") is viewed as active, while other users are viewed as inactive. The problem of influence maximization is studied with the use of this status-based dynamics. Initially all users are considered inactive. Then, the selected users are activated who may further influence their friends (neighbor nodes) to be active as well. The simplest model is to quantify the influence of each node with some heuristics.

Several examples are given as follows.

(1) High-degree heuristic. It selects seed nodes according to their degree d_v. The strategy is very simple and also natural because the nodes with more neighbors would arguably tend to impose more influence upon their direct neighbors. This consideration of high-degree nodes is also known in the sociology literature as "degree centrality".
(2) Low-distance heuristic. Another commonly used influence measure in sociology is distance centrality, which considers the nodes with the shortest paths to others as seed nodes. This strategy is based on the intuition that individuals are more likely to be influenced by those who are closely related to them [23].
(3) Degree discount heuristic. The general idea of this approach is that if u has been selected as a seed, then when selecting v as a new seed based on its degree, we should not count the edge $\overrightarrow{uv}$ towards its degree. This is referred to as single discount. More specifically, for a node v with d_v neighbors, of which t_v neighbors are selected as seeds already, we should discount v's degree by $2t_v + (d_v - t_v)t_v p$, where p denotes propagation probability.
(4) Linear threshold model. In this family of models, whether a given node v will be active or not depends on an arbitrary monotone function of the set of neighbors of v that are already active. We associate a monotone threshold function f_v which maps subsets of v 's neighbors to real numbers in $[0, 1]$.

    Then, each node v is given a threshold θ_v , and v will turn active in step t if $f_v(S) > \theta_v$, where S is the set of neighbors of v that are active in step $t - 1$.

    Specifically, in [79] the threshold function $f_v(S)$ is instantiated as $f_v(S) = \sum_{u \in S} b_{v,u}$ where $b_{v,u}$ can be seen as a fixed weight, subject to the following constraint:

$$\sum_{\text{neighbors of } v} b_{v,u} \leq 1. \tag{11.16}$$

(5) General cascade model. Define an incremental function $p_v(u, S) \in [0, 1]$ as the success probability of user u activating user v, i.e. user u tries to activate v and finally succeeds, where S are those of v's neighbors that have already attempted but failed to drive v active. A special version of this model used in [79] is called the independent cascade model (ICM), in which $p_v(u, S)$ is a constant. This means that whether v is active does not depend on the order in which v's neighbors try to activate it. A special case of ICM is weighted cascade model (WCM), where each edge from node u to v is assigned probability $1/d_v$ for activating v.

One challenge in the diffusion influence model is the evaluation of its effectiveness and efficiency. From a theoretical perspective, Kempe *et al.* [79] proved that the optimizations of their two proposed models (i.e. linear threshold model and general cascade model) were NP-hard. Their proposed approximation algorithms can also theoretically guarantee that the influence spread is within $(1-1/e)$ of the optimal influence spread. From an empirical perspective, Kempe *et al.* [79] showed that their proposed models could outperform the traditional heuristics model in terms of the maximization of social influence. Recent studies have been mainly focused on the improvement of the algorithm efficiency. For example, Leskovec *et al.* [82] presented an optimization strategy referred to as "cost-effective lazy forward" or "CELF", which could accelerate the procedure by up to 700 times without loss of effectiveness. Chen *et al.* [80] further improved the efficiency by employing a new heuristics and in [81] they extended the algorithm to handle large-scale datasets. Another problem is the evaluation of the effectiveness of the models for influence maximization. However, since these methods were designed only for small datasets, it is still a challenging problem to extend these methods to large datasets.

Maximizing the spread of influence

Kempe *et al.* [79] proposed the linear threshold model (LTM) and independent cascade model (ICM). Finding the optimal solution to either model is NP-hard. The solution is to use a submodular function to approximate the influence function. Submodular functions have a number of very nice tractability properties in terms of designing approximation algorithms. One important property is shown as follows. Given a function f that is submodular and takes only nonnegative values, we have $f(S \cup \{v\}) \geq f(S)$ for all elements v and sets S. Thus, the problem can be transformed into finding a k-element set S with which $f(S)$ is maximized. The problem can be solved by using a greedy hill-climbing algorithm which approximates the optimal solution within a factor of $(1-1/e)$. The following theorem formally defines the problem [83, 84].

For a non-negative monotone submodular function f, let S be a set of size k obtained by selecting elements one at a time, each time choosing an element that provides the largest marginal increment in the function value. Let S be a set that maximizes the value of f over all k-element sets. Then $f(S) \geq (1-1/e) \bullet f(s^*)$.

In other words, S provides a $(1-1/e)$-approximation. The model can be further extended to assume that each node v has an associated non-negative weight w_v, which can be used to capture the human prior knowledge to the task at hand such as how important it is that v can be activated in the final outcome.

To adapt the model to a more realistic scenario, we may have a number of different marketing actions M_i available, each of which may affect some subsets of nodes by increasing their probabilities of being active. However, different nodes may respond to marketing actions in different ways. Thus a more general model was considered in [79]. Specifically, we can introduce investment t_i for each marketing action M_i. Thus the goal is to reach a maximum profit lift given that the total investments do not exceed the budget. A marketing strategy is then an m-dimensional vector $\mathbf{t}$ of investments. The probability that node v will become active is determined by the strategy and denoted by $h_v(\mathbf{t})$.

It is assumed that the function is non-decreasing and satisfies the "diminishing returns" property for all $t \geq t'$ and $a \geq 0$:

$$h_v(t + a) - h_v(t) \leq h_v(t' + a) - h_v(t'). \tag{11.17}$$

Satisfying the above inequality corresponds to an interesting marketing intuition: marketing actions would be more effective when the targeted individual is less "marketing-saturated" at that point. Finally, the objective of the model is to maximize the expected size of the final active set. Given an initial set A, then the expected revenue of the marketing strategy **t** can be defined as:

$$g(\mathbf{t}) = \sum_{A \subset V} \sigma(A) \bullet \Pi_{v \in A} h_v(t) \bullet \Pi_{u \in (V-A)}(1 - h_v(t)). \tag{11.18}$$

Note that if A is the active set of nodes, the inactive set of nodes is denoted as $V - A$. We can use the submodular property in order to optimize the function corresponding to the revenue of the marketing strategy. We can design a greedy hill-climbing algorithm, which can still guarantee an approximation within a constant factor. The proof of this result can be found in [79].

Viral Marketing aims to increase brand awareness and marketer revenue with the help of social networks and social influence. Direct marketing is an important application of influence maximization that attempts to market only to a selective set of potentially profitable customers. In reality, a person's decision to buy a product is often influenced by their friends and acquaintances. However, it is not desirable to ignore the networking influence, because this can lead to severely suboptimal decisions.

We introduce a model that tries to combine the network value with the customer intrinsic value [18]. Here, the intrinsic value represents attributes (e.g. customer behavior history) that are directly associated with a customer. Such attributes might affect the likelihood of the customer to buy a product, while the network value represents the social network (e.g. customers' friends), which may influence the customer's buying decisions.

The basic idea is to formulate the social network as Markov random fields, where each customer's probability of buying is modeled as a function of both the intrinsic desirability of the product for the customer and the influence of other customers. Formally, the input can be defined as follows. Consider a social network $G = (V, E)$, with n potential customers and their relationships recorded in E, and let x_i indicate the attributes associated with each customer v_i. We assign a Boolean variable y_i to each customer that takes the value 1 if the customer v_i buys the product being marketed and 0 otherwise. Furthermore, let NB^i be the neighbors of v_i in the social network and z_i be a variable representing the marketing action that is taken for customer v_i. Note that z_i can be a Boolean variable, with $z_i = 1$ if the customer is selected to market (e.g. offered a free product), and $z_i = 0$ otherwise. Alternatively, z_i could be a continuous variable indicating a discount offered to the customer. Given this, we can define the marketing process for customer v_i in a Markov random field as follows:

$$\begin{aligned} P(y_i|\mathbf{y_{NB^i}}, \mathbf{x_i}, \mathbf{z}) &= \sum_{C(NB^i)} P(y_i, \mathbf{y_{NB^i}}|\mathbf{x_i}, \mathbf{z}) \\ &= \sum_{C(NB^i)} P(y_i|\mathbf{y_{NB^i}}, \mathbf{x_i}, \mathbf{z}) P(\mathbf{y_{NB^i}}|\mathbf{X_i}, \mathbf{z}), \end{aligned} \tag{11.19}$$

where $C(NB^i)$ is the set of all possible configuration of the neighbors of v_i; and $\mathbf{X}$ represents the attributes of all customers. To estimate $P(y_{NB^i}|\mathbf{X_i}, \mathbf{z})$, Domingos and Richardson [18] employed the maximum entropy estimation to approximate the probability based on the assumption of independence, i.e.

$$P(y_{NB^i}|\mathbf{X_i}, \mathbf{z}) = \Pi_{v_j \in (NB^i)} P(y_j|\mathbf{X}, \mathbf{z}). \tag{11.20}$$

The marketing action M is modeled as a Boolean variable. The cost of marketing to a customer is further considered in the Markov model. Let r_0 be the revenue from selling the product to the customer if no marketing action is performed, and r_1 be the revenue if marketing is performed. The cost can be considered as offering a discount to the marketed customer. Thus the expected lift in profit from marketing to customer v_i in isolation (without influence) can be defined as follows:

$$ELP_i^1(Y, M_i) = r_i P(y_i = 1|Y, f_i^1(M)) - r_0 P(y_i = 1|Y, f_i^0(M_i)) - c, \tag{11.21}$$

where $f_i^0(M_i)$ is the result of setting M_i to 1 and leaving the rest of M unchanged, and similarly for $f_i^0(M_i)$. Thus the global lift in profit for a particular choice M is

$$ELP_i^1(Y, \mathbf{M}) = \sum_{i=1}^{n} [r_i P(X_i = 1|Y, \mathbf{M}) - r_0 P(X_i = 1|Y, M_0) - c_i]. \tag{11.22}$$

A customer's total value is the global lift in profit from marketing to him/her:

$$ELP\big(Y, f_i^1(M_i)\big) - ELP\big(Y, f_i^0(M_i)\big), \tag{11.23}$$

and his/her network value is the difference between his/her total and intrinsic values. The model can be also adjusted to a continuous version with no qualitative difference.

In the context of marketing, the goal of modeling the value of a customer is to find M that can maximize the lift in profit. Richardson and Domingos proposed several approximation algorithms in [18] to solve this problem. They made further contribution to this field in their later paper [20] through showing a tractable way to calculate the optimal M by directly solving the equation

$$r\Delta_i \frac{d\Delta P_i(z, Y)}{dz} = \frac{dc(z)}{dz}, \tag{11.24}$$

where z denotes the market action. The network effect $\Delta_i(Y) = \sum_{j=1}^{n} w_{ji} \Delta_j(Y)$ is the total increment in probability of purchasing in the network (including y_i) that results from a unit change in $P_0(y_i)$ when w_{ij} indicates how much v_j can influence v_i. $\Delta P_i(z, Y) = \beta_i[P_0(X_i = 1|Y, M)]$ denotes the immediate change in customer v_i's probability of purchasing when he is marketed to with action z.

Some other work aimed to find the optimal marketing strategy by directly maximizing the revenue rather than the social influence. Some investigations have been made in this field [85]. The basic idea is given as follows: since a customer who owns a product can have an impact on potential buyers, it is important to decide the sequence of marketing, as well as the price offered to the buyers. Thus a simple marketing strategy, called influence-and-exploit strategy, is introduced, which basically consists of an influence step and an exploit step. In the influence step, the seller starts by giving some products for free to some specially selected customers (hopefully the most influential ones), and in the exploit step, the seller trys to sell products to the remaining customers with a fixed optimal price. Hartline *et al.* [85] also proved that the influence-and-exploit method works as a reasonable approximation of the NP-hard problem of finding the optimal marketing strategy.

11.3.3 Diffusion network inference

In the above, we focused on discussing and introducing social influence from the perspective of measurements and influence propagation over social networks. In this subsection, we will introduce an interesting research issue which becomes increasingly important in information diffusion recently: inferring the hidden diffusion network based on information propagation logs. In many real scenarios, we can only observe the events that a node is activated by a contagion, such as the fact that she/he learns about a piece of information, makes a decision, adopts a new behavior, or becomes infected by a disease. However, the underlying network connectivity among these users on which the information, knowledge, and disease propagate is unknown and needs to be inferred. Inferring the underlying diffusion dynamics is important to facilitate us to gain new insights on information diffusion and enables us to forecast the potential influence of a new cascade.

The problem of diffusion network inference was first studied by Adar and Adamic [86]. It is a problem where dynamic information propagation processes are unfolded over an unobserved network, and the goal is to reconstruct the unobserved network based on the temporal dynamics of the propagation processes, namely cascades.

We formally describe this problem as follows. Given a hidden directed network G^*, we observe a set of cascades C spreading over it. Each propagation step of cascade C in the hidden network can be denoted as $(u_i, t_i, \phi_i)_c$, which means that cascade C infects node u_i at time t_i with a set of features or attributes ϕ_i. Note that we only observe the time t_u when cascade C infected node u without explicitly observing which node infects node u. Now, given a large volume of such cascades, the problem is how to infer the unobserved directed network G^* over which the cascades propagate (Figure 11.5) [87].

There are generally two types of algorithms proposed to solve this problem, namely the survival-theory-based model [88] and the submodularity, based model [42, 87]. Since the survival-theory-based model is much more popular and widely used in many different models, in this chapter we focus on introducing this method in detail [89–91].

The survival-theory-based model assumes that infections can occur at different rates over different edges of a network. Denote $f(t_i|t_j, \alpha_{j,i})$ as the conditional likelihood of

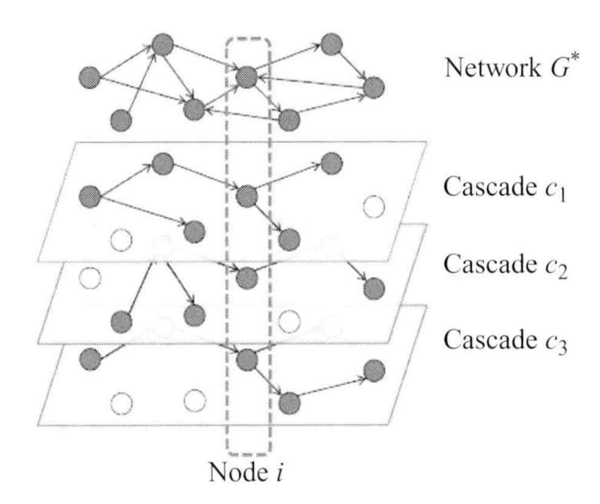

Figure 11.5 An illustration of the studied problem in [87]. The underlying diffusion network over which contagions spread is illustrated at the top of the figure. Each subsequent layer depicts a cascade formed by the diffusion of a particular contagion. For each cascade, the gray nodes are the "infected" nodes and the edges among these "infected" nodes form the propagation paths of the contagion. Given only the number of node infections in each cascade, the problem is how to infer the connectivity of the underlying network G^*.

transmission between node j and node i. From the formulation, we can see that the transmission likelihood between node i and j depends on their infection times (t_i, t_j) and the pairwise transmission rate $\alpha_{j,i}$. A node cannot be infected by a node infected later in time. In other words, a node j that has been infected at time t_j may infect a node i at time t_i only if $t_j < t_i$. Although in some scenarios it may be possible to estimate a non-parametric likelihood empirically, for simplicity we consider three well-known parametric models: exponential, power law, and Rayleigh (see Table 11.5). Transmission rates are denoted as $\alpha_{j,i} \geq 0$. As $\alpha_{j,i} \to 0$, the likelihood of infection tends to zero and the expected transmission time becomes arbitrarily long.

Exponential and power laws are monotonic models that have been previously used in modeling diffusion networks and social networks [42, 92]. Power-law model infections with long-tails. The Rayleigh model is a non-monotonic parametric model previously used in epidemiology [93]. It is well-adapted to model fads, where infection likelihood rises to a peak and then drops extremely rapidly.

Gomez-Rodriguez *et al.* recalled some additional standard notations [90]. The cumulative density function, denoted as $F(t_i|t_j;\alpha_{j,i})$, was computed from the transmission likelihoods. Given that a node j was infected at time t_j, the survival function of edge $j \to i$ was the probability that node i was not infected by the node j at time t_i:

$$S(t_i|t_j;\alpha_{j,i}) = 1 - F(t_i|t_j;\alpha_{j,i}). \tag{11.25}$$

The hazard function, or instantaneous infection rate, of edge $j \to i$ is the ratio

$$H(t_i|t_j;\alpha_{j,i}) = \frac{f(t_i|t_j;\alpha_{j,i})}{S(t_i|t_j;\alpha_{j,i})}. \tag{11.26}$$

For clarity, the transmission likelihood, log survival function, and hazard functions of the three models are given in Table 11.5.

Table 11.5 The transmission likelihood, log survival function, and hazard functions of the exponential, power law and Rayleigh models

Model	Transmission likelihood $f(t_i\|t_j;a_{j,i})$	Log survival function $logS(t_i\|t_j;a_{j,i})$	Hazard function $H(t_i\|t_j;a_{j,i})$
Exponential	$\begin{cases} \alpha_{j,i} \bullet e^{-\alpha_{j,i}(t_i-t_j)} & if\ t_j < t_i \\ 0 & otherwise \end{cases}$	$-\alpha_{j,i}(t_i - t_j)$	$\alpha_{j,i}$
Power law	$\begin{cases} \frac{\alpha_{j,i}}{\delta}\left(\frac{t_i-t_j}{\delta}\right)^{-1-\alpha_{j,i}} & if\ t_j + \delta < t_i \\ 0 & otherwise \end{cases}$	$-\alpha_{j,i}log\left(\frac{t_i-t_j}{\delta}\right)$	$\alpha_{j,i} \bullet \frac{1}{t_i-t_j}$
Rayleigh	$\begin{cases} \alpha_{j,i}(t_i - t_j)e^{-\frac{1}{2}\alpha_{j,i}(t_i-t_j)^2} & if\ t_j < t_i \\ 0 & otherwise \end{cases}$	$-\alpha_{j,i}\frac{(t_i-t_j)^2}{2}$	$\alpha_{j,i}(t_i - t_j)$

Based on the above definitions and formulations, the probability that a node remains uninfected until time T can be computed by Eqn (11.27). Then the conditional likelihoods can be computed by Eqn (11.28):

$$\Pi_{t_k \leq T} S(T|t_j; \alpha_{k,i}), \tag{11.27}$$

$$f(t^{\leq T}; A) = \Pi_{t_i \leq T} f(t_i|t_1, \ldots, t_N \backslash t_i; A). \tag{11.28}$$

The goal of this approach is to find the transmission rate $\alpha_{j,i}$ of each pair of nodes such that the likelihood of an observed set of cascades $C = \{t^1, \ldots, t^{|C|}\}$ is maximized. This goal can be achieved by solving the following optimization formula:

$$\begin{aligned} &min_A - \sum_{c \in C} logf(t^c; A) \\ &\text{subject to } \alpha_{j,i} \geq 0, i, j = 1, \cdots, N, i \neq j, \end{aligned} \tag{11.29}$$

where $A := \{\alpha_{j,i}|i, j = 1, \ldots, n, i \neq j\}$ are the variables of the information transmission rates on edges. The edges of the network are those pairs of nodes with transmission rates $\alpha_{j,i} > 0$.

There are also some studies using a submodularity-based approach to address this problem. As finding the optimal solution to the submodularity-based approach is NP-hard, Gomez-Rodriguez *et al.* proposed a heuristic algorithm using the greedy strategy to solve the problem [42, 87]. The algorithm starts with an empty graph, and iteratively, in step i, adds the edge e_i which maximizes the marginal gain. The submodularity property guarantees the efficiency of finding a provably near-optimal solution to this otherwise NP-hard optimization problem.

We introduced the basic formulation of the network inference problem and briefly presented its solutions. Recently, an increasing effort was made to uncover, model, and understand a broad range of propagation processes over a wide variety of network structures: propagation of information [42, 62, 86, 95], adoption of new products [21, 29, 35], diffusion of technical innovations [5], spread of chain letters [96], promotion of products via viral marketing [79, 82, 91, 97], spread of computer viruses [98], different topic diffusion underlying network [89], infectious diseases [93, 99, 100], and even the diffusion of human travel [101].

11.3.4 Challenges of IP&M

Existing approaches for opinion leader identification vary due to the differences among social networks. For example, the identification of opinion leaders in a Yahoo blog is quite different from that of Facebook. Each approach may have its own advantages for a particular network, but the advantages are limited. However, the big data era brings a variety of data, and more columns become available. Therefore, the problem of selecting proper columns will become a prominent challenge. The challenge has many aspects. For example, how to improve the accuracy when the computing resources are limited? How to adapt the approach to various social networks? Whether there exist any, and if so, what are these general rules for attribute selection?

Another challenge is the scalability of the algorithms and models in this field. Scalability is a key factor for extending them to large-scale social networks with millions or even billions of nodes. Previous solutions, such as the greedy algorithm proposed by Kempe *et al.* [79] and its improvements (e.g. high-degree heuristic, low-distance heuristic, degree discount heuristic algorithms and linear threshold model) are generally time consuming. Most algorithms for diffusion network inference also face the same challenges. The good news is that more and more researchers start to pay attention to this problem. Chen *et al.* [80] designed a heuristic algorithm that is easily scalable to millions of nodes and edges. However, with the explosive increase of network size, more efficient algorithms will be still in an urgent need to balance between the running time and the accuracy. Some researchers are considering the direction of parallel computation, and thus, increasing models and algorithms are designed on the parallel computing platform such as Hadoop and Spark.

11.4 Challenges in big data

Social influence researchers are starting to grapple with massive data sets, encountering challenges with handling, processing and analyzing the data that was once desired by social scientists. Big data has been characterized by "4*V*s," volume, velocity, and variety; the fourth "*V*" could be variability, virtual, or value. Therefore, the challenges of social influence research, especially for the computing aspect, are generally focused on the scalability, velocity and parallel computing.

There are various challenges of social influence measures. For network-based measures, since those models and algorithms are based on the original topology of social networks, one will have to handle the problems of computing velocity. Firstly, with the extension of social networks, the computing time of shortest path and random-walk-based measures are increasing dramatically. Secondly, for the design of parallel computing, researchers designed several parallel computing on large sparse matrix recently.

Such new algorithms are suitable for the distributed computing on parallel systems. The efficiency in turn motivates more researches and applications.

The interaction and topic based measures are also faced with the scalability problems. Since social media users interact in different ways, it is really challenging to measure influence and identify opinion leaders with heterogenous data. More column attributes will increase the computing cost, and the column selection will be another problem.

Nowadays, diffusion network inference is generally based on small networks, and researchers cannot guarantee the efficiency of these methods on large-scale social networks. The scalability of these approaches accordingly emerges as yet another new challenge.

11.5 Summary

Traditionally, social influence study aims at analyzing and modeling the influence phenomena over social networks qualitatively and quantitatively. In the big data era, with social medias deeply rooted in social lives day by day, the scale of social networks grows faster, and the content over social networks becomes increasingly diversified. The huge size, multiple attributes, and continuous growth of social networks bring formidable challenges. However, from another perspective, there will be more opportunities for both theoretical studies and practical applications. For the former, one needs to investigate new intriguing phenomena, so as to develop new theories and models. For the latter, social network services need more effective and efficient influence applications urgently. Generally, big data bring a spring for social influence research.

In this chapter, we focused on the quantitative research of social influence, and introduced the corresponding models and algorithms. Moreover, we analyzed their pros and cons in the context of big data. Our review covers most of the popular and latest research results. Firstly, we introduced the basic statistical measures of networks such as centrality, closeness and betweenness. Secondly, we presented the social influence and selection models. These covered the fundamental concepts on influence. Thirdly, we presented the influence maximization and its application in viral marketing and advertising. In the future, we expect that the challenges or opportunities of social influence will inspire more theoretical and practical contributions in this field. Specifically, a valuable research is the development of efficient and effective social influence mechanisms to enable various applications in social media. As to models and algorithms, researchers ought to develop more models and algorithms by improving the level of feature selection, algorithm design and evaluation approaches.

This area lies in the intersection of several disciplines such as computer science, sociology, and physics. In particular, scalable and parallel data mining algorithms, scalable database and web technology have been revolutionizing how sociologists approach this problem. Instead of building conceptual models and conducting small-scale simulations and user studies, large-scale data mining algorithms that analyze social network data have significantly impacted on this field. This chapter presented the history of social influence analysis, and discussed the challenges and opportunities of such research in the context of big data. This area is still in its infancy, and we anticipate that more techniques will be developed for this problem in the big data era.

Acknowledgement

The authors appreciate all the contributors of cited papers which our work is based upon, and would like to thank Dr. Zhaoyun Ding and the editors for their excellent work.

References

[1] D. M. Romero, W. Galuba, S. Asur, and B. A. Huberman, "Influence and passivity in social media," in *Machine Learning and Knowledge Discovery in Databases*, Springer, 2011, pp. 18–33.

[2] D. Easley and J. Kleinberg, *Networks, Crowds, and Markets: Reasoning about a Highly Connected World*, Cambridge University Press, 2010.

[3] L. Katz, "A new status index derived from sociometric analysis," *Psychometrika*, vol. **18**, no. 1, pp. 39–43, 1953.

[4] E. Katz and P. F. Lazarsfeld, *Personal Influence, The Part Played by People in the Flow of Mass Communications*, Transaction Publishers, 1970.

[5] E. M. Rogers, *Diffusion of Innovations*, Simon and Schuster, 2010.

[6] E. Keller and J. Berry, *The Influentials: One American in Ten Tells the Other Nine How to Vote, Where to Eat, and What to Buy*, Simon and Schuster, 2003.

[7] G. K. Al Mamunur Rashid and J. Riedl, "Influence in ratings-based recommender systems: an algorithm-independent approach," in *Proceedings of the SIAM International Conference on Data Mining*, SIAM, 2005.

[8] E. Bakshy, J. M. Hofman, W. A. Mason, and D. J. Watts, "Everyone's an influencer: quantifying influence on twitter," in *Proceedings of the Fourth ACM International Conference on Web Search and Data Mining*, ACM, 2011, pp. 65–74.

[9] M. Cha, H. Haddadi, F. Benevenuto, and P. K. Gummadi, "Measuring user influence in twitter: the million follower fallacy," *ICWSM*, vol. **10**, pp. 10–17, 2010.

[10] J. Weng, E.-P. Lim, J. Jiang, and Q. He, "Twitterrank: finding topic-sensitive influential twitterers," in *Proceedings of the Third ACM International Conference on Web Search and Data Mining*, ACM, 2010, pp. 261–270.

[11] J. Yang and J. Leskovec, "Modeling information diffusion in implicit networks," in *Data Mining (ICDM), 2010 IEEE 10th International Conference on*, IEEE, 2010, pp. 599–608.

[12] L. Backstrom, D. Huttenlocher, J. Kleinberg, and X. Lan, "Group formation in large social networks: membership, growth, and evolution," in *Proceedings of the 12th ACM SIGKDD International Conference on Knowledge Discovery and Data Mining*, ACM, 2006, pp. 44–54.

[13] D. Crandall, D. Cosley, D. Huttenlocher, J. Kleinberg, and S. Suri, "Feedback effects between similarity and social influence in online communities," in *Proceedings of the 14th ACM SIGKDD International Conference on Knowledge Discovery and Data Mining*, ACM, 2008, pp. 160–168.

[14] L. M. Aiello, A. Barrat, R. Schifanella, *et al.*, "Friendship prediction and homophily in social media," *ACM Transactions on the Web (TWEB)*, vol. **6**, no. 2, p. 9, 2012.

[15] D. Kempe, J. Kleinberg, and É. Tardos, "Influential nodes in a diffusion model for social networks," in *Automata, Languages and Programming*, Springer, 2005, pp. 1127–1138.

[16] J. J. Brown and P. H. Reingen, "Social ties and word-of-mouth referral behavior," *Journal of Consumer Research*, pp. 350–362, 1987.

[17] V. Mahajan, E. Muller, and F. M. Bass, "New product diffusion models in marketing: A review and directions for research," *The Journal of Marketing*, pp. 1–26, 1990.

[18] P. Domingos and M. Richardson, "Mining the network value of customers," in *Proceedings of the Seventh ACM SIGKDD International Conference on Knowledge Discovery and Data Mining*, ACM, 2001, pp. 57–66.

[19] J. Goldenberg, B. Libai, and E. Muller, "Talk of the network: a complex systems look at the underlying process of word-of-mouth," *Marketing Letters*, vol. **12**, no. 3, pp. 211–223, 2001.

[20] M. Richardson and P. Domingos, "Mining knowledge-sharing sites for viral marketing," in *Proceedings of the Eighth ACM SIGKDD International Conference on Knowledge Discovery and Data Mining*, ACM, 2002, pp. 61–70.

[21] J. Leskovec, L. A. Adamic, and B. A. Huberman, "The dynamics of viral marketing," *ACM Transactions on the Web (TWEB)*, vol. **1**, no. 1, p. 5, 2007.

[22] N. A. Christakis and J. H. Fowler, "The spread of obesity in a large social network over 32 years," *New England Journal of Medicine*, vol. **357**, no. 4, pp. 370–379, 2007.

[23] J. H. Fowler and N. A. Christakis, "Dynamic spread of happiness in a large social network: longitudinal analysis over 20 years in the framingham heart study," *BMJ*, vol. **337**, 2008.

[24] W. Dong and A. Pentland, "Modeling influence between experts," in *Artifical Intelligence for Human Computing*, Springer, 2007, pp. 170–189.

[25] J. Tang, J. Sun, C. Wang, and Z. Yang, "Social influence analysis in large-scale networks," in *Proceedings of the 15th ACM SIGKDD International Conference on Knowledge Discovery and Data Mining*, ACM, 2009, pp. 807–816.

[26] T. Sakaki, M. Okazaki, and Y. Matsuo, "Earthquake shakes twitter users: real-time event detection by social sensors," in *Proceedings of the 19th International Conference on World Wide Web*, ACM, 2010, pp. 851–860.

[27] E. Bakshy, D. Eckles, R. Yan, and I. Rosenn, "Social influence in social advertising: evidence from field experiments," in *Proceedings of the 13th ACM Conference on Electronic Commerce*, ACM, 2012, pp. 146–161.

[28] L. Rashotte, "Social influence," *The Blackwell Encyclopedia of Social Psychology*, vol. **9**, pp. 562–563, 2007.

[29] D. J. Watts and P. S. Dodds, "Influentials, networks, and public opinion formation," *Journal of Consumer Research*, vol. **34**, no. 4, pp. 441–458, 2007.

[30] R. A. Coulter, L. F. Feick, and L. L. Price, "Changing faces: cosmetics opinion leadership among women in the new hungary," *European Journal of Marketing*, vol. **36**, no. 11/12, pp. 1287–1308, 2002.

[31] W. Fei-Yue, "Study on cyber-enabled social movement organizations based on social computing and parallel systems," *Journal of University of Shanghai for Science and Technology*, vol. **1**, p. 003, 2011.

[32] D. Laney, "3D data management: controlling data volume, velocity," and variety. Technical report, META Group, Tech. Rep., 2001.

[33] F.-Y. Wang, "A big-data perspective on AI: Newton, Merton, and analytics intelligence," *Intelligent Systems, IEEE*, vol. **27**, no. 5, pp. 2–4, 2012.

[34] D. Centola, "An experimental study of homophily in the adoption of health behavior," *Science*, vol. **334**, no. 6060, pp. 1269–1272, 2011.
[35] S. Aral and D. Walker, "Identifying influential and susceptible members of social networks," *Science*, vol. **337**, no. 6092, pp. 337–341, 2012.
[36] S. P. Borgatti and M. G. Everett, "A graph-theoretic perspective on centrality," *Social Networks*, vol. **28**, no. 4, pp. 466–484, 2006.
[37] L. C. Freeman, "Centrality in social networks conceptual clarification," *Social Networks*, vol. **1**, no. 3, pp. 215–239, 1979.
[38] P. Bonacich, "Power and centrality: A family of measures," *American Journal of Sociology*, pp. 1170–1182, 1987.
[39] ——, "Factoring and weighting approaches to status scores and clique identification," *Journal of Mathematical Sociology*, vol. **2**, no. 1, pp. 113–120, 1972.
[40] M. E. Newman, "The structure and function of complex networks," *SIAM Review*, vol. **45**, no. 2, pp. 167–256, 2003.
[41] Xindong Wu, Yi Li, and Lei Li, "Influence analysis of online social networks," *Chinese Journal of Computers*, vol. **37**, no. 4, pp. 1–18, 2014.
[42] M. Gomez Rodriguez, J. Leskovec, and A. Krause, "Inferring networks of diffusion and influence," in *Proceedings of the 16th ACM SIGKDD International Conference on Knowledge Discovery and Data Mining*, ACM, 2010, pp. 1019–1028.
[43] A.-L. Barabasi, "The origin of bursts and heavy tails in human dynamics," *Nature*, vol. **435**, no. 7039, pp. 207–211, 2005.
[44] R. D. Malmgren, D. B. Stouffer, A. E. Motter, and L. A. Amaral, "A poissonian explanation for heavy tails in e-mail communication," *Proceedings of the National Academy of Sciences*, vol. **105**, no. 47, pp. 18 153–18 158, 2008.
[45] M. S. Granovetter, "The strength of weak ties," *American Journal of Sociology*, pp. 1360–1380, 1973.
[46] M. Granovetter, "Economic action and social structure: the problem of embeddedness," *American Journal of Sociology*, pp. 481–510, 1985.
[47] P. W. Holland and S. Leinhardt, "Transitivity in structural models of small groups." *Comparative Group Studies*, 1971.
[48] D. J. Watts and S. H. Strogatz, "Collective dynamics of 'small-world' networks," *Nature*, vol. **393**, no. 6684, pp. 440–442, 1998.
[49] C. E. Tsourakakis, "Fast counting of triangles in large real networks without counting: algorithms and laws," in *Data Mining, 2008. ICDM'08. Eighth IEEE International Conference on*, IEEE, 2008, pp. 608–617.
[50] L. C. Freeman, "A set of measures of centrality based on betweenness," *Sociometry*, pp. 35–41, 1977.
[51] M. Girvan and M. E. Newman, "Community structure in social and biological networks," *Proceedings of the National Academy of Sciences*, vol. **99**, no. 12, pp. 7821–7826, 2002.
[52] A. Java, P. Kolari, T. Finin, and T. Oates, "Modeling the spread of influence on the blogosphere," in *Proceedings of the 15th International World Wide Web Conference*, 2006, pp. 22–26.
[53] R. Xiang, J. Neville, and M. Rogati, "Modeling relationship strength in online social networks," in *Proceedings of the 19th International Conference on World Wide Web*, ACM, 2010, pp. 981–990.
[54] A. Goyal, F. Bonchi, and L. V. Lakshmanan, "Learning influence probabilities in social networks," in *Proceedings of the third ACM International Conference on Web Search and Data Mining*, ACM, 2010, pp. 241–250.

[55] P. Singla and M. Richardson, "Yes, there is a correlation:-from social networks to personal behavior on the web," in *Proceedings of the 17th International Conference on World Wide Web*, ACM, 2008, pp. 655–664.
[56] R. T. A. Leenders, "Modeling social influence through network autocorrelation: constructing the weight matrix," *Social Networks*, vol. **24**, no. 1, pp. 21–47, 2002.
[57] K. Saito, R. Nakano, and M. Kimura, "Prediction of information diffusion probabilities for independent cascade model," in *Knowledge-Based Intelligent Information and Engineering Systems*, Springer, 2008, pp. 67–75.
[58] K. Saito, M. Kimura, K. Ohara, and H. Motoda, "Selecting information diffusion models over social networks for behavioral analysis," in *Machine Learning and Knowledge Discovery in Databases*, Springer, 2010, pp. 180–195.
[59] M. Trusov, A. V. Bodapati, and R. E. Bucklin, "Determining influential users in internet social networks," *Journal of Marketing Research*, vol. **47**, no. 4, pp. 643–658, 2010.
[60] C. Tan, J. Tang, J. Sun, Q. Lin, and F. Wang, "Social action tracking via noise tolerant time-varying factor graphs," in *Proceedings of the 16th ACM SIGKDD International Conference on Knowledge Discovery and Data Mining*, ACM, 2010, pp. 1049–1058.
[61] H. Kwak, C. Lee, H. Park, and S. Moon, "What is twitter, a social network or a news media?" in *Proceedings of the 19th International Conference on World Wide Web*, ACM, 2010, pp. 591–600.
[62] D. M. Romero, B. Meeder, and J. Kleinberg, "Differences in the mechanics of information diffusion across topics: idioms, political hashtags, and complex contagion on twitter," in *Proceedings of the 20th International Conference on World Wide Web*, ACM, 2011, pp. 695–704.
[63] D. Gruhl, R. Guha, D. Liben-Nowell, and A. Tomkins, "Information diffusion through blogspace," in *Proceedings of the 13th International Conference on World Wide Web*, ACM, 2004, pp. 491–501.
[64] D. Krackhardt, "The strength of strong ties: the importance of philos in organizations," *Networks and Organizations: Structure, Form, and Action*, vol. **216**, p. 239, 1992.
[65] L. Liu, J. Tang, J. Han, M. Jiang, and S. Yang, "Mining topic-level influence in heterogeneous networks," in *Proceedings of the 19th ACM International Conference on Information and Knowledge Management*, ACM, 2010, pp. 199–208.
[66] P. Cui, F. Wang, S. Liu, *et al.*, "Who should share what?: item-level social influence prediction for users and posts ranking," in *Proceedings of the 34th International ACM SIGIR Conference on Research and Development in Information Retrieval*, ACM, 2011, pp. 185–194.
[67] S. A. Macskassy and M. Michelson, "Why do people retweet? Anti-homophily wins the day!" in *ICWSM*, 2011.
[68] C. Lee, H. Kwak, H. Park, and S. Moon, "Finding influentials based on the temporal order of information adoption in twitter," in *Proceedings of the 19th International Conference on World Wide Web*, ACM, 2010, pp. 1137–1138.
[69] E. Bakshy, B. Karrer, and L. A. Adamic, "Social influence and the diffusion of user-created content," in *Proceedings of the 10th ACM Conference on Electronic Commerce*, ACM, 2009, pp. 325–334.
[70] W. Pan, W. Dong, M. Cebrian, *et al.*, "Modeling dynamical influence in human interaction: Using data to make better inferences about influence within social systems," *Signal Processing Magazine, IEEE*, vol. **29**, no. 2, pp. 77–86, 2012.
[71] G. Ver Steeg and A. Galstyan, "Information transfer in social media," in *Proceedings of the 21st International Conference on World Wide Web*, ACM, 2012, pp. 509–518.

[72] ——, "Information-theoretic measures of influence based on content dynamics," in *Proceedings of the Sixth ACM International Conference on Web Search and Data Mining*, ACM, 2013, pp. 3–12.

[73] J. Huang, X.-Q. Cheng, H.-W. Shen, T. Zhou, and X. Jin, "Exploring social influence via posterior effect of word-of-mouth recommendations," in *Proceedings of the Fifth ACM International Conference on Web Search and Data Mining*, ACM, 2012, pp. 573–582.

[74] F. Provost, B. Dalessandro, R. Hook, X. Zhang, and A. Murray, "Audience selection for on-line brand advertising: privacy-friendly social network targeting," in *Proceedings of the 15th ACM SIGKDD International Conference on Knowledge Discovery and Data Mining*, ACM, 2009, pp. 707–716.

[75] A. Goyal, F. Bonchi, and L. V. Lakshmanan, "Discovering leaders from community actions," in *Proceedings of the 17th ACM Conference on Information and Knowledge Management*, ACM, 2008, pp. 499–508.

[76] N. Agarwal, H. Liu, L. Tang, and P. S. Yu, "Identifying the influential bloggers in a community," in *Proceedings of the 2008 International Conference on Web Search and Data Mining*, ACM, 2008, pp. 207–218.

[77] X. Song, Y. Chi, K. Hino, *et al.*, "Identifying opinion leaders in the blogosphere," in *Proceedings of the Sixteenth ACM Conference on Information and Knowledge Management*, ACM, pp. 971–974, 2007.

[78] F. M. Bass, "A new product growth for model consumer durables," *Management Science*, vol. **15**, no. 5, pp. 215–227, 1969.

[79] D. Kempe, J. Kleinberg, and É. Tardos, "Maximizing the spread of influence through a social network," in *Proceedings of the Ninth ACM SIGKDD International Conference on Knowledge Discovery and Data Mining*, ACM, 2003, pp. 137–146.

[80] W. Chen, Y. Wang, and S. Yang, "Efficient influence maximization in social networks," in *Proceedings of the 15th ACM SIGKDD International Conference on Knowledge Discovery and Data Mining*, ACM, 2009, pp. 199–208.

[81] W. Chen, C. Wang, and Y. Wang, "Scalable influence maximization for prevalent viral marketing in large-scale social networks," in *Proceedings of the 16th ACM SIGKDD International Conference on Knowledge Discovery and Data Mining*, ACM, 2010, pp. 1029–1038.

[82] J. Leskovec, A. Krause, C. Guestrin, *et al.*, "Cost-effective outbreak detection in networks," in *Proceedings of the 13th ACM SIGKDD International Conference on Knowledge Discovery and Data Mining*, ACM, 2007, pp. 420–429.

[83] G. Cornuejols, M. L. Fisher, and G. L. Nemhauser, "Exceptional paper-location of bank accounts to optimize float: An analytic study of exact and approximate algorithms," *Management Science*, vol. **23**, no. 8, pp. 789–810, 1977.

[84] G. L. Nemhauser, L. A. Wolsey, and M. L. Fisher, "An analysis of approximations for maximizing submodular set functions," *Mathematical Programming*, vol. **14**, no. 1, pp. 265–294, 1978.

[85] J. Hartline, V. Mirrokni, and M. Sundararajan, "Optimal marketing strategies over social networks," in *Proceedings of the 17th International Conference on World Wide Web*, ACM, 2008, pp. 189–198.

[86] E. Adar and L. A. Adamic, "Tracking information epidemics in blogspace," in *Web Intelligence, 2005. Proceedings, the 2005 IEEE/WIC/ACM International Conference on*, IEEE, 2005, pp. 207–214.

[87] M. Gomez-Rodriguez, J. Leskovec, and A. Krause, "Inferring networks of diffusion and influence," *ACM Transactions on Knowledge Discovery from Data (TKDD)*, vol. **5**, no. 4, p. 21, 2012.

[88] M. G. Rodriguez, D. Balduzzi, and B. Schölkopf, "Uncovering the temporal dynamics of diffusion networks," arXiv preprint arXiv:1105.0697, 2011.

[89] S. Wang, X. Hu, P. S. Yu, and Z. Li, "Mmrate: inferring multi-aspect diffusion networks with multi-pattern cascades," in *Proceedings of the 20th ACM SIGKDD International Conference on Knowledge Discovery and Data Mining*, ACM, 2014, pp. 1246–1255.

[90] M. Gomez Rodriguez, J. Leskovec, and B. Schölkopf, "Structure and dynamics of information pathways in online media," in *Proceedings of the Sixth ACM International Conference on Web Search and Data Mining*, ACM, 2013, pp. 23–32.

[91] N. Du, L. Song, M. Gomez-Rodriguez, and H. Zha, "Scalable influence estimation in continuous-time diffusion networks," in *Advances in Neural Information Processing Systems*, 2013, pp. 3147–3155.

[92] S. Myers and J. Leskovec, "On the convexity of latent social network inference," in *Advances in Neural Information Processing Systems*, 2010, pp. 1741–1749.

[93] J. Wallinga and P. Teunis, "Different epidemic curves for severe acute respiratory syndrome reveal similar impacts of control measures," *American Journal of Epidemiology*, vol. **160**, no. 6, pp. 509–516, 2004.

[94] J. F. Lawless, *Statistical Models and Methods for Lifetime Data*, John Wiley & Sons, 2011, vol. 362.

[95] J. Leskovec, M. McGlohon, C. Faloutsos, N. S. Glance, and M. Hurst, "Patterns of cascading behavior in large blog graphs." in *SDM*, vol. **7**, SIAM, 2007, pp. 551–556.

[96] D. Liben-Nowell and J. Kleinberg, "Tracing information flow on a global scale using internet chain-letter data," *Proceedings of the National Academy of Sciences*, vol. **105**, no. 12, pp. 4633–4638, 2008.

[97] T. Lappas, E. Terzi, D. Gunopulos, and H. Mannila, "Finding effectors in social networks," in *Proceedings of the 16th ACM SIGKDD International Conference on Knowledge Discovery and Data Mining*, ACM, 2010, pp. 1059–1068.

[98] C. Wang, J. C. Knight, and M. C. Elder, "On computer viral infection and the effect of immunization," in *Computer Security Applications, 2000, ACSAC'00, 16th Annual Conference*, IEEE, 2000, pp. 246–256.

[99] M. Lipsitch, T. Cohen, B. Cooper, *et al.*, "Transmission dynamics and control of severe acute respiratory syndrome," *Science*, vol. **300**, no. 5627, pp. 1966–1970, 2003.

[100] L. Hufnagel, D. Brockmann, and T. Geisel, "Forecast and control of epidemics in a globalized world," *Proceedings of the National Academy of Sciences*, vol. **101**, no. 42, pp. 15 124–15 129, 2004.

[101] D. Brockmann, L. Hufnagel, and T. Geisel, "The scaling laws of human travel," *Nature*, vol. **439**, no. 7075, pp. 462–465, 2006.

Part IV

Big data over biological networks

12 Inference of gene regulatory networks: validation and uncertainty

Xiaoning Qian, Byung-Jun Yoon, and Edward R. Dougherty

A fundamental problem of biology is to construct gene regulatory networks that characterize the operational interaction among genes. The term "gene" is used generically because such networks could involve gene products. Numerous inference algorithms have been proposed. The validity, or accuracy, of such algorithms is of central concern. Given data generated by a ground-truth network, how well does a model network inferred from the data match the data-generating network? This chapter discusses a general paradigm for inference validation based on defining a distance between networks and judging validity according to the distance between the original network and the inferred network. Such a distance will typically be based on some network characteristics, such as connectivity, rule structure, or steady-state distribution. It can also be based on some objective for which the model network is being employed, such as deriving an intervention strategy to apply to the original network with the aim of correcting aberrant behavior. Rather than assuming that a single network is inferred, one can take the perspective that the inference procedure leads to an "uncertainty class" of networks, to which belongs the ground-truth network. In this case, we define a measure of uncertainty in terms of the cost that uncertainty imposes on the objective, for which the model network is to be employed, the example discussed in the current chapter involving intervention in the yeast cell cycle network.

12.1 Introduction

From a translational perspective, we are interested in gene regulatory networks (GRNs) as a vehicle to derive optimal intervention strategies for regulatory pathologies, cancer being the salient example (see [1–3] for reviews and [4] for extensive coverage). Two basic intervention approaches have been considered for gene regulatory networks in the context of probabilistic Boolean networks (PBNs), external control and structural intervention [4], a key to intervention being that the dynamic behavior of a PBN can be modeled by a Markov chain, thereby making intervention in PBNs amenable to the theory of Markov decision processes. Perhaps we should note that the ability of Markov chains to model GRNs has a long history in translational genomics [5]. *External control*

Big Data over Networks, ed. Shuguang Cui, Alfred O. Hero III, Zhi-Quan Luo, and José M. F. Moura.
Published by Cambridge University Press.

is based on externally manipulating the value of a control gene to beneficially alter the steady-state distribution, either indirectly via a one-step cost function [6] or directly via an objective function based on the steady-state distribution [7, 8]. *Structural intervention* involves a one-time change of the network structure (wiring) to beneficially alter the steady-state distribution [9, 10].

A pre-requisite to the application of control to a GRN is construction of a GRN that provides sufficient modeling of the actual regulatory regime so that an intervention strategy derived from the GRN performs well in practice. The availability of high-throughput genomic data has motivated a host of algorithms to infer gene regulatory networks, from coarse-grained discrete networks to stochastic differential equations [11–15]. A naive perspective might lead one to believe that the flood of data might make the inference problem easy; on the contrary, even binary networks are sufficiently complex that their successful inference requires efficient generation and use of data. Inference validity (accuracy) concerns the degree to which an inferred "model network" matches, in some defined manner, the "ground-truth network" from which the data have been generated. Validation is defined via a distance between networks, or the distance between two structures of the same kind deduced from the networks, such as their steady-state distributions [16, 17]. As a function from a sample data set to a class of network models, an inference procedure is a mathematical operator and its performance must be evaluated within a mathematical framework, in this case, distance functions. Unfortunately, there has been very little study devoted to network validation. There are many subtle statistical issues. If we are to be able to judge the worth of proposed algorithms, then these issues need to be addressed within a formal mathematical framework.

The organization of this chapter is as follows: Section 12.2 provides a review on modeling, controlling, and inference of regulatory networks based on Markovian logical network models, which include PBNs. Several network distance functions are introduced in Section 12.3 for validation of network inference algorithms. Based on these introduced network distances, the performance evaluation of network inference algorithms is presented in Section 12.4. Practical issues of inferring and validation of gene regulatory networks based on time-course observations, including consistency (Section 12.5) and distance estimates through approximation (Section 12.6) as we do not know the actual or "true" gene regulatory networks, are discussed with an inference example based on metastatic melanoma gene expression profiles in Section 12.7. Finally, Section 12.8 presents an objective-based uncertainty quantification (UQ) framework to provide guidelines on network inference and experiment design for data collection when the available data are not sufficient to infer high-quality network models for corresponding operational objectives, including disease prognosis or gene-based therapeutic design.

12.2 Background

There are many kinds of regulatory networks, the common aspect being that the state of the network at time T is characterized (possibly stochastically) in terms of the states of the network at times $t < T$. We confine ourselves to discrete networks. The network is composed of a finite node (gene) set, $V = \{X_1, X_2, \ldots, X_n\}$, each node taking values in $\{0, 1, \ldots, d-1\}$. The state space possesses $D = d^n$ states, denoted by $\mathbf{x}_1, \mathbf{x}_2, \ldots, \mathbf{x}_D$, with $\mathbf{x}_j = (x_{j1}, x_{j2}, \ldots, x_{jn})$. For notational ease, we write vectors in row form but treat them as columns when multiplied by a matrix. The corresponding dynamical system is based on the state-vector transition $\mathbf{X}(t) \to \mathbf{X}(t+1)$, $t = 0, 1, 2, \ldots$. The state $\mathbf{X} = (X_1, X_2, \ldots, X_n)$ is often referred to as a gene activity profile (GAP).

We focus on logical regulatory network models based on Markov chains. The state space and transition probability structure of the underlying Markov chains for corresponding regulatory network models are introduced in Section 12.2.1. Specific examples, including Boolean networks (BNs), Boolean networks with perturbation (BNps), and probabilistic Boolean networks (PBNs) are presented in Section 12.2.2. Both intervention (Section 12.2.3) and inference (Section 12.2.4) of these types of regulatory networks are briefly reviewed.

12.2.1 Markov chains

We assume that the process $\mathbf{X}(t)$ is a Markov chain, meaning that the probability of $\mathbf{X}(t)$ conditioned on $\mathbf{X}$ at $t_1 < t_2 < \cdots < t_s < t$ is equal to the probability of $\mathbf{X}(t)$ conditioned on $\mathbf{X}(t_s)$. We also assume the process is *homogeneous*, meaning that the transition probabilities depend only on the time difference; that is, for any t and u, the u-step transition probability, $p_{jk}(u) = P(\mathbf{X}(t+u) = \mathbf{x}_k | \mathbf{X}(t) = \mathbf{x}_j)$, depends only on u. Under these assumptions, we need to only consider the one-step *transition probability matrix*,

$$\mathbf{P} = \begin{pmatrix} p_{11} & p_{12} & \cdots & P_{1D} \\ p_{21} & p_{22} & \cdots & P_{2D} \\ \vdots & \vdots & \ddots & \vdots \\ p_{D1} & p_{D2} & \cdots & P_{DD} \end{pmatrix},$$

where $p_{jk} = p_{jk}(1)$ is the one-step transition probability. For $t = 0, 1, 2, \ldots,$ the state probability structure of the network is given by the t-state-probability vector $\mathbf{p}(t) = (p_1(t), p_2(t), \ldots, p_D(t))$, where $p_j(t) = P(\mathbf{X}(t) = \mathbf{x}_j)$. $\mathbf{p}(0)$ is the initial-state probability vector.

The gene one-step probabilities are given by $p_i(j, r) = P(X_i(t+1) = r | \mathbf{X}(t) = \mathbf{x}_j)$. If, given the GAP at t, the conditional probabilities of the genes are independent, then $p_{jk} = p_1(j, x_{k1}) p_2(j, x_{k2}) \cdots p_n(j, x_{kn})$. Suppose that gene X_i at time $t+1$ depends only on values of genes in a regulatory set, $R_i \subset V$, at time t, the dependency being independent of t. Then the gene one-step probabilities are given by

$$p_i(j, r) = P(X_i(t+1) = r | X_l(t) = x_{jl} \text{ for } X_l \in R_i).$$

In this form, we see that the Markov dependencies are restricted to regulatory genes.

The network has a *regulatory graph* consisting of the n genes and a directed edge from gene X_i to gene X_j if $X_i \in R_j$. There is also a *state-transition graph* whose nodes are the D state vectors. There is a directed edge from state $\mathbf{x}_j$ to state $\mathbf{x}_k$ if and only if $\mathbf{x}_j = \mathbf{X}(t)$ implies $\mathbf{x}_k = \mathbf{X}(t+1)$.

A homogeneous, discrete-time Markov chain with state space $\{\mathbf{x}_1, \mathbf{x}_2, \ldots, \mathbf{x}_D\}$ possesses a steady-state distribution $(\pi_1, \pi_2, \ldots, \pi_D)$ if, for all pairs of states $\mathbf{x}_k$ and $\mathbf{x}_j$, $p_{jk}(u) \to \pi_k$ as $u \to \infty$. If there exists a steady-state distribution, then, for any k, the probability of the Markov chain being in state $\mathbf{x}_k$ in the long run is π_k: specifically, for any initial distribution $\mathbf{p}(0)$, $p_k(t) \to \pi_k$ as $t \to \infty$.

12.2.2 Logical regulatory networks

A basic type of regulatory model occurs when the transition $\mathbf{X}(t) \to \mathbf{X}(t+1)$ is governed by a logic structure, meaning there exists a logical state function $\mathbf{f} = (f_1, f_2, \ldots, f_n)$ such that $X_i(t+1) = f_i(R_i(t))$. The archetypal example is a Boolean network (BN), where the values are binary, 0 or 1, and the function f_i can be defined via a logic expression or a truth table consisting of 2^n rows, with each row assigning a 0 or 1 as the value for the GAP defined by the row [18]. A BN is deterministic and the entries in its transition probability matrix are either 0 or 1. The *connectivity* of the BN is the maximum number of predictors for a gene. If each gene has the same number of predictors, then we say that the network has *uniform connectivity*. The long-run behavior of a deterministic BN depends on the initial state and the network will eventually settle down and cycle endlessly through a set of states called an *attractor cycle*. The set of all initial states that reach a particular attractor cycle forms the *basin of attraction* for the cycle. Attractor cycles are disjoint.

A *Boolean network with perturbation* (BNp) results from assuming that there is an i.i.d. random perturbation vector $\gamma \in \{0, 1\}^n$ such that the ith gene flips if the ith component of γ is equal to 1, with perturbation probability $p = P(\gamma_i = 1)$ for all the genes. With perturbation, $\mathbf{X}(t+1) = \mathbf{f}(\mathbf{X}(t))$ with probability of $(1-p)^n$ and $\mathbf{X}(t+1) = \mathbf{X}(t) \oplus \gamma$ with probability $1-(1-p)^n$, where $\oplus$ is component-wise addition modulo 2. Perturbation makes the corresponding Markov chain of a BNp irreducible and ergodic. Hence, the network possesses a steady-state distribution. A BNp inherits the attractor structure from the original Boolean network without perturbation, the difference being that a random perturbation can cause a BNp to jump out of an attractor cycle, perhaps then transitioning to a different attractor cycle prior to another perturbation.

A binary *probabilistic Boolean network* (PBN) is composed of a family of m BNps together with probabilities governing the selection of a BNp at each time point (a PBN can be defined using BNs but we restrict our attention to BNps in order to guarantee a steady-state distribution) [19]. The m constituent BNps are characterized by m network functions, $\mathbf{f}^1, \mathbf{f}^2, \ldots, \mathbf{f}^m$. At any time point there is a positive probability q of switching from the current governing constituent BNp to another. By definition, a PBN inherits the attractor cycles of its constituent BNps. If $q = 1$, the PBN is said to be *instantaneously random*; if $q < 1$, the PBN is said to be *context-sensitive*, each constituent BNp being referred to as a context (see [2] for computation of the transition probability matrix for a context-sensitive PBN). If a context switch is called for, there are probabilities $c_1, c_2, \ldots, c_m$ governing context selection. Although here we have defined PBNs as having binary values, the general definition assumes that each gene can take a finite number of values, say, in the set $\{0, 1, \ldots, d-1\}$.

12.2.3 Control policy for maximal steady-state alteration

For network intervention for regulatory pathologies, we consider an infinite-horizon policy that achieves maximal steady-state alteration [7] for the maximum shift of steady-state mass from undesirable to desirable states. To define desirable and undesirable network states, one way is based on the values of given genetic markers. For example, the up-regulation of gene WNT5A has been associated with increased risk of metastasis in melanoma [20, 21]. WNT5A can be considered as a genetic marker and it is reasonable to consider network states with WNT5A up-regulated as undesirable states while the desirable states would be those in which WNT5A is down-regulated. With defined desirable and undesirable states denoted as $\mathcal{D}$ and $\mathcal{U}$ respectively, we can derive an external control policy to change the steady-state distribution. Let $\{\mathbf{x}_t, t = 0, 1, \ldots\}$ be the sequence of network state transitions without control and let $\{a_t, t = 0, 1, \ldots\}$ be the control actions which, for example, can admit no intervention with $a_t = 0$ or flip the expression level of a control gene when $a_t = 1$. We aim to find an intervention policy to maximally shift the steady-state mass of undesirable states to desirable ones. The derivation of an intervention policy is to prescribe for taking actions at each time point t in accordance with a random mechanism, possibly a function of the entire history of the system up to time t. Let $h_t = (\mathbf{x}_0, a_0, \mathbf{x}_1, a_1, \ldots, \mathbf{x}_t, a_t)$ denote the observed history till time t. Let $\mathcal{A} = \mathcal{A}(\mathbf{x}_t) = \{0, 1\}$ be the set of intervention actions for any state $\mathbf{x}_t$ in the state space. If the history h_{t-1} is observed up to time t, then the decision maker chooses an action $a \in \mathcal{A}(\mathbf{x}_t)$ with probability $\upsilon_t(a \mid h_{t-1}, \mathbf{x}_t)$, which characterizes the stochasticity of the underlying constrained Markov decision process (CMDP). The amount of shift in the aggregated probability of undesirable states for a PBN controlled under υ is defined as

$$\Delta\pi_{\mathcal{U}}(\upsilon) = \sum_{\mathbf{x}_j \in \mathcal{U}} \pi_{\mathbf{x}_j} - \sum_{\mathbf{x}_j \in \mathcal{U}} \pi_{\mathbf{x}_j}(\upsilon), \tag{12.1}$$

where π and $\pi(\upsilon)$ are the steady-state vectors for the Markov chains governed by the original and controlled PBNs, respectively. The goal is to maximize $\Delta\pi_{\mathcal{U}}(\upsilon)$. This problem has a stationary (time-invariant) and deterministic optimal solution through solving a linear programming problem, which we refer to as the maximal steady-state alteration (MSSA) algorithm [7].

12.2.4 Inference algorithms

We are interested in inferring regulatory relationships among genes so that we can design effective intervention strategies based on learned regulatory networks [22]. We focus on network inference algorithms for PBNs from one or several time series of observed gene expression states $\mathbf{x}_t$, which find both regulatory relationships among genes and the corresponding Boolean functions for each gene to "consistently" explain the observed state transitions in time series data.

For example, REVEAL [23] identifies predictors for each gene by estimating the mutual information between the temporal profile of each gene and all the combination profiles of potential regulators, starting from one regulator per gene. The algorithm requires an exponential number of state transitions in the observed time course data, with respect to the number of genes n in the network, to find a unique solution in the worst case. However, as most of biological networks are sparse [24, 25], REVEAL works effectively in practice, and Akutsu *et al.* [26] also have proven that only $O(\log n)$ state transitions are required when the maximum size of the regulatory sets ($K = \max_{i=1}^{n} |R_i|$) or the connectivity of the network is small. However, the original REVEAL algorithm and the exhaustive algorithm in [26] focus on inferring BNs instead of PBNs and require finding "consistent" Boolean functions for each gene. They assume that the observed time course data themselves are completely consistent based on underlying Boolean functions without errors.

With random perturbations introduced in PBNs, instead of finding consistent Boolean functions, the inference algorithm Best-Fit [27–29] searches for the best-fit function for each gene by exhaustively searching for all combinations of potential regulator sets. Similarly, with small regulatory sets, the algorithm is feasible with a given number of state transitions and is efficient with the time complexity $O(M \log M \text{poly}(n))$ with M state transitions, in which $\text{poly}(n)$ is time to compute the minimum error for one given state transition [28]. For our implementations [27, 28, 30] based on both REVEAL and Best-Fit algorithms, we have modified the algorithms to get both regulator sets and corresponding best-fit functions. Finally, with a limited number of observed state transitions and potential random perturbations, the inferred regulatory functions may be *partially defined Boolean functions* [28]. To obtain a unique solution, we can further impose other biologically motivating constraints based on the attractor structure of network dynamics as in [31, 32]. Another direction to constrain the solution space is to adopt the MDL-based network inference algorithm [33–35] to penalize the model complexity of inferred networks. We have modified the algorithm proposed in [34] to identify the best regulatory set with the minimum combination of network coding length, capturing the model complexity, and data coding length, which is similar to REVEAL based on mutual information. The MDL network coding length in [34] has similar asymptotic performance to the Bayesian information criterion (BIC) model complexity, which we also have implemented in our set of inference algorithms. Finally, both MDL [34] and BIC [30] adopt ad-hoc measures of model description length that necessitate tuning parameters as weighting coefficients to balance the model and data coding lengths [33, 35] and inference performances or validity measures may change with different tuning parameters. The universal MDL (uMDL) network inference algorithm [35] derives a theoretical measure for the model and data coding length together from a universal normalized maximum likelihood model and no tuning parameters are needed [33].

12.3 Network distance functions

In this section, we first state the required properties of distance functions to measure differences between networks in Section 12.3.1. Different network distance functions focusing on different aspects of network models, including network topology, dynamics, and intervention, are then described in detail.

12.3.1 Semi-metrics

Given networks $\mathcal{H}$ and $\mathcal{M}$, we require a function $\mu(\mathcal{M}, \mathcal{H})$ quantifying the difference between them to be a semi-metric, meaning that it satisfies the following four properties:

(1) $\mu(\mathcal{M}, \mathcal{H}) \geq 0$,
(2) $\mu(\mathcal{M}, \mathcal{M}) = 0$,
(3) $\mu(\mathcal{M}, \mathcal{H}) = \mu(\mathcal{H}, \mathcal{M})$ [symmetry],
(4) $\mu(\mathcal{M}, \mathcal{H}) \leq \mu(\mathcal{M}, \mathcal{N}) + \mu(\mathcal{N}, \mathcal{H})$ [triangle inequality].

As a semi-metric, μ is called a distance function. If μ satisfies a fifth condition,

(5) $\mu(\mathcal{M}, \mathcal{H}) = 0 \Rightarrow \mathcal{M} = \mathcal{H}$,

then it is a metric. A distance function is often defined in terms of some characteristic, by which we mean some structure associated with a network, such as its regulatory graph, steady-state distribution, or probability transition matrix. This is why we do not require the fifth condition, $\mu(\mathcal{M}, \mathcal{H}) = 0 \Rightarrow \mathcal{M} = \mathcal{H}$, for a network distance function.

12.3.2 Rule-based distance

For Boolean networks (with or without perturbation) possessing the same gene set, a distance is given by the proportion of incorrect rows in the function-defining truth tables. Denoting the state functions for networks $\mathcal{H}$ and $\mathcal{M}$ by $\mathbf{f} = (f_1, f_2, \ldots, f_n)$ and $\mathbf{g} = (g_1, g_2, \ldots, g_n)$, respectively, since there are n truth tables consisting of $D = 2^n$ rows each, this distance is given by

$$\mu_{fun}(\mathcal{M}, \mathcal{H}) = \frac{1}{n2^n} \sum_{i=1}^{n} \sum_{k=1}^{D} I[f_i(\mathbf{x}_k) \neq g_i(\mathbf{x}_k)], \tag{12.2}$$

where I denotes the indicator function, $I[A] = 1$ if A is a true statement and $I[A] = 0$ otherwise. If we wish to give more weight to those states more likely to be observed in the steady state, then we can weight the inner sums in (12.2) by the corresponding terms in the steady-state distribution, $\pi = (\pi_1, \pi_2, \ldots, \pi_D)$. For BNs without perturbation, μ_{fun} is a metric. If there is perturbation, then μ_{fun} is not a metric because two distinct networks may be identical with regard to the rules but possess different perturbation probabilities.

The definition of μ_{fun} is modified for PBNs by including another summation over the m contexts:

$$\mu_{fun}(\mathcal{M},\mathcal{H}) = \frac{1}{mn2^n}\sum_{l=1}^{m}\sum_{i=1}^{n}\sum_{k=1}^{D} I\left[f_i^l(\mathbf{x}_k) \neq g_i^l(\mathbf{x}_k)\right], \tag{12.3}$$

where the network functions now take the forms $\mathbf{f}^l$ and $\mathbf{g}^l$, $l = 1, 2, \ldots, m$.

12.3.3 Topology-based distance

To focus on network topology, we construct the adjacency matrix. For $i, j = 1, 2, \ldots, n$, the (i, j) matrix entry is 1 if there is a directed edge from the ith to the jth gene; otherwise, the (i, j) entry is 0. If $A = (a_{ij})$ and $B = (b_{ij})$ are the adjacency matrices for networks $\mathcal{H}$ and $\mathcal{M}$, respectively, where $\mathcal{H}$ and $\mathcal{M}$ possess the same gene set, then the *Hamming distance* between $\mathcal{H}$ and $\mathcal{M}$ is

$$\mu_{ham}(\mathcal{M},\mathcal{H}) = \sum_{i,j=1}^{n} |a_{ij} - b_{ij}|. \tag{12.4}$$

Alternatively, the Hamming distance may be computed by normalizing the sum, such as by the number of genes or the number of edges in one of the networks, for instance, when one of the networks is considered as representing ground truth.

As defined, we are considering directed graphs. If we consider undirected graphs, then an edge is either present or absent. In this case, the Hamming distance is still defined by (12.4) but the adjacency matrix is symmetric.

12.3.4 Transition-probability-based distance

Distances for Markov networks can be defined via their probability transition matrices by considering matrix norms. A norm is a function $\|\bullet\|$ on a linear (vector) space, $\mathcal{L}$, such that:

(1) $\|v\| \geq 0$,
(2) $\|v\| = 0 \Rightarrow v = 0$,
(3) $\|av\| = |a| \cdot \|v\|$ [homogeneity],
(4) $\|v + w\| \leq \|v\| + \|w\|$ [triangle inequality].

Given a norm on $\mathcal{L}$, a metric is defined on $\mathcal{L}$ by $\|v - w\|$.

For an $n \times n$ matrix and $r \geq 1$, the r-norm is defined by

$$\|\mathbf{P}\|_r = \left(\sum_{i,j=1}^{D} |p_{ij}|^r\right)^{\frac{1}{r}}. \tag{12.5}$$

The *supremum norm* is defined by $\|\mathbf{P}\|^{\infty} = \max\{|p_{ij}|; i, j = 1, 2, \ldots, D\}$. A norm induces a metric via $\|\mathbf{P} - \mathbf{Q}\|_r$. If $\mathbf{P} = (p_{ij})$ and $\mathbf{Q} = (q_{ij})$ are the probability transition matrices for networks $\mathcal{H}$ and $\mathcal{M}$, respectively, then a network distance function is defined by

$$\mu^r_{prob}(\mathcal{M}, \mathcal{H}) = \|\mathbf{P} - \mathbf{Q}\|_r. \quad (12.6)$$

Whereas $\| \bullet \|_r$ defines a matrix metric, μ^r_{prob} is only a network semi-metric because two distinct networks may have the same transition probability matrix.

12.3.5 Steady-state distance

Since steady-state behavior is of particular interest, for instance, being associated with phenotypes, a natural choice for a network distance is to measure the difference between steady-state distributions [15]. If $\pi = (\pi_1, \pi_2, \ldots, \pi_D)$ is a probability vector, then its r-norm is defined by

$$\|\pi\|_r = \left(\sum_{i=1}^{D} |\pi_i|^r\right)^{\frac{1}{r}} \quad (12.7)$$

for $r \geq 1$, and its supremum norm is defined by $\|\pi\|_\infty = \max\{|\pi_i| : i = 1, 2, \ldots, D\}$. If $\pi = (\pi_1, \pi_2, \ldots, \pi_D)$ and $\omega = (\omega_1, \omega_2, \ldots, \omega_D)$ are the steady-state distributions for networks $\mathcal{H}$ and $\mathcal{M}$, respectively, then a network distance is defined by

$$\mu^r_{ss}(\mathcal{M}, \mathcal{H}) = \|\pi - \omega\|_r. \quad (12.8)$$

Other norms can be used to define the distance function.

Other long-run distances can be defined, including distances for networks that do not possess a steady-state distribution or distances defined in terms of trajectories [16].

12.3.6 Control-based distance

Assuming that we are interested in an impaired biological system that has a higher risk of entering into aberrant phenotypes, from the collected measurements, our goal is to design effective stationary control policies to reduce the risk of entering into these undesirable or bad states. Previous distance functions are solely interested in the network itself. However, in this scenario, inference procedures should be evaluated in regard to our final objective of effectively reducing the undesirable risk by evaluating the control performance of intervention strategies derived using the network model inferred from observed data. In fact, in a real-world scenario, we typically do not have the ground truth of the underlying system and this type of objective-based validity measure may be the only reasonable framework for network inference validation.

The definition of an objective-based validity measure relative to controllability is not a semi-metric; nonetheless, for consistency we call it a distance function. Let $\mathcal{H}$ and $\mathcal{M}$ denote two networks, with intervention strategies $\upsilon_{\mathcal{H}}$ and $\upsilon_{\mathcal{M}}$ being the maximal steady-state alteration policies for $\mathcal{H}$ and $\mathcal{M}$, respectively, and $\pi^{\mathcal{H}}$ and $\pi^{\mathcal{M}}$ being the steady-state vectors for $\mathcal{H}$ controlled by $\upsilon_{\mathcal{H}}$ and $\upsilon_{\mathcal{M}}$, respectively. Then the distance function relative to controllability is defined by

$$\mu_{ctrl}(\mathcal{M}, \mathcal{H}) = \sum_{i \in \mathcal{U}} \pi_i^{\mathcal{M}} - \sum_{i \in \mathcal{U}} \pi_i^{\mathcal{H}}, \tag{12.9}$$

where $\mathcal{U}$ is again the class of undesirable states.

12.4 Inference performance

An inference procedure operates on data generated by a network $\mathcal{H}$ and constructs an inferred network $\mathcal{M}$ to serve as an estimate of $\mathcal{H}$, or it constructs a characteristic to serve as an estimate of the corresponding characteristic of $\mathcal{H}$. For instance, the data may be used to infer a distribution that estimates the steady-state distribution of $\mathcal{H}$. The data could be dynamical, consisting of time-course observations, or it might be taken from the steady state. In the latter case, it makes sense to consider inference accuracy relative to the steady-state distribution of $\mathcal{H}$, rather than $\mathcal{H}$ itself. For full network inference, the inference procedure is a mathematical operation, a mapping from a space of samples to a space of networks. There is a sample data set S and the inference procedure is of the form $\mathcal{M} = \psi(S)$. If a related network characteristic is being estimated, then $\psi(S)$ is a characteristic of interest, for instance, $\psi(S) = F$, a probability distribution.

We present a mathematical framework to measure inference performance based on network distance functions in Section 12.4.1. Both analytic and empirical illustrations of network inference performance evaluation are provided in Section 12.4.2 and Section 12.4.3, respectively.

12.4.1 Measuring inference performance using distance functions

Focusing on full network inference, the goodness of an inference procedure ψ is measured relative to some distance $\mu(\mathcal{M}, \mathcal{H}) = \mu(\psi(S), \mathcal{H})$, which is a function of the sample S, which is itself a realization of a random process, Σ, governing data generation. The data might be directly generated by $\mathcal{H}$ or might result from directly generated data corrupted by noise; $\mu(\psi(\Sigma), \mathcal{H})$ is a random variable and the performance of ψ is characterized by the distribution of $\mu(\psi(\Sigma), \mathcal{H})$, which depends on the distribution of Σ. The salient statistic is the expectation, $E_\Sigma[\mu(\psi(\Sigma), \mathcal{H})]$, taken with respect to Σ.

Rather than considering a single network, we can consider a distribution, H, of random networks, where, by definition, the occurrences of realizations $\mathcal{H}$ of H are governed by a probability distribution. Averaging over the class of random networks, our interest focuses on

$$\mu^*(\mathrm{H}, \Sigma, \psi) = E_{\mathrm{H}}[E_\Sigma[\mu(\psi(\Sigma), \mathcal{H})]].$$

Inference procedure ψ_1 is better than inference procedure ψ_2 relative to the distance μ, the random network H, and the sampling procedure Σ if $\mu^*(\mathrm{H}, \Sigma, \psi_1) < \mu^*(\mathrm{H}, \Sigma, \psi_2)$. Whether an inference procedure is "good" is relative to the distance function and how one views the value of the expected distance. It is not really possible to determine an absolute notion of goodness.

In practice, the expectation is estimated by an average,

$$\mu^*(\mathrm{H}, \Sigma, \psi) = \frac{1}{N}\sum_{j=1}^{N} \mu(\psi(S_j), \mathcal{H}_j), \tag{12.10}$$

where $S_1, S_2, \ldots, S_N$ are sample point sets generated according to Σ from networks $\mathcal{H}_1, \mathcal{H}_2, \ldots, \mathcal{H}_N$ randomly chosen from H.

When estimating a characteristic, replace $\mathcal{H}$ and H by λ and Λ, where λ and Λ are a characteristic and a random characteristic, respectively, and replace the network distance μ by the characteristic distance.

12.4.2 Analytic example

In the case of a BNp, $\mathcal{H}$, using best-fit inference, it is possible to give an exact expression for $E_\Sigma[\mu(\psi(\Sigma), \mathcal{H})]$. Referring to (12.2), let **f** and **g** be the true and inferred state functions, respectively. For gene X_i, in the sample data, let $U_0(k; i)$ and $U_1(k; i)$ denote the number of instances in which $X_i(t+1) = 0$ and $X_i(t+1) = 1$, given state $\mathbf{x}_k$. Assuming no connectivity restriction, the best-fit function is the one that agrees the best with the data. Hence, it is defined by $g_i(\mathbf{x}_k) = 0$ if $U_0(k; i) > U_1(k; i)$, $g_i(\mathbf{x}_k) = 1$ if $U_0(k; i) < U_1(k; i)$, and $g_i(\mathbf{x}_k)$ is determined by a random flip if $U_0(k; i) = U_1(k; i)$.

During the sampling procedure, if the network is in state $\mathbf{x}_k$ at time t, then, $X_i(t+1) \neq f_i(\mathbf{x}_k)$ if and only if there is a perturbation. Hence,

$$P(X_i(t+1) \neq f_i(\mathbf{x}_k) | \mathbf{x}_k \text{ at time } t) = p.$$

Referring to (12.2), $g_i(\mathbf{x}_k) \neq f_i(\mathbf{x}_k)$ if and only if during the sampling procedure the frequency of perturbations for gene X_i exceeds the frequency of non-perturbations. Assuming that $\mathbf{x}_k$ is observed N_k times, N_k odd, then the probability of this occurring (conditioned on N_k) is

$$\xi(k, i) = \sum_{j=\frac{N_k+1}{2}}^{N_k} \binom{N_k}{j} p^j (1-p)^{N_k - j}.$$

Should N_k be even, we need to account for the fact that the frequencies might be equal, in which case a random flip determines if $g_i(\mathbf{x}_k) \neq f_i(\mathbf{x}_k)$. Hence, for N_k even,

$$\xi(k, i) = \sum_{j=\frac{N_k+2}{2}}^{N_k} \binom{N_k}{j} p^j (1-p)^{N_k - j} + \frac{1}{2}\binom{N_k}{N_k/2} p^{\frac{N_k}{2}} (1-p)^{\frac{N_k}{2}}.$$

The expected value of $\mu_{fun}(\mathcal{M}, \mathcal{H})$ is given by

$$E[\mu_{fun}(\mathcal{M}, \mathcal{H})] = \frac{1}{n2^n}\sum_{i=1}^{n}\sum_{k=1}^{2^n} E_{N_k}[\xi(k, i)].$$

Note that $\xi(k, i)$ will be very small when N_k is large. The difficulty is that N_k may be very small unless the sample is very large.

As noted following (12.2), we can weight μ_{fun} by the steady-state probabilities, in which case, for random sampling in the steady state, we obtain:

$$E[\mu_{fun}(\mathcal{M}, \mathcal{H})] = \frac{1}{n2^n} \sum_{i=1}^{n} \sum_{k=1}^{2^n} \pi_k E_{N_k}[\xi(k, i)]. \tag{12.11}$$

Large values of $E_{N_k}[\xi(k, i)]$ (resulting from the distribution of N_k being concentrated near 0) are associated with small probabilities π_k and these receive less weight in (12.11), which is reasonable since they are less likely to occur. This formula also applies in the limit when a single realization is observed over a very long period of time, but not for any finite-length observation.

The preceding analysis assumes connectivity n. If the connectivity is restricted, then the construction must be restricted and the evaluation of $E[\mu_{fun}(\mathcal{M}, \mathcal{H})]$ becomes much more tedious.

12.4.3 Synthetic examples

With given randomly generated synthetic networks and simulated time series of network states based on synthetic networks, we can investigate inference performances based on the aforementioned network distances for different network inference algorithms. Following the experiment design in [22], we have implemented REVEAL, MDL, BIC, uMDL, and Best-Fit algorithms based on the PBN Toolbox (http://code.google.com/p/pbn-matlab-toolbox/), the Bayes Net Toolbox (https://code.google.com/p/bnt/), as well as the source code provided by the authors of [35]. The detailed descriptions of these different algorithms can be found in the corresponding papers [23, 27, 28, 30, 34, 35]. For BIC and MDL, we set the regularization coefficients to values previously reported to have good performance in [34], $\lambda = 0.5$ for BIC and $\lambda = 0.3$ for MDL.

We generate 500 random BNps with $n = 8$ genes, maximum input degree $K = \max_{i=1}^{n} k_i = 3$, and perturbation probability $p = 0.01$. For each node, we uniformly assign 1 to K regulators based on the assumption that real-world biological networks are typically sparse. The sparse topology enables feasible network inference for all the selected algorithms with reasonable computational complexity [26, 28]. In addition, to address the possible ambiguity of inferred networks when they can not be uniquely determined based on available time series [23, 26, 28, 34, 36], we always choose the most sparse network among all the feasible networks as the final solution. To generate random synthetic BNps, we uniformly randomly assign regulators for each gene and the binary values in the corresponding truth tables for all the regulatory functions are then filled in based on Bernoulli random numbers with the bias following a beta distribution with mean 0.5 and standard deviation 0.01. For each random BNp, we simulate time series of state transitions based on its underlying Markov chain. The number of simulated state transitions M ranges from 10 to 50 to reflect the difficulty level of network inference. For measuring control-based distance, we choose the first node as the marker gene without loss of generality and define the undesirable states as these network states with the first node down-regulated. In the binary representation of network states, $\mathcal{U} = \{\mathbf{x}|x_1 = 0\}$. We further assume that we can either knock up or down the last gene to derive control policies. In our simulated random BNps, we have the original average undesirable steady-state mass $\pi_{\mathcal{U}}^{org} = 0.481$ with standard deviation 0.277, with $\pi_{\mathcal{U}}^{org} \approx 0.5$ because we set the bias to 0.5. When we apply the MSSA algorithm to derive the optimal stationary control policies for these random BNps, the average controlled undesirable steady state mass is $\pi_{\mathcal{U}} = 0.301$ with the standard deviation 0.282.

Figure 12.1 plots all five network distances for inferential validation averaged over all generated random BNps: μ_{fun}, μ_{ham}, μ_{prob}^{1}, μ_{ss}^{1}, and μ_{ctrl} from Figure 12.1a–e. The trends of these validation indices are similar to the results provided in [22]. First, BIC and MDL perform similarly for all indices based on different distances as we discussed in [22, 34]. The distances μ_{fun}, μ_{prob}^{1}, and μ_{ss}^{1} measure inference performance based on how well the algorithms recover network dynamics. Hence, the trends for different inference algorithms are similar for these three network distances as shown in Figure 12.1a, c, and d. From these figures, Best-Fit performs consistently better than the other algorithms since it aims to find the network models that best fit the observed state transitions. REVEAL performs similarly as Best-Fit with respect to μ_{prob}^{1} and μ_{ss}^{1} when potential perturbations are considered. With regularization on model complexity by BIC, MDL, and uMDL, the distances are greater, especially when we have a small number of observed state transitions. REVEAL and Best-Fit reconstruct networks with more edges to explain the observed data, which leads to smaller μ_{ss}. However, as we have empirically observed [22], REVEAL often has more false positive edges and more complex regulatory functions and hence is more susceptible to overfitting as shown in Figure 12.1a, where REVEAL performs similarly as the other algorithms compared to Best-Fit.

Considering the accurate recovery of regulatory relationships or network topology, it is interesting to see that Best-Fit still achieves the best performance with respect to μ_{ham} while REVEAL does not perform very well. This again confirms that REVEAL may introduce many false positives. For uMDL, the performance with respect to μ_{ham} improves quickly with increasing sample size compared to other complexity regularization algorithms BIC and MDL as uMDL consistently generates very low false-positive edges (close to zero) as shown in the original paper [35].

When we investigate the inferential validity with respect to controllability, μ_{ctrl}, we see interesting changes of tendency between the five algorithms. Especially with very few state transitions, $M = 10$, BIC, MDL, and uMDL algorithms perform better than REVEAL and Best-Fit, which indicates that the regularization on model complexity with a limited number of observations helps reconstruct network models that yield better controllers. With more observations, REVEAL and Best-Fit gradually perform better than BIC, MDL, and uMDL due to introduced bias by model complexity regularization.

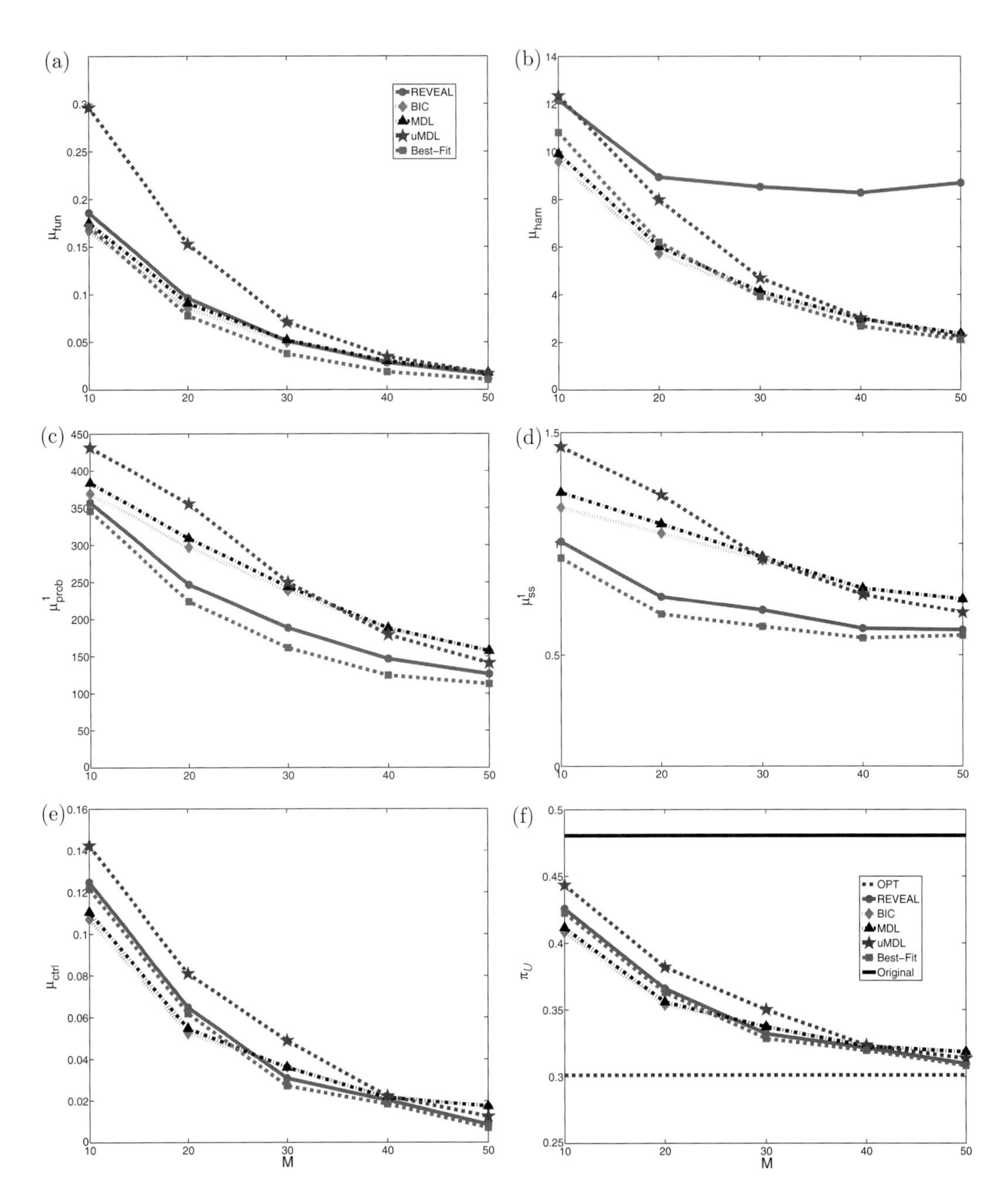

Figure 12.1 Performance of five network inference algorithms measured by five different network distances as inferential validity indices based on simulated BNps with eight genes and $K = 3$: (a) average rule-based distance μ_{fun}; (b) average Hamming distance μ_{ham}; (c) average transition-probability-based distance μ^r_{prob} with $r = 1$; (d) average steady-state distance μ^1_{ss}; (e) average control-based distance μ_{ctrl}; and (f) average undesirable steady-state mass $\pi_{\mathcal{U}}$ after applying derived stationary control policies based on inferred networks to the original ground truth BNps, compared to the average undesirable mass obtained by the optimal control policy (OPT) based on the complete knowledge of original BNps and the average undesirable mass before intervention (Original).

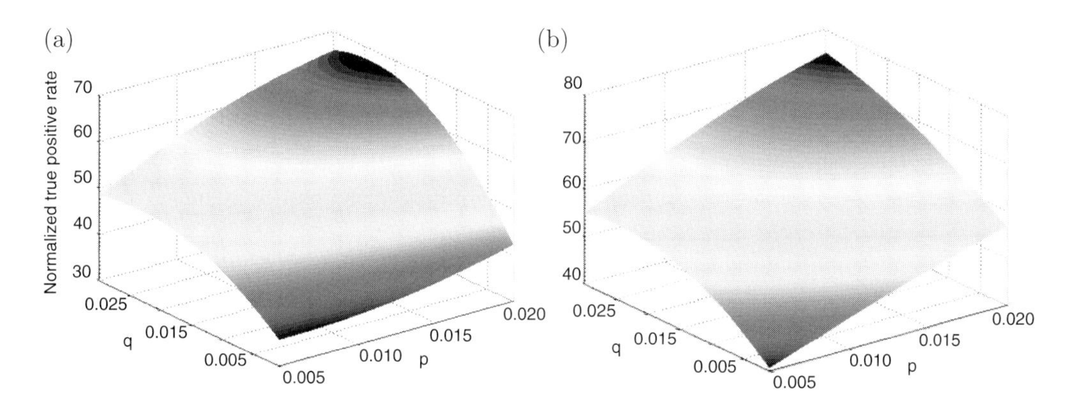

Figure 12.2 Network topology recovery as a function of switching (q) and perturbation probabilities (p) for context-dependent PBNs with 10 genes based on the Best-Fit algorithm [29]: (a) $M = 6000$, (b) $M = 10\,000$.

To further demonstrate that it is possible to derive effective control policies with the inferred networks based on a limited number of observations, we also plot the average undesirable steady-state mass using the control policy designed on the inferred networks via the MSSA algorithm in Figure12.1f. Compared to the average original undesirable mass and the average undesirable mass following application of the MMSA control policy designed on the original network, we can see that we can reduce the undesirable steady-state mass on average even when we have very few observations ($M = 10$). In addition, all algorithms show improved performance when M increases.

From these synthetic examples, we can see that although perfect network inference may require a large number of observations, we may still achieve desirable operation with partially recovered networks from a limited number of observed state transitions. The previously introduced network distances provide inferential validation indices with respect to different operational objectives for thorough performance evaluation.

With sparse connectivity assumptions, inference of regulatory functions in small-size networks is feasible. With increasing model complexity, for example, in context-sensitive PBNs, the data requirement can significantly increase. In [29], 80 context-sensitive PBNs with 10 genes are randomly generated from four-context BNs with maximum connectivity $K = 3$. Figure 12.2 plots the percentage of recovered regulators based on the Best-Fit algorithm with respect to perturbation probability p and switching probability q, with 6000 (Figure 12.2a) and 10 000 (Figure 12.2b) observed state transitions. Regulatory recovery is a function of the Hamming distance μ_{ham} and false positives. It is the same as the normalized true positive rate considering network topology. From the figure, it is clear that with more observations, we can recover the network topology better as shown by the increased surface height in Figure 12.2b. As discussed in [29], more observations are required for similar inference performance if we have networks with more genes. Finally, it is interesting to see the inference performance generally improves for both increasing p and q, as this leads to the likely observation of more state transitions as the observed times series increases in length.

12.5 Consistency

The greater the amount of data, the better inference one can expect. The hope is that, for large data sets, the inferred network will be close to the generating network. We define an inference procedure, ψ, to be consistent if $\mu^*(\mathrm{H}, \Sigma, \psi) \to 0$ as $|\Sigma| \to \infty$. We illustrate consistency using Boolean networks with perturbation and best-fit inference applied to a single observed time series and the distance function μ_{fun}.

Assuming a perturbation probability $p < 0.5$ and letting $\eta_j(\mathbf{x}_k)$ denote the number times we observe the transition $\mathbf{x}_k \to \mathbf{x}_j$, if $\mathbf{f}(\mathbf{x}_k) = \mathbf{x}_{(k)}$ is the function-defined transition, then

$$P(\eta_{(k)}(\mathbf{x}_k) > \max\{\eta_1(\mathbf{x}_k), \ldots, \eta_{(k)-1}(\mathbf{x}_k), \eta_{(k)+1}(\mathbf{x}_k), \ldots, \eta_N(\mathbf{x}_k)\}) \to 1$$

as $|\Sigma| \to \infty$. Thus, if $\hat{\mathbf{f}}$ denotes the inferred state function, then $P(\hat{\mathbf{f}} = \mathbf{f}) \to 1$ as $|\Sigma| \to \infty$. Similar asymptotic statements hold for $f_1, f_2, \ldots, f_n$. This insures that, for any $\tau > 0$, $P(\mu_{fun}(\psi(\Sigma), \mathcal{H}) > \tau) \to 0$ as $|\Sigma| \to \infty$ for any Boolean network $\mathcal{H}$. Since $\mu_{fun}(\psi(\Sigma), \mathcal{H}) \leq 1$, this is equivalent to $E_\Sigma[\mu_{fun}(\psi(\Sigma), \mathcal{H})] \to 0$ as $|\Sigma| \to \infty$. Finally, if H is the class of all Boolean networks on n genes with perturbation probability p, then, since H is a finite set,

$$\mu^*(\mathrm{H}, \Sigma, \psi) = E_\mathrm{H}[E_\Sigma[\mu_{fun}(\psi(\Sigma), \mathcal{H})]] \to 0 \quad as \quad |\Sigma| \to \infty, \tag{12.12}$$

and the inference procedure is consistent relative to μ_{fun}.

12.6 Approximation

The performance measures we are considering require that the true network is known. Suppose we do not know the random network, H, generating the data but know a network $\mathcal{N}$ that we believe to be a good approximation to the networks in H. We might then compare the inferred network to $\mathcal{N}$. In effect, such a comparison is approximating $\mu^*(\mathrm{H}, \Sigma, \psi)$ by $E_\Sigma[\mu(\psi(\Sigma), \mathcal{N})]$. Following [16], we consider the accuracy of this approximation.

The triangle inequality implies

$$\begin{aligned} \mu(\psi(S), \mathcal{N}) - \mu(\mathcal{N}, \mathcal{H}) &\leq \mu(\psi(S), \mathcal{H}) \\ &\leq \mu(\psi(S), \mathcal{N}) + \mu(\mathcal{N}, \mathcal{H}) \end{aligned} \tag{12.13}$$

for any sample set S and $\mathcal{H} \in \mathrm{H}$. Hence,

$$\begin{aligned} &E_\Sigma[\mu(\psi(\Sigma), \mathcal{N})] - E_\mathrm{H}[\mu(\mathcal{N}, \mathrm{H})] \\ &\leq E_\Sigma[E_\mathrm{H}[\mu(\psi(\Sigma), \mathrm{H})]] \\ &\leq E_\Sigma[\mu(\psi(\Sigma), \mathcal{N})] + E_\mathrm{H}[\mu(\mathcal{N}, \mathrm{H})]. \end{aligned} \tag{12.14}$$

If $E_\mathrm{H}[\mu(\mathcal{N}, \mathrm{H})] \approx 0$, then the preceding inequality leads to the approximate inequality

$$E_\Sigma[\mu(\psi(\Sigma), \mathcal{N})] \leq_\approx \mu^*(\mathrm{H}, \Sigma, \psi) \leq_\approx E_\Sigma[\mu(\psi(\Sigma), \mathcal{N})] \tag{12.15}$$

and

$$\mu^*(\mathrm{H}, \Sigma, \psi) \approx E_\Sigma[\mu(\psi(\Sigma), \mathcal{N})], \tag{12.16}$$

in which case it is reasonable to judge the performance of ψ relative to H by $E_\Sigma[\mu(\psi(\Sigma), \mathcal{N})]$; however, if $E_\mathrm{H}[\mu(\mathcal{N}, \mathrm{H})]$ is not small, then both bounds in (12.14) are loose and nothing can be asserted regarding the performance of ψ on H. Moreover, if $E_\mathrm{H}[\mu(\mathcal{N}, \mathrm{H})] \approx 0$, one still has to have sufficient data to estimate $E_\Sigma[\mu(\psi(\Sigma), \mathcal{N})]$ well using the average distance.

The preceding approximation methodology can occur in the following way. A network is inferred from one or more real datasets and compared, not to the unknown random network generating the data, but to a model network humanly constructed from the literature (and implicitly assumed to approximate the data-generating network). For instance, a directed graph (adjacency matrix), **A**, is constructed from relations found in the literature and the Hamming distance is used in the approximating expectation, $E_\Sigma[\mu_{ham}(\psi(\Sigma), \mathbf{A})]$ in (12.16). The aim is to compare the result of the inference procedure to some characteristic related to existing biological knowledge. But is the constructed regulatory graph a good approximation to the real system generating the data? In the following section, we will consider this issue in more detail.

12.7 Validation from experimental data

Consider a scenario in which real data from training-data sampling procedure Σ are used to infer a network, a characteristic is deduced from the inferred network, and an independent test-sampling procedure Ω generates real data to validate the designed network by direct construction of the characteristic. Validation is then via the random variable $\mu(\psi(\Sigma), \zeta(\Omega))$. To simplify the notation we consider a single underlying network $\mathcal{H}$ rather than a random network H. In this situation, $E_\mathrm{H}[\mu(\mathcal{N}, \mathrm{H})]$ in (12.14) is replaced by $\mu(\zeta(\Omega), \lambda_\mathcal{H})$, where $\lambda_\mathcal{H}$ is the characteristic for $\mathcal{H}$, and Eqn (12.14) takes the form

$$\begin{aligned} &E_\Omega[E_\Sigma[\mu(\psi(\Sigma), \zeta(\Omega))]] - E_\Omega[\mu(\zeta(\Omega), \lambda_\mathcal{H})] \\ &\quad \leq E_\Sigma[\mu(\psi(\Sigma), \lambda_\mathcal{H})] \\ &\quad \leq E_\Omega[E_\Sigma[\mu(\psi(\Sigma), \zeta(\Omega))]] + E_\Omega[\mu(\zeta(\Omega), \lambda_\mathcal{H})]. \end{aligned} \tag{12.17}$$

If $E_\Omega[\mu(\zeta(\Omega), \lambda_\mathcal{H})] \approx 0$, then

$$\mu^*(\mathrm{H}, \Sigma, \psi) \approx E_\Omega[E_\Sigma[\mu(\psi(\Sigma), \zeta(\Omega))]]. \tag{12.18}$$

If ζ is a consistent estimator of $\lambda_\mathcal{H}$, so that $E_\Omega[\mu(\zeta(\Omega), \lambda_\mathcal{H})] \approx 0$ for large samples, then, on average, the approximation is good.

If there are insufficient data to split between training and testing, one must test on the training data, so that $\Omega = \Sigma$ in (12.17) and we obtain the resubstitution estimate, $E_\Sigma[\mu(\psi(\Sigma), \zeta(\Sigma))]$, in (12.18). If ζ is a consistent estimator of $\lambda_\mathcal{H}$ and the single training sample is large, then the conclusion of Eqn (12.18) again holds. But we do not have a large sample. Hence, Eqn (12.17) cannot be used to insure good average performance. With independent test data, we are concerned with

$$\Delta_{test} = |E_\Sigma[\mu(\psi(\Sigma), \lambda_{\mathcal{H}})] - E_\Omega[E_\Sigma[\mu(\psi(\Sigma), \zeta(\Omega))]]|.$$

When the same data are used for training and testing, our interest is with

$$\Delta_{train} = |E_\Sigma[\mu(\psi(\Sigma), \lambda_{\mathcal{H}})] - E_\Sigma[\mu(\psi(\Sigma), \zeta(\Sigma))]|.$$

As with classification, resubstitution is risky because the characteristic of the designed network is being compared to a characteristic inferred from the same data with which the network has been designed. This can be a serious problem for small samples because overfitting can cause Δ_{train} to be much less than Δ_{test}. Whereas recently there has been an effort to understand these kinds of issues in pattern recognition – for instance, finding the joint distribution of the true and estimated errors [37] or characterizing the mean-square-error of the error estimate relative to the true error [38, 39] – there appears to be an absence of such effort for network validation.

12.7.1 Metastatic melanoma network inference

Based on cDNA microarray data as the steady-state data S collected from experiments for studying melanoma metastasis [20, 21], corresponding PBNs can been inferred based on an attractor-preserving inference method [31]. A PBN has been inferred from microarray data considering seven critical genes: WNT5A, pirin, S100P, RET1, MART1, HADHB, and STC2. The steady-state distribution of the inferred PBN has been compared to the actual histogram of the data, the histogram serving as an estimate of the steady-state distribution of the underlying physical network. Figure 12.3 illustrates the comparison of the portion of the steady-state distribution corresponding to the data states with the data histogram. Referring to (12.8), the 1-norm and 2-norm yield the resubstitution error distances $\mu^1_{ss}(\psi(S), \zeta(S)) = 0.45$ and $\mu^2_{ss}(\psi(S), \zeta(S)) = 0.13$, respectively.

12.8 Uncertainty quantification

When dealing with a complex real-world system it is often unrealistic to assume that it can be accurately modeled. In particular, a gene regulatory network that governs the behavior of a cell consists of tens of thousands of genes that regulate and are regulated by each other. These intertwined dynamical interactions are responsible for the complexities and characteristics of living organisms. Considering the size and complexity of a GRN, it is obvious that we cannot perfectly describe the network and its dynamics using a mathematical model. Thus, we limit ourselves to simple models that are hopefully adequate for the applications at hand; however, even for simple models, the complexity quickly grows with increasing numbers of genes, thereby making it difficult to accurately infer these models. In such circumstances we are better off not assuming a

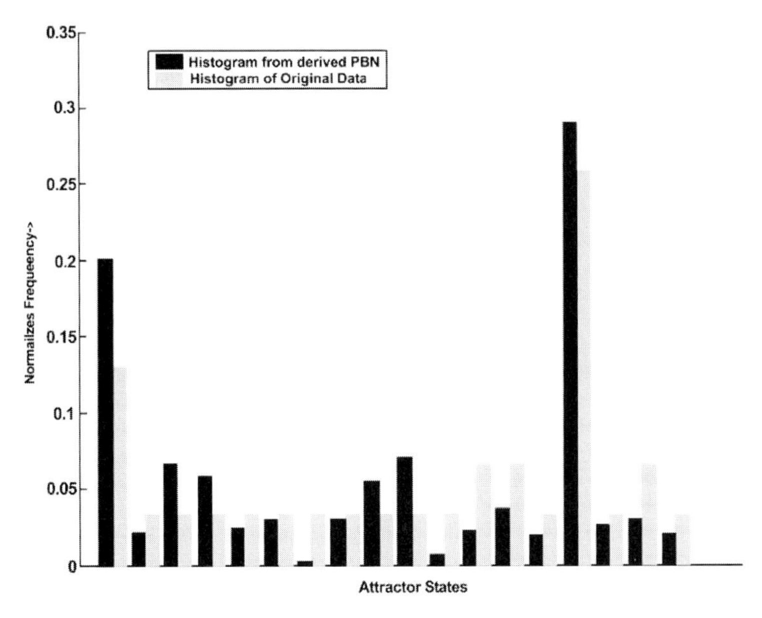

Figure 12.3 The network difference based on steady-state distributions illustrated as the comparison of the histogram of the steady-state distribution from the inferred metastatic melanoma network and the data histogram from actual experiments [16].

uniquely specified model, but instead assuming that the "true" model belongs to an *uncertainty class* of models.

Consider, for example, the rule-based distance in (12.2). Omitting the normalization, the distance simply counts the incorrect functional values. Assuming that these rules are to be inferred from data, those that are observed a large number of times are more likely to be correct than those that are rarely observed. Suppose we restrict our inference to those state-to-gene mappings that are observed some minimal number of times and we leave the others as "uncertain". The net result is an uncertainty class of networks, each one depending on how the uncertain relations are defined. In a binary network given r uncertain mappings, the resulting uncertainty class will contain 2^r networks. If our goal is optimal intervention, then we must define optimality relative to the uncertainty.

The design of optimal operations in the presence of uncertain system identification has been addressed in various engineering problems, including filtering, classification, and control. A well-known example is robust filtering. Qualitatively, a filter is said to be *robust* when its performance degradation is acceptable for signals that are statistically close to those for which it has been designed. The problem of robustness arises when optimizing a filter given a signal model, which in this case is uncertain. The problem was first addressed in the 1970s for signal detection [40], Wiener filtering with uncertain spectra [41], and then more generally for optimal linear filtering via minimax robustness [42–44]. Around 2000, optimal robust filtering was examined in a Bayesian framework for morphological [45] and linear [46] filtering. At first, these attempts resulted in suboptimal solutions. Recently, a general procedure to obtain fully optimal

Bayesian robust filters for both linear and morphological filtering has been discovered [47]. Bayesian robust classification took a similar route, first with suboptimal [48] and then with fully optimal solutions [49]. It is interesting to note that there is a much longer history of fully optimal Bayesian control [50–52]. Relative to GRNs, suboptimal control in the presence of uncertainty has been considered from both minimax [53] and Bayesian [54] perspectives. Recently, fully optimal Bayesian control of GRNs has been considered, with extreme computational complexity being severely limiting for practical applications [55].

12.8.1 Mean objective cost of uncertainty

A fundamental issue when considering a model with uncertain characteristics is the quantification of the uncertainty present in the model. Traditionally, such uncertainty has been often measured in terms of entropy. For example, several studies on GRN inference have considered the problem of designing optimal gene perturbation experiments that can maximally reduce the entropy of the potential network models [56, 57]. Here we discuss recent work quantifying model uncertainty in an objective-oriented manner based on the expected increase of the operational cost that it induces [58]. The resulting objective-based UQ (uncertainty quantification) framework provides a general mathematical basis for designing robust operators and can be applied to various applications, including robust filtering, classification, and control.

In the context of robust operator design, we are not as much concerned about the uncertainty of the system itself as we are with the effect of the uncertainty on operator performance. For example, in the case of optimal filtering as long as unknown model parameters do not affect overall filter performance, the model uncertainty is of no concern. On the other hand, when there exist parameters that may significantly degrade filter performance in case of small mismatch, the model uncertainty is consequential. Hence, (i) model uncertainty should be understood based on the *objective* of employing the model (e.g. signal estimation); and (ii) the uncertainty should be quantified in terms of the expected *cost* it induces (e.g. estimation error) with respect to an operator class.

Formally, consider a model defined by a parameter vector θ from a class Θ of parameter vectors – or equivalently, a class of models – governed by an *uncertainty distribution* $f(\theta)$. This distribution may come from observation data and/or prior knowledge regarding the model. Consider also a family Ψ of operators and a family $\Xi = \{\xi_\theta | \theta \in \Theta\}$ of cost functions, $\xi_\theta : \Psi \to [0, \infty)$. In the presence of uncertainty in θ, an *intrinsically Bayesian robust (IBR)* operator is defined by

$$\psi_{\text{IBR}} = \arg\min_{\psi \in \Psi} E_\theta[\xi_\theta(\psi)]. \tag{12.19}$$

An IBR operator minimizes, among all operators, the expected cost across the uncertainty class relative to the uncertainty distribution.

Suboptimality relative to (12.19) results from taking the minimum over all operators that are optimal for some model in Θ, which results in a *model-constrained Bayesian robust (MCBR)* operator defined by

$$\psi_{\mathrm{MCBR}} = \arg\min_{\psi_\varphi:\varphi\in\Theta} E_\theta[\xi_\theta(\psi_\varphi)], \tag{12.20}$$

where ψ_φ is optimal for model $\varphi \in \Theta$ [45, 48, 54].

Although it is possible to measure the uncertainty in terms of the entropy of the model θ based on its distribution $f(\theta)$ – with the entropy being 0 when there is no uncertainty – the entropy may not tell us much about operator performance. The entropy could be large, but if the networks in Θ are very close to each other with regard to whatever network characteristics determine the operators, then the fact that the entropy is large is of little consequence. In the other direction, the entropy could be small, but the networks be far apart with regard to the characteristics determining the operators, which would be of significant consequence.

Recognizing the significance of operator performance, the *objective cost of uncertainty (OCU)* relative to $\theta \in \Theta$ is defined as the differential cost of using the IBR operator instead of the operator ψ_θ that is optimal for the given model [58]:

$$U_{\Psi,\Xi,f}(\Theta,\theta) = \xi_\theta(\psi_{\mathrm{IBR}}) - \xi_\theta(\psi_\theta). \tag{12.21}$$

The overall cost of uncertainty can be computed by taking its expectation over $f(\theta)$,

$$\mathrm{M}_{\Psi,\Xi,f}(\Theta) = E_\theta[U_{\Psi,\Xi,f}(\Theta,\theta)], \tag{12.22}$$

which is called the *mean objective cost of uncertainty (MOCU)*. The objective cost of uncertainty is relative to the operator family Ψ and the family Ξ of cost functions, in addition to the uncertainty distribution $f(\theta)$, thereby taking into account the objective of operator design, be it signal estimation, classification, or control.

The concept of MOCU has connections to existing techniques and concepts used in various fields. For example, ψ_{IBR} can be viewed as a Bayesian estimator that minimizes the expected cost over all potential models $\theta \in \Theta$ [59] and the MOCU can be viewed as the minimum expected value of a Bayesian loss function, where the Bayesian loss function maps an operator to its differential cost (for using the given operator instead of the optimal operator) and its minimum expectation is attained by the optimal robust operator that minimizes the average differential cost. This differential cost has been referred to as the *regret* in decision theory [60], which is defined as the difference between the maximum payoff (for making the optimal decision) and the actual payoff (for the decision that has been made). From this perspective, the MOCU can also be viewed as the minimum expected regret for using the robust operator.

Here we are interested in validation and the MOCU can be employed in this light. The rule-based distance of (12.2) counts the incorrectly inferred model parameters. Indeed, such a count provides a general form of validation measure taking the form

$$\mu_{par}(\mathcal{M},\mathcal{H}) = \sum_{j=1}^{J} c_j I\left[\theta_j^{\mathcal{M}} - \theta_j^{\mathcal{H}}\right|, \tag{12.23}$$

where $\theta_j^{\mathcal{M}}$ and $\theta_j^{\mathcal{H}}$, $j = 1, 2, \ldots, J$, are the model parameters for $\mathcal{M}$ and $\mathcal{H}$, respectively, and $c_1, c_2, \ldots, c_J$ are nonnegative weights corresponding to the importance of the parameters. Now, suppose we utilize an inference process under the assumption that a parameter will only be inferred if there is very strong confidence that it is correct – for instance, a regulatory relation has been observed some (large) minimal number of times; otherwise, we declare the parameter to be uncertain. This will leave us with with some set of uncertain parameters and, given an objective, we can compute the MOCU. If we incorporate the weights of the unknown parameters into the costs, ξ_θ in (12.22), then the MOCU can be taken as a measure of validation. It measures the expected cost incurred by inadequate inference relative to the objective.

If we restrict optimization to network-specific optimal operators by using an MCBR operator instead of an IBR operator for computing the objective cost of uncertainty, then we get the following *model-constrained OCU* relative to θ:

$$U^{\mathrm{mc}}_{\Psi,\Xi,f}(\Theta,\theta) = \xi_\theta(\psi_{\mathrm{MCBR}}) - \xi_\theta(\psi_\theta). \tag{12.24}$$

Taking the expectation over $f(\theta)$ yields the model-constrained MOCU, namely,

$$\mathrm{M}^{\mathrm{mc}}_{\Psi,\Xi,f}(\Theta) = E_\theta\left[U^{\mathrm{mc}}_{\Psi,\Xi,f}(\Theta,\theta))\right], \tag{12.25}$$

which gives us an upper bound on the true MOCU,

$$\mathrm{M}_{\Psi,\Xi,f}(\Theta) \leq \mathrm{M}^{\mathrm{mc}}_{\Psi,\Xi,f}(\Theta). \tag{12.26}$$

An IBR operator is preferable to an MCBR operator because the minimum is taken over a larger operator class. We would certainly prefer to know the MOCU rather than an upper bound on it. Nonetheless, there are two reasons why MCBR operators might be useful. First, it may be practically easier to find a MCBR operator. For instance, in the case of stationary network control, MCBR operators can be found by using dynamic programming for each $\theta \in \Theta$ as shown in [54], whereas IBR operators cannot be found in such a way. In the case of finite-state discrete-time networks, there is a finite number of stationary controllers, but their number is huge once the network is beyond several genes. Whereas one may in principle be able to find an IBR operator by exhaustive search, this may be computationally unfeasible. Thus, we may have to be content with MCBR operators and the model-constrained MOCU as an approximation to the true MOCU.

12.8.2 Intervention in yeast cell cycle network with uncertainty

In [58], we have assessed the performance of the IBR structural intervention ψ^*, as defined in (12.19), and compared it with that of the MCBR structural intervention $\psi_{\hat{\varphi}}$, which is designed based on a specific network $\hat{\varphi} \in \Theta$ that minimizes the cost as defined in (12.20).

The experiments in [58] are based on a baseline network model for the yeast (*Saccharomyces cerevisiae*) cell cycle [61–63] (Figure 12.4). A specific class of regulatory

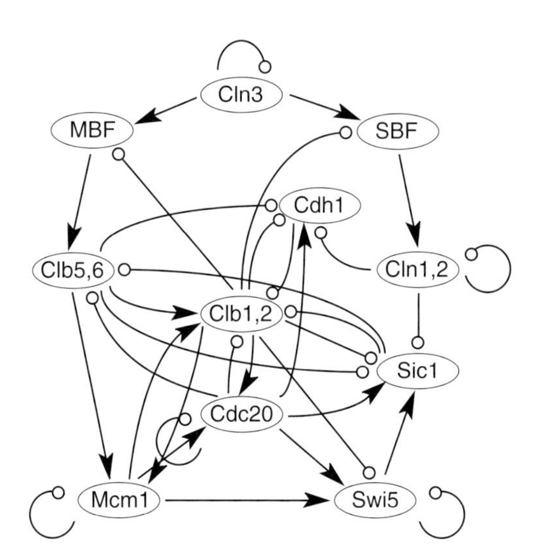

Figure 12.4 *Saccharomyces cerevisiae* cell cycle network modified from [61]: the normal arrows represent activation and the arrows with hollow circles represent suppression [58].

functions using the "majority vote" rules is adopted for simulating network dynamics:

$$f_i(\mathbf{x}) = \begin{cases} 1 & \text{if} \sum_j R_{j\to i}x_j > 0 \\ x_i & \text{if} \sum_j R_{j\to i}x_j = 0 \\ -1 & \text{if} \sum_j R_{j\to i}x_j < 0 \end{cases}, \forall i,$$

in which $R_{j\to i} \in \{-1, 0, 1\}$ denotes the possible "suppressing", "none", or "activating" regulatory relationships from gene X_j to gene X_i. This baseline yeast cell cycle network model has 11 genes ($n = 11$), 15 activating regulations, and 19 suppressing regulations. Based on this regulatory model, there is a state transition trajectory (subsequence) corresponding to the essential cell cycle process [61, 63]: in the order of genes Cln3, MBF, SBF, Cln1,2, Clb5,6, Clb1,2, Mcm1, Cdc20, Swi5, Sic1, and Cdh1, the network transitions from the excited G1 phase (10000000011 → 01100000011 → 01110000011 → 01110000000), going through the S state (01111000000), the G2 state (01111110000), then to the M phase (00011111000 → 000001111000 → 000001111100 → 00000001111), and finally returning to the biological G1 stationary state (00000000011).

We generated a random ensemble of 100 BNps based on this model, where each random BNp contains uncertainty regarding regulatory relationships among genes $R_{j\to i}$, and investigated the chances that the IBR and MCBR operators will lead to different expected costs for a given uncertainty class. These BNps were generated by following the procedure in [63] to randomly add edges between pairs of genes and fill in 1 or −1 for the regulatory matrix $\mathbf{R} = (R_{j\to i})$ to form a combined structural and functional ensemble that has the same number of nodes, the same number of activating edges, the same number of suppressing edges, and the same essential state transition trajectory as in

Table 12.1 The average number m_d of random networks with different ψ^* from any model specific structural intervention ψ_θ and its standard deviation σ_d in randomly generated functional and structural ensemble based on the yeast cell cycle network [58]

R	1	2	3	4
m_d	0.90	1.80	2.10	2.20
σ_d	0.88	0.79	0.99	2.04

the previously described original cell cycle network. To study the performances of IBR and MCBR structural interventions on random BNps with uncertainty, we randomly picked 10 independent sets of regulations and assumed their activating/suppressing relationships to be uncertain, each set containing $1 \leq R \leq 4$ uncertain regulations. For all of these random networks, we set the perturbation probability to $p = 0.01$. In Table 12.1, we show the average number m_d (and the standard deviation σ_d) of random networks with uncertainty for which the IBR intervention was different from the MCBR intervention (i.e. $\psi^* \neq \psi_{\hat{\varphi}}$). From Table 12.1, the average number m_d increases gradually with the number of uncertain regulations R. This is reasonable, since the optimal robust strategy ψ^* will be more likely to be different from any network-specific intervention ψ_θ as the model uncertainty increases with increasing R.

It also is interesting to see that with different levels of uncertainty, we in fact can still achieve similar robust intervention strategies. As shown by other examples in [58], the objective-based UQ framework and the concept of MOCU can effectively capture the critical properties and dynamics of gene regulatory networks that are pertinent to our aim for network intervention in the presence of substantial uncertainty. Together with previous synthetic examples, it is clear that with reasonably inferred networks with uncertainty, we are still able to effectively steer the network dynamics away from undesirable states, which is our ultimate operational objective in this set of experiments.

References

[1] A. Datta, R. Pal., A. Choudhary, and E. Dougherty, "Control approaches for probabilistic gene regulatory networks," *IEEE Signal Processing Magazine*, vol. **24**, no. 1, pp. 54–63, 2007.

[2] B. Faryabi, G. Vahedi, A. Datta, J.-F. Chamberland, and E. Dougherty, "Recent advances in the external control of Markovian gene regulatory networks," *Current Genomics*, vol. **10**, no. 7, pp. 463–477, 2009.

[3] E. Dougherty, R. Pal, X. Qian, M. Bittner, and A. Datta, "Stationary and structural control in gene regulatory networks: Basic concepts," *International Journal of Systems Science*, vol. **41**, no. 1, pp. 5–16, 2010.

[4] I. Shmulevich and E. Dougherty, *Probabilistic Boolean Networks: The Modeling and Control of Gene Regulatory Networks*, New York: SIAM Press, 2010.

[5] S. Kim, H. Li, E. Dougherty, *et al.*, "Can Markov chain models mimic biological regulation?" *J. Biol. Syst.*, vol. **10**, no. 4, pp. 337–357, 2002.
[6] R. Pal, A. Datta, and E. Dougherty, "Optimal infinite horizon control for probabilistic Boolean networks," *IEEE Trans. Signal Processing*, vol. **54**, no. 6-2, pp. 2375–2387, 2006.
[7] N. Yousefi and E. Dougherty, "Intervention in gene regulatory networks with maximal phenotype alteration," *Bioinformatics*, vol. **29**, no. 14, pp. 1758–1767, 2013.
[8] B. Faryabi, G. Vahedi, J.-F. Chamberland, A. Datta, and E. Dougherty, "Optimal constrained intervention in genetic regulatory networks," *EURASIP J. Bioinformatics and Systems Biology*, vol. **620767**, p. 10 pages, 2008.
[9] Y. Xiao and E. Dougherty, "The impact of function perturbations in Boolean networks," *Bioinformatics*, vol. **23**, no. 10, pp. 1265–1273, 2007.
[10] X. Qian and E. Dougherty, "Effect of function perturbation on the steady-state distribution of genetic regulatory networks: Optimal structural intervention," *IEEE Trans. Signal Processing*, vol. **56**, no. 10-1, pp. 4966–4975, 2008.
[11] H. de Jong, "Modeling and simulation of genetic regulatory systems: A literature review," *Computational Biology*, vol. **9**, no. 1, pp. 67–103, 2002.
[12] I. Shmulevich and E. Dougherty, *Genomic Signal Processing*, Princeton: Princeton University Press, 2007.
[13] K. Cho, S. Choo, S. Jung, *et al.*, "Reverse engineering of gene regulatory networks," *IET Systems Biology*, vol. **1**, no. 3, pp. 149–163, 2007.
[14] T. Schlitt and A. Brazma, "Current approaches to gene regulatory network modeling," *BMC Bioinformatics*, vol. **8**, no. Suppl 6, p. S9, 2007.
[15] C. Sima, J. Hua, and S. Jung, "Inference of gene regulatory network using time-series data: A survey," *Current Genomics*, vol. **10**, no. 6, pp. 416–429, 2009.
[16] E. Dougherty, "Validation of inference procedures for gene regulatory networks," *Current Genomics*, vol. **8**, no. 6, pp. 351–359, 2007.
[17] ——, "Validation of gene regulatory networks: scientific and inferential," *Briefings in Bioinformaticss*, vol. **12**, no. 3, pp. 245–252, 2011.
[18] S. Kauffman, "Metabolic stability and epigenesis in randomly constructed genetic nets," *Theoretical Biology*, vol. **22**, pp. 437–467, 1969.
[19] I. Shmulevich, E. Dougherty, and W. Zhang, "From Boolean to probabilistic Boolean networks as models of genetic regulatory networks," *Proceedings of the IEEE*, vol. **90**, no. 11, pp. 1778–1792, 2002.
[20] M. Bittner, P. Meltzer, Y. Chen, *et al.*, "Molecular classification of cutaneous malignant melanoma by gene expression profiling," *Nature*, vol. **406**, no. 6795, pp. 536–540, 2000.
[21] A. Weeraratna, Y. Jiang, G. Hostetter, *et al.*, "Wnt5a signalling directly affects cell motility and invasion of metastatic melanoma," *Cancer Cell*, vol. **1**, pp. 279–288, 2002.
[22] X. Qian and E. Dougherty, "Validation of gene regulatory network inference based on controllability," *Frontiers in Genetics*, vol. **4**, p. 272, 2013.
[23] S. Liang, S. Fuhrman, and R. Somogyi, "REVEAL: A general reverse engineering algorithm for inference of genetic network architectures," in *Pacific Symposium on Biocomputing*, vol. **3**, 1998, pp. 18–29.
[24] M. Arnone and E. Davidson, "The hardwiring of development: organization and function of genomic regulatory systems," *Development*, vol. **124**, pp. 1851–1864, 1997.

[25] D. Thieffry, A. Huerta, E. Pèrez-Rueda, and J. Collado-Vides, "From specific gene regulation to genomic networks: A global analysis of transcriptional regulation in *Escherichia coli*," *BioEssays*, vol. **20**, pp. 433–440, 1998.

[26] T. Akutsu, S. Miyano, and S. Kuhara, "Identification of genetic networks from a small number of gene expression patterns under the Boolean network model," in *Pacific Symposium on Biocomputing*, vol. **4**, 1999, pp. 17–28.

[27] I. Shmulevich, A. Saarinen, O. Yli-Harja, and J. Astola, "Inference of genetic regulatory networks under the Best-Fit Extension paradigm," in *Computational And Statistical Approaches To Genomics*, W. Zhang and I. Shmulevich, Eds, Boston: Kluwer Academic Publishers, 2002.

[28] H. Lähdesmäki, I. Shmulevich, and O. Yli-Harja, "On learning gene regulatory networks under the Boolean network model," *Machine Learning*, vol. **52**, pp. 147–167, 2003.

[29] S. Marshall, L. Yu, Y. Xiao, and E. Dougherty, "Inference of probabilistic Boolean networks from a single observed temporal sequence," *EURASIP J. Bioinformatics and Systems Biology*, vol. **2007**, 32454, 2007.

[30] K. Murphy and S. Mian, "Modelling gene expression data using dynamic Bayesian networks," University of California, Berkeley, Tech. Rep., 1999.

[31] R. Pal, I. Ivanov, A. Datta, M. Bittner, and E. Dougherty, "Generating Boolean networks with a prescribed attractor structure," *Bioinformatics*, vol. **21**, pp. 4021–4025, 2005.

[32] X. Qian and E. Dougherty, "Phenotypically constrained boolean network inference with prescribed steady states," in *IEEE International Workshop on Genomic Signal Processing and Statistics*, Houston, TX, 2013.

[33] I. Tabus and J. Astola, "On the use of MDL principle in gene expression prediction," *Journal of Applied Signal Processing*, vol. **4**, pp. 297–303, 2001.

[34] W. Zhao, E. Serpedin, and E. Dougherty, "Inferring gene regulatory networks from time series data using the minimum description length principle," *Bioinformatics*, vol. **22**, no. 17, pp. 2129–2135, 2006.

[35] J. Dougherty, I. Tabus, and J. Astola, "Inference of gene regulatory networks based on a universal minimum description length," *EURASIP J. Bioinform. Syst. Biol.*, vol. **1**, p. 482090, 2008.

[36] S. Martin, Z. Zhang, A. Martino, and J.-L. Faulon, "Boolean dynamics of genetic regulatory networks inferred from microarray time series data," *Bioinformatics*, vol. **23**, no. 7, pp. 866–874, 2007.

[37] A. Zollanvari, U. Braga-Neto, and E. Dougherty, "On the joint sampling distribution between the actual classification error and the resubstitution and leave-one-out error estimators for linear classifiers," *IEEE Trans. Information Theory*, vol. **56**, no. 2, pp. 784–804, 2010.

[38] ——, "Analytic study of performance of error estimators for linear discriminant analysis," *IEEE Trans. Signal Processing*, vol. **59**, no. 9, pp. 4238–4255, 2011.

[39] ——, "Exact representation of the second-order moments for resubstitution and leave-one-out error estimation for linear discriminant analysis in the univariate heteroskedastic Gaussian model," *Pattern Recognition*, vol. **45**, no. 2, pp. 908–917, 2012.

[40] V. Kuznetsov, "Stable detection when the signal and spectrum of normal noise are inaccurately known," *Telecommun. Radio Eng*, vol. **30**, p. 31, 1976.

[41] S. Kassam and T. Lim, "Robust Wiener filters," *Journal of the Franklin Institute*, vol. **304**, no. 4-5, pp. 171–185, 1977.

[42] H. Poor, "On robust Wiener filtering," *IEEE Transactions on Automatic Control*, vol. **25**, no. 3, pp. 531–536, 1980.
[43] K. Vastola and H. Poor, "Robust Wiener–Kolmogorov theory," *IEEE Transactions on Information Theory*, vol. **30**, no. 2, pp. 316–327, 1984.
[44] S. Verdu and H. Poor, "Minimax linear observers and regulators for stochastic systems with uncertain second-order statistics," *IEEE Transactions on Automatic Control*, vol. **29**, no. 6, pp. 499–511, 1984.
[45] E. Dougherty and Y. Chen, "Robust optimal granulometric bandpass filters," *Signal Processing*, vol. **81**, no. 7, pp. 1357–1372, 2001.
[46] A. Grygorian and E. Dougherty, "Bayesian robust optimal linear filters," *Signal Processing*, vol. **81**, no. 12, pp. 2503–2521, 2001.
[47] L. Dalton and E. Dougherty, "Intrinsically optimal bayesian robust filtering," *IEEE Transactions on Signal Processing*, vol. **62**, no. 3, pp. 657–670, 2014.
[48] E. Dougherty, J. Hua, Z. Xiong, and Y. Chen, "Optimal robust classifiers," *Pattern Recognition*, vol. **38**, no. 10, pp. 1520–1532, 2005.
[49] L. Dalton and E. Dougherty, "Optimal classifiers with minimum expected error within a Bayesian framework – part II: properties and performance analysis," *Pattern Recognition*, vol. **46**, no. 5, pp. 1301–1314, 2013.
[50] E. Silver, "Markovian decision processes with uncertain transition probabilities or rewards," DTIC Document, Tech. Rep., 1963.
[51] J. Gozzolino, R. Gonzalez-Zubieta, and R. Miller, "Markovian decision processes with uncertain transition probabilities," DTIC Document, Tech. Rep., 1965.
[52] J. Martin, *Bayesian Decision Problems and Markov Chains, Publications in Operations Research*, New York: Wiley, 1967.
[53] R. Pal, A. Datta, and E. Dougherty, "Robust intervention in probabilistic boolean networks," *IEEE Transactions on Signal Processing*, vol. **56**, no. 3, pp. 1280–1294, 2008.
[54] ——, "Bayesian robustness in the control of gene regulatory networks," *IEEE Transactions on Signal Processing*, vol. **57**, no. 9, pp. 3667–3678, 2009.
[55] M. Yousefi and E. R. Dougherty, "A comparison study of optimal and suboptimal intervention policies for gene regulatory networks in the presence of uncertainty," *EURASIP Journal on Bioinformatics and Systems Biology*, vol. **2014**, p. 6, 2014.
[56] T. Ideker, V. Thorsson, and R. Karp, "Discovery of regulatory interactions through perturbation: inference and experimental design," in *Pacific Symposium on Biocomputing*, vol. **5**, 2000, pp. 302–313.
[57] A. Almudevar and P. Salzman, "Using a Bayesian posterior density in the design of perturbation experiments for network reconstruction," in *Proceedings of the 2005 IEEE Symposium on Computational Intelligence in Bioinformatics and Computational Biology, 2005, CIBCB'05*, 2005, pp. 1–7.
[58] B.-J. Yoon, X. Qian, and E. Dougherty, "Quantifying the objective cost of uncertainty in complex dynamical systems," *IEEE Transactions on Signal Processing*, vol. **61**, no. 9, pp. 2256–2266, 2013.
[59] E. Lehmann and G. Casella, *Theory of Point Estimation*, Springer, 1998.
[60] D. Berry and B. Fristedt, *Bandit Problems: Sequential Allocation of Experiments*, Chapman and Hall, 1985.
[61] F. Li, T. Long, Y. Lu, Q. Ouyang, and C. Tang, "The yeast cell-cycle network is robustly designed," *Proc. Natl. Acad. Sci. USA*, vol. **101**, no. 14, pp. 4781–4786, 2009.

[62] Y. Wu, X. Zhang, Y. Yu, and Q. Ouyang, "Identification of a topological characteristic responsible for the biological robustness of regulatory networks," *PLoS Computational Biology*, vol. **5**, p. 7, 2009.

[63] K. Lau, S. Ganguli, and C. Tang, "Function constrains network architecture and dynamics: a case study on the yeast cell cycle Boolean network," *Phys. Rev. E*, vol. **75**, p. 051907, 2007.

13 Inference of gene networks associated with the host response to infectious disease

Zhe Gan, Xin Yuan, Ricardo Henao, Ephraim L. Tsalik, and Lawrence Carin

Inspired by the problem of inferring gene networks associated with the host response to infectious diseases, a new framework for discriminative factor models is developed. Bayesian shrinkage priors are employed to impose (near) sparsity on the factor loadings, while non-parametric techniques are utilized to infer the number of factors needed to represent the data. Two discriminative Bayesian loss functions are investigated, i.e. the logistic log-loss and the max-margin hinge loss. Efficient mean-field variational Bayesian inference and Gibbs sampling are implemented. To address large-scale datasets, an online version of variational Bayes is also developed. Experimental results on two real-world microarray-based gene expression datasets show that the proposed framework achieves comparatively superior classification performance, with model interpretation delivered via pathway association analysis.

13.1 Background

From a statistical-modeling perspective, gene expression analysis can be roughly divided into two phases: exploration and prediction. In the former, the practitioner attempts to get a general understanding of a dataset by modeling its variability in an interpretable way, such that the inferred model can serve as a feature extractor and hypotheses-generating mechanism of the underlying biological processes. Factor models are among the most widely employed techniques for exploratory gene expression analysis [1, 2], with principal component analysis a popular special case [3]. Predictive modeling, on the other hand, is concerned with finding a relationship between gene expression and phenotypes, that can be generalized to unseen samples. Examples of predictive models include classification methods like logistic regression and support vector machines [4, 5].

Factor models infer a latent covariance structure among the genes or biomarkers, with data modeled as generated from a noisy low-rank matrix factorization, manifested in terms of a *loadings matrix* and a *factor scores* matrix. Different specifications for these matrices give rise to special cases of factor models, such as principal components analysis [6], nonnegative matrix factorization [7], independent component analysis [8],

Big Data over Networks, ed. Shuguang Cui, Alfred O. Hero III, Zhi-Quan Luo, and José M. F. Moura. Published by Cambridge University Press.

and sparse factor models [1]. Factor models employing a sparse factor loadings matrix are of significant interest in gene-expression analysis, as the nonzero elements in the loadings matrix may be interpreted as correlated gene networks [1, 2, 9].

Discriminative models, in particular binary linear classification models, aim to find a linear combination of input features or covariates to separate observed data into two groups or phenotypes (using kernel techniques, such approaches are readily extended to nonlinear classifiers [10]). One proceeds by first learning the parameters of the classifier by using labeled data, in which one knows the phenotype of every data point (in a semi-supervised learning procedure, some data are labeled and others non-labeled [11]). Once so learned, the model parameters are fixed and used to predict the phenotype of unlabeled data; this is often termed the inference step [12]. In the work reported here we jointly learn the factor model (feature learning) and classifier, and we develop a framework that scales well to high-dimensional data.

13.2 Factor models in gene expression analysis

Gene expression analysis typically involves considering a relatively small number of observations, each of which is composed of expression values from tens of thousands of genes. In this setting, called the "large p, small N" problem [13], the number of biomarkers p is much larger than the number of observations N ($N \ll p$). In this regime direct analysis either using factor models or discriminative models is infeasible, because the problem is ill-posed [14]. For factor models, in order to yield reliable modeling, two key assumptions are widely imposed: (i) the number of factors needed to explain the data is small (low-rank assumption of the dataset), and (ii) each factor is responsible for explaining only a small subset of variables. The latter also applies to discriminative models, in the sense that only a small subset of variables is necessary for classification. From an application point of view, this *sparsity* assumption yields results that can be interpreted, e.g. the small subset of correlated genes associated with a factor correspond ideally to biologically meaningful *pathways*, *modules* or *gene networks*. Factor models can be used as a general feature extraction tool in the context of gene expression analysis [1].

Under the Bayesian paradigm, the sparsity assumption is specified via prior distributions, such as spike-and-slab [1, 15, 16] or shrinkage priors [10, 17–20]. The key difference between these two prior distribution families is that the former assumes signal (slab) and no-signal (spike) as coming from a bi-modal distribution, whereas in the latter both signal and no-signal are modeled as a unimodal heavy-tailed distribution, in which values close to zero are designated as no-signal (i.e. as noise). Common choices for continuous shrinkage priors include Student's-t [10], double-exponential (Laplace) [17], the horseshoe [18], and the three-parameter beta normal (TPBN) [20]. The TPBN is an example of global-local shrinkage priors as defined in [19], that has demonstrated superior performance in terms of mixing when compared to other shrinkage specifications and spike-and-slab priors.

Factor models and discriminative models have been successfully used to identify host responses in infectious disease studies [21–26]. Such analyses are usually performed by first using a factor model to obtain a low-dimensional representation of the data, encoded by the factor scores. A discriminative model is next used to characterize the phenotype of interest from the factor scores. This two-step procedure can be seen as the exploration and prediction phases described above. Provided that both models are equipped with sparsity priors, we can use the discriminative model to identify a subset of factor scores responsible for the classification rule. Since these should correlate with the phenotype, they are thus proxies for the corresponding host response. Subsequently, we can use the factor loadings to identify the correlated subset of genes responsible for the relevant factor scores. As a result, we can build a classification model and learn about the gene network that contributes to the predictor's outcome, in a systematic manner.

A drawback of the two-step procedure described above is that feature extraction (factor modeling) and classification are performed separately, thus the factor model is not informed of the ultimate use of the factor scores in a classifier (it simply tries to fit the data in a generative sense, and may ignore subtle features of the data that are critical for the subsequent classification task). In a discriminative factor model [27], also known as supervised dictionary learning, the two-steps are performed jointly.

In a Bayesian setting, classification models have always been troubled with complications in terms of learning and inference, owing to the lack of conditional conjugacy between the likelihood function implied by the decision function and the prior distributions for the parameters of the model. Probit regression is an interesting case, because it is perhaps the first classification model provided with efficient inference, as a consequence of variable augmentation [28]. Although augmentation schemes for logistic regression [29] and support vector machines [30] have been proposed recently, they have not yet been combined with factor models. Since discriminative factor models using probit regression as the classifier have been investigated previously [31, 32], we focus this work on effective inference for discriminative factor models using logistic regression and support vector machine classifiers.

The remainder of this chapter is organized as follows. Sections 13.3 and 13.4 describe factor models and discriminative models, respectively. Section 13.5 introduces our discriminative factor model, while Section 13.6 describes the learning and inference procedures employed. Section 13.7 presents extensive results on real gene-expression data, and Section 13.8 provides summary comments and observations about the proposed framework. Finally, inference details are given in Section 13.9.

13.3 Factor models

Assume a matrix of observed data $\mathbf{X} \in \mathbb{R}^{p \times N}$, where each column corresponds to one of $n \in \{1, \ldots, N\}$ samples, $\mathbf{x}_n \in \mathbb{R}^p$, each of which contains expression values for p genes (while we focus the discussion on gene expression analysis, the basic modeling structure developed here is applicable to other biomarkers, such as proteomics, metabolomics, etc.). We consider a linear factor model of the form

$$\begin{aligned} \mathbf{x}_n &= \mathbf{A}\boldsymbol{\Lambda}\mathbf{s}_n + \boldsymbol{\epsilon}_n, \\ \mathbf{s}_n &\sim \mathcal{N}(\mathbf{0}, \mathbf{I}), \\ \boldsymbol{\epsilon}_n &\sim \mathcal{N}(\mathbf{0}, \boldsymbol{\Psi}^{-1}), \end{aligned} \tag{13.1}$$

where a_{jk} is an element of loadings matrix $\mathbf{A} \in \mathbb{R}^{p \times K}$, K is the total number of factors, $\boldsymbol{\epsilon}_n$ is additive noise (or model residual) for the nth sample, ψ_j is an element of the noise precision matrix $\boldsymbol{\Psi} = \text{diag}(\psi_1, \ldots, \psi_p)$, λ_k is one of the factor specific scaling coefficients in matrix $\boldsymbol{\Lambda} = \text{diag}(\lambda_1, \ldots, \lambda_K)$, and $\mathbf{s}_n \in \mathbb{R}^K$ is a vector of Gaussian distributed factor scores for the nth observation. While we truncate the model to a (large) upper bound of K of factors, the shrinkage/sparsity imposed on the factor loadings allows us to infer the subset of factors (typically less than K) actually needed to represent the data.

Provided that factor scores $\mathbf{s}_n$ and noise terms $\boldsymbol{\epsilon}_n$ are independent and Gaussian distributed, from (13.1) it follows that $\mathbf{x}_n \sim \mathcal{N}(\mathbf{0}, \mathbf{A}\boldsymbol{\Lambda}\boldsymbol{\Lambda}\mathbf{A}^\top + \boldsymbol{\Psi}^{-1})$, i.e. the factor model estimates the covariance structure of data $\mathbf{x}_n$ as a low-rank representation if $K < p$, once the factor scores $\mathbf{s}_n$ are analytically marginalized out.

The generative model encoded by (13.1) implies that the expression value for a particular gene is a linear combination of columns of $\mathbf{A}$, with the columns weighted by the factor scores $\mathbf{s}_n$. We can thus see the columns of $\mathbf{A}$ as constituting a dictionary with K gene-related dictionary elements (columns); since $\mathbf{A}$ is shared among all data samples $\{\mathbf{x}_n\}$, the factor scores $\mathbf{s}_n$ serve as a low-dimensional representation of each $\mathbf{x}_n$. Prior distributions for $\mathbf{A}$ and $\boldsymbol{\Lambda}$ are specified to encourage (near) sparsity, to ease interpretability of the features encoded by columns of $\mathbf{A}$, while also being able to automatically estimate the usually unknown number of factors needed by the model to effectively explain the data (to infer the subset of K columns actually needed to represent the observed data).

13.3.1 Shrinkage prior

To impose (near) sparseness on the loadings matrix $\mathbf{A}$, we employ the three-parameter beta normal (TPBN) prior, a general prior distribution that can be expressed as scale mixtures of normals [20]. Specifically, if $a_{jk} \sim \text{TPBN}(a, b, \phi_k)$, where $j = 1, \ldots, p$, $k = 1, \ldots, K$, we can write

$$\begin{aligned} a_{jk} &\sim \mathcal{N}(0, \zeta_{jk}), \\ \zeta_{jk} &\sim \text{Gamma}(a, \xi_{jk}), \\ \xi_{jk} &\sim \text{Gamma}(b, \phi_k). \end{aligned} \tag{13.2}$$

When $a = b = 1/2$, the TPBN prior reduces to the horseshoe prior [18]. For fixed values of a and b, ϕ_k controls the shrinkage level of the kth column of $\mathbf{A}$, so that smaller values of ϕ_k yield stronger shrinkage. Furthermore, we can set a prior on ϕ_k to allow the model to learn individual shrinkage levels for each column of $\mathbf{A}$. For instance,

$$\phi_k \sim \text{Gamma}\left(\tfrac{1}{2}, w\right),$$
$$w \sim \text{Gamma}\left(\tfrac{1}{2}, 1\right),$$

where w is a latent variable whose distribution serves as support for the shrinkage levels of $\mathbf{A}$. Alternatively, when prior knowledge about the expected sparsity level of $\mathbf{A}$ is available, ϕ_k can be set accordingly, rather than inferred from data.

The prior in (13.2) also encompasses as special cases well-known shrinkage priors, such as double-exponential [17, 33], Strawderman–Berger [34], normal-exponential-gamma [35], and horseshoe [18] priors. Further, it can be seen as an example of a yet larger family of continuous shrinkage hierarchies, known as global-local shrinkage priors [19, 36].

One of the most appealing features of (13.2) is its excellent mixing properties, which stem from the fact of it being marginally a continuous unimodal distribution with closed-form conditional posterior. Bimodal sparsity, including distributions such as spike-and-slab [1, 15] and the Indian buffet process [37], often face mixing difficulties, whereas commonly used unimodal shrinkage priors (such as double-exponential and Student's-t) are not flexible enough as they tend to over-shrink coefficients with values far away from zero, as previously shown in [18].

13.3.2 Multiplicative gamma process

The multiplicative gamma process (MGP), originally proposed in [38], is imposed on the factor loadings matrix, $\mathbf{A}$, as a global shrinkage prior to estimate the effective number of factors. We choose it over more-involved approaches, such as reversible jump Markov chain Monte Carlo [39] or discrete variable selection priors such as the Indian buffet process [37], beta processes [40] or evolutionary stochastic model search [1]. The multiplicative gamma process introduces infinitely many factors to the model in a way that the variance of each column of the loadings $\mathbf{A}$ will stochastically shrink towards zero, as the index of the column increases. Alternatively, we can impose the MGP on the scaling coefficients of the factors denoted by $\mathbf{\Lambda}$ in (13.1) similar to [41], in which the factorization $\mathbf{A}\mathbf{\Lambda}\mathbf{s}_n$ is seen as a sum of K rank-one matrices weighted by the elements of $\mathbf{\Lambda}$, such that large indices k will have negligible impact on the full factorization, which is in essence similar to a singular value decomposition. From the model in (13.1), if $\mathbf{\Lambda} = \text{diag}(\lambda_1, \ldots, \lambda_K) \sim \text{MGP}(a_1, a_2)$, then

$$\begin{aligned} &\lambda_k \sim \mathcal{N}(0, \tau_k^{-1}), \qquad && \tau_k = \prod_{l=1}^{k} \delta_l, \\ &\delta_1 \sim \text{Gamma}(a_1, 1), && \delta_l \sim \text{Gamma}(a_2, 1), \quad \text{for } l \geq 2, \end{aligned} \tag{13.3}$$

where δ_l for $l = 1, \ldots, \infty$ are independent. Each term of $\prod_{l=1}^{k} \delta_l$ is stochastically increasing provided that $a_2 > 1$, therefore the precision of the Gaussian distribution, τ_k, will shrink λ_k towards zero as k increases. As described in [38], inference for the MGP prior in (13.3) can be done in two ways: (i) approximate the potentially infinite number of columns of $\mathbf{\Lambda}$ by setting K to a reasonably large truncation level; (ii) selecting the number of factors adaptively. Here we choose the former because, as stated in [38], it produces accurate estimates of the effective number of factors as long as the selected truncation level is large enough; further, it is computationally simpler than the adaptive approach.

One additional benefit of the MGP prior is that it helps alleviate one of the sources of lack of identifiability in factor models. Having stochastically ordered columns for $\mathbf{A}$ helps mitigate factor switching during inference, which happens due to the well-known permutation ambiguity of factor models; i.e. the factor models $\mathbf{A}\mathbf{\Lambda}\mathbf{s}_n$ and $\mathbf{A}\mathbf{\Lambda}\mathbf{P}^{-1}\mathbf{P}\mathbf{s}_n$, where $\mathbf{P}$ is an arbitrary permutation matrix, have the same likelihood [14]. In the case of (13.3), $\mathbf{P}$ is no longer arbitrary as $\{\tau_k\}_{k=1}^{K}$ couples the elements of $\mathbf{\Lambda}$, which as a result locks the ordering of columns of $\mathbf{A}$ and elements of $\mathbf{s}_n$.

13.4 Discriminative models

Consider a set of K covariates $\mathbf{s}_n$ with an associated binary label y_n. The goal of a discriminative model is to predict the label $y^\star$ of unseen covariates $\mathbf{s}^\star$. In a probabilistic model this usually amounts to estimating the joint distribution $p(y, \mathbf{s})$ from a training set $\{y_n, \mathbf{s}_n\}_{n=1}^{N}$, then using the predictive distribution $p(y^\star|\mathbf{s}^\star)$ to estimate $y^\star$. In the linear case, we can parameterize the model as

$$y_n = g\left(\mathbf{h}^\top \mathbf{s}_n\right), \tag{13.4}$$

where $g(\cdot) : \mathbb{R}^K \rightarrow \{-1, 1\}$ or $g(\cdot) : \mathbb{R}^K \rightarrow \{0, 1\}$ are mapping functions and $\mathbf{h} \in \mathbb{R}^K$ is a vector of classification coefficients weighting the relative contribution of each of the K covariates to the decision process.

One of the most common choices for function $g(\cdot)$ in (13.4), in Bayesian modeling, is the Heaviside step function that gives rise to probit regression, in which case it can be shown that $p(y_n = 1|\mathbf{s}_n) = \Phi(\mathbf{h}^\top \mathbf{s}_n)$, where $\Phi(\cdot)$ is the cumulative density function of a standard Gaussian distribution [12]. The popularity of the probit function is likely due to the fact that an effective and easy-to-implement inference procedure based on variable augmentation exists [28]. Another two alternatives inspired by the machine learning community are logistic regression and support vector classification, each of which has an associated loss function: log-loss and hinge-loss, respectively [12]. Since Bayesian probit regression is a well-understood model, we focus the remainder of this section on recently proposed variable augmentation approaches for logistic regression and support vector classification.

13.4.1 Bayesian log-loss

The log-loss is defined as the logarithm of a Bernoulli likelihood, so for $y_n \in \{0, 1\}$ and the model in (13.4) we can write

$$\ell(y_n, \mathbf{s}_n, \mathbf{h}) = y_n \log\left(g\left(\mathbf{h}^\top \mathbf{s}_n\right)\right) + (1 - y_n) \log\left(1 - g\left(\mathbf{h}^\top \mathbf{s}_n\right)\right). \tag{13.5}$$

From a Bayesian perspective, instead of optimizing for $\mathbf{h}$ by using the loss function in (13.5), we estimate the posterior distribution $p(\mathbf{h}|y, \mathbf{s})$ using the variable-augmentation approach proposed in [29]; to do this, we first introduce Pólya–Gamma random variables.

A random variable X has a Pólya–Gamma distribution with parameters $b > 0$ and $c \in \mathbb{R}$, denoted $X \sim \mathrm{PG}(b, c)$, if

$$X = \frac{1}{2\pi^2} \sum_{k=1}^{\infty} \frac{g_k}{(k - 1/2)^2 + c^2/(4\pi^2)},$$

where each $g_k \sim \mathrm{Gamma}(b, 1)$ is an independent gamma random variable [29].

A key result of [29] is that Bernoulli likelihoods, $y_n \sim \mathrm{Bernoulli}(\sigma(\mathbf{h}^\top \mathbf{s}_n))$, parameterized by the log-odds, $\mathbf{h}^\top \mathbf{s}_n$, and the logistic function, $\sigma(x) = 1/(1 + \exp(-x))$, can be written as scale mixtures of Gaussians w.r.t Pólya–Gamma distributions; this is, if $\gamma \sim \mathrm{PG}(b, 0)$, then

$$\frac{\exp(\psi)^a}{(1 + \exp(\psi))^b} = 2^{-b} \exp(\kappa\psi) \int_0^{\infty} \exp\left(-\frac{\gamma\psi^2}{2}\right) p(\gamma) d\gamma, \tag{13.6}$$

where $\kappa = a - b/2$ and $\gamma|\psi \sim \mathrm{PG}(b, \psi)$. For the logistic regression model, the likelihood function can be written as

$$L(y_n|\mathbf{h}, \mathbf{s}_n) = \frac{\exp\left(y_n \mathbf{h}^\top \mathbf{s}_n\right)}{1 + \exp\left(\mathbf{h}^\top \mathbf{s}_n\right)}. \tag{13.7}$$

Let $a = y_n$, $b = 1$, $\psi = \mathbf{h}^\top \mathbf{s}_n$, using the Pólya–Gamma data augmentation in (13.6), we can rewrite the likelihood in (13.7) as

$$\begin{aligned} L(y_n|\mathbf{h}, \mathbf{s}_n, \gamma_n) &\propto \exp\left\{\kappa_n \mathbf{h}^\top \mathbf{s}_n - \frac{1}{2}\gamma_n \left(\mathbf{h}^\top \mathbf{s}_n\right)^2\right\} \\ &\propto \exp\left\{-\frac{\gamma_n}{2}\left(\mathbf{h}^\top \mathbf{s}_n - \frac{\kappa_n}{\gamma_n}\right)^2\right\} \\ &\propto \exp\left\{-\frac{1}{2}\left(z_n - \mathbf{h}^\top \mathbf{s}_n\right)^\top \gamma_n \left(z_n - \mathbf{h}^\top \mathbf{s}_n\right)\right\}, \end{aligned}$$

where $\kappa_n = y_n - \frac{1}{2}$, $z_n = \kappa_n/\gamma_n$. Therefore, the augmented model for $\mathbf{h}$ can be expressed as

$$\begin{aligned} \mathbf{h}|\mathbf{y}, \mathbf{S}, \boldsymbol{\gamma} &\propto p(\mathbf{h}) \prod_{n=1}^{N} L(y_n|\mathbf{h}, \mathbf{s}_n, \gamma_n), \\ \gamma_n &\sim \mathrm{PG}(1, 0), \end{aligned} \tag{13.8}$$

where $\mathbf{y} = [y_1 \ \dots \ y_N]$, $\mathbf{S} = [\mathbf{s}_1 \ \dots \ \mathbf{s}_N]$ and $p(\mathbf{h})$ is the prior for $\mathbf{h}$. Note that if $p(\mathbf{h})$ is Gaussian, due to conjugacy, the conditional posterior $p(\mathbf{h}|\mathbf{y}, \mathbf{S})$ will have the same distribution because the likelihood $L(y_n|\mathbf{h}, \mathbf{s}_n, \gamma_n)$ is conditionally Gaussian w.r.t. $\mathbf{h}$.

13.4.2 Bayesian hinge-loss

The hinge-loss, most popular for its role in support vector machines (SVMs) [42], can be written for labels $y_n \in \{-1, 1\}$ as

$$\ell(y_n, \mathbf{s}_n, \mathbf{h}) = \max\left(1 - y_n \mathbf{h}^\top \mathbf{s}_n, 0\right). \tag{13.9}$$

Minimizing (13.9) amounts to finding a decision boundary $\{\mathbf{s}_n : \mathbf{h}^\top \mathbf{s}_n = 0\}$ with associated decision function, $\text{sign}(\mathbf{h}^\top \mathbf{s}_n)$, such that the distance between the so called *margin* boundaries defined as $\{\mathbf{s}_n : \mathbf{h}^\top \mathbf{s}_n = \pm 1\}$ is as large as possible, hence discriminative models based on the hinge-loss are called max-margin classifiers [42]. Unlike the log-loss, the hinge-loss only penalizes misclassifications and margin violations, which stems from the fact that $\ell(y_n, \mathbf{s}_n, \mathbf{h}) = 0$ whenever $y_n \mathbf{h}^\top \mathbf{s}_n > 1$.

Within the Bayesian framework, we can use (13.9) to form a pseudo-likelihood function that can be written as

$$L(y_n|\mathbf{h}, \mathbf{s}_n) = \exp\left\{-2\max\left(1 - y_n \mathbf{h}^\top \mathbf{s}_n, 0\right)\right\}, \tag{13.10}$$

which admits a location-scale mixture of Gaussian representation, by making use of the following integral identity introduced in [43]

$$\exp\{-2\max(u, 0)\} = \int_0^\infty \mathcal{N}(u| - \gamma, \gamma)d\gamma. \tag{13.11}$$

From (13.10) and (13.11) we obtain an augmented likelihood expressed as

$$L(y_n|\mathbf{h}, \mathbf{s}_n, \gamma_n) \propto \gamma_n^{-1/2} \exp\left(-\frac{1}{2}\frac{\left(1 + \gamma_n - y_n \mathbf{h}^\top \mathbf{s}_n\right)^2}{\gamma_n}\right),$$

which we recognize as the core of a Gaussian density with variance γ_n [44]. Interestingly, we get a similar augmented model to that of (13.8) but with conditional posterior $\gamma_n^{-1} \sim \text{IG}(|1 - y_n \mathbf{h}^\top \mathbf{s}_n|, 1)$ instead of $\gamma_n \sim \text{PG}(1, \mathbf{h}^\top \mathbf{s}_n)$ [29], where $\text{IG}(\cdot, \cdot)$ is the inverse Gaussian distribution [43].

13.5 Discriminative factor model

In the previous two sections, we introduced factor models and discriminative models. We next consider the problem of jointly learning both, by connecting them through the factor scores matrix $\mathbf{S}$. Intuitively, instead of doing dimensionality reduction and classification separately as a two-step process, we use the discriminative model as a way to inform the factor model about the fact that its low-dimensional representation of data, $\mathbf{S}$, should be biased towards discriminative power. As a result, we obtain a factor model with two goals: explain the data via covariance structure estimation and produce a classification rule via dimensionality reduction, using factor scores as proxy for data.

Table 13.1 Variable augmentation specifications for classification. $\mathcal{N}(x; \mu, \sigma^2)$ is the density of a Gaussian distribution with parameters μ and σ^2

Classifier	$L(y_n \mid \mathbf{h}, \mathbf{s}_n, \gamma_n)$	$p(\gamma_n)$	$g(\cdot)$
Probit	$I(\gamma_n > 0)$	$\mathcal{N}\left(\mathbf{h}^\top \mathbf{s}_n, 1\right)$	$I\left(\mathbf{h}^\top \mathbf{s}_n > 0\right)$
Logistic	$\mathcal{N}\left(\mathbf{h}^\top \mathbf{s}_n; \left(y_n - \frac{1}{2}\right)\gamma_n^{-1}, \gamma_n^{-1}\right)$	PG(1, 0)	$I\left(\mathbf{h}^\top \mathbf{s}_n > 0\right)$
Max-margin	$\mathcal{N}\left(1 - y_n \mathbf{h}^\top \mathbf{s}_n; -\gamma_n, \gamma_n\right)$	Uniform$(0, \infty)$	sign $\left(\mathbf{h}^\top \mathbf{s}_n\right)$

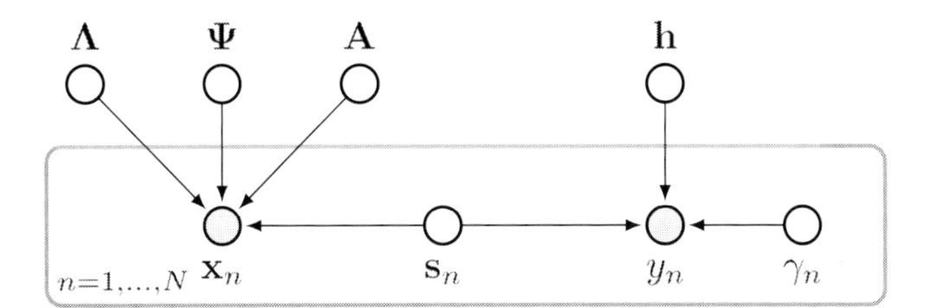

Figure 13.1 Graphical model for the discriminative factor model in (13.12). Note that $\boldsymbol{\epsilon}_n$ has been marginalized out.

Our discriminative factor-model specification with graphical model shown in Figure 13.1 can be written as

$$\begin{aligned}
\mathbf{x}_n &= \mathbf{A}\boldsymbol{\Lambda}\mathbf{s}_n + \boldsymbol{\epsilon}_n, \\
y_n &= g\left(\mathbf{h}^\top \mathbf{s}_n\right), \\
a_{jk} &\sim \text{TPBN}(a, b, \phi_k), \\
\boldsymbol{\Lambda} &\sim \text{MGP}(a_1, a_2), \\
\mathbf{s}_n &\sim \mathcal{N}(\mathbf{0}, \mathbf{I}), \\
\boldsymbol{\epsilon}_n &\sim \mathcal{N}\left(\mathbf{0}, \boldsymbol{\Psi}^{-1}\right), \\
h_k &\sim \text{TPBN}(a, b, \phi),
\end{aligned} \tag{13.12}$$

where $g(\mathbf{h}^\top \mathbf{s}_n)$ is the decision function corresponding to either log-loss or hinge-loss, as described in the previous section and summarized in Table 13.1.

Note that we have provided the vector of classifier weights, $\mathbf{h}$, with the same shrinkage prior imposed on the elements of the factor loadings matrix, $\mathbf{A}$. This is done to acknowledge that only a subset of the factors learned by the model will likely be geared towards discrimination, while the remaining ones may be entirely (or primarily) focused on explaining the data.

From Figure 13.1 we see that conditioned on the factor scores, $\mathbf{S}$, data and labels are conditionally independent, meaning that inference-wise $\mathbf{S}$ is the only variable in the model whose conditional posterior is both dependent on $\mathbf{X}$ and $\mathbf{y}$, data and labels, respectively. It also implies that inference for all the other variables of the model remains conveniently unchanged, when compared to, say, just the factor model in (13.1).

13.5.1 Multi-task learning

The proposed discriminative factor model can be readily extended to a multi-task learning setting, in which multiple binary classification problems are performed jointly. This mechanism has the benefit of information sharing across tasks, thus enhancing the representation learned by the model [23, 45, 46]. For instance, in our proposed model, the classifier coefficients may be different in multiple tasks, but the matrix factorization mechanism, i.e. the inferred matrices $\mathbf{A}$, $\boldsymbol{\Lambda}$ and $\mathbf{S}$ are shared, thereby utilizing all label information to make the model discriminative, but assuming that the original high-dimensional gene data lies in a common low-dimensional subspace for all tasks.

Consider the situation in which $c \in \{1, \ldots, C\}$ tasks performed together. The complete model can be expressed as

$$\begin{aligned} \mathbf{x}_n &= \mathbf{A}\boldsymbol{\Lambda}\mathbf{s}_n + \boldsymbol{\epsilon}_n, \\ y_n^{(c)} &= g\left(\left(\mathbf{h}^{(c)}\right)^\top \mathbf{s}_n\right), \\ \gamma_n^{(c)} &\sim p\left(\gamma_n^{(c)}\right), \end{aligned} \tag{13.13}$$

where superscript (c) represents a specific task, and $g(\cdot)$ and $p(\gamma^{(c)})$ are one of the choices in Table 13.1, depending on the classification model to be used. By comparing (13.12) and (13.13), we see that the factor model part of the hierarchy remains unchanged, and only the discriminative component reflects that we are feeding more information to the model. As a result of this, we expect the model to learn a richer low-dimensional representation, $\mathbf{S}$, thus a larger number of factors K are likely to be used.

The model in (13.13) can also be used for multi-class tasks, if we encode $\tilde{y}_n \in \{1, \ldots, C\}$, where C is the number of classes, as separate binary "tasks". For this purpose we use a *one-vs.-all* encoding, in which we learn a multi-class model by learning C binary classifiers, each of which attempts to differentiate one of the classes from all the others. Once built, the classification rule is

$$\tilde{y}_n = \operatorname{argmax}_c \left(\mathbf{h}^{(c)}\right)^\top \mathbf{s}_n.$$

In biologically motivated applications, learning multiple binary classifiers, as opposed to a single multi-class model, is beneficial in the sense that we can use individual classification coefficient vectors, $\mathbf{h}^{(c)}$, to drive the interpretation of the model one phenotype at the time.

13.6 Inference

Learning the parameters of the model in (13.12) (or (13.13)) with classification specifications in Table 13.1 can be done either via Markov chain Monte Carlo (MCMC) [47, 48] or a mean-field variational Bayes (VB) approximation [49]. The MCMC algorithm involves a sequence of Gibbs updates, where each latent variable is iteratively resampled conditioned on instances of all the others.

Variational Bayes

The VB framework attempts to approximate the full posterior distribution of the model by a simpler distribution, $q(\Theta) \approx p(\Theta|\boldsymbol{y}, \mathbf{X})$, where $\Theta = \{a_{jk}, \lambda_k, s_{kn}, \psi_j, \zeta_{jk}, \xi_{jk}, \phi_k, w, \delta_k, h_k, \gamma_n\}$, for $j = 1, \ldots, p$, $n = 1, \ldots, N$, denotes the set of independent latent variables in the model, and $\mathbf{X} = [\mathbf{x}_1, \ldots, \mathbf{x}_N]$, $\boldsymbol{y} = [y_1, \ldots, y_N]$ represent observed data and labels, respectively. Specifically, VB assumes a complete factorization across latent variables, $q(\Theta) = \prod_i q_i(\boldsymbol{\theta}_i)$, where $\boldsymbol{\theta}_i$ is an element of Θ.

Solving for the optimal distribution $q^\star(\Theta)$ that minimizes the distance between p and q effectively estimates the conditional posterior distribution $p(\Theta|\boldsymbol{y}, \mathbf{X})$. A commonly used distance metric between the two distributions functions is the Kullback–Leibler (KL) divergence [50]. We write the KL-divergence of p from q as follows:

$$\mathrm{KL}(q \| p) = \int_\Theta q(\Theta) \log \frac{q(\Theta)}{p(\Theta|\boldsymbol{y}, \mathbf{X})} d\Theta,$$

through which

$$\begin{aligned} \log p(\mathbf{y}, \mathbf{X}) &= \mathrm{KL}(q \| p) + \mathcal{L}(q), \\ \mathcal{L}(q) &= -\int_\Theta q(\Theta) \log \frac{q(\Theta)}{p(\Theta, \boldsymbol{y}, \mathbf{X})} d\Theta. \end{aligned}$$

Observe that $\log p(\boldsymbol{y}, \mathbf{X})$ is fixed w.r.t. the variations in $q(\Theta)$. Therefore, maximizing the evidence lower bound (ELBO), $\mathcal{L}(q)$, is equivalent to minimizing the KL-divergence between the two distributions. This minimal distance occurs when

$$\log q^\star(\Theta) = \mathbb{E}[\log p(\boldsymbol{y}, \mathbf{X}, \Theta)] + \text{const}.$$

Assuming a complete factorization across the latent variables $q(\Theta) = \prod_i q_i(\boldsymbol{\theta}_i)$, each parameter in a variational Bayes approximation is independently updated according to

$$q_j^\star(\Theta_j) \propto \exp\{\mathbb{E}_{i \neq j}[\log p(\boldsymbol{y}, \mathbf{X}, \Theta)]\}.$$

For the models considered here, both MCMC and VB are straightforward to implement, because local conjugacy of all the parameters of the model allows us to write corresponding conditional posteriors in closed form. A sketch of the inference procedures, and details of conditional posterior distributions of all the parameters of the model, are found in Section 13.9.

Scaling up

Every iteration of the VB inference algorithm requires a full pass through the dataset, which can be time consuming when applied to large datasets. Therefore, an online version of VB inference can be developed, building upon a recent online implementation for latent Dirichlet allocation [51]. Online VB exploits the difference between *local* variables, which are observation dependent, and *global* variables, which are shared among the entire dataset. In the context of factor models, the data are the gene expression values of an individual, the global variables are the factor loadings (shared among all data samples), and the local variables are individual-specific factor scores.

In practice, stochastic optimization is applied to the variational objective function in the online VB representation [52]. The key observation is that the coordinate ascent updates in VB precisely correspond to the natural gradient of the variational objective function. Considering that we split the entire dataset with N observations into D mini-batches, the following steps are repeated to implement the online VB.

(1) Randomly select one mini-batch; optimize its local variational parameters.
(2) Obtain the current estimate of the global variational parameters, as though we were running classical coordinate ascent update on the dataset formed by repeating N/D times of the selected mini-batches.
(3) Update the global variational parameters to be a weighted average of the current estimate and previous estimate.

To be more specific, the global variables are denoted as $\Theta_{\mathbf{g}} = \{a_{jk}, \lambda_k, \psi_j, \zeta_{jk}, \xi_{jk}, \phi_k, w, \delta_k, h_k\}$ and the local variables are denoted as $\Theta_{\mathbf{l}} = \{s_{kn}, \gamma_n\}$, for $j = 1, \ldots, p$, $k = 1, \ldots, K$ and $n = 1, \ldots, N$. The update of one global variable θ_g, after seeing the mini-batch indexed by l, becomes

$$\theta_g^{(l)} = (1 - \rho_l)\theta_g^{(l-1)} + \rho_l \theta_g^*,$$

where θ_g^* is the current estimate and $\rho_l = (\rho_0 + l)^{-\kappa}$ is an appropriately decreasing learning rate. To ensure that the global parameters converge to a stationary point, we set $\kappa \in (0.5, 1]$ and $\rho_0 > 0$. The update of local variational parameters are the same as the batch VB inference. This modeling and inference framework allows one to scale up the analysis to "big data", with a large number of biomarkers (large p) as well as a large number of sample N, although in practice we still typically deal with $N < p$.

Computational cost

The computational cost of the factor model with or without the classifier is roughly $\mathcal{O}(pK^2)$ per iteration. In our experience, we find that between 50 and 100 VB iterations are enough for the model to stabilize. It is important to take into account that the time needed to run any of the models proposed here is significantly smaller than the time needed to generate the data, thus statistical analyses performed using our models will not constitute a bottleneck in real-world applications.

Implementation and availability

All the code used for the experiments including implementations of the models considered was written in Matlab and can be found at: http://www.duke.edu/~rh137/host.html.

13.7 Experiments

In this section, we present extensive experiments on two real-world microarray-based gene expression datasets, from two studies about infectious diseases. In the following, we briefly describe the data and performance measures used to evaluate the models being compared. Extensive numerical results are provided, with model interpretation delivered via pathway association analysis.

ARI dataset

The acute respiratory infection (ARI) dataset is developed with the goal of differentiating bacterial from viral infections in the context of a relatively heterogeneous cohort, also containing subjects with non-infectious illnesses. It is composed of intensities of $p = 22\,277$ probes from Affymetrix HG-U133A 2.0 arrays with $N = 280$ subjects categorized into the following three groups: bacterial (70), viral (115) and non-infectious illness (88). For the analysis, we keep the top 25% (5569) GCRMA normalized [53] probes with largest intensity profiles.

TB dataset

This particular tuberculosis (TB) dataset[1] is the result of a recently published study [54], which consists of gene expression intensities for 47323 genes and $N = 491$ subjects measured using Illumina HumanHT-12 V4.0 expression bead-chips, categorized in four phenotypes: active TB (190), latent TB (68), other diseases (233) and HIV positive (161). The raw data were preprocessed using background correction, quantile normalized signal intensities, log-transformation and gene filtering. For the analysis we keep the top $p = 4732$ genes with largest intensity profiles.

13.7.1 Performance measures

The performance of factor models is evaluated by the mean squared error (MSE) between the original observation matrix $\mathbf{X}$, and the reconstructed matrix $\hat{\mathbf{X}} = \mathbf{A}\mathbf{\Lambda}\mathbf{S}$, where $\mathbf{A}$, $\mathbf{\Lambda}$ and $\mathbf{S}$ are the inferred factor loadings, factor scaling matrix and factor scores, respectively. To be more precise, MSE is defined as $\frac{1}{N}\sum_{n=1}^{N}\sqrt{\sum_{j=1}^{p}(X_{jn} - \hat{X}_{jn})^2}$. Classification is done within a 10-fold cross validation (CV) framework and the receiver operating characteristic (ROC) [55], area under curve (AUC), classification accuracy (ACC), true positive rate (TPR), and true negative rate (TNR) are reported as quantitative performance measures. CPU time is also recorded as proxy for computational cost.

13.7.2 Experimental setup

Inference is performed via VB and the maximum number of iterations is set to 60. We verified empirically that further increasing iterations does not significantly change the outcome of any of our models. A Gibbs sampler is implemented as well, which is set to 2000 burn-in iterations and 1000 posterior collection samples. In order to address potential large-scale datasets, an online version of the VB inference is further developed. All computations are carried out on a 3.40 GHz desktop with 12 GB RAM.

[1] Available at: http://www.ncbi.nlm.nih.gov/geo/query/acc.cgi?acc=GSE39941.

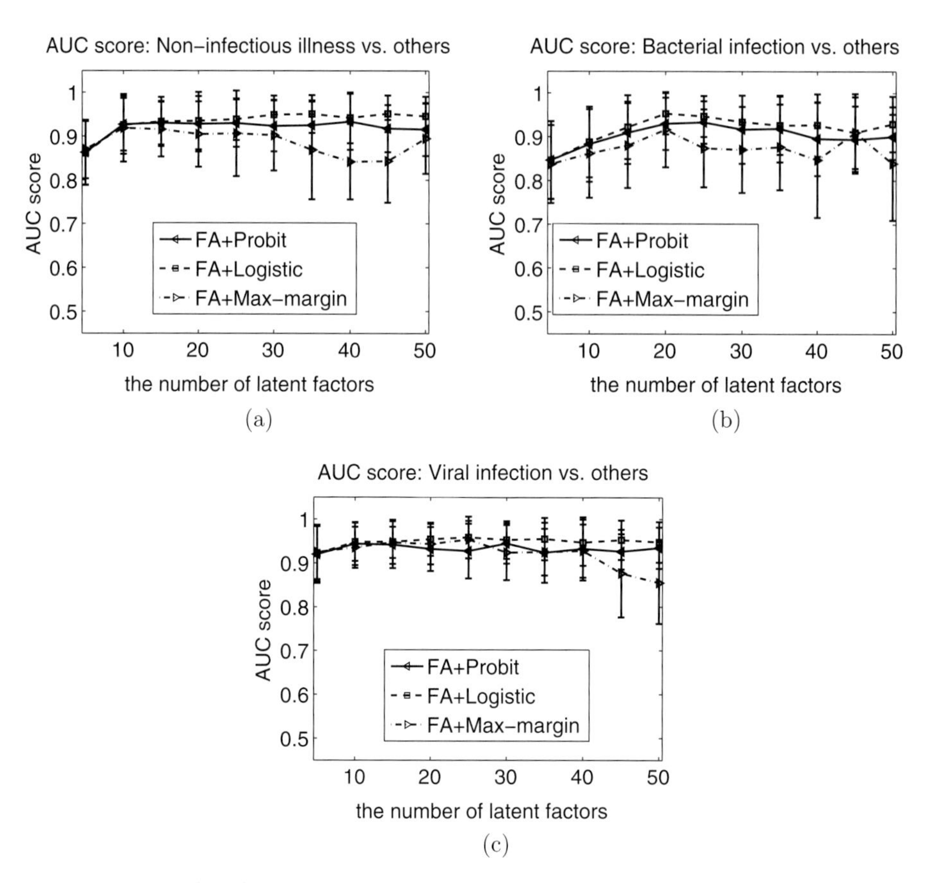

Figure 13.2 AUC values from 10-fold cross-validation on the ARI dataset using the two-step approach: (a) non-infectious illness; (b) bacterial infection; and (c) viral infection.

When implementing the inference algorithms, we initialize **A** and **S** randomly using isotropic normals with standard deviation 1, and set $\mathbf{\Lambda}$ to be the identity matrix. To impose strong sparsity, the hyper-parameters ϕ_k, $k = 1, \ldots, K$ in the TPBN prior are all set to be 10^{-4} rather than inferred from the data. The hyper-parameters of the precision ψ_j, $j = 1, \ldots, p$ are $a_\psi = 1.1$ and $b_\psi = 10^{-3}$. For the ARI dataset, the hyper-parameters of the MGP are set to $a_1 = 2.1$ and $a_2 = 3.1$, and the truncation level of the MGP to $K = 50$. For the TB dataset, we set $a_1 = 1.1$, $a_2 = 2.1$ and $K = 100$.

13.7.3 Classification results

Recall that our model is built using a multi-task learning scheme, in which a factor model jointly learns several predictors, e.g. four in the TB dataset: HIV positive vs. HIV negative, active TB vs. others, latent TB vs. others, and other diseases vs. all TB. The discriminative models are implemented either by probit, logistic or max-margin classifiers as previously shown in Table 13.1.

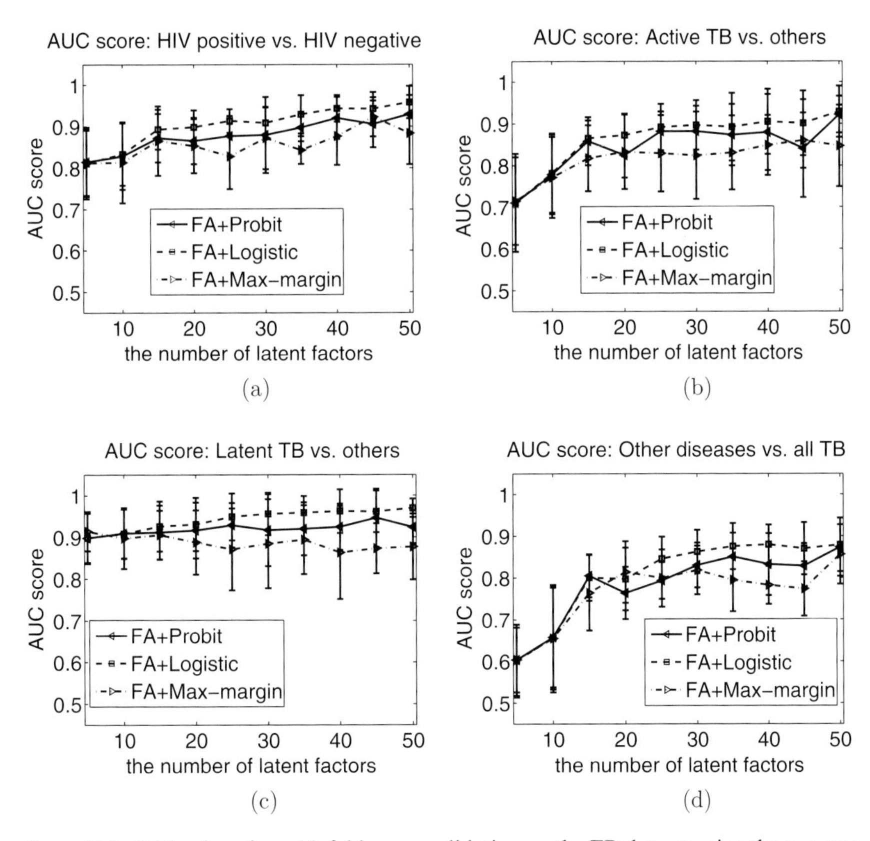

Figure 13.3 AUC values from 10-fold cross-validation on the TB dataset using the two-step approach: (a) HIV positive; (b) active TB; (c) latent TB; and (d) other diseases.

Two-step approach

For illustration, we first train the factor model and the one-vs.-all classifiers separately in a two-step approach. The MGP is not utilized in the factor model in order to investigate the impact of the number of factors on the classification performance. Figures 13.2 and 13.3 show AUCs resulting from this two-step approach for the two datasets, ARI and TB, respectively. It can be seen that the logistic classifier consistently achieves a better classification performance when compared with the other two methods. Using a different number of factors results in different classification accuracies. The number of factors that yields the best performance is close 20 and 40 for ARI and TB data, respectively.

Discriminative factor model

We show the results using the proposed discriminative factor model from Section 13.5 to verify that exploiting the label information during the process of factor modeling can further improve the classification performance. Further, that using the MGP prior can sidestep the model selection issue, and provides us with reasonable choices of the number of factors.

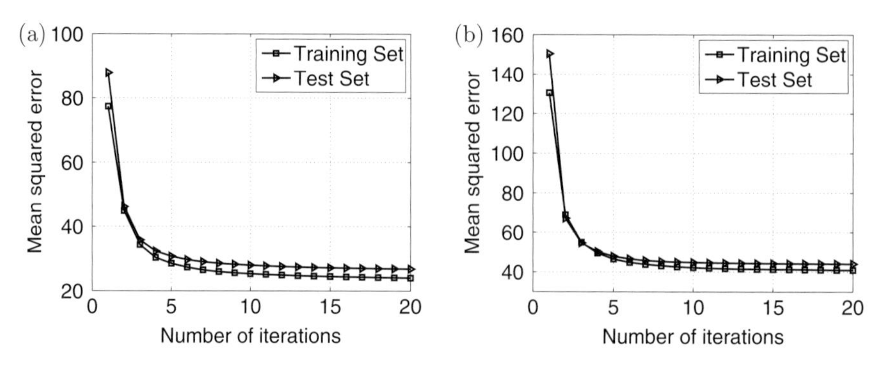

Figure 13.4 Mean squared error: (a) ARI dataset; and (b) TB dataset. Training and test sets are displayed separately.

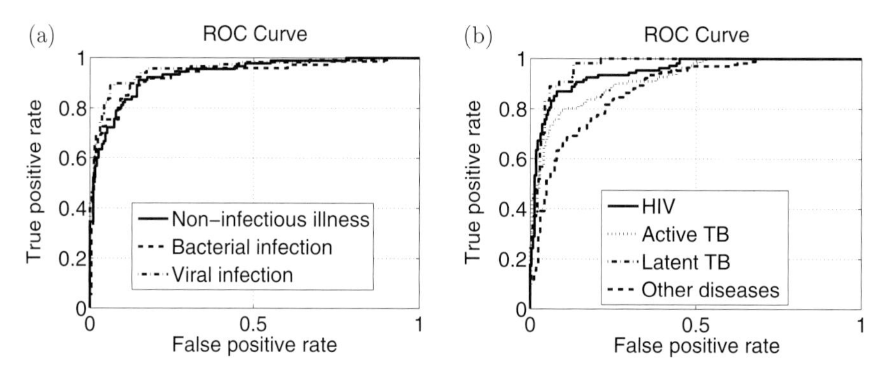

Figure 13.5 ROC curves from 10-fold cross validation: (a) ARI dataset; and (b) TB dataset.

Figure 13.4 plots the mean squared error (MSE) between the model reconstruction and the original data. It also shows that VB inference converges quickly, i.e. ten iterations provides a stable result. Furthermore, the MSE for training and test sets does not differ considerably, which indicates that our discriminative factor model has the ability to successfully prevent overfitting, due to the utilization of the TPBN shrinkage priors.

Since the logistic classifier empirically achieves the best performance, as discussed below, only the ROC curves for the logistic classifier are plotted in Figure 13.5. The AUCs, ACCs, TPRs, and TNRs are detailed in Table 13.2, both for ARI and TB data. For each fold, we calculate AUC, ACC and MSE values, and report averaged results with corresponding error bars (standard deviations). The TPRs and TNRs are calculated on the full set of cross-validated predictions, thus no error bars are available.

From Table 13.2 we see that, on average, 20 and 42 factors are inferred from ARI and TB data, respectively, which is consistent with the model selection results by choosing the model with the highest AUC scores in Figures 13.2 and 13.3. This implies our MGP prior has the ability to choose an adequate number of factors. In terms of classification, all the classifiers we have presented achieve comparable performances, among which

Table 13.2 The 10-fold cross-validation results for TB (left) and ARI (right) datasets

Measure	Logistic	Max-margin	Probit	Logistic	Max-margin	Probit
		HIV positive vs. HIV negative			*Non-infectious illness vs. others*	
AUC (%)	95.18± 1.63	93.67± 3.42	92.48± 4.93	93.57± 5.39	91.35± 5.91	92.81± 5.90
ACC (%)	89.84± 3.71	87.10± 4.80	86.26± 5.90	87.50± 6.57	86.43± 6.48	85.36± 6.83
TPR (%)	85.98	82.24	79.44	90.00	81.11	84.44
TNR (%)	92.07	87.22	88.55	85.26	87.89	85.26
		Active TB vs. others			*Bacterial infection vs. others*	
AUC (%)	91.52± 6.50	86.89± 8.08	91.44± 4.87	93.85± 5.68	91.47± 9.32	92.51± 6.92
ACC (%)	86.27± 5.50	82.64± 5.26	85.95± 7.58	90.00± 6.48	84.64± 8.59	88.57± 4.39
TPR (%)	79.28	76.58	85.59	87.67	80.82	84.93
TNR (%)	89.69	82.96	78.48	87.92	87.92	88.89
		Latent TB vs. others			*Viral infection vs. others*	
AUC (%)	96.18± 4.17	93.47± 6.55	94.02± 5.28	95.61± 4.38	94.19± 5.05	94.90± 4.46
ACC (%)	93.13± 5.76	90.73± 5.56	90.42± 4.45	91.07± 5.12	89.64± 4.89	88.93± 7.23
TPR (%)	87.04	87.04	90.74	88.89	85.47	83.76
TNR (%)	94.29	85.00	84.29	92.64	92.02	95.09
		Other diseases vs. all TB				
AUC (%)	88.11± 3.88	84.93± 5.89	86.79± 4.61			
ACC (%)	79.64± 4.88	75.74± 5.96	79.05± 4.41			
TPR (%)	81.07	78.11	77.51			
TNR (%)	77.58	74.55	80.61			
Factors	42.0± 8.43	40.4± 8.88	43.6± 8.88	20.40± 3.27	19.80± 2.44	20.50± 1.96
Train MSE	39.84± 1.26	40.03± 1.42	39.59± 1.07	23.02± 0.82	23.19± 0.61	22.99± 0.56
Test MSE	45.01± 1.86	45.21± 1.86	44.95± 1.92	24.98± 1.08	25.09± 0.86	24.93± 0.78
Time (s)	3031.2± 44.2	3114.0± 25.5	3039.6± 43.4	2161.6± 15.2	2097.7± 8.2	2105.2± 22.5

the logistic classifier performs best. The good prediction performance indicates that our model has the ability to find the potentially important factors, thus it is able to successfully characterize the host response to the stimuli considered. Furthermore, since the TB dataset has more categories to predict, the number of inferred factors is larger compared to the ARI dataset as one may expect.

Inference via Gibbs sampling results in similar performances, thus those results are not presented, for brevity. Nevertheless, we show trace plots in Figure 13.6 by randomly selecting one element from the factor loadings $\mathbf{A}$, factor scores $\mathbf{S}$, noise precision Ψ, and classifier coefficients $\mathbf{h}$, respectively. It is observed that the Gibbs sampler yields in general good mixing; the trace plot for $\mathbf{A}$ clearly demonstrates the shrinkage imposed on the factor loadings through the TPBN prior.

Online learning

To demonstrate the ability of our proposed model to scale up to large datasets, we present results on the ARI dataset, using our implementation of online VB with the logistic classifier. Local parameters are updated using 20 iterations per mini-batch,

Table 13.3 The 10-fold cross-validation results on the ARI dataset using online VB. The first column indicates the mini-batch size. Results in the second row are taken from Table 13.2. AUC scores correspond from left to right to non-infectious illness, bacterial, and viral classifiers

Size	CPU time (s)	MSE		AUC (%)	
252	2161.6$\pm$ 15.2	24.98$\pm$ 1.08	93.57$\pm$ 5.39	93.85$\pm$ 5.68	95.61$\pm$ 4.38
126	1087.5$\pm$ 3.9	25.72$\pm$ 1.22	92.16$\pm$ 7.51	90.62$\pm$ 6.98	94.67$\pm$ 4.16
63	881.3$\pm$ 3.1	25.83$\pm$ 1.11	92.81$\pm$ 6.90	90.87$\pm$ 7.41	94.63$\pm$ 4.57
36	857.4$\pm$ 3.7	26.68$\pm$ 1.15	91.29$\pm$ 6.08	90.14$\pm$ 9.60	93.13$\pm$ 5.89

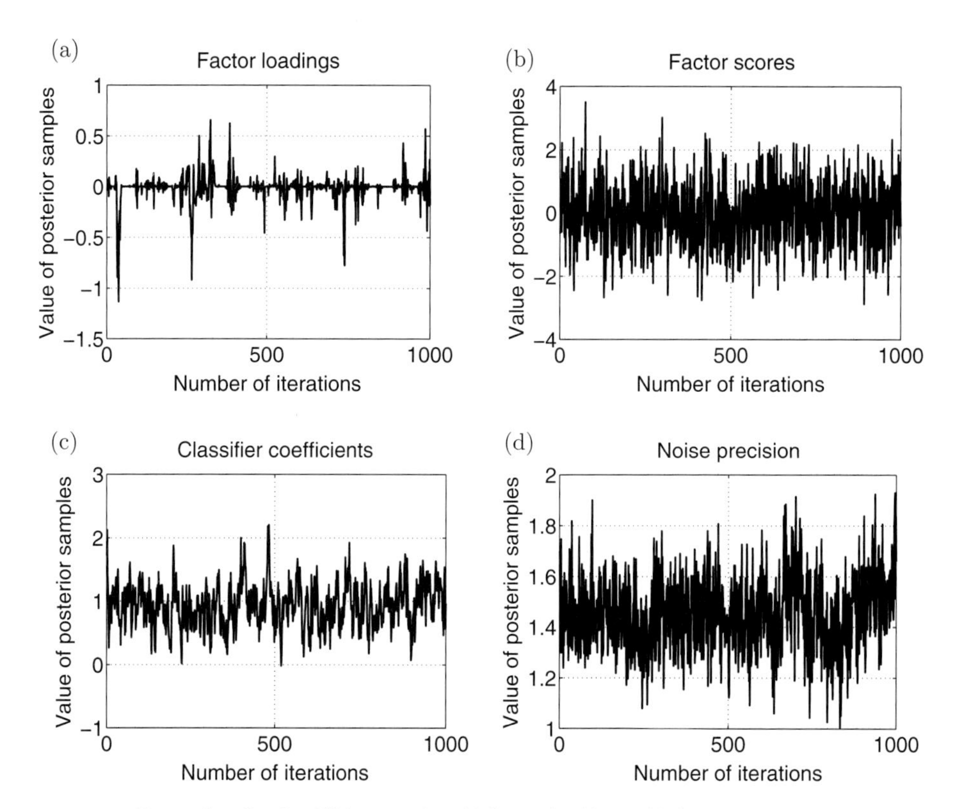

Figure 13.6 Trace plots for the Gibbs sampler: (a) factor loadings; (b) factor scores; (c) noise precision; and (d) classifier coefficients.

and results are shown for 4 epochs. The learning rate is $\rho_l = (\rho_0 + l)^{-\kappa}$, and we set $\rho_0 = 1$ and $\kappa = 0.5$. Results are summarized in Table 13.3. We see that online VB can achieve almost the same performance as batch VB, in terms of both factor modeling and classification; at the same time, online VB is considerably less computationally expensive.

13.7.4 Interpretation

Interpretation of our discriminative factor model is based on the factor loadings and classification weights, $\mathbf{A}$ and $\mathbf{h}$, respectively. As previously discussed, we can relate

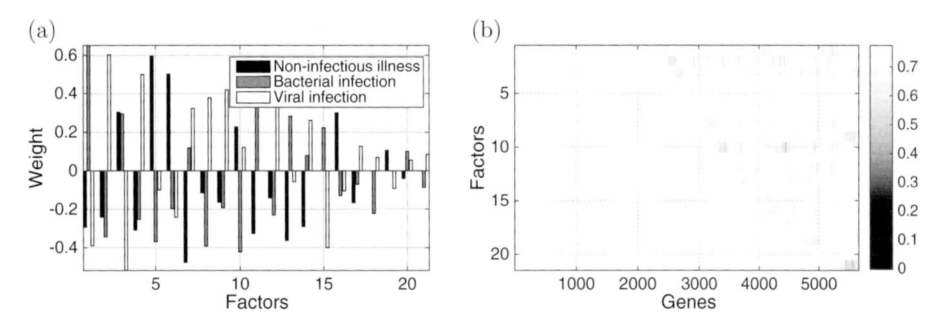

Figure 13.7 Model features for ARI data: (a) classification weights $\mathbf{h}$; (b) thresholded loading matrix $\mathbf{A}$.

columns in $\mathbf{A}$, the factors, to networks of correlated genes and the magnitude of the weights in $\mathbf{h}$ as a measure of the relative importance of each factor to the classification outcome. Figure 13.7a shows classification weights for each of the three classifiers built for the ARI dataset. We see that individual factors contrast different phenotypes, for example factor 1 is specific to bacterial infection, factors 2 and 3 to viral infection, and so on. Figure 13.7b shows a thresholded version of loadings matrix $\mathbf{A}$, where we have set to zero any of its elements such that $\{a_{jk} : |a_{jk}| < 3\text{std}(a_{1k}, \ldots, a_{pk})\}$. This procedure not only eases visualization but helps subsequent interpretation by only focusing on the genes that have an important contribution to the factor scores, and thus to the discriminative models. Note that nearly half of the genes are not present in the thresholded loading matrix in Figure 13.7b, meaning that they are for the most part considered by the model as "noise", thus explained by $\boldsymbol{\epsilon}_n$ rather than $\mathbf{A}\mathbf{s}_n$. After thresholding each of the 21 factors found by the model, they constitute gene networks of approximately 300 genes each. Interpretation of the gene networks encoded by the loadings matrix is done by means of a pathway association analysis using DAVID [56]. The idea is to statistically quantify the association between a set of genes (factor) to a biological theme, function or pathway. In our analysis we particularly focus on gene ontology (GO) terms related to molecular function; however, if the application requires it one may rather target it to pathway databases or medical terms. Table 13.4 shows the size of each gene network and the top GO associations corresponding to the top 7 factors with largest absolute classification weights, as shown in Figure 13.7a. In parentheses are the number of genes in the network associated with a particular pathway. All associations are significant at the 0.05 level with p-values corrected for multiple testing using the work of Benjamin and Hochberg [57]. From Table 13.4 we see GO terms intimately related to host response to infectious diseases, such as immune response, defense response, and inflammatory response, but more interestingly we see some factors targeting particular biological functions, for example, factors 1 and 3 confirm our interpretation of the classifier weights from Figure 13.7a, in terms of them being specific of bacterial and viral infections, respectively, whereas factor 6 has apoptosis as its main theme. Complete tables containing all associations, not only the statistically significant ones, p-values with different correction methods and gene lists for each GO term can be found at http://www.duke.edu/~rh137/host.html.

Table 13.4 Pathway analysis for ARI data. GO terms associated with gene networks encoded by the columns of factor loadings matrix. All associations are significant at the 0.05 level

ID	N	GO terms
1	252	Defense response (43), immune response (45), response to wounding (36), inflammatory response (26), positive regulation of immune system process (20), response to bacterium (18), innate immune response (15)
2	447	Golgi apparatus (46), nuclear lumen (65), intracellular organelle lumen (74), organelle lumen (75), endoplasmic reticulum (47)
3	286	Immune response (43), response to virus (13), nucleic acid transport (12), RNA transport (12), establishment of RNA localization (12), RNA localization (12)
4	301	Immune response (32), programmed cell death (27), death (28), defense response (26), positive regulation of immune system process (15)
5	291	Immune response (49), defense response (33), lymphocyte activation (16), innate immune response (13), leukocyte activation (17), cell activation (17), response to virus (11), response to vitamin (8), response to organic substance (28)
6	295	Negative regulation of cell death (21), negative regulation of programmed cell death (21), macromolecular complex subunit organization (31), immune response (31), nucleosome assembly (10), negative regulation of apoptosis (21)
7	302	Immune response (47), defense response (33), antigen processing and presentation (11), regulation of apoptosis (34), regulation of programmed cell death (34), regulation of cell death (34), vesicle-mediated transport (27)

We also performed a thorough pathway association analysis for the TB dataset with results as biologically relevant as those in Table 13.4. Complete tables of association and plots similar to those in Figure 13.7 can be found at http://www.duke.edu/˜rh137/host.html.

13.8 Closing remarks

This chapter highlights the importance of Bayesian modeling for gene expression analysis. Discriminative factor models, which are the particular theme of this chapter, are presented within a principled framework to jointly build factor models and multiple classifiers. As an alternative to Bayesian classifiers based on the traditional probit link, we have integrated logistic regression and support vector classification into our modeling scheme, using novel variable augmentation techniques. We have equipped the factor models with global-local shrinkage priors, recently proposed within the machine learning community. Our model can infer the number of factors automatically from the data. We also provide extensions to multi-task learning. Inference is developed using both MCMC and variational Bayes algorithms, while online learning is further investigated to scale the model to large datasets.

It is understood that real-world datasets are getting larger and also more complex. One important research direction is to develop factor models with built-in nonlinear classifiers, since the data may not be always separable in the linear subspace implied by our factor model specification. We have made some progress with a discriminative factor model using nonlinear classifiers, based on mixtures of locally linear classifiers or support vector machines [58]. However, this new methodology still needs a considerable amount of study, thus is left as future work. Another possibility we have been considering is to relax the Gaussianity assumption implied by the likelihood function of our factor model, which may be useful in other types of omics data, such as proteomics, metabolomics, and RNA sequencing. In particular, we are considering replacing the Gaussian likelihood in our model with a likelihood function for ranked data, so we can seamlessly treat ordinal, continuous and discrete data all within the same family of generalized discriminative factor models [59].

13.9 Inference details

Given all the conditional posterior distributions for all the parameters involved in (13.12), MCMC based on Gibbs sampling and mean-field VB inference can be easily implemented as briefly described in Section 13.6.

Gibbs sampling

We sample all parameters of the model from their corresponding conditional posterior distributions one at the time. We repeat this cycle of samples enough times for the model to mix; this is usually known as the "burn-in" period. Then, we keep sampling for a specified number of iterations in order to summarize the distributions of the parameters of interest ("collection" period).

Variational Bayes

Instead of sampling from the parameters of the model, we introduce a fully factorized approximation $q(\Theta|\mathbf{x}_n, y_n)$ to the exact posterior $p(\Theta|\mathbf{x}_n, y_n)$, for $n = 1, \ldots, N$, where $\Theta = \{a_{jk}, \lambda_k, s_{kn}, \psi_j, \zeta_{jk}, \xi_{jk}, \phi_k, w, \delta_k, h_k, \gamma_n\}$, for $j = 1, \ldots, p$ and $k = 1, \ldots, K$. Inference proceeds by repeatedly updating the moments of the variational distributions until convergence is observed. In the case of local conjugacy, the moments of the variational approximation match the moments of the analytically computed conditional posteriors.

Conditional posteriors

In the remainder of this section we present the relevant conditional posteriors. For reference, $j = 1, \ldots, p$ indexes variables (genes), $n = 1, \ldots, N$ indexes observations and $k = 1, \ldots, K$ indexes factors.

Factor loadings

- Define $X_{jn}^{-k} = x_{jn} - \sum_{l \neq k}^{K} \lambda_l A_{jl} s_{ln}$, then $A_{jk}|- \sim \mathcal{N}(\mu_{jk}, \Sigma_{jk})$, where

$$\Sigma_{jk} = \left(\sum_{n=1}^{N} \psi_j \lambda_k^2 s_{kn}^2 + \zeta_{jk}^{-1} \right)^{-1}, \quad \mu_{jk} = \Sigma_{jk} \left(\sum_{n=1}^{N} \psi_j \lambda_k s_{kn} X_{jn}^{-k} \right).$$

- $\zeta_{jk}|- \sim \text{GIG}\left(0, 2\xi_{jk}, A_{jk}^2\right)$ and $\zeta_{jk}^{-1}|- \sim \text{GIG}\left(0, A_{jk}^2, 2\xi_{jk}\right)$.
- $\xi_{jk}|- \sim \text{Gamma}\left(1, \zeta_{jk} + \phi_k\right)$.
- $\phi_k|- \sim \text{Gamma}\left(\frac{1}{2}p + \frac{1}{2}, w + \sum_{j=1}^{p} \xi_{jk}\right)$.
- $w|- \sim \text{Gamma}\left(\frac{1}{2}K + \frac{1}{2}, 1 + \sum_{k=1}^{K} \phi_k\right)$.

Factor scalings

- $\lambda_k|- \sim \mathcal{N}\left(\mu_k, \sigma_k^2\right)$, where

$$\sigma_k^2 = \left(\sum_{n=1}^{N}\sum_{j=1}^{p} \psi_j A_{jk}^2 s_{kn}^2 + \tau_k\right)^{-1}, \quad \mu_k = \sigma_k^2 \left(\sum_{n=1}^{N}\sum_{j=1}^{p} \psi_j A_{jk} s_{kn} X_{jn}^{-k}\right).$$

- $\delta_1|- \sim \text{Gamma}(\hat{a}_1, \hat{b}_1)$, where

$$\hat{a}_1 = a_1 + \frac{K}{2}, \quad \hat{b}_1 = 1 + \frac{1}{2}\sum_{k=1}^{K} \tau_k^{(1)} \lambda_k^2.$$

- $\delta_h|- \sim \text{Gamma}(\hat{a}_h, \hat{b}_h)$, where $h \geq 2$ and

$$\hat{a}_h = a_2 + \frac{K - h + 1}{2}, \quad \hat{b}_h = 1 + \frac{1}{2}\sum_{k=h}^{K} \tau_k^{(h)} \lambda_k^2, \quad \tau_k^{(h)} = \prod_{l=1, l \neq h}^{k} \delta_l = \frac{\tau_k}{\delta_h}.$$

Noise variance

- $\psi_j|- \sim \text{Gamma}(g_j, h_j)$, where

$$g_j = g_0 + \frac{N}{2}, \quad h_j = h_0 + \frac{1}{2}\sum_{n=1}^{N} \left(x_{jn} - \mathbf{A}_j^\top \boldsymbol{\Lambda} \mathbf{s}_n\right)^2.$$

When utilizing various classifiers considered, we have different update equations for $\mathbf{S}$ and $\mathbf{h}$, factor scores and classifier coefficients, respectively.

Max-margin classifier

- Define $T_{kn} = 1 - \sum_{l \neq k} y_n s_{ln} h_l$, then $s_{kn}|- \sim \mathcal{N}(\mu_{kn}, \Sigma_{kn})$, where

$$\Sigma_{kn} = \left(\sum_{j=1}^{p} \psi_j \lambda_k^2 A_{jk}^2 + 1 + \frac{h_k^2}{\gamma_n}\right)^{-1},$$

$$\mu_{kn} = \Sigma_{kn} \left(\sum_{j=1}^{p} \psi_j \lambda_k A_{jk} X_{jn}^{-k} + \left(1 + T_{kn}\gamma_n^{-1}\right) y_n h_k\right).$$

- $h_k|- \sim \mathcal{N}\left(\mu_k, \sigma_k^2\right)$, where

$$\sigma_k^2 = \left(\sum_{n=1}^{N} \frac{s_{kn}^2}{\gamma_n} + \omega_k^{-1}\right)^{-1}, \quad \mu_k = \sigma_k^2 \left(\sum_{n=1}^{N} (1 + T_{kn}\gamma_n^{-1}) y_n s_{kn}\right).$$

- $\gamma_n^{-1}|- \sim \text{IG}\left(\left|1 - y_n \mathbf{h}^\top \mathbf{s}_n\right|^{-1}, 1\right)$.

Logistic regression

- Define $T_{kn} = \gamma_n^{-1}\left(y_n - \frac{1}{2}\right) - \sum_{l \neq k} s_{ln} h_l$, then $s_{kn}|- \sim \mathcal{N}(\mu_{kn}, \Sigma_{kn})$, where

$$\Sigma_{kn} = \left(\sum_{j=1}^{p} \psi_j \lambda_k^2 A_{jk}^2 + 1 + \gamma_n h_k^2\right)^{-1},$$

$$\mu_{kn} = \Sigma_{kn}\left(\sum_{j=1}^{p} \psi_j \lambda_k A_{jk} X_{jn}^{-k} + \gamma_n h_k T_{kn}\right).$$

- $h_k|- \sim \mathcal{N}\left(\mu_k, \sigma_k^2\right)$, where

$$\sigma_k^2 = \left(\sum_{n=1}^{N} \gamma_n s_{kn}^2 + \omega_k^{-1}\right)^{-1}, \quad \mu_k = \sigma_k^2\left(\sum_{n=1}^{N} \gamma_n s_{kn} T_{kn}\right).$$

- $\gamma_n|- \sim \mathrm{PG}\left(1, \mathbf{h}^\top \mathbf{s}_n\right)$.

Probit regression

- Define $T_{kn} = \gamma_n - \sum_{l \neq k} s_{ln} h_l$, then $s_{kn}|- \sim \mathcal{N}(\mu_{kn}, \Sigma_{kn})$, where

$$\Sigma_{kn} = \left(\sum_{j=1}^{p} \psi_j \lambda_k^2 A_{jk}^2 + 1 + h_k^2\right)^{-1},$$

$$\mu_{kn} = \Sigma_{kn}\left(\sum_{j=1}^{p} \psi_j \lambda_k A_{jk} X_{jn}^{-k} + h_k T_{kn}\right).$$

- $h_k|- \sim \mathcal{N}\left(\mu_k, \sigma_k^2\right)$, where

$$\sigma_k^2 = \left(\sum_{n=1}^{N} s_{kn}^2 + \omega_k^{-1}\right)^{-1}, \quad \mu_k = \sigma_k^2\left(\sum_{n=1}^{N} T_{kn} s_{kn}\right).$$

- γ_n, where

$$\gamma_n | y_n = 1, - \sim \mathcal{N}\left(\mathbf{h}^\top \mathbf{s}_n, 1\right) I(\gamma_n \geq 0),$$
$$\gamma_n | y_n = -1, - \sim \mathcal{N}\left(\mathbf{h}^\top \mathbf{s}_n, 1\right) I(\gamma_n < 0).$$

Acknowledgements

The research reported here was supported by the Defense Advanced Research Projects Agency (DARPA), under the Predicting Health and Disease (PHD) program. Ephraim L. Tsalik was supported by Award Number 1IK2CX000530 from the Clinical Science Research and Development Service of the VA Office of Research and Development. The views expressed in this article are those of the authors and do not necessarily represent the views of the Department of Veterans Affairs.

References

[1] C. M. Carvalho, J. Chang, J. E. Lucas, *et al.*, "High-dimensional sparse factor modeling: applications in gene expression genomics," *Journal of the American Statistical Association*, vol. **103**, no. 484, pp. 1438–1456, 2008.

[2] J. Lucas, C. Carvalho, and M. West, "A Bayesian analysis strategy for cross-study translation of gene expression biomarkers," *Statistical Applications in Genetics and Molecular Biology*, vol. **8**, no. 1, pp. 1–26, 2009.

[3] T. Speed, *Statistical Analysis of Gene Expression Microarray Data*, CRC Press, 2003.

[4] S. Dudoit, J. Fridlyand, and T. P. Speed, "Comparison of discrimination methods for the classification of tumors using gene expression data," *Journal of the American Statistical Association*, vol. **97**, no. 457, pp. 77–87, 2002.

[5] I. Guyon, J. Weston, S. Barnhill, and V. Vapnik, "Gene selection for cancer classification using support vector machines," *Machine Learning*, vol. **46**, pp. 389–422, 2002.

[6] I. Jolliffe, *Principal Component Analysis*, Wiley Online Library, 2005.

[7] D. D. Lee and H. S. Seung, "Algorithms for non-negative matrix factorization," in *Advances in Neural Information Processing Systems*, 2001.

[8] A. Hyvärinen, J. Karhunen, and E. Oja, *Independent Component Analysis*, John Wiley & Sons, 2004.

[9] L. Carin, J. L. Alfred Hero III, D. Dunson, *et al.*, "High-dimensional longitudinal genomic data: an analysis used for monitoring viral infections," *IEEE Signal Processing Magazine*, vol. **29**, no. 1, pp. 108–123, 2012.

[10] M. E. Tipping, "Sparse Bayesian learning and the relevance vector machine," *The Journal of Machine Learning Research*, vol. **1**, pp. 211–244, 2001.

[11] B. Krishnapuram, D. Williams, Y. Xue, *et al.*, "On semi-supervised classification," in *Advances in Neural Information Processing Systems*, 2004.

[12] C. M. Bishop *et al.*, *Pattern Recognition and Machine Learning*, Springer, New York, 2006.

[13] J. Bernardo, M. Bayarri, J. Berger, *et al.*, "Bayesian factor regression models in the 'large p, small n' paradigm," *Bayesian Statistics*, vol. **7**, pp. 733–742, 2003.

[14] A. M. Kagan, C. R. Rao, and Y. V. Linnik, *Characterization Problems in Mathematical Statistics*, Wiley, 1973.

[15] H. Ishwaran and L. F. James, "Gibbs sampling methods for stick-breaking priors," *Journal of the American Statistical Association*, vol. **96**, no. 453, pp. 1–23, 2001.

[16] R. Henao and O. Winther, "Sparse linear identifiable multivariate modeling," *The Journal of Machine Learning Research*, vol. **12**, pp. 863–905, 2011.

[17] T. Park and G. Casella, "The Bayesian lasso," *Journal of the American Statistical Association*, vol. **103**, no. 482, pp. 681–686, 2008.

[18] C. M. Carvalho, N. G. Polson, and J. G. Scott, "Handling sparsity via the horseshoe," in *International Conference on Artificial Intelligence and Statistics*, 2009, pp. 73–80.

[19] N. G. Polson and J. G. Scott, "Shrink globally, act locally: sparse Bayesian regularization and prediction," *Bayesian Statistics*, vol. **9**, pp. 501–538, 2010.

[20] A. Armagan, D. B. Dunson, and M. Clyde, "Generalized beta mixtures of Gaussians," in *Advances in Neural Information Processing Systems*, 2011.

[21] A. K. Zaas, M. Chen, J. Varkey, *et al.*, "Gene expression signatures diagnose influenza and other symptomatic respiratory viral infections in humans," *Cell Host & Microbe*, vol. **6**, no. 3, pp. 207–217, 2009.

[22] B. Chen, M. Chen, J. Paisley, *et al.*, "Bayesian inference of the number of factors in gene-expression analysis: application to human virus challenge studies," *BMC Bioinformatics*, vol. **11**, no. 1, pp. 1–16, 2010.

[23] M. Chen, D. Carlson, A. Zaas, *et al.*, "Detection of viruses via statistical gene expression analysis," *Biomedical Engineering, IEEE Transactions on*, vol. **58**, no. 3, pp. 468–479, 2011.

[24] C. W. Woods, M. T. McClain, M. Chen, *et al.*, "A host transcriptional signature for presymptomatic detection of infection in humans exposed to influenza H1N1 or H3N2," *PloS One*, vol. **8**, no. 1, pp. e52 198, 1–9, 2013.

[25] R. Henao, J. W. Thompson, M. A. Moseley, *et al.*, "Latent protein trees," *Annals of Applied Statistics*, vol. **7**, no. 2, pp. 691–713, 2013.

[26] A. K. Zaas, B. H. Garner, E. L. Tsalik, *et al.*, "The current epidemiology and clinical decisions surrounding acute respiratory infections," *Trends in Molecular Medicine*, vol. **20**, no. 10, pp. 579–588, 2014.

[27] J. Mairal, F. Bach, J. Ponce, G. Sapiro, and A. Zisserman, "Supervised dictionary learning," in *Advances in Neural Information Processing Systems*, 2009.

[28] J. H. Albert and S. Chib, "Bayesian analysis of binary and polychotomous response data," *Journal of the American Statistical Association*, vol. **88**, no. 422, pp. 669–679, 1993.

[29] N. G. Polson, J. G. Scott, and J. Windle, "Bayesian inference for logistic models using Pólya–gamma latent variables," *Journal of the American Statistical Association*, vol. **108**, no. 504, pp. 1339–1349, 2013.

[30] N. G. Polson and S. L. Scott, "Data augmentation for support vector machines," *Bayesian Analysis*, vol. **6**, no. 1, pp. 1–23, 2011.

[31] N. Quadrianto, V. Sharmanska, D. A. Knowles, and Z. Ghahramani, "The supervised IBP: neighbourhood preserving infinite latent feature models." in *Uncertainty in Artificial Intelligence*, 2013.

[32] E. Salazar, M. S. Cain, S. R. Mitroff, and L. Carin, "Inferring latent structure from mixed real and categorical relational data," in *International Conference on Machine Learning*, 2012.

[33] C. Hans, "Bayesian lasso regression," *Biometrika*, vol. **96**, no. 4, pp. 835–845, 2009.

[34] J. O. Berger and W. E. Strawderman, "Choice of hierarchical priors: admissibility in estimation of normal means," *The Annals of Statistics*, pp. 931–951, 1996.

[35] J. E. Griffin and P. J. Brown, "Bayesian adaptive lassos with non-convex penalization," University of Warwick. Centre for Research in Statistical Methodology, Tech. Rep., 2007.

[36] N. G. Polson and J. G. Scott, "Local shrinkage rules, Lévy processes and regularized regression," *Journal of the Royal Statistical Society: Series B (Statistical Methodology)*, vol. **74**, no. 2, pp. 287–311, 2012.

[37] T. L. Griffiths and Z. Ghahramani, "Infinite latent feature models and the Indian buffet process," in *Advances in Neural Information Processing Systems*, 2005.

[38] A. Bhattacharya and D. B. Dunson, "Sparse Bayesian infinite factor models," *Biometrika*, vol. **98**, no. 2, pp. 291–306, 2011.

[39] H. F. Lopes and M. West, "Bayesian model assessment in factor analysis," *Statistica Sinica*, vol. **14**, no. 1, pp. 41–68, 2004.

[40] J. Paisley and L. Carin, "Nonparametric factor analysis with beta process priors," in *International Conference on Machine Learning*, 2009.

[41] X. Zhang and L. Carin, "Joint modeling of a matrix with associated text via latent binary features," in *Advances in Neural Information Processing Systems*, 2012.

[42] C. Cortes and V. Vapnik, "Support-vector networks," *Machine Learning*, vol. **20**, no. 3, pp. 273–297, 1995.
[43] D. F. Andrews and C. L. Mallows, "Scale mixtures of normal distributions," *Journal of the Royal Statistical Society. Series B (Methodological)*, pp. 99–102, 1974.
[44] N. G. Polson, S. L. Scott *et al.*, "Data augmentation for support vector machines," *Bayesian Analysis*, vol. **6**, no. 1, pp. 1–23, 2011.
[45] Y. Xue, X. Liao, L. Carin, and B. Krishnapuram, "Multi-task learning for classification with Dirichlet process priors," *The Journal of Machine Learning Research*, vol. **8**, pp. 35–63, 2007.
[46] S. Ji, D. Dunson, and L. Carin, "Multitask compressive sensing," *IEEE Transactions on Signal Processing*, vol. **57**, no. 1, pp. 92–106, 2009.
[47] S. Geman and D. Geman, "Stochastic relaxation, Gibbs distributions, and the Bayesian restoration of images," *IEEE Transactions on Pattern Analysis and Machine Intelligence*, vol. **6**, no. 6, pp. 721–741, 1984.
[48] C. Andrieu, N. De Freitas, A. Doucet, and M. I. Jordan, "An introduction to MCMC for machine learning," *Machine Learning*, vol. **50**, pp. 5–43, 2003.
[49] M. J. Beal, "Variational algorithms for approximate Bayesian inference," Ph.D. dissertation, University College London, 2003.
[50] S. Kullback and R. A. Leibler, "On information and sufficiency," *The Annals of Mathematical Statistics*, vol. **22**, no. 1, pp. 79–86, 1951.
[51] M. Hoffman, F. R. Bach, and D. M. Blei, "Online learning for latent Dirichlet allocation," in *Advances in Neural Information Processing Systems*, 2010.
[52] M. D. Hoffman, D. M. Blei, C. Wang, and J. Paisley, "Stochastic variational inference," *The Journal of Machine Learning Research*, vol. **14**, no. 1, pp. 1303–1347, 2013.
[53] Z. Wu, R. A. Irizarry, R. Gentleman, F. Martinez-Murillo, and F. Spencer, "A model-based background adjustment for oligonucleotide expression arrays," *Journal of the American Statistical Association*, vol. **99**, no. 468, pp. 909–917, 2004.
[54] S. T. Anderson, M. Kaforou, A. J. Brent, *et al.*, "Diagnosis of childhood tuberculosis and host RNA expression in Africa," *New England Journal of Medicine*, vol. **370**, no. 18, pp. 1712–1723, 2014.
[55] T. Fawcett, "An introduction to ROC analysis," *Pattern Recognition Letters*, vol. **27**, no. 8, pp. 861–874, 2006.
[56] B. T. S. Da Wei Huang and R. A. Lempicki, "Systematic and integrative analysis of large gene lists using DAVID bioinformatics resources," *Nature Protocols*, vol. **4**, no. 1, pp. 44–57, 2008.
[57] Y. Benjamini and Y. Hochberg, "Controlling the false discovery rate: a practical and powerful approach to multiple testing," *Journal of the Royal Statistical Society. Series B (Methodological)*, pp. 289–300, 1995.
[58] R. Henao, X. Yuan, and L. Carin, "Bayesian nonlinear support vector machines and discriminative factor modeling," in *Advances in Neural Information Processing Systems*, 2014.
[59] X. Yuan, R. Henao, E. L. Tsalik, R. J. Longley, and L. Carin, "Non-Gaussian discriminative factor models via the max-margin rank-likelihood," in *International Conference on Machine Learning*, 2015.

14 Gene-set-based inference of biological network topologies from big molecular profiling data

Lipi Acharya and Dongxiao Zhu

Network discovery is often of primary interest in many scientific domains. It becomes much more challenging in biological domain because: (1) such networks are not directly observable in the experiments; (2) such networks are dynamic, i.e. different parts of the network are activated from time to time and from condition to condition; and (3) the increasingly available biological data are often big (volume), heterogeneous (variety), and error prone (veracity). There is an urgent need for the new methods, algorithms and tools to discover networks from big biological data. In this chapter, we make two assumptions that lead to two approaches to network discovery from big biological data. (1) The true network topology is a distribution of candidate topologies. The challenge is that an exponential number of possible topologies are computational intractable to characterize. Our strategy, i.e. gene set Gibbs sampling (GSGS), is to draw sample topologies and use them to infer the true topology – an approximate learning falling into stochastic algorithm framework. (2) The true network topology is deterministic. The challenge is the large search space, where we design an artificial intelligence algorithm, i.e. gene set simulated annealing (GSSA), to efficiently and intelligently explore the search space of network structures. We use both simulation data and real-world data to demonstrate the performance of our approaches compared to the selected competing approaches.

14.1 Introduction

The past decade has witnessed a tremendous explosion in the amount of data generated through high-throughput molecular profiling technologies such as microarrays and next-generation sequencing. Big molecular profiling datasets are enabling a high-resolution view of biological systems and allowing scientists to interrogate the biomolecular activities of tens of thousands of genes simultaneously. However, challenges remain in analyzing big molecular profiling data and gaining meaningful insights into the biomolecular interaction and regulation mechanisms. These mechanisms are often understood through the inference of biological networks using computational systems biology approaches. A wide range of methods have been proposed in the literature for inferring the structure

Big Data over Networks, ed. Shuguang Cui, Alfred O. Hero III, Zhi-Quan Luo, and José M. F. Moura. Published by Cambridge University Press.

of different types of biological networks, such as gene regulatory networks, protein–protein interaction networks, and signaling networks in the form of Bayesian networks [1, 2], probabilistic Boolean networks (PBNs) [3, 4], mutual information networks [5–7], graphical Gaussian models [8–11], and other approaches [12–16].

Big molecular profiling datasets are becoming increasingly available, and are feathered by both high dimension and large sample size. Intelligent computational approaches are urgently needed to perform big data analytics. For a reliable inference of biomolecular interaction and regulation mechanisms, it is necessary to find the group of molecules involved in a biological network, which we refer to as a *network component*, as well as the underlying network structure in a given network component. Although many annotated network components and tools for their analysis have become available in recent years [17–21], our current knowledge about biological networks is still limited. Prior-known biological networks may not represent the complete picture of the underlying biomolecular activities. There might exist additional relationships among molecules present in the corresponding network components. Moreover, biological networks in the databases are usually generic in nature, whereas scientists are often interested in understanding context-specific mechanisms in a network component.

For inferring biological networks from big molecular profiling data, one school of methods may be Markov chain Monte Carlo (MCMC)-based approaches where samples are drawn from big data to approximate the true distributions, which is otherwise not probable. Another school of methods is based on stochastic search, such as simulated annealing, genetic algorithm, and cultural algorithm. In this chapter, we present case studies from the two schools of methods in the context of discovering network topology from gene sets. More specifically, we consider the computational framework presented in Figure 14.1 for inferring biological networks from big molecular profiling data. Throughout, we hypothesize a biological network as an ensemble of several overlapping sub-networks and utilize the gene sets corresponding to sub-networks for reconstructing the underlying network topology. In the present context, gene sets refer to the sets of genes appearing in the sub-networks without any information about the sub-network structures. Thus, the underlying biological network can be reconstructed by inferring the sub-network structures from the overlapping gene sets related to a given network component and combining the inferred sub-networks into a single unit. We divide our discussion into three different stages. The first stage focuses on obtaining biologically meaningful network components from big molecular profiling data. Specifically, we discuss approaches such as utilizing database knowledge and dimensionality reduction algorithms for deriving network components. In the second stage, we discuss two approaches, utilizing database knowledge in simulation studies and discretizing molecular profiling data related to network components, for deriving gene sets corresponding to sub-networks. In the final stage, we present a general setting for gene-set-based inference of biological networks with a detailed review of the sampling-based approach (GSGS) [22] and the discrete optimization approach (GSSA) [23].

Specifically, GSGS is an MCMC-based stochastic algorithm, where the goal of biological network inference is translated into drawing network samples sequentially from the joint distribution of gene sets related to the given network component and summarizing the most likely structure from the sampled structures. Indeed, the number

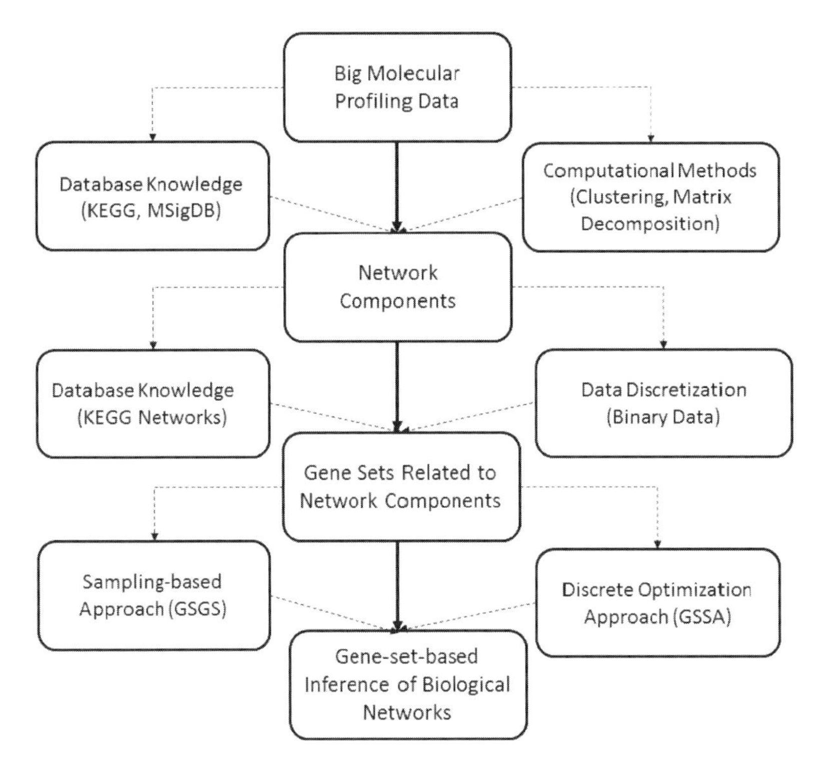

Figure 14.1 A computational framework for inferring biological network topologies from big molecular profiling data. Network components are first derived by either utilizing database knowledge or a dimensionality-reduction approach. For a given network component, gene sets corresponding to sub-networks in the component are either derived directly from a prior-known biological network structure for proof-of-concept studies or by discretizing molecular profiling data related to the component using binary labels. The underlying network structure is inferred from the compendium of gene sets related to the given network component by applying gene-set-based sampling (example: GSGS) or discrete optimization approach (example: GSSA).

of biological networks consistent with a gene set compendium grows exponentially with increase in the number and lengths of gene sets in the compendium. Therefore, it might be computationally intractable to characterize the underlying network topology by exhaustive enumeration of all possible network structures. GSGS offers an approximate learning approach developed under the framework of Gibbs sampling [22, 24, 25]. On the other hand, GSSA is designed under the setting of artificial intelligence algorithms and utilizes a strategy based on simulated annealing to search for the underlying network topology in the space of network structures defined by gene sets [23, 26, 27].

14.2 Big data to network components

In this section, we discuss approaches for deriving network components from big molecular profiling data. A network component comprises of a set of molecules present in the underlying biological network topology.

Database knowledge Network analysis has received much attention in recent years and there has been a rapid accumulation of network components, identified via computational or experimental approaches, in biomedical databases. Prior-known network components and network structures available in databases could be utilized for validating the results from a proof-of-concept study, performing enrichment analyses, incorporating prior knowledge in a study, or inferring context-specific biological networks. Some of the popular databases for obtaining network components and the underlying biological network topologies include

- KEGG (www.genome.jp/kegg),
- BioCarta (www.biocarta.com),
- NCBI BioSystems (www.ncbi.nlm.nih.gov/biosystems),
- MSigDB (www.broadinstitute.org/gsea/msigdb).

A more comprehensive list is presented in [28]. In the present context, subsets of big molecular profiling data corresponding to prior known network components can be selected for inferring context-specific biological networks.

Computational approaches Network components can be identified by applying dimensionality reduction approaches on big molecular data. Indeed, analysis of big molecular profiling data is challenging due to the curse of dimensionality. Therefore, an initial characterization of such datasets is first required to arrange genes into smaller groups representing potential network components, where genes with similar expression profiles belong to the same group. Clustering [29–31] is often one of the initial steps used for identifying these groups known as clusters. Hierarchial clustering [32], K-means clustering [33], and model-based clustering [34] are some of the popular gene clustering algorithms. Besides clustering, matrix factorization [35, 36] and pattern recognition techniques [37] can also be used for identification of network components. For example, the module discovery problem is addressed by combining the estimation of correlation structure with matrix factorization in [38], whereas the approach presented in [37] identifies a sequence of dominant patterns present in high-dimensional data. Network components obtained using any of the above methods are statistically tested for their biological significance using over-representation analysis [39, 40]. Note that enrichment analysis gives an indication of the presence or absence of an underlying network structure in a computationally derived network component. Enriched network components can be utilized for the inference of biological network topologies.

14.3 Gene sets related to network components

This section focuses on approaches for deriving gene sets corresponding to sub-networks in a network component, and writing a gene set compendium as a matrix of discrete values.

Simulation In the proof-of-concept studies, where the underlying biological network structure is available, specific algorithms can be developed to sample gene sets from the given network topology. For instance, a path sampling algorithm is presented in [22] to generate a collection of linear overlapping sub-networks from a directed biological network structure, where gene sets are derived by randomly permuting the linear ordering of genes in each sub-network. Note that, if there are m gene sets and n distinct genes in a gene set compendium related to a network component, then the compendium can be as an $m \times n$ matrix. If there are k genes in the ith gene set, then the k locations in the ith row can be set to nonzero indices representing these genes, and the remaining $n - k$ locations can be set to 0. Using a matrix representation, a gene set compendium can be easily accommodated in a computational framework [22, 23]. It is well-known that binary discrete measurements are used in the inference of Bayesian networks in order to achieve a manageable computational complexity. From a given gene set compendium, a binary $m \times n$ matrix can be constructed by considering presence/absence of genes in each gene set. If there are k genes in the ith gene set, then the corresponding k locations in the ith row of data are set to 1 and the remaining $n - k$ locations are set to 0. Representation of a gene set compendium as a binary matrix can be utilized to compare the performance of gene-set-based network inference algorithms and Bayesian network methods.

Discretization When molecular profiling data corresponding to a network component is given, we derive a gene set compendium by discretizing molecular profiling measurements using binary labels. Discretization scheme aims to achieve an approximation to the gene sets related to the network structure. Some of the common data discretization methods include binning (equal-width, equal-frequency) or simple thresholding, where a fixed percentage of top measurements in the given data are set to 1 and the remaining measurements are set to 0. This approach has been used to compare the performance of gene-set-based network inference algorithms and Bayesian network methods by utilizing *E. coli* datasets from the DREAM initiative (see Figure 14.7) as well as reconstruction of context-specific biological networks by utilizing breast-cancer data in [22, 23].

14.4 Reconstructing biological network topologies using gene sets

In this section, we focus on the inference of biological networks using the gene set compendiums related to network components. Throughout, the underlying network topology is assumed to be directed. We present a general setting describing the problem of inferring biological network structures from overlapping gene sets with a detailed review of gene set Gibbs sampling (GSGS) [22] and gene set simulated annealing (GSSA) [23]. The major advantage of working with gene-set-based approaches is their ability to naturally incorporate higher-order interaction and regulation mechanisms as most of the other existing network inference approaches typically rely on pairwise similarities or statistical causal interactions [1, 2, 5, 41–43].

14.4.1 A general setting

Structural assumptions Throughout, we hypothesize a biological network as an ensemble of many overlapping sub-networks with linear arrangement of genes in each subnetwork. These sub-networks are referred to as information flows (IFs). In other words, IFs represent directed linear paths in a biological network and uniquely determine its structure. We treat IFs as the basic building blocks of a biological network. An information flow gene set (IFGS) is defined as the set of genes in a given IF without gene ordering information. From a compendium of IFGSs, the underlying biological network can be reconstructed by inferring the order of genes in each IFGS and combining the inferred IFs into a single unit. Indeed, biological networks, such as signaling networks, are characterized by linear signal transduction activities. Also, the above assumptions provide a tractable approximation to the underlying network topology, which otherwise may be infeasible to achieve.

The problem Since there are $L!$ different gene ordering permutations for an IFGS containing L genes, the number of biological networks consistent with a gene set compendium containing m IFGSs is of the order of $L!^m$. However, it may not be computationally feasible to exhaustively enumerate all $L!^m$ networks to find the underlying structure, even when the values of m and L are small. Consequently, the goal of biological network inference can be translated into a sampling or optimization problem. For instance, if we treat the ordering of genes in each IFGS as a random variable, network structures can be either sampled from the joint distribution of IFGSs or searched in the space of feasible biological networks. Algorithms such as simulated annealing, genetic algorithm, Gibbs sampling and Metropolis–Hastings are commonly used in such scenarios.

Markov chain parameters IFGSs are treated as random samples from a first-order Markov chain model, where the state of a node only depends on the state of its previous node. From a given set of m IFs containing n distinct genes, the model parameters, initial probability vector π_0 and transition probability matrix Π, are calculated as below:

$$\pi_0 = \left(\frac{c_1}{m}, \ldots, \frac{c_n}{m}\right) \tag{14.1}$$

and

$$\Pi = [p_{rs}]_{n \times n}, \tag{14.2}$$

where $p_{rs} = c_{rs} / \sum_{s=1}^{n} c_{rs}$, $r, s = 1, \ldots, n$, c_l is the total number of times the lth gene appears as the first node among m IFs, for each $l = 1, \ldots, n$, and c_{rs} is the total number of times the rth gene transits to the sth gene among m IFs. Note that the overlapping information among IFs is captured in Π. The parameters π_0 and Π can be calculated for each of the $\prod_{i=1}^{m} L_i!$ network structures, where each network is a collection of m IFs, and can be used to assign a score to each network. The score of a biological network is defined as the product of the scores of IFs in the compendium. An IF is scored using the parameters π_0 and Π defined above. For instance, the score for the IF $z \to y \to x$ is calculated as

$$\mathcal{P}(z \to y \to x) = P(z) \times P(y|z) \times P(x|y), \tag{14.3}$$

where the terms on the right-hand side of the above equation are obtained from Π. When the network inference is formulated as an optimization problem, the above scoring can be used to search for the optimal biological network structure. In the case where network inference is treated as a sampling problem, the scores for all $\prod_{i=1}^{m} L_i!$ network structures can be normalized to represent the joint distribution of IFGSs.

14.4.2 Gene set Gibbs sampling

In this section, we review the gene set Gibbs sampling (GSGS) approach from [23]. GSGS is developed under the Gibbs sampling framework [24, 25], where the goal of inferring directed biological networks is translated into drawing samples from the joint distribution of IFGSs followed by summarizing the most likely network from the sampled networks. GSGS is a stochastic algorithm which samples network structures by sampling the ordering for each IFGS from the conditional distribution defined by the remaining IFGSs in the given gene set compendium.

Given an IFGS compendium related to a network component, GSGS begins by assigning random ordering to genes in each IFGS. These orderings are updated sequentially by sampling an order for each IFGS conditioned on the known orders of remaining $m - 1$ IFGSs. For sampling an order for an IFGS X_i containing L_i genes from the conditional distribution, Markov chain parameters π_{-i} and Π_{-i} are calculated using $m - 1$ IFs obtained by leaving X_i out of the compendium. The parameters π_{-i} and Π_{-i} are used to score all $L_i!$ orderings for X_i. These scores are normalized and used for sampling an order for X_i from the conditional distribution using the inverse cumulative density function (CDF). This procedure is performed for each IFGS in the compendium in a sequential manner to generate a sample from the underlying distribution. The next sample is generated using the IFs in the previous sample. In the end, the most frequent ordering for each IFGS among the sampled IFs is used to construct the final biological network.

Figure 14.2 shows a comparison between GSGS and two Bayesian network approaches K2 and Metropolis–Hastings in terms of F-score. A significantly higher F-score is observed in the case of GSGS than K2 and MH. In Figure 14.3, performance of GSGS in inferring the underlying biomolecular mechanisms is presented using *In silico* network available from the DREAM initiative [47, 48].

14.4.3 Gene set simulated annealing

This section reviews a search strategy from [23] to learn the optimal network structure from IFGSs related to a network component. In [23], network inference from IFGSs is formulated as a discrete optimization problem and a simulated annealing algorithm [26], GSSA, is designed to search for the optimal network structure in the space of feasible networks. GSSA mimics the physical process of heating and then cooling down a substance slowly to obtain a strong crystalline structure.

In GSSA, the feasible space is defined by networks that possess the degree distribution of the true network structure. In simulation studies, where the underlying network structure and true IFs are known, feasible networks are derived by randomly permuting

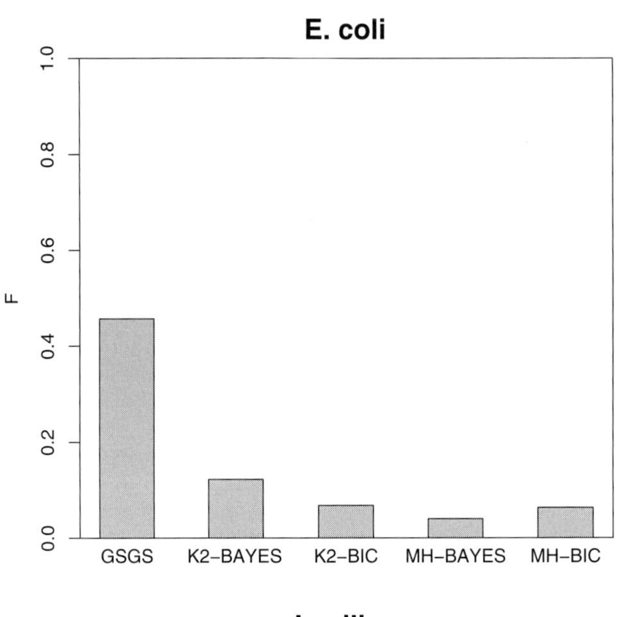

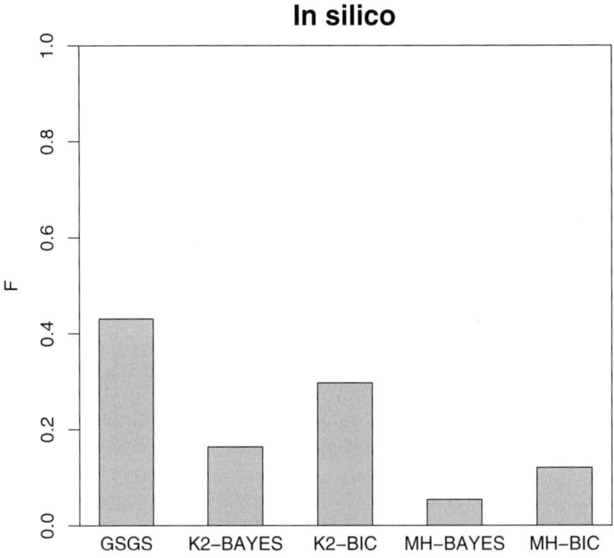

Figure 14.2 Comparison of the GSGS approach with K2 and Metropolis-Hastings (MH) in terms of F-score. Two gene set compendiums derived from *E. coli* [44–46] and *In silico* [47, 48] networks available from the DREAM initiative were used in the comparison. In the case of K2 and MH, binary representation of gene sets was utilized, and both BIC and Bayesian scoring functions were used. There were a total of 125 IFGSs in the case of *E. coli* network, and 57 in the case of *In silico* network. A total of 1000 networks were sampled in the case of GSGS and MH.

the orders of intermediate nodes in true IFs by keeping the pair of end nodes fixed. More specifically, we assume that the source and sink nodes in each gene set are known a priori, but not the order in which the intermediate genes appear in the true IF. Since the incoming and outgoing degrees of all intermediate nodes in IFs are 1, the feasible

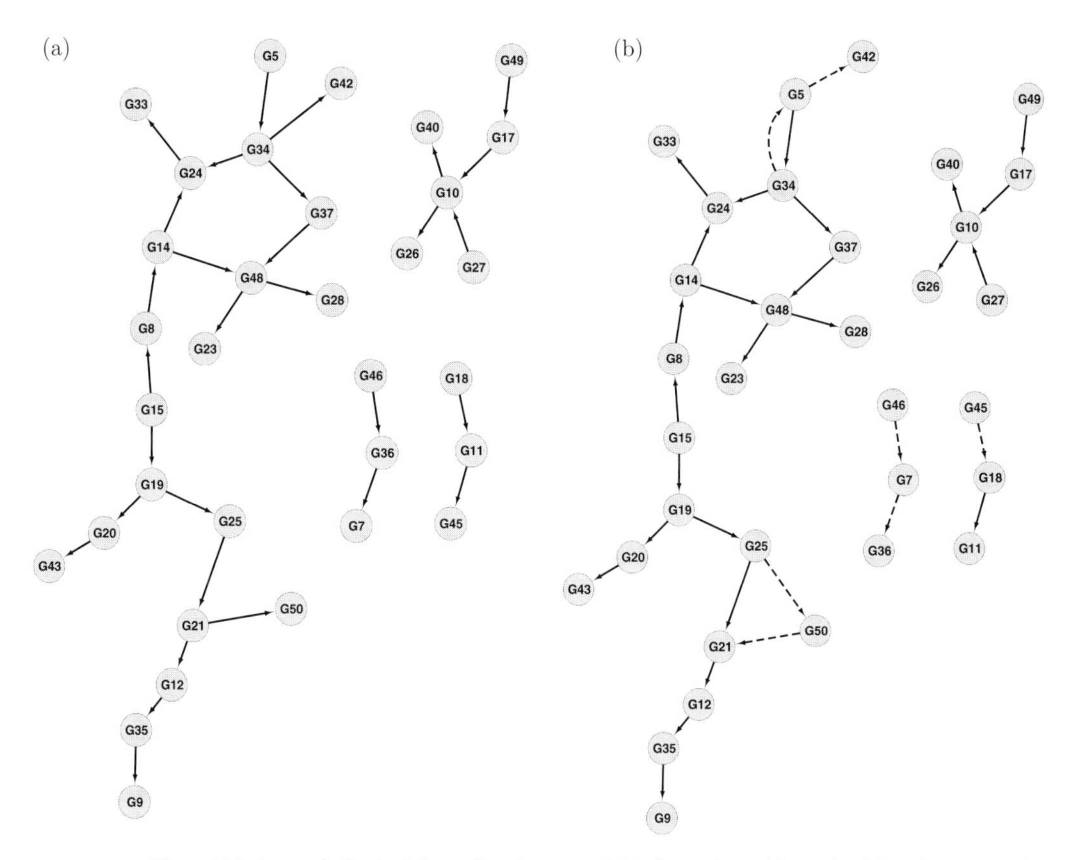

Figure 14.3 A proof of principle study where panel (a) shows the gold standard *In silico* network available from the DREAM initiative [47, 48] and panel (b) represents the network predicted by GSGS. Solid edges correspond to true positives and dashed edges represent false positives.

networks share the same degree distribution. In real-world studies, feasible space is approximated by using database knowledge. This is achieved by inferring the pair of end nodes in each IFGS followed by generating the feasible networks as described above. In the studies conducted in [22] and [23], the end nodes for each IFGS were obtained by considering the hierarchical representation of genes in the pathway structures available in the KEGG database.

The GSSA algorithm proceeds by randomly sampling a network from the neighborhood of the current network. Note that a network structure is represented as a compendium of IFs. The neighbor of a network is defined as the set of structures obtained by randomly permuting the orders of $L_i - 2$ intermediate genes in the ith IF by keeping the remaining $m - 1$ IFs fixed, for each $i = 1, \ldots, m$. Only one IF is randomly selected and perturbed at a time for defining the set of networks to sample from. If the initial network structure belongs to the feasible space, GSSA is guaranteed to take jumps within the feasible set of candidate networks sharing the same degree distribution as the true network. The network sampled from the neighborhood of the current network is scored using the procedure discussed in Section 14.4.1 with the only difference that negative logarithm is applied on the score presented in 14.4.1 to

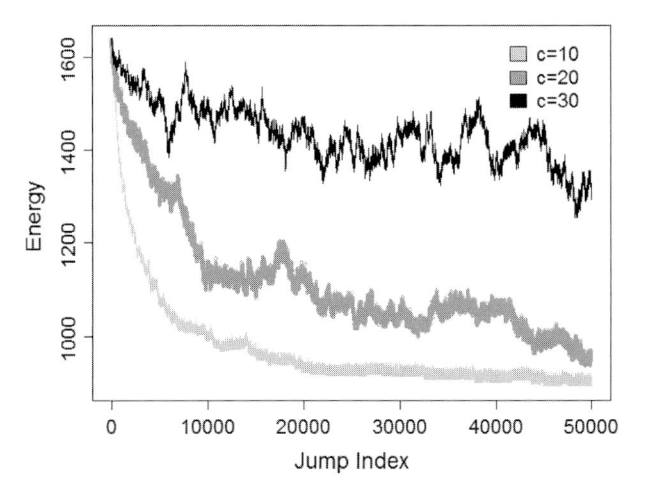

Figure 14.4 Energy values calculated by using different cooling schedule constants in the case of IFGS compendium derived from the generic vascular smooth muscle contraction network in KEGG [20].

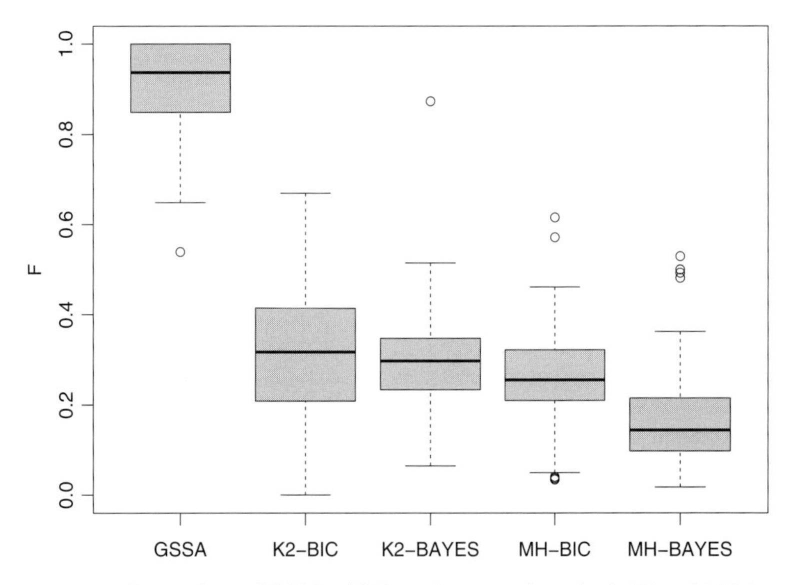

Figure 14.5 Comparison of GSSA with Bayesian network methods K2 and MH in terms of F-score. Both BIC and Bayesian scoring metrics were used in the case of Bayesian network methods. A total of 83 IFGS compendiums derived from KEGG database were used in this study. The number and lengths of IFGSs varied in the ranges 5–723 and 4–13, respectively.

present the problem as a minimization problem. This score is referred to as "energy". If the neighboring network has a lower score, the move to the neighboring network is accepted, otherwise a Bernoulli sample is drawn with the probability of success as $\min\{1, \exp(\text{score(current network)} - \text{score(neighboring network)}/T)\}$, where T represents cooling schedule, which at the kth stage is defined as

$$T_k = \frac{c}{\log(k+1)}, \quad k = 1, 2, \ldots. \tag{14.4}$$

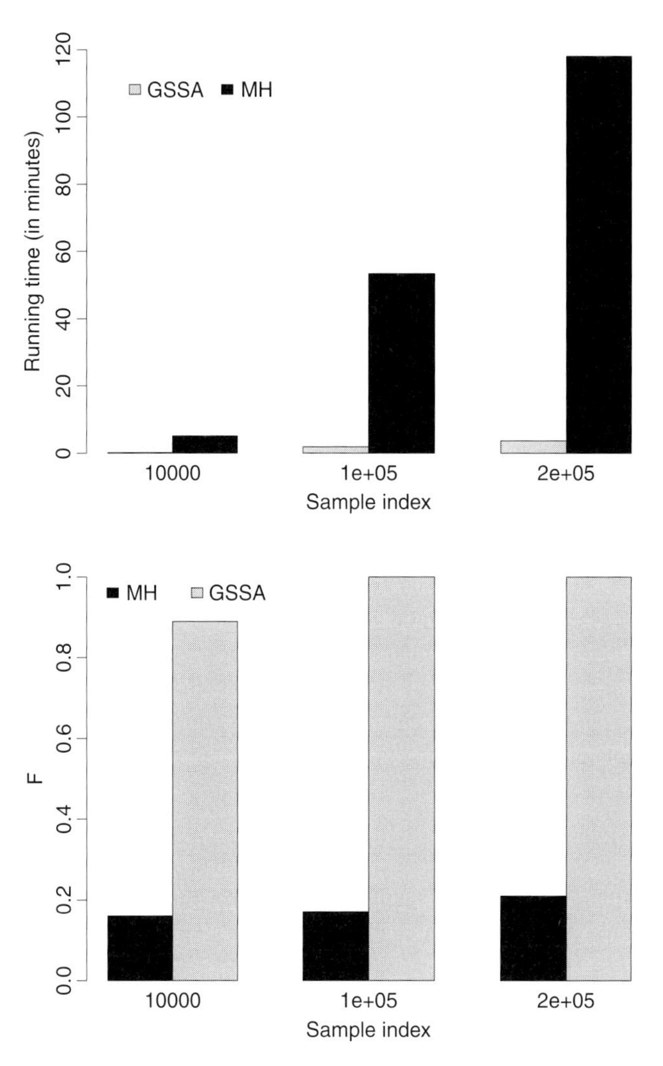

Figure 14.6 Performance of GSSA and MH in terms of running time and F-score at different jump indices. An IFGS compendium derived from the KEGG database containing 54 gene sets has been used in the comparison.

Here $c > 0$ is constant that is referred to as the *cooling schedule constant*. The choice of c is often problem specific. A small value of c may lead to a local solution, whereas a large value may slow down the speed of convergence of GSSA (see Figure 14.4). The above cooling schedule has been used to study the convergence properties of simulated annealing approach in a general setting [49].

Figure 14.5 compares the performance of GSSA and the Bayesian network methods K2 and MH by using gene set compendiums derived from the KEGG database [20]. In Figure 14.6, the performance of GSSA and MH is compared in terms of running time and F-score at different jump or sample indices. We observe a significantly better performance of GSSA in recovering the underlying network topologies compared with K2 and MH, and a much reduced computational cost. Figure 14.7 compares the performance

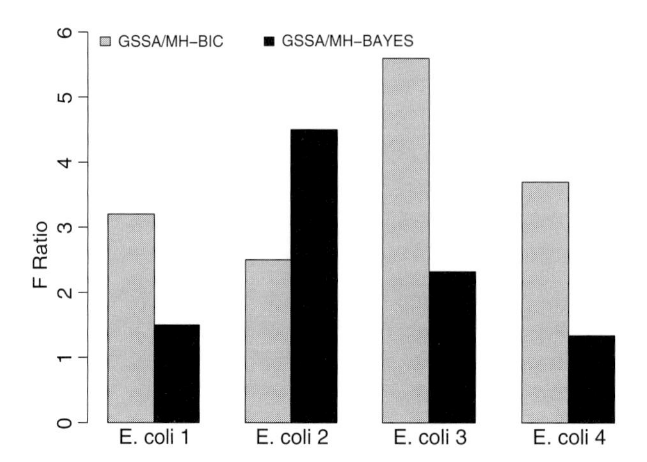

Figure 14.7 Comparison of GSSA and MH in terms of F-score ratio. A ratio greater than 1 indicates a better performance by GSSA. Four *E.coli* benchmark datasets available from the DREAM initiative were used in this study.

of GSSA and MH in terms of the ratio of F-scores, where four gene set compendiums derived using *E. coli* benchmark datasets available from the DREAM initiative were used [44–46]. The first two compendiums were derived from two datasets comprising 50 genes and 51 samples, whereas the remaining two compendiums were obtained by using two datasets containing 100 genes and 101 samples. The corresponding gold standard networks comprised 62, 82, 125, and 119 edges, respectively. The inferred structures were compared with the corresponding gold standards. The IFGS compendiums were derived by declaring the top 10% of the measurements in each dataset as 1 and the remaining measurements as 0, and considering IFGSs with lengths in the range 3–9. The final compendiums comprised 47, 45, 45, and 49 IFGSs, respectively. In [22, 23], GSGS and GSSA were also used to infer breast-cancer-specific biological networks corresponding to ERBB network component available from the KEGG database. A partial view of the inferred subnetworks and final network is presented in Figures 14.8 and 14.9, respectively. Note that the network represented in Figure 14.9 is specific to breast cancer. This network was inferred using IFGSs derived from molecular profiling data corresponding to the ERBB pathway in the KEGG database. On the other hand, the network structure available in the KEGG database is generic in nature. Moreover, the information available in public databases in often incomplete. Thus, instead of utilizing a performance measure to compare the two structures, we have identified activities that are common to both the generic network and the breast-cancer-specific network (solid edges), and the ones that may be considered as novel edges (dashed edges).

It is important to note that there are two key factors that help with the inference of edge directions in the GSGS and GSSA approaches: (1) overlapping among gene sets; and (2) prior-knowledge, which in the case of GSSA is defined as the knowledge of two end nodes in each gene set, and in the case of GSGS, it may be the availability of prior-known edges. A combination of these two factors determines the accuracy of the final inferred network. For instance, in the case of GSSA, increase in the overlapping

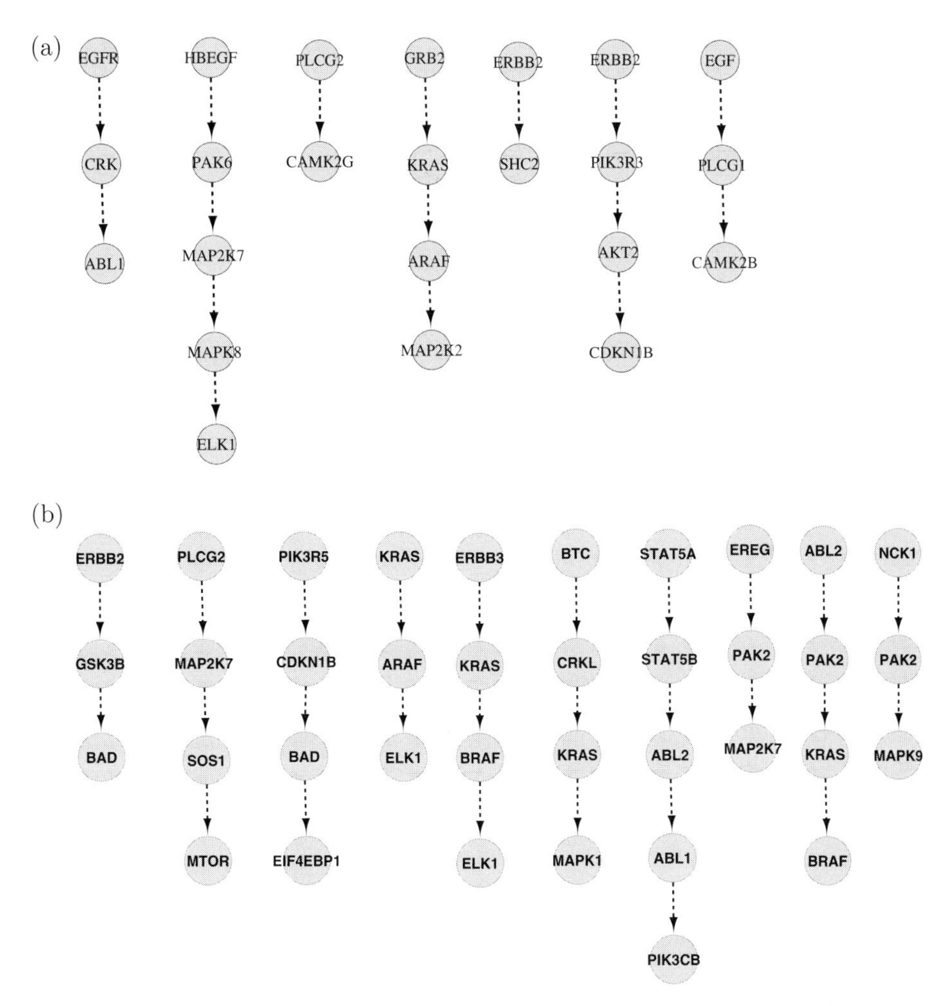

Figure 14.8 Linear subnetworks inferred by GSSA (a) and GSGS (b), which correspond to complete or partial linear sub-networks in the ERBB network available in the KEGG database.

information, i.e. increase in the number of gene sets with prior-known end nodes for each gene set, also means the increase in the knowledge of source and sink nodes, which increases the accuracy of the inferred network. Although, this scenario was not discussed for the GSGS approach, the impact of including prior knowledge on the accuracy of the inferred network has been discussed in [22].

14.5 Discussion and future work

Recent advances in high-throughput data acquisition technologies are resulting in the accumulation of vast amounts of big molecular profiling data. Big data are a rich source of information for gaining deeper insights into the biomolecular interaction

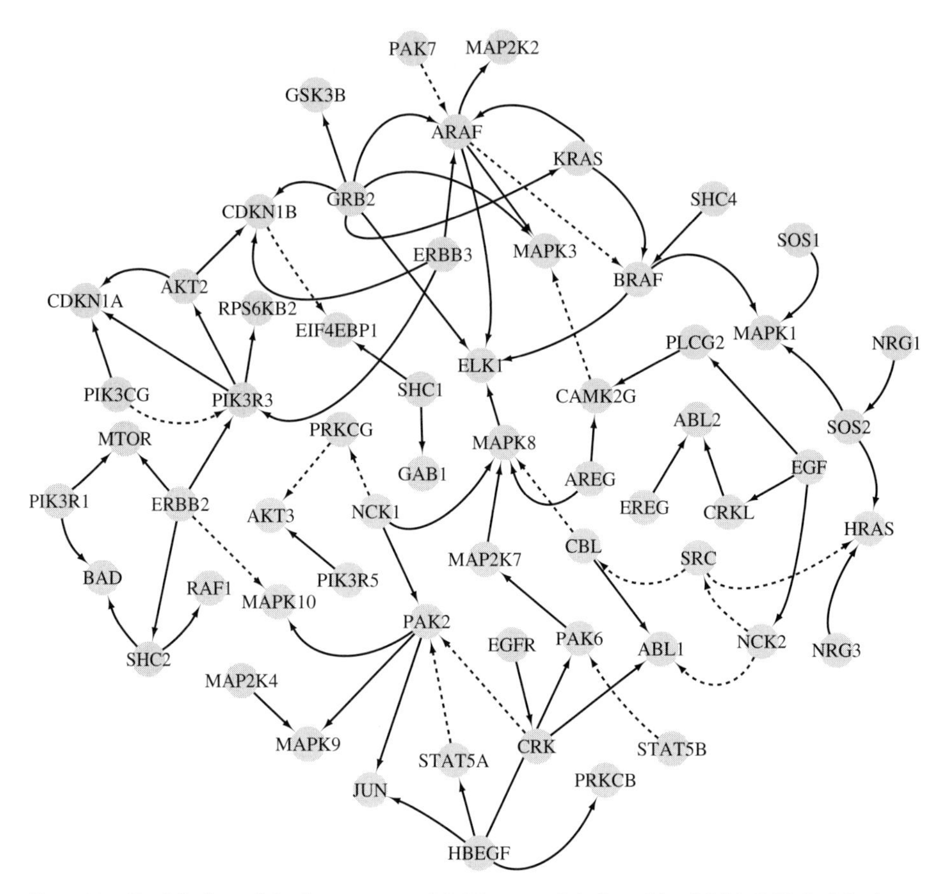

Figure 14.9 Partial view of the breast cancer ERBB network inferred by GSSA [23]. Solid edges represent complete or partial sub-networks in the ERBB network available from the KEGG database, whereas dashed edges follow the hierarchial arrangement of the structure in the KEGG database.

and regulation mechanisms. In this chapter, we focused on the inference of biological network topologies from big molecular profiling data. We divided our discussion into three stages: identification of network components from big molecular profiling data, derivation of gene sets related to a network component, and gene-set-based inference of the underlying network topology in the given network component.

Network components can be obtained from public databases or derived by applying computational techniques such as clustering (K-means, hierarchial clustering, etc.) or matrix factorization (singular value decomposition, nonnegative matrix factorization, etc.) on big molecular profiling data followed by enrichment analysis. Enriched network components can be investigated for the underlying network structure.

For deriving overlapping gene sets related to a network component, we discussed two approaches: simulation and data discretization. In simulation studies, where the underlying network structure is known, gene sets can be derived by listing all directed linear paths containing three or more genes followed by random permutation of the ordering of genes in each path. In the case of molecular profiling measurements, gene sets can be derived by discretizing molecular profiling data related to the given network component using binary labels.

In the final stage, we reviewed gene-set-based approaches for inferring biological network topologies in a given network component. Throughout, a biological network structure was hypothesized as an ensemble of overlapping sub-networks with a directed and linear arrangement of genes in each sub-network. Gene sets referred to the sets of genes in sub-networks without ordering information. We first presented a general setting describing the problem and then reviewed two approaches GSGS [22] and GSSA [23] for inferring the underlying network topology by inferring the order of genes in each gene set and combining the inferred subnetworks into a single unit. Throughout, gene sets were treated as random samples from a first-order Markov chain model.

Gene-set-based approaches are relatively new in the field of systems biology and provide a fresh perspective to understand the structure of biological networks with several possibilities of extensions. For instance, the framework presented in Section 14.4.1 could be used with Bayesian network methods. Bayesian network methods are commonly used in the inference of directed network topologies, however, they suffer from non-trivial computational complexity [43, 50, 51]. Heuristics, such as K2 [43], MCMC [52], simulated annealing [53] and others [54, 55], are often used to tackle the computational cost associated with the inference of Bayesian networks, but even in these cases a number of unrealistic assumptions are made by restricting the size of parent set for each node, choosing a specific scoring criteria and other parameter values. Bayesian networks also assume acyclicity in the underlying topology which prohibits the inclusion of feedback loops, a common biological feature, in the network. Gene-set-based approaches do not have these restrictions. Although an individual gene set is treated as a linear Markov chain without any loop, the final structure obtained by combining the Markov chains may contain loops. Gene-set-based approaches incorporate higher-order interaction and regulation mechanisms in a more natural way and allow more flexibility in incorporating prior knowledge in the inference procedure with a much reduced computational cost in several cases.

Further, gene-set-based approaches discussed in this chapter could be extended to the setting of dynamic Bayesian networks (DBNs). Since Bayesian networks do not incorporate cyclic behavior, they are unfolded in time for inferring cyclic mechanisms and are referred to as dynamic Bayesian networks. Similar to the case of Bayesian networks, methods used for inferring dynamic Bayesian networks suffer from an inevitable increase in the model size, large computational time and memory requirements. Gene-set-based approaches for inferring DBNs using time series data may lead to the discovery of more complex interactions in biological networks.

It is clear from the discussion in this chapter that gene-set-based approaches hold much promise for understanding the structural organization of biological networks. Therefore, future researches in this direction would benefit from developing an easy-to-use computational and visualization tool by seamlessly integrating the three stages: identification of network components, derivation of gene sets, and inference of biological network topologies by sufficiently exploiting the potential of big molecular profiling data.

References

[1] N. Friedman, M. Linial, I. Nachman, and D. Peer, "Using Bayesian networks to analyze expression data," *J. Comput. Biol.*, vol. **7**, pp. 601–620, 2000.

[2] E. Segal, M. Shapira, A. Regev, D. Peer, D. Botstein, D. Koller, and N. Friedman, "Module networks: identifying regulatory modules and their condition-specific regulators from gene expression data," *Nat. Genet.*, vol. **34**, pp. 166–176, 2003.

[3] I. Shmulevich, E. R. Dougherty, S. Kim, and W. Zhang, "Probabilistic boolean networks: a rule-based uncertainty model for gene regulatory networks," *Bioinformatics*, vol. **18**, pp. 261–274, 2002.

[4] I. Shmulevich, I. Gluhovsky, R. Hashimoto, E. R. Dougherty, and W. Zhang, "Probabilistic boolean networks: a rule-based uncertainty model for gene regulatory networks," *Comp. Funct. Genomics.*, vol. **4**, pp. 601–608, 2003.

[5] A. J. Butte and I. S. Kohane, "Mutual information relevance networks: functional genomic clustering using pairwise entropy measurements," *Pac. Symp. Biocomput.*, vol. **5**, pp. 415–426, 2000.

[6] G. Altay and F. Emmert-Streib, "Revealing differences in gene network inference algorithms on the network-level by ensemble methods," *Bioinformatics*, vol. **26**, no. 14, pp. 1738–1744, 2010.

[7] P. E. Meyer, K. Kontos, and G. Bontempi, "Information-theoretic inference of large transcriptional regulatory networks," *EUROSIP J. Bioinform. Syst. Biol.*, 2007.

[8] H. Kishino and P. J. Waddell, "Correspondence analysis of genes and tissue types and finding genetic links from microarray data," *Genome Informatics*, vol. **11**, pp. 83–95, 2000.

[9] A. Dobra, C. Hans, B. Jones, J. R. Nevins, and M. West, "Sparse graphical models for exploring gene expression data," *J. Multiv. Anal.*, vol. **90**, pp. 196–212, 2004.

[10] J. Schäfer and K. Strimmer, "An empirical Bayes approach to inferring large-scale gene association networks," *Bioinformatics*, vol. **21**, pp. 756–764, 2005.

[11] ——, "A shrinkage approach to large-scale covariance matrix estimation and implications for functional genomics," *Stat. Appl. Genet. Mol. Biol.*, vol. **4**, 2005.

[12] T. S. Gardner, D. di Bernardo, D. Lorenz, and J. J. Collins, "Inferring genetic networks and identifying compound mode of action via expression profiling," *Science*, vol. **301**, no. 5629, pp. 102–105, 2003.

[13] J. Tegner, M. K. S. Yeung, J. Hasty, and J. J. Collins, "Reverse engineering gene networks: integrating genetic perturbations with dynamical modeling," *Proc. Natl Acad. Sci. USA*, vol. **100**, pp. 5944–5949, 2003.

[14] D. Zhu, A. O. Hero, Z. S. Qin, and A. Swaroop, "High throughput screening of co-expressed gene pairs with controlled False Discovery Rate (FDR) and Minimum Acceptable Strength (MAS)," *J. Comput. Biol.*, vol. **12**, pp. 1027–1043, 2005.

[15] A. L. Tarca, S. Draghici, P. Khatri, *et al.*, "A novel signaling pathway impact analysis," *Bioinformatics*, vol. **25**, pp. 75–82, 2009.

[16] C. J. Vaske, S. C. Benz, J. Z. Sanborn, *et al.*, "Inference of patient-specific pathway activities from multi-dimensional cancer genomics data using PARADIGM," *Bioinformatics*, vol. **26**, pp. 237–245, 2010.

[17] A. Subramanian, P. Tamayo, V. K. Mootha, *et al.*, "Gene set enrichment analysis: a knowledge-based approach for interpreting genome-wide expression profiles," *Proc. Natl Acad. Sci. USA*, vol. **102**, pp. 15 545–15 550, 2005.

[18] L. Tian, S. A. Greenberg, S. W. Kong, *et al.*, "Discovering statistically significant pathways in expression profiling studies," *Proc. Natl Acad. Sci. USA*, vol. **102**, pp. 13 544–13 559, 2005.

[19] D. W. Huang, B. T. Sherman, and R. A. Lempicki, "Systematic and integrative analysis of large gene lists using DAVID Bioinformatics Resources," *Nat Protoc.*, vol. **4**, no. 1, pp. 44–57, 2009.

[20] M. Kanehisa, S. Goto, M. Furumichi, M. Tanabe, and M. Hirakawa, "KEGG for representation and analysis of molecular networks involving diseases and drugs," *Nucleic Acids Res.*, vol. **38**, pp. 355–360, 2010.

[21] G. J. Dennis, B. T. Sherman, D. A. Hosack, *et al.*, "DAVID: database for annotation, visualization and integrated discovery," *Genome Biol.*, vol. **4**, no. 5, 2003.

[22] L. Acharya, T. Judeh, Z. Duan, M. Rabbat, and D. Zhu, "GSGS: a computational approach to reconstruct signaling pathway structures from gene sets," *IEEE/ACM Trans. Comput. Biology Bioinform.*, vol. **9**, no. 2, pp. 438–450, 2012.

[23] L. Acharya, T. Judeh, G. Wang, and D. Zhu, "Optimal structural inference of signaling pathways from unordered and overlapping gene sets," *Bioinformatics*, vol. **28**, no. 4, pp. 546–556, 2012.

[24] A. Gelman, J. B. Carlin, H. S. Stern, and D. B. Rubin, *Bayesian Data Analysis*, 2nd edition, Chapman & Hall, 2003.

[25] G. H. Givens and J. A. Hoeting, *Computational Statistics*, Wiley Series in Proabbility and Statistics, 2005.

[26] S. Kirkpatrick, C. D. J. Gelatt, and M. P. Vecchi, "Optimization by simulated annealing," *Science*, vol. **220**, pp. 671–680, 1983.

[27] T. Judeh, T. Jayyousi, L. Acharya, R. G. Reynolds, and D. Zhu, "Gene set cultural algorithm: A cultural algorithm approach to reconstruct networks from gene sets," in *Proceedings of the ACM Conference on Bioinformatics, Computational Biology and Biomedical Informatics (BCB)*, 2013.

[28] G. D. Bader, M. P. Cary, and S. Chris, "Pathguide: a pathway resource list," *Nucleic Acids Research*, vol. **34**, pp. 504–506, 2006.

[29] K. Y. Yeung, M. Medvedovic, and R. E. Bumgarner, "Clustering gene-expression data with repeated measurements," *Genome Biol.*, vol. **4**, no. 5, 2003.

[30] M. Medvedovic and S. Sivaganesan, "Bayesian infinite mixture model based clustering of gene expression profiles," *Bioinformatics*, vol. **18**, pp. 1194–1206, 2002.

[31] M. Medvedovic, K. Y. Yeung, and R. E. Bumgarner, "Bayesian mixtures for clustering replicated microarray data," *Bioinformatics*, vol. **20**, pp. 1222–1232, 2004.

[32] M. Eisen, P. Spellman, P. O. Brown, and D. Botstein, "Cluster analysis and display of genome-wide expression patterns," *Proc. Natl Acad. Sci. USA*, vol. **95**, pp. 14 863–14 868, 1998.

[33] J. A. Hartigan and M. A. Wong, "A k-means clustering algorithm," *Applied Stat.*, vol. **28**, pp. 100–108, 1979.

[34] G. J. McLachlan and D. Peel, *Finite Mixture Models*, Wiley Series in Probability and Mathematical Statistics, Applied Probability and Statistics Section, John Wiley & Sons, 2000.

[35] A. Ben-Hur and I. Guyon, *Detecting Stable Clusters Using Principal Component Analysis in Methods in Molecular Biology*, Humana Press, 2003.

[36] P. M. Kim and B. Tidor, "Subsystem identification through dimensionality reduction of large-scale gene expression data," *Genome Res.*, vol. **13**, no. 7, pp. 1706–1718, 2003.

[37] M. Koyuturk, A. Grama, and N. Ramakrishnan, "Compression, clustering and pattern discovery in very high dimensional discrete-attribute datasets," *IEEE Transactions on Knowledge and Data Engineering*, vol. **17**, no. 4, 2005.

[38] X. Yang, Y. Zhou, R. Jin, and C. Chan, "Reconstruct modular phenotype-specific gene networks by knowledge-driven matrix factorization," *Bioinformatics*, vol. **25**, pp. 2236–2243, 2009.

[39] S. Draghici, P. Khatri, R. P. Martins, G. C. Ostermeier, and S. A. Krawetz, "Global functional profiling of gene expression," *Genomics*, vol. **81**, no. 2, pp. 98–104, 2003.

[40] P. Khatri and S. Draghici, "Ontological analysis of gene expression data: current tools, limitations, and open problems," *Bioinformatics*, vol. **21**, pp. 3587–3595, 2005.

[41] A. Margolin, T. Nemenman, K. Basso, *et al.*, "Aracne: an algorithm for the reconstruction of gene regulatory networks in a mammalian cellular context," *BMC Bioinform.*, 2006.

[42] J. J. Faith, B. Hayete, J. T. Thaden, *et al.*, "Large-scale mapping and validation of escherichia coli transcriptional regulation from a compendium of expression profiles," *PLoS Biol.*, vol. **5**, no. 1, 2007.

[43] G. F. Cooper and E. Herskovits, "A Bayesian method for the induction of probabilistic networks from data," *Machine Learning*, vol. **9**, no. 4, pp. 309–347, 1992.

[44] D. Marbach, T. Schaffter, C. Mattiussi, and D. Floreano, "Generating realistic in silico gene networks for performance assessment of reverse engineering methods," *J. Comput. Biol.*, vol. **16**, no. 2, pp. 229–239, 2009.

[45] D. Marbach, R. J. Prill, T. Schaffter, *et al.*, "Revealing strengths and weaknesses of methods for gene network inference," *Proc. Natl Acad. Sci. USA*, vol. **107**, no. 14, pp. 6286–6291, 2010.

[46] R. J. Prill, D. Marbach, J. Saez-Rodriguez, *et al.*, "Towards a rigorous assessment of systems biology models: the DREAM3 challenges," *PLoS ONE*, vol. **5**, 2010.

[47] P. Mendes, *Framework for Comparative Assessment of Parameter Estimation and Inference Methods in Systems Biology*, MIT Press, Cambridge, MA, 2009.

[48] G. Stolovitzky, R. J. Prill, and A. Califano, *Lessons from the DREAM2 Challenges*, Annals of the New York Academy of Sciences, G. Stolovitzky and P. Kahlem and A. Califano, Eds., 1158, pp. 159–195, 2009.

[49] B. Hajek, "Cooling schedules for optimal annealing," *Mathematics of Operations Research*, vol. **13**, no. 2, pp. 311–329, 1998.

[50] D. Chickering, "Optimal structure identification with greedy search," *J. Mach. Learn. Res.*, vol. **3**, pp. 507–554, 2002.

[51] R. W. Robinson, *Counting Unlabeled Acyclic Digraphs*, Springer Lecture Notes in Mathematics, 622, pp. 28–43, 1977.

[52] K. Murphy, "Active learning of causal Bayes net structure," UC Berkeley, Tech. Rep., 2001.

[53] B. W. Kernighan and S. Lin, "An efficient heuristic procedure for partitioning graphs," *Bell Systen Technical Journal*, vol. **49**, pp. 291–307, 1970.

[54] J. H. Holland, *Adaptation in Natural and Artificial Systems: An Introductory Analysis with Applications to Biology, Control, and Artificial Intelligence*, MIT Press, Cambridge, MA, 1992.

[55] F. Glover, "Tabu Search – Part I," *ORSA J. Comp.*, vol. **1**, no. 3, pp. 190–206, 1989.

15 Large-scale correlation mining for biomolecular network discovery

Alfred Hero and Bala Rajaratnam

Continuing advances in high-throughput mRNA probing, gene sequencing, and microscopic imaging technology is producing a wealth of biomarker data on many different living organisms and conditions. Scientists hope that increasing amounts of relevant data will eventually lead to better understanding of the network of interactions between the thousands of molecules that regulate these organisms. Thus progress in understanding the biological science has become increasingly dependent on progress in understanding the data science. Data-mining tools have been of particular relevance since they can sometimes be used to effectively separate the "wheat" from the "chaff", winnowing the massive amount of data down to a few important data dimensions. Correlation mining is a data-mining tool that is particularly useful for probing statistical correlations between biomarkers and recovering properties of their correlation networks. However, since the number of correlations between biomarkers is quadratically larger than the number biomarkers, the scalability of correlation mining in the big data setting becomes an issue. Furthermore, there are phase transitions that govern the correlation mining discoveries that must be understood in order for these discoveries to be reliable and of high confidence. This is especially important to understand at big data scales where the number of samples is fixed and the number of biomarkers becomes unbounded, a sampling regime referred to as the "purely high-dimensional setting". In this chapter, we will discuss some of the main advances and challenges in correlation mining in the context of large scale biomolecular networks with a focus on medicine. A new correlation mining application will be introduced: discovery of correlation sign flips between edges in a pair of correlation or partial correlation networks. The pair of networks could respectively correspond to a disease (or treatment) group and a control group.

15.1 Introduction

Data mining at a large scale has matured over the past 50 years to a point where, every minute, millions of searches over billions of data dimensions are routinely handled by search engines at Google, Yahoo, LinkedIn, Facebook, Twitter, and other media. Similarly, large ontological databases like GO [1] and DAVID [2] have enabled

Big Data over Networks, ed. Shuguang Cui, Alfred O. Hero III, Zhi-Quan Luo, and José M. F. Moura. Published by Cambridge University Press.

large-scale text data mining for researchers in the life sciences [3]. Curated repositories in the NCBI databases [4], such as the NCI pathway interaction database [5], or aggregated repository search engines, such as Pathway Commons [6], can be used to search over the network of interactions between genes and proteins as reported in scientific publications. These reported interactions are sometimes based on causal analysis, e.g. the result of gene-knockout studies that identify causal interactions between a knocked-out gene and some other set of downstream genes. However, with increasing frequency, life-science researchers have been reporting networks of associations between thousands of variables measured in a high-throughput assay like gene chip, RNAseq, or chromosomal conformal capture (3C, 4C, HiC). These networks can yield information on direct and indirect interactions in very large dimensions. Indeed there exist many algorithms for reconstructing gene interaction networks over several thousand gene microarray probes. Several of these algorithms have been in compared [7].

Data mining is based on computing a set of scores, indexed over the variables in the database, that are ranked in decreasing order of magnitude to yield a rank-ordered list. Variables at the top of the list are considered to be the best match to the data-mining criterion. The matching criterion will depend on the objective of the experimenter and the nature of the data. For "big data" applications there are two major issues: computational challenges and false positives, where the latter are defined as the occurrence of spurious scores near the top of the list that should be lower down on the list. Correlation mining[1] is a kind of data mining that searches for patterns of statistical correlation, or partial correlation, in an interaction network or across several interaction networks. These correlations can usually only be empirically estimated from multiple samples of the variables, as in gene microarray or RNAseq data collected over a population of subjects. From the estimated correlations one obtains an empirical correlation network, or partial correlation network, that can be mined for interconnectivity patterns such as edges, connected nodes, hub nodes, or sub-graphs. Correlation mining was introduced by the authors in [9] for mining connected nodes from correlation networks and in [10] for mining hub nodes from correlation or partial correlation networks.

In this chapter we focus on the related problem of mining edges from correlation or partial correlation networks. In particular, we will emphasize significance testing for network edges and hubs in the emerging "big data" setting where there are an exceedingly large number of variables. We will provide theory for reliable recovery of edges that applies to arbitrarily large numbers of variables and we will illustrate this theory by implementing correlation mining on a gene–gene correlation network of over 12 000 variables (gene probes), which can in principle have on the order of 7 million edges. We will also provide perspectives on future challenges in correlation mining biological data with a special focus on health and medical applications.

Accurate estimation of correlation requires acquiring multiple samples, e.g. technical replication of a gene chip assay n times or biological replication of the assay on n different members of the population. With the emergence of increasingly low-cost and

[1] Our definition of correlation mining is not to be confused with "correlated graph pattern mining" that seeks to find co-occurring subgraphs in graph databases [8].

higher-throughput technology, e.g. oligonucleotide gene microchips and RNAseq assays, the expression of increasing numbers of biomarkers can be determined from a single biological sample at increasingly lower cost. On the other hand, the cost of acquiring additional reliable samples does not appear to be decreasing at the same rate as the cost of including additional biomarkers in the assay. For example, running a controlled experiment or challenge study with animals or human volunteers is labor-intensive and very costly, and will likely remain so for the foreseeable future.

Therefore, at least for experimental biology applications, correlation mining practitioners face a deluge of biomarkers (variables) with a relative paucity of biological replicates (samples). This situation poses great difficulty in correlation mining due to the multiple comparisons problem: with so few samples one is bound to find spurious correlations between some pairs of the many variables. Controlling, or at least predicting, the number of spurious correlations must be the principal scientific objective in such situations. Achieving this objective lies in the realm of high-dimensional statistical inference in the "large p small n regime". Statisticians have studied several sub-regimes, e.g. those characterized as "high-dimensional", "very-high-dimensional", or "ultra-high-dimensional" settings [11]. These settings, however, still require that both the dimension p and the sample size n go to infinity. Consequently they may not be very useful for biological experimentalists who lack the budget to collect an increasing number of samples, especially given how large p is. A more useful and relevant regime, which will be the focus of this chapter, is the "purely high-dimensional" setting where the number of samples n remains fixed while the number of variables p is allowed to grow without bound. This high-dimensional regime is, in fact, the highest-possible-dimensional regime, short of having no samples at all. Thus it is appropriate to call this purely high-dimensional regime the "ultimately high-dimensional regime", and we shall use these two terms interchangeably in the rest of this chapter. A table illustrating our asymptotic framework is given below (Table 15.1), and also serves to compare and contrast our framework to previously proposed asymptotic regimes.

The purely high-dimensional regime of large p and fixed n poses several challenges to those seeking to perform correlation mining on biological data. These include both computational challenges and the challenge of error control and performance prediction. Yet this regime also holds some rather pleasant surprises. Remarkably there are modest benefits to having few samples in terms of computation and scaling laws. There is a scalable computational complexity advantage relative to other high-dimensional regimes where both n and p are large. In particular, correlation mining algorithms can take advantage of numerical linear algebra shortcuts and approximate k nearest neighbor search to compute large sparse correlation or partial correlation networks. Another benefit of purely high dimensionality is an advantageous scaling law of the false positive rates as a function of n and p. Even small increases in sample size can provide significant gains in this regime. For example, when the dimension is $p = 10\,000$ and the number of samples is $n = 100$, the experimenter only needs to double the number of samples in order to accommodate an increase in dimension by six orders of magnitude ($p = 10\,000\,000\,000$) without increasing the false positive rate [10].

Table 15.1 Overview of different asymptotic regimes of statistical data analysis. These regimes are determined by the relation between the number n of samples drawn from the population and the number p, called the dimension, of variables or biomarker probes. In the classical asymptotic regime the number p is fixed while n goes to infinity. This is the regime where most of the well known classical statistical testing procedures, such as student t tests of the mean, Fisher F tests of the variance, and Spearman tests of the correlation, can be applied reliably. Mixed asymptotic regimes where n and p both go to infinity have received much attention in modern statistics. However, in this era of big data where p is exceedingly large, the mixed asymptotic regime is inadequate since it still requires that n go to infinity. The recently introduced "purely high-dimensional regime" [9, 10], which is the regime addressed in this paper, is more suitable to big data problems where n is finite

Asymptotic framework	Terminology	Sample size n	Dimension p	Application setting	References
Classical (or sample increasing)	small-dimensional	$\longrightarrow \infty$	fixed	"small data"	Fisher [12, 13], Rao [14, 15], Neyman and Pearson [16], Wilks [17], Wald [18–21], Cramér [22, 23], Le Cam [24, 25], Chernoff [26], Kiefer and Wolfowitz [27], Bahadur [28], Efron [29]
Mixed asymptotics	high-dimensional very-high-dimensional ultra-high-dimensional	$\longrightarrow \infty$ $\longrightarrow \infty$ $\longrightarrow \infty$	$\longrightarrow \infty$ $\longrightarrow \infty$ $\longrightarrow \infty$	"medium sized" data (mega or giga scales)	Donoho [30], Zhao and Yu [31], Meinshausen and Bühlmann [32], Candès and Tao [33], Bickel, Ritov, and Tsybakov [34], Peng, Wang, Zhou, and Zhu [35], Wainwright [36, 37], Khare, Oh, and Rajaratnam [38]
Purely high-dimensional	purely high-dimensional	fixed	$\longrightarrow \infty$	"Big Data" (tera, peta and exascales)	Hero and Rajaratnam [9], Hero and Rajaratnam [10], Firouzi, Hero, and Rajaratnam [39]

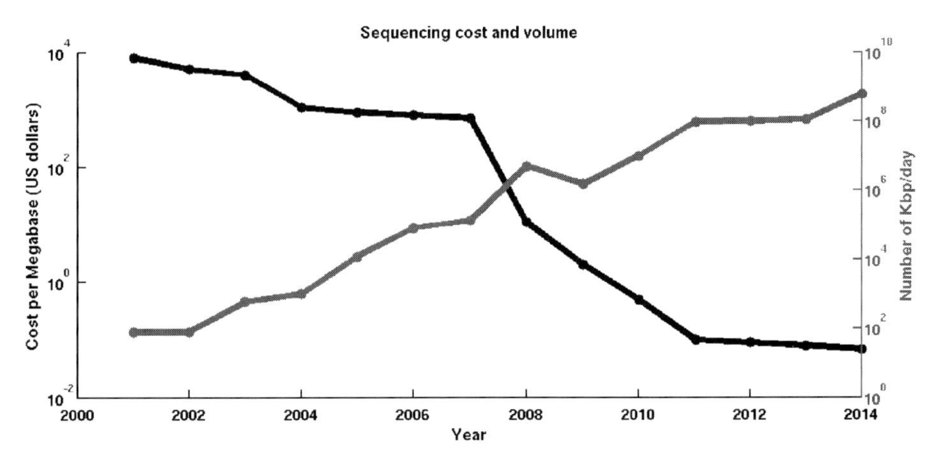

Figure 15.1 The cost in US dollars per megabase (black curve) and the number of kilobases (gray curve) of DNA sequenced per day over the past 14 years. The cost of determining one megabase of DNA has fallen at a very rapid rate since the transition in early 2008 from the Sanger-based technology to second-generation technology. At the same time the total volume of DNA sequenced has risen at an even more rapid rate as the price has fallen, demand for sequencing has grown, and sequencing centers have proliferated. (Note: volume is a product of sequencing depth and number of DNA samples sequenced.) *Source*: NCBI.

The relevance of the purely high-dimensional regime in biology can be understood more concretely in the context of developments in the technology of DNA sequencing and gene microarrays. The cost of sequencing has dropped very rapidly to the point where, as of this writing, the sequencing of a individual's full genome (3000 megabases) is approaching $1000 US. Recent RNAseq technology allows DNA sequencing to be used to measure levels of gene expression. Figure 15.1 shows how the cost of sequencing (in dollars per megabase) vs. the volume of DNA sequenced (in kilobases per day) as a function of time [40, 41]. While there has been a flattening of the cost curve over the past two years, the cost is still decreasing, albeit more slowly than before. Despite the recent flattening out of cost, the volume sequenced has been increasing at a rapid rate. This suggests that the usage of high-throughput sequencing technology will expand, creating a demand for efficient correlation mining methods like the ones discussed in this chapter.

The trends are similar in gene microarray technology, a competitor to sequencing based gene expression technologies such as RNAseq. One such technology is the 60mer oligonucleotide gene expression technology by Agilent, which is similar to the shorter 25mer oligonucleotide technology sold by Affymetrix as the "genechip". In Figure 15.2 is shown the price per slide of the Agilent Custom Microarray as a function of the number of biomarkers (probes) included on the slide. The price increases sublinearly in the number of probes. As the probe density increases the price will come down further at the high-density end of the scale. Thus it is becoming far less costly to collect more biomarker probes (p variables) than it is to collect more biological replicates

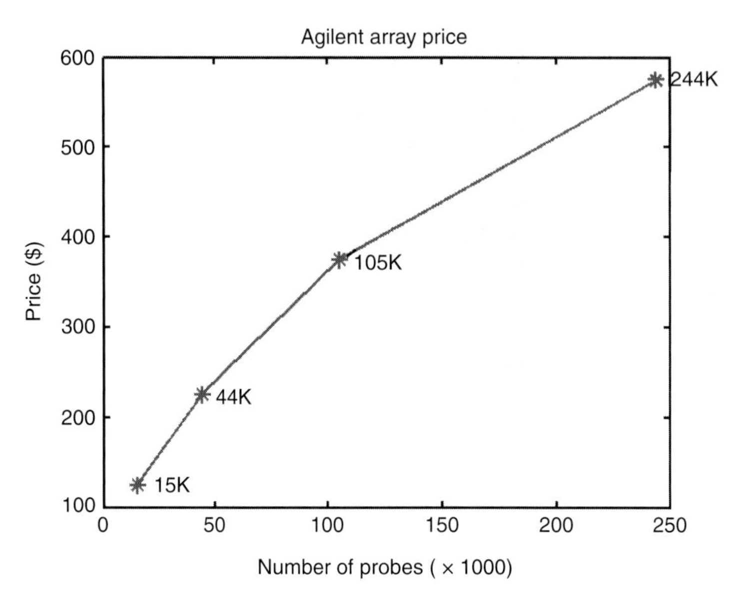

Figure 15.2 The price in US dollars of an Agilent array as a function of the number of gene probes per array. The price increases sublinearly in the number of probes in the array. Data are from the Agilent Custom Gene Expression Micorarrays G2309F, G2513F, G4503A, G4502A (February 2013). *Source*: BMC RNA Profiling Core.

(n samples). One can therefore expect an accelerated demand for high-dimensional analysis techniques such as the correlation mining methods discussed in this chapter.

The outline of the chapter is as follows. In Section 15.2 we give an illustrative example to motivate the utility of correlation measures for mining biological data. In Section 15.2.2 we formally introduce correlation and partial correlation in the context of recovering networks of biomarker interactions. In Section 15.3, we briefly describe rules of thumb for setting correlation screening thresholds to protect against these errors. In Section 15.4 we describe some future challenges in correlation mining of biological data, In Section 15.5 we provide some concluding remarks.

15.2 Illustrative example

We illustrate the utility of our correlation mining framework in the context of gene chip data collected from an acute respiratory infection (ARI) flu study [42, 43]. In this ARI flu study 17 human male and female volunteers were infected with a virus and their peripheral blood gene expression was assayed at 16 time points pre- and post-infection. The study resulted in a matrix of 12 023 gene probes by 267 samples was formed from the 267 available Affymetrix HU133 chips.

Roughly half of the volunteers became symptomatically ill (Sx) while the rest (Asx) did not become ill despite exposure to the virus. Each volunteer had two reference samples collected before viral exposure. The samples of the Sx volunteers were subdivided

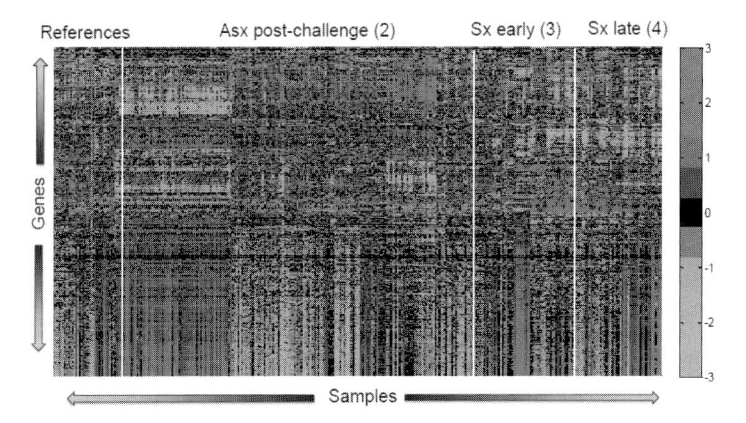

Figure 15.3 A heatmap of the 12 023 × 267 matrix (6.5 megapixels) of gene expression from a viral flu challenge study involving samples taken before viral exposure (reference), subjects that did not become ill after exposure (Asx post-challenge), subjects that became symptomatic (Sx early and Sx late). These are four different sub-population classes, called phenotypes, that can be mined for genes whose expression patterns change between pairs of classes. There is no easily discernible discriminating gene pattern that one can see just by looking at the raw data heatmap. The objective of correlation mining is to extract genes whose pattern of correlation changes over different sub-populations.

into early and late stages of infection using an unsupervised Bayesian factor analysis method [43]. Figure 15.3 shows the 12 023 × 267 matrix with columns arranged over the four categories: *reference, Asx post-challenge, Sx early, Sx late*. From these raw data one can see some patterns but they do not clearly differentiate the classes of individuals. This lack of definition is due to the fact that these classes are not homogeneous. The immune response of the volunteers evolves over time, the temporal evolution is not synchonronous across the population, and some genes exhibit different responses between the men and women in the study.

To better discriminate the genes that differentiate between the classes, a score function can be designed to rank and select the genes with highest scores. Since the number of samples is limited, the score function and the cutoff threshold should be carefully selected according to statistical principles to avoid false positives. Two classes of score functions will be discussed here: first order and second order.

Let us say that an experimenter wishes to find genes that have very different mean expression values μ when averaged over two sub-populations A and B. Let $\{X_k^A(g)\}_{k=1}^{n_A}$ be the measured expression levels of gene g over a population of n_A samples in population A. The sample mean $\mu_A(g)$ is

$$\mu_A(g) = n_A^{-1} \sum_{k=1}^{n_A} X_k^A(g).$$

Similarly define the sample mean μ_B over sub-population B. The difference $\mu_B - \mu_A$ between the sample means over the sub-populations is an example of a first-order score function. Thresholding this score function will produce a list of genes that have the highest contrast in their means. More generally, a first-order score function is any function of the data that is designed to constrast the sample means across sub-populations. The student-t test statistic and the Welch test statistic [44] are also first-order score functions that are variance normalized sample mean differences between two populations.

15.2.1 Pairwise correlation

A different objective for the experimenter, and the motivation for this chapter, is to find pairs of genes that have very different correlation coefficients ρ over two sub-populations. Let $\left\{X_k^A(g), X_k^A(\gamma)\right\}_{k=1}^{n_A}$ and $\left\{X_k^B(g), X_k^B(\gamma)\right\}_{k=1}^{n_B}$ be the measured expression levels of two genes g and γ over two sub-populations A and B. The standard Pearson product moment correlation coefficient between these two genes for sub-population A is

$$\rho_A(g,\gamma) = \frac{\sum_{k=1}^{n_A}\left(X_k^A(g) - \mu_A(g)\right)\left(X_k^A(\gamma) - \mu_A(\gamma)\right)}{\sqrt{\sum_{k=1}^{n_A}\left(X_k^A(g) - \mu_A(g)\right)^2}\sqrt{\sum_{k=1}^{n_A}\left(X_k^A(\gamma) - \mu_A(\gamma)\right)^2}} \in [-1.1],$$

and similarly for $\rho_B(g,\gamma)$. The difference between sample correlations $\Delta\rho_{A,B} = \rho_A(g,\gamma) - \rho_B(g,\gamma)$ is a contrast that is an example of a higher-order score function. The magnitude of $\Delta\rho_{A,B}$ takes a maximum value of 2, which it approaches when the correlation is high but of opposite sign in both sub-populations. Genes that have high contrast $\Delta\rho_{A,B}$ in their correlations may not have high contrast in their means. Furthermore, unlike the mean, the correlation is directly related to the predictability of one gene from another gene, since the mean squared error of the optimal linear predictor is proportional to $(1-\rho^2)$: a variable that is highly correlated to another variable is easy to predict from that other variable. Finally, it has been observed by many researchers that, while the mean often fluctuates over sub-populations, the correlation between biomarkers is preserved. Thus the sample correlation can be more stable than the sample mean, especially when used to compare multiple populations or multiple species [45–48], which often makes it a better discovery tool.

For an illustration of the differences between mining with the mean vs. mining with the correlation the reader is referred to Figure 15.4. In this figure, two pairs of genes are shown from the ARI challenge study whose sample correlation coefficients flip from a highly positive coefficient in the Asx sub-population to a highly negative coefficient in the Sx sub-population. The HMBOX1 and NRLP2 genes shown in Figure 15.4a are known transcription factors that are involved in immune response. HMBOX1, a homeobox transcription factor, negatively regulates natural killer cells (NK) in the immune system defense mechanism [49]. NRLP2, an intracellular nod-like receptor, regulates the level of immune response to the level of threat [50]. The flip from positive correlation (Asx) to negative correlation (Sx) between NRLP2 and HMBOX1 is an interesting and discriminating biomarker pattern between Sx and Asx phenotypes. Note that neither of these genes exhibit a significant change in mean over these sub-populations. Hence, while a second-order contrast function assigns these genes very high scores, revealing

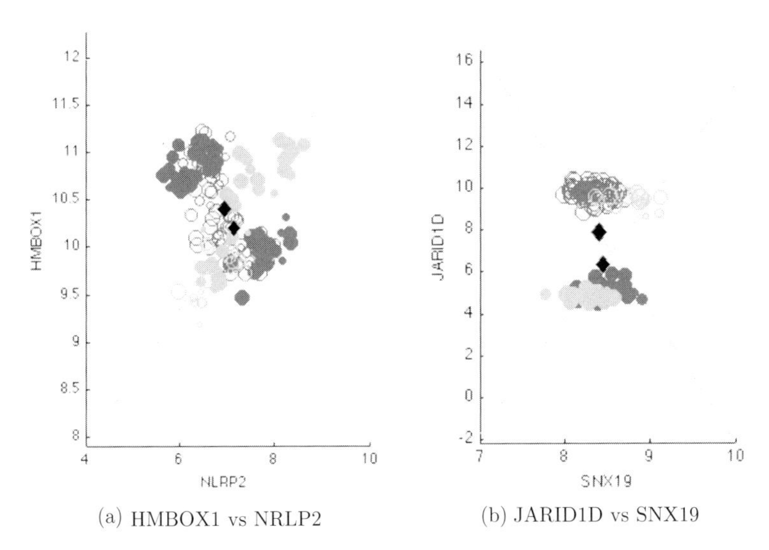

(a) HMBOX1 vs NRLP2 (b) JARID1D vs SNX19

Figure 15.4 Two pairs of genes (rows of matrix in Figure 15.3) discovered by a second-order analysis (polarity-flip correlation screening) between Asx and Sx sub-populations in a acute respiratory infection (ARI) dataset (expression levels of 12 023 genes measured over 267 samples). Solid circles denote women, hollow circles denote men. Light gray symbols denote Asx subjects and dark gray symbols denote Sx subjects. Size of symbol encodes time index of sample (smaller size and larger size symbols encode early and late post-exposure times, respectively). In both left and right panels there is a flip from high positive correlation to high negative correlation (indicated by the diagonal and antidiagonal lines in each scatterplot). The sample mean would not reveal these genes as differentiating the Sx and Asx subjects.

them as highly differentiated over Sx and Asx, a first order contrast function would miss them entirely.

Figure 15.4b shows a correlation flip between JARID1D and SNX19 that is capturing differences between the immune response of women (solid circles) and men (hollow circles). These differences would not be revealed by a first-order analysis without stratifying the population into male and female volunteers. Without such stratification, the first-order analysis would detect no significant change in mean over the Sx and Asx sub-populations.

15.2.2 From pairwise correlation to networks of correlations

Assume that there are p genes that are assayed in a population and let $\rho(g, \gamma)$ be the sample correlation between the pair g, γ. Define the $p \times p$ matrix of sample correlation coefficients $\mathbf{R} = [[\rho(g, \gamma)]]_{g,\gamma=1}^{p}$. By thresholding the magnitude of $\mathbf{R}$ one obtains an adjacency matrix whose nonzero entries specify the edges between pairs of genes in a correlation graph. The higher the correlation threshold the sparser will be the adjacency matrix and the fewer edges will be contained in the graph. The correlation graph, also called a correlation network, has p nodes and up to $\binom{p}{2}$ edges. To avoid confusion between the correlation network and the partial correlation network, discussed below, the former is often called the *marginal correlation network*.

Another network associated with the interactions between genes is the *partial correlation network*. The partial correlation network is computed by thresholding the diagonally normalized Moore–Penrose generalized inverse of the sample correlation, which we denote $\mathbf{P}$, where

$$\mathbf{P} = \text{diag}\left(\mathbf{R}^{\dagger}\right)^{-1/2} \mathbf{R}^{\dagger} \text{diag}\left(\mathbf{R}^{\dagger}\right)^{-1/2} \tag{15.1}$$

and $\mathbf{R}^{\dagger}$ is the Moore–Penrose generalized inverse and diag($\mathbf{R}^{\dagger}$) is the matrix obtained by zeroing out the off diagonal entries of $\mathbf{R}^{\dagger}$. Thus edges in the partial correlation graph depend on the correlation matrix only through its inverse.

Correlation graphs and partial correlation graphs have different properties, a fact that is perhaps easily understood in the context of sparse Gaussian graphical models [51], also known as Gauss–Markov random fields. For a centered Gaussian graphical model, the joint distribution $f(\mathbf{X})$ of the variables $\mathbf{X} = [X(1), \ldots, X(p)]^T$ is multivariate Gaussian with mean zero mean and covariance parameter $\mathbf{\Sigma}$. Consider samples $\{\mathbf{X}_k\}_{k=1}^n$ which are assumed to be independent and identically distributed (i.i.d.). For this Gaussian model the sparsity of (i, j)th element of the inverse correlation matrix, and hence sparsity of the partial correlation network, implies that components i and j are independent given the remaining variables. This conditional independence property is referred to as the "pairwise" Markov property. Another kind of Markov property, the so-called local Markov property, states that a specified variable given its nearest neighboring variables (in the partial correlation network) is conditionally independent of all the remaining variables. Yet another Markov property is the global Markov property which states that two blocks of variables A and B in a graph are conditionally independent given another set of variables C when the third block C separates in the graph the original two blocks of variables (see [51] for more details). Here "separate" means than every path between a pair of nodes, one in block A and another in block B, has to traverse C. It can be shown that the pairwise, local, and global Markov properties are equivalent under very mild conditions [51]. In other words, a Markov network constructed from the pairwise Markov property allows one to read off complex multivariate relationships at the level of groups of variables by simply using the partial correlation graph. Thus the sparsity of the inverse correlation captures Markov properties between variables in the dataset. Since the marginal correlation graph encodes pairwise or bivariate relationships, the partial correlation network is often at least as sparse, and usually much sparser, than the marginal correlation network. This key property makes partial correlation networks more useful for obtaining a parsimonious description of high-dimensional data.

One of the principal challenges faced by practitioners of data mining is the problem of phase transitions. We illustrate this problem in the context of correlation mining as a prequel to the theory presented in Section 15.3. As explained in Section 15.3, as one reduces the threshold used to recover a correlation or partial correlation network, one eventually encounters an abrupt phase transition point where we start to see an increasingly large number of false positive edges or false positive nodes connected in the recovered graph. This phase transition threshold can be mathematically approximated and, by using the theory below, a threshold can be selected that guarantees a prescribed false positive rate under an assumption of sparsity.

We illustrate the behavior of partial correlation and marginal correlation networks on the reference samples of the ARI challenge study dataset. In this data set there were 34 reference samples taken at two time points before viral exposure from the 17 volunteers enrolled in the study. Using theory discussed in Section 15.3 we selected and applied a threshold (0.92) to the correlation and partial correlation matrices. This threshold was determined using Theorem 15.1 to approximate the false positive rate P_e with the expression (15.2) and using this expression to select the threshold ρ that gives $P_e = 10^{-6}$ when there are $p = 12\,023$ variables and $n = 34$ samples. This false positive rate constraint is equivalent to a constraint that on the average there are fewer than 0.22 nodes that are mistakenly connected in the graph.

The application of the threshold $\rho = 0.92$ to the sample correlation resulted in a correlation network having 8718 edges connecting 1658 nodes (genes). On the other hand, using the same threshold on the sample partial correlation, which according to Theorem 15.1 also guarantees 10^{-6} false positive rate, the recovered partial correlation network had only 111 edges connecting 39 genes. This later network is shown in Figure 15.5. The fact that the partial correlation network is significantly sparser strongly suggests that the genes behave in a Markovian manner with Markov structure specified by the graph shown in Figure 15.5. This graph reveals four connected components that are conditionally independent of each other since there are no edges between them.

By investigating the genes in the partial correlation network's connected components in Figure 15.5, the four modules can be putatively associated with different biological functions. In particular, one of the connected components at the top right corresponds to sentinel genes like HERC5, MX1, RSAD2, and OAS3 in the immune system. One of the components in the top middle is composed of genes like UTY and and PRKY that are involved in protein–protein interactions and kinase production. One of the large connected components at bottom center is a module of housekeeping biomarkers, e.g. AFFX-BioDn-T_at, that are used by the gene chip manufacturer (Affymetrix) for calibration purposes. While all of these genes are also present in the connected components of the much less sparse marginal correlation network, the picture is not nearly as simple and clear.

15.3 Principles of correlation mining for big data

In big data collection regimes one is often stuck with an exceedingly large number of variables (large p) and a relatively small and fixed number of samples (small n). This regime is especially relevant for biomolecular applications where size p of the genome, proteome or interactome can range from tens of thousands to millions of variables, while the number n of samples in a given sub-population is fixed and only on the order of tens or hundreds. Recently an asymptotic theory has been developed expressly for correlation mining in this purely high-dimensional regime [9, 10]. Unlike other high-dimensional regimes where both p and n go to infinity, the theory of [9, 10] only requires p to go to

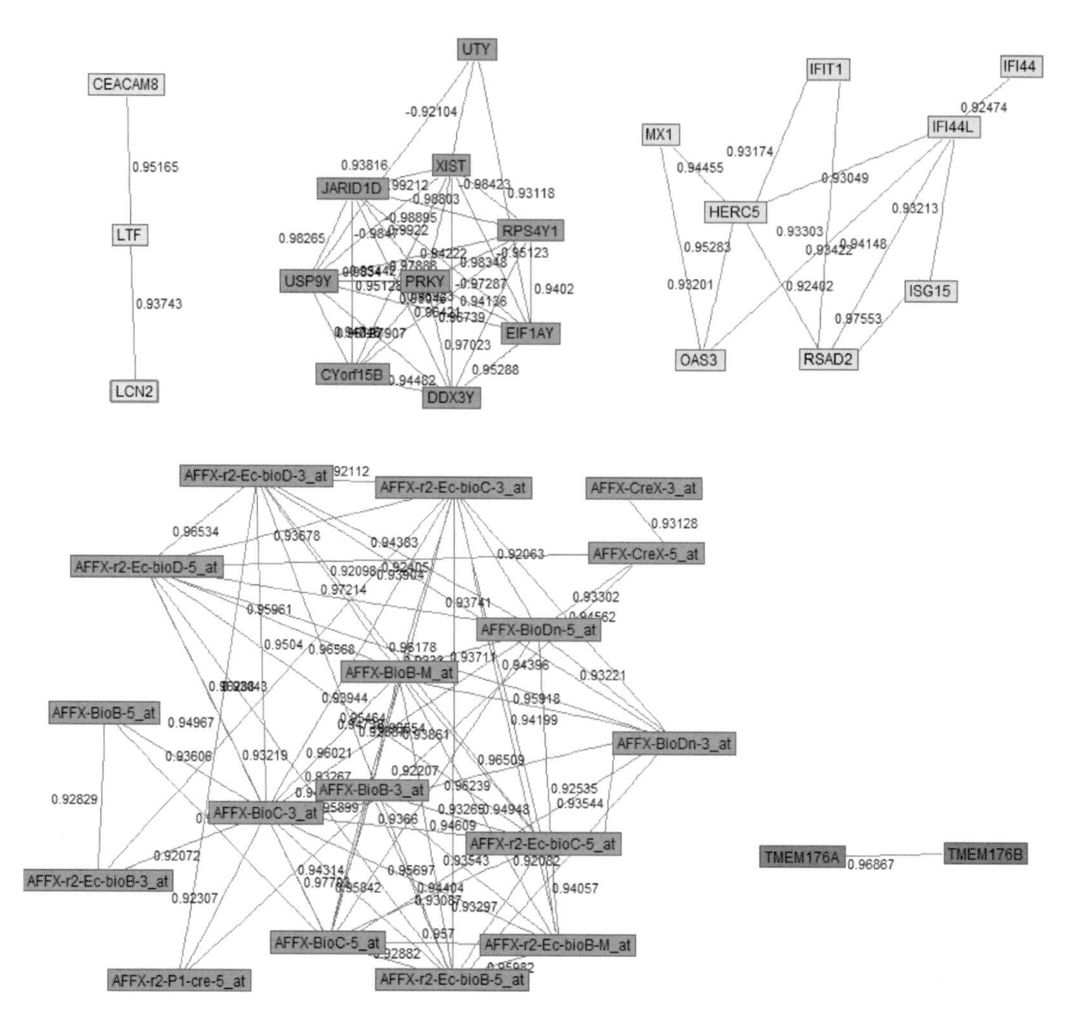

Figure 15.5 Recovered partial correlation gene network over the 12 023 genes and 34 reference samples assayed in the ARI challenge study data. The applied partial correlation threshold used to obtain this network was determined using the theory in Section 15.3 in order to guarantee a 10^{-6} false positive rate or less. There are only 39 nodes (genes) connected by 111 edges in the displayed partial correlation network as compared to over 1658 nodes and 8718 edges recovered in the (marginal) correlation network (not shown).

infinity while n remains fixed. This property makes this purely high-dimensional regime more relevant to limited sampling applications, which characterize big data. Before discussing the theory of correlation mining we contrast it to other correlation recovery methods.

We define correlation mining in a very specific manner in order to differentiate it from the many other methods that have been proposed to recover properties of correlation matrices. Most of these methods are concerned with the so-called covariance selection problem [52]. The objective of covariance selection is to find nonzero entries in the inverse covariance matrix for which many different approaches have been proposed [32, 35, 38, 53–66] and some have been applied to bioinformatics applications [67–69]. Covariance selection adopts an estimation framework where one attempts to fit a sparse covariance (or inverse covariance) model to the data using a goodness of fit criterion. For example, one can minimize residual fitting error, such as penalized Frobenius norm squared, or maximize likelihood. The resulting optimization problem is usually solved by iterative algorithms with a stopping criterion. These methods are not immediately scalable to large numbers of variables except in very sparse situations [59].

In contrast to covariance selection methods, which seek to recover the entire covariance or inverse covariance including the zero values, correlation mining methods use thresholding to identify variables and edges connected in the correlation or partial correlation network that have the the strongest correlations. Thus correlation mining does not penalize for estimation error nor does it recover zero correlations. Correlation mining is fundamentally a testing problem as contrasted to an approximation problem – a characteristic that clearly differentiates correlation mining from covariance selection. Correlation mining can be looked at as complementary to covariance selection and has several advantages in terms of computation and error control. Since it involves only simple thresholding operations, correlation mining is highly scalable. Since correlation mining filters out all but the highest correlations, a certain kind of extreme value theory applies. This theory specifies the asymptotic distribution of the false positive rate in the large p small n big data regime. This can then be used to accurately predict the onset of phase transitions and to select the applied threshold to ensure a prescribed level of error control. Inverse covariance estimation problems do not in general admit this property.

In [9, 10] three different categories of correlation mining problems were defined. These correlation mining problems aimed to identify variables having various degrees of connectivity in the correlation or partial correlation network. These were: correlation screening over a single population, correlation screening over multiple populations, and hub screening over a single population. In each of these problems theory was developed for the purely high-dimensional regime of large p and fixed n under the assumption of block sparse covariance. A block sparse $p \times p$ covariance matrix is a positive definite symmetric matrix $\mathbf{\Sigma}$ for which there exists a row–column permutation matrix Π such that $\Pi \mathbf{\Sigma} \Pi^T$ is block diagonal with a single block of size $m \times m$ with $m = o(p)$. For more details the reader is referred to the original papers [9, 10]. While not specifically emphasized in these papers, the asymptotic theory also directly applies to screening edges in the network. Indeed the proofs of the asymptotic limits in [9, 10] use an obvious equivalence between the incidence of false edges and false connected nodes in the recovered network. In particular, we have the following theorem that is proved as an immediate corollary to [10, Prop. 2].

Theorem 15.1 *Assume that the n samples $\{\mathbf{X}_k\}_{k=1}^n$ are i.i.d. random vectors in $\mathbb{R}^p$ with bounded elliptically contoured density and block sparse $p \times p$ covariance matrix. Let ρ be the threshold applied to either the sample correlation matrix or the sample*

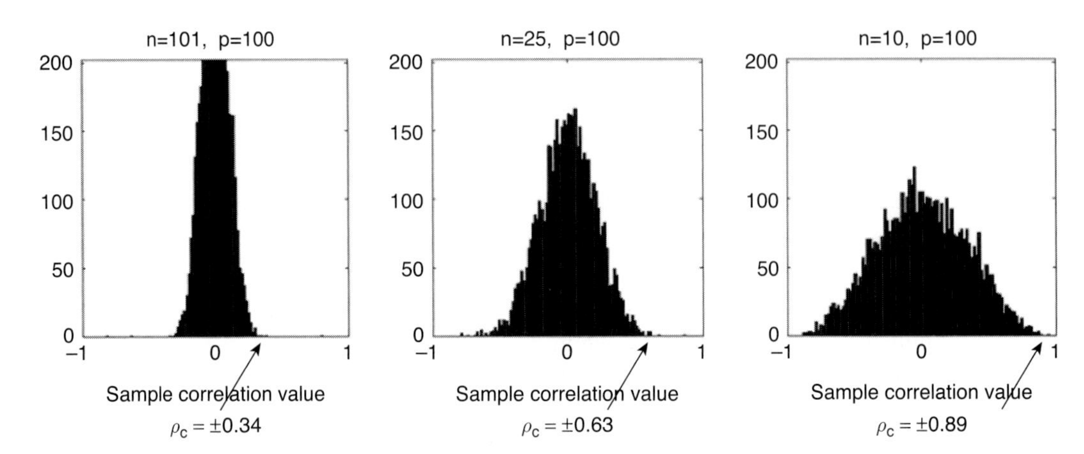

Figure 15.6 Histograms of the number of edge discoveries in a sample correlation graph over $p = 100$ biomarkers as a function of the threshold ρ applied to the magnitude of each entry of the correlation matrix. In this simulation the true covariance matrix is diagonal so every edge discovery is a false discovery. In the three panels from left to right the number of samples is reduced from $n = 101$ to $n = 10$. The theoretically determined critical phase transition threshold ρ_c, given by expression (15.5), closely tracks the observed phase transition threshold in all cases. (Figure adapted from [9, Fig. 1].)

partial correlation matrix. Assume that p goes to infinity and $\rho = \rho_p$ goes to one according to the relation $\lim_{p\to\infty} p(p-1)(1-\rho_p^2)^{(n-2)/2} = e_n$. Then the probability P_e that there exists at least one false edge in the correlation or partial correlation network satisfies

$$\lim_{p\to\infty} P_e = 1 - \exp(-\kappa_n/2), \tag{15.2}$$

where

$$\kappa_n = e_n a_n (n-2)^{-1},$$

where a_n is the volume of the $n-2$ dimensional unit sphere in $\mathbb{R}^{n-1}$ and is given by $a_n = \frac{\Gamma((n-1)/2)}{\sqrt{\pi}\Gamma((n-2)/2)} = 2B((n-2)/2, 1/2)$.

We now provide an intuitive way to understand the flavor of our results. Consider a null model where the covariance matrix is diagonal, i.e. when there are no true nonzero correlations, and so any edge that is detected as nonzero will be a false edge. Under this setting one can generate the distribution of the sample correlation coefficients for various dimensions. Figure 15.6 illustrates these distributions in the setting when $p = 100$ for various values of n. It is clear that when the sample size n is low relative to the dimension p, the probability of obtaining false edges are higher, and consequently a higher threshold level ρ_p is required in order to avoid detecting false edges. In such situations consider once more the quantity P_e, the probability of obtaining at least one false edge. It is clear that for a fixed sample size n and fixed threshold ρ, as the dimension $p \longrightarrow \infty$ the probability of detecting a false edge P_e tends to 1. So it makes sense to let the threshold ρ tend to 1 in such settings so that the probability of detecting a false edge is small. However, for a fixed sample size n and fixed dimension p, if the threshold ρ tends to 1, then the probability of detecting a false edge goes to zero. This is because as ρ gets larger and tends to 1, it will eventually surpass the largest sample correlation. Thus increasing ρ to 1 will eventually threshold all sample correlations to zero, and will result in the probability of detecting a false edge tending to 0. The two scenarios described above therefore lead to degenerate or trivial limits. The theory above resolves this degeneracy by letting the threshold ρ_p tend to 1 at the correct rate as $p \longrightarrow \infty$ to obtain a non-degenerate limit. This in turn leads to a very useful expression for the probability of detecting a false edge P_e in the "purely high-dimensional" setting, when only the dimension $p \longrightarrow \infty$, while the sample size n is fixed.

Theorem 15.1 gives a limit for the probability P_e that is universal in the following senses: (1) it applies equally to correlation and partial correlation networks; (2) the limit does not depend on the true covariance.

The quantity e_n in Theorem 15.2 does not depend on p. However, we can remove the limit from the definition of e_n and substitute it into the expression for κ to obtain a useful large p approximation to P_e

$$P_e = 1 - \exp(-\lambda(\rho, n)/2), \tag{15.3}$$

with

$$\lambda_{\rho,n} = p(p-1)P_0(\rho, n),$$

and, P_0 is the normalized incomplete Beta function

$$P_0(\rho, n) = 2B((n-2)/2, 1/2) \int_{\rho}^{1} (1-u^2)^{\frac{n-4}{2}} du, \tag{15.4}$$

with $B(a, b)$ the Beta function. A bound on the fidelity of the approximation $1 - \exp(-\lambda_{\rho,n}/2)$ to P_e was obtained in [10, Sec. 3.2] for hub screening and it also applies to the case of edge screening treated here.

The large p approximation (15.3) to P_e resembles the probability that a Poisson random variable with rate function $\lambda_{\rho,n}/2$ exceeds zero. It can be shown that $\lambda_{\rho,n}/2$ is asymptotically equal to the expected number of false edges. Hence, in plain words, Theorem 15.1 says that the incidence of false edges in the thresholded sample correlation or sample partial correlation network asymptotically behave as if the edges were Poisson distributed.

Remarkably, the Poisson rate $\lambda_{\rho,n}/2$ in the large p approximation (15.3) does not depend on the true covariance matrix. This is a consequence of the block sparse assumption on the covariance. When the covariance is only row sparse, i.e. the number of nonzeros in each row increases only as order $o(p)$, then the same theorem holds except that now $\lambda_{\rho,n}$ will generally depend on the true covariance matrix.

The behavior of $\lambda_{\rho,n}$ as a function of ρ, n, and p specifies the behavior of the approximation (15.3) to P_e. In particular for fixed n and p there is an abrupt phase transition in P_e as the applied correlation threshold ρ decreases from 1 to 0 (see Figure 15.6). An asymptotic analysis of the incomplete Beta function yields a sharp approximation to the phase transition threshold ρ_c [9]. This threshold, defined as the

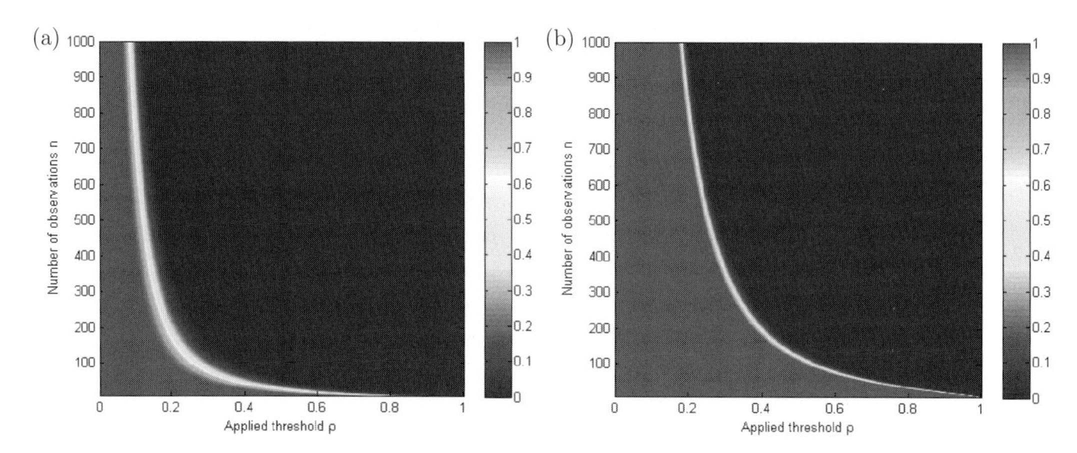

Figure 15.7 Heatmap of the large p approximation to the false edge discovery probability P_e in (15.2) as a function of the number of samples n and the applied threshold ρ for $p = 10$ (a) and for $p = 10\,000$ (b). The phase transition clearly delineates the region of low P_e from the region of high P_e and becomes increasingly abrupt as p increases to $p = 10\,000$.

knee in the curve $\lambda(\rho, n)$ defining the asymptotic mean number of edges surviving the threshold ρ, has the form [10, Eq. (10)]:

$$\rho_c = \sqrt{1 - (c_n(p-1))^{-2/(n-4)}}, \tag{15.5}$$

where $c_n = 2B((n-2)/2, 1/2)$.

In Figure 15.7 the large p approximation (15.3) to the false discovery probability P_e is rendered as a heatmap over n and ρ for the cases $p = 10$ and $p = 10\,000$. The phase transition clearly delineates the region of low P_e from the region of high P_e and becomes more abrupt for the larger value of p. The critical phase transition point increases much less slowly in n than it does in p. For example, from the heatmap we can see that if the biologist wanted to reliably detect edges of magnitude greater then 0.5 with $p = 10$ variables she would need more than $n = 60$ samples. However, if the number of variables were to increase by a factor of 1000 to $p = 10\,000$ she would need only to increase the number of samples by a factor of 2 to $n = 120$ samples. This represents a very high return on investment into additional samples. However, as can be seen from the heatmaps, there are rapidly diminishing returns as the number of samples increase for fixed p. Though we do not undertake a detailed demonstration of Theorem 15.1 and further development thereof for use in graphical model selection, we note that this topic is the subject of ongoing and future work.

15.3.1 Correlation mining for correlation flips between two populations

In Figure 15.4 of Section 15.2.1 we were shown a pair of genes whose correlation flipped from a positive value in one population to a negative value in another population. The correlation mining theory of the previous subsection is easily extended to this case using the results of [9, Prop. 3] in persistency correlation screening of multiple independent populations. For two populations of samples $\{\mathbf{X}_k\}_{k=1}^{n_A}$ and $\{\mathbf{Y}_k\}_{k=1}^{n_B}$ on the same domain

$\mathbb{R}^p$ let ρ_A and ρ_B be two thresholds in [0, 1]. When each of these thresholds is applied to the respective correlation matrices one obtains correlation networks G_A and G_B, respectively. We define the p node *correlation flip network* as the p node network obtained by placing an edge between two nodes if there exists a corresponding edge in G_A and G_B and the correlations associated with these two edges are of opposite sign. The number of false positive edges will depend on the two thresholds and the number of samples in each population. Specifically, we have the following theorem that formalizes this statement.

Theorem 15.2 *Assume that the samples $\{\mathbf{X}_k\}_{k=1}^{n_A}$ and $\{\mathbf{Y}_k\}_{k=1}^{n_B}$ are both i.i.d. and mutually independent random vectors in $\mathbb{R}^p$ with bounded elliptically contoured densities and block sparse $p \times p$ covariance matrices. Let ρ_A and ρ_B be the associated correlation thresholds. Assume that p goes to ∞ and ρ_A and ρ_B go to one according to the relations $\lim_{p\to\infty} p^{1/2}(p-1)(1-\rho_A^2)^{(n_A-2)/2} = e_{n_A}$ and $\lim_{p\to\infty} p^{1/2}(p-1)(1-\rho_B^2)^{(n_B-2)/2} = e_{n_B}$. Then the probability P_e that there exists at least one false edge in the correlation flip network satisfies*

$$\lim_{p\to\infty} P_e = 1 - \exp(-\kappa_{n_A,n_B}), \tag{15.6}$$

where

$$\kappa_{n_A,n_B} = e_{n_A} e_{n_B} a_{n_A} a_{n_B} (n_A - 2)^{-1} (n_B - 2)^{-1}/2.$$

As in Theorem 15.1, Theorem 15.2 can be used to obtain a large p approximation to P_e

$$P_e = 1 - \exp(-\lambda_{\rho_A,\rho_B,n_A,n_B}), \tag{15.7}$$

with

$$\lambda_{\rho_A,\rho_B,n_A,n_B} = e_{n_A} e_{n_B} p(p-1)^2 P_0(\rho_A, n_A) P_0(\rho_B, n_B)/2,$$

and $P_0(\cdot,\cdot)$ as given in (15.4).

Again the form of the limit (15.7) is the probability that Poisson random variable is not equal to zero, where the rate of the Poisson random variable is $\lambda_{\rho_A,\rho_B,n_A,n_B}$. This expression for the Poisson rate differs by a factor of $1/2$ from the expression for the Poisson rate in the persistency-screening case considered in [9, Prop. 3]. This is simply due to the fact that the version of persistency screening in [9, Prop. 3] does not carry the restriction that the correlations be of opposite sign in the two populations. It can be shown, using the results of [10], that the theorem equally applies to mining edges in a partial correlation flip network.

To apply the theorem to correlation flip mining, the two thresholds ρ_A and ρ_B will not be chosen independently. Rather, as in persistency screening [9], they should be selected in a coupled manner according to the relation [9, Sec. 3.3] so as to equalize the mean number of false positive edges in the marginal networks G_A and G_B.

For illustration, we apply the large p approximation (15.7) to the ARI challenge study example shown in Figure 15.4. There were $n_A = 170$ Sx samples and $n_B = 152$ Asx samples collected in the study and the number of genes is $p = 12\,023$. The large p approximation to P_e specifies the two (coupled) thresholds that guarantee at most 10^{-6}

false edges in the correlation flip network:

$$\rho_A = 0.44, \qquad \rho_B = 0.47.$$

This can be compared to the thresholds that guarantee the same error rate on false edges if the individual correlation networks were screening independently:

$$\rho_A = 0.54, \qquad \rho_B = 0.57,$$

which simply reflects the fact that we can reliably detect lower-value correlations in the correlation flip network since, for equal thresholds, false edges are rarer than in the individual correlation networks.

In the correlation flip example above we have deliberately not compared the results of our proposed threshold-based approach to other more complex optimization-based methods in the literature such as glasso, elastic net, or SPACE. These methods are simply not scalable to the dimension ($p = 12\,023$) that is considered in the above example, at least not without making additional restrictive assumptions.

15.3.2 Large-scale implementation of correlation mining

Only a subset of the correlations is required to construct a correlation network whose edges correspond to correlations exceeding the applied correlation threshold ρ. This fact makes large-scale implementation of correlation mining scalable to large-dimension p. In particular, it is not necessary to either store or compute the full correlation matrix $\mathbf{R}$ in order to find the correlation graph. This is for to the following two reasons.

(1) **Z-score representation of R and P** The sample correlation matrix $\mathbf{R}$ and the partial correlation matrix $\mathbf{P}$, as defined in (15.1), have Gram product representations [10, Sec. 2.3]

$$\mathbf{R} = \mathbb{T}^T\mathbb{T}, \qquad \mathbf{P} = \mathbb{Y}^T\mathbb{Y},$$

where $\mathbb{T} = [\mathbf{T}_1, \ldots, \mathbf{T}_p]$ is an $n \times p$ matrix and $\mathbb{Y} = [\mathbf{Y}_1, \ldots, \mathbf{Y}_p]$ is an $(n-1) \times p$ matrix. The columns of these matrices are Z-scores that are defined in [10, Eq. (4)] and [10, Eq. (9)]. Therefore, all that needs to be stored are the smaller matrices $\mathbb{T}$ and $\mathbb{Y}$ (recall that $n \ll p$). Furthermore, computation of the Z-score matrix $\mathbb{Y}$ only requires a single $(n-1) \times (n-1)$ matrix inversion. This property is especially useful in typical cases where $n \ll p$.

(2) **Ball graph representation of correlation network** Owing to the Z-score representations above, the correlation and partial correlation networks are equivalent to ball graphs, also called a Euclidean proximity graphs, for which fast and scalable algorithms exist [70]. We explain this equivalence for the correlation graph that thresholds $\mathbf{R}$; the partial correlation graph equivalence is similarly explained.

A Euclidean proximity graph with ball radius r places an edge between vectors $\mathbf{T}_i$ and $\mathbf{T}_j$ in $\mathbb{R}^n$ iff the Euclidean distance $\|\mathbf{T}_i - \mathbf{T}_j\|$ does not exceed r. The ball graph is easily converted to a correlation network since two vectors $\mathbf{T}_i$, $\mathbf{T}_j$ in $\mathbb{R}^n$ with sample correlation ρ_{ij} are at mutual Euclidean distance $\|\mathbf{T}_i - \mathbf{T}_j\| = \sqrt{2(1-\rho_{ij})}$. Hence if a ball graph with ball radius $r = \sqrt{2(1-\rho)}$ is constructed on the

Z-score sample $\{\mathbf{T}_i\}_{i=1}^p$ the resulting graph will be a one-sided positive correlation network, i.e. a network whose edges represent postive correlations exceeding ρ. A one-sided negative correlation network is obtained by applying the same ball graph construction with pairwise distances $\|(-\mathbf{T}_i) - \mathbf{T}_j\| = \|\mathbf{T}_i + \mathbf{T}_j\|$, resulting in a network whose edges represent negative correlations less than $-\rho$. Merging the two one-sided correlation networks yields the correlation network.

By exploiting the Z-score and ball graph representations of the sample correlation and sample partial correlation, correlation mining algorithms can be implemented on very-high-dimensional biomarker spaces as long as the number of samples is small. Matlab and R code for implementing correlation mining algorithms for edge recovery and hub recovery is currently being developed for online access.

15.4 Perspectives and future challenges

As discussed above, one of the basic challenges in gleaning information from big data is the large p small n problem: a deluge of variables and a paucity of samples of these variables to adequately represent the population. This is of critical importance when one searches for correlations among the variables as correlations requires multiple samples. That this challenge is especially relevant to molecular biology applications is due to advances in high-throughput sequencing technology: the cost of collecting additional variables (p) is dropping much faster than the cost of obtaining additional independent samples (n). Thus the theory and practice of correlation mining is well placed to tackle big data problems in the medical and biological sciences. However, not surprisingly, there exist challenges that will require future work. Before discussing these challenges, we summarize the current state of the art of knowledge in correlation mining.

15.4.1 State-of-the-art in correlation mining

As discussed in Section 15.1, the objective and formulation of correlation mining are different from those of covariance selection. Correlation mining seeks to locate a few nodes, edges or hubs associated with high (partial) correlation correlation while covariance selection seeks to estimate a sparse (inverse) covariance matrix, thereby localizing the zero-valued (partial) correlations. Put more succinctly, correlation mining is founded on testing a few high (partial) correlations while covariance selection is focused on minimization of approximation error for the entire sample (inverse) covariance. This difference leads to sharp finite sample theory and simplified implementation of correlation mining as compared to covariance selection.

To illustrate we consider covariance selection for recovering the inverse covariance matrix. As mentioned above, partial correlation networks arise naturally when the experimenter assumes a Gaussian graphical model for the data [11]. Under a sparse inverse covariance hypothesis, maximum likelihood and other parametric estimation methods can then be used to recover an approximation to the true network. There are many methods for partial correlation network recovery that are based on the GGM. These include, among others: (1) GeneNet [71] which uses a regularized generalized-inverse of the covariance and uses the bootstrap to estimate the most suitable regularization parameter; (2) the work of [58], which proposes two algorithms: a block coordinate descent method and another based on Nesterov's first-order method to maximize the ℓ_1-regularized Gaussian log-likelihood; and (3) the block coordinatewise method g-lasso of [53]. In recent years, a fast proximal Newton-like method (QUIC) was proposed by [59]. This work was followed up by even faster proximal gradients methods, G-ISTA and G-AMA, in [60] and [61] respectively. Currently, the state-of-the-art method in the field, G-AMA, is able to handle problem sizes as high as 5000, though handling problem sizes of 10 000 or 20 000 is more challenging. Some of these methods are compared in [7].

There have also been efforts to go beyond the ℓ_1-regularized Gaussian likelihood approach to ℓ_1-regularized pseudo-likelihood and regression based approaches. These include methods such as "neighborhood selection" (NS) [32], SPACE [35], SPLICE [62], and SYMLASSO [72] (see also [73]). In a recent major contribution to this area, a convex pseudo-likelihood framework for high-dimensional partial correlation estimation with convergence guarantees was recently proposed in [38]. This method, CONCORD, overcomes the many hurdles faced by existing pseudo-likelihood based methods but at the same time retains all their attractive properties. In particular, it yields an approach which is provably convergent, respects symmetry, enjoys asymptotic statistical guarantees, is computationally tractable, and has excellent finite sample performance. The CONCORD pseudo-likelihood in [38] is solved by coordinatewise minimization. In even more recent work, ISTA- and FISTA-type proximal gradient methods have been proposed in [63] to scale up the implementation of CONCORD. In particular, the authors in [63] demonstrate convincingly that pseudo-likelihood methods can be also benefit from leveraging recent advances in convex optimization theory.

Despite the tremendous advances described above, ℓ_1-regularized and other proposed methods are not currently as easily scalable due to the iterative nature of the corresponding algorithms. Unlike correlation mining methods described in this manuscript, the methods above also do not have the tight theory provided by Theorem 15.1 to control false positives, and neither do they have the results obtained in [10].

The following summarizes some of the main features of the proposed correlation mining framework for biology applications.

- Correlation mining has its niche in the purely high-dimensional regime characterizing big biological datasets where n is fixed while p is large and increasing. This is where the asymptotic theory provides tight bounds on edge, node, and hub recovery performance.
- In this purely high-dimensional regime correlation mining algorithms are scalable to very large p. This is because they are non-iterative, involve only inversion of $n \times n$ matrices, and can be implemented using approximate nearest-neighbor search algorithms.
- Correlation mining algorithms and theory apply equally to correlation mining in both (marginal) correlation networks and partial correlation networks.

- The theory provides a mathematical expression for the critical phase transition ρ_c in the mean number of false positive edges in the recovered correlation network as a function of the applied threshold ρ.
- In the purely high-dimensional regime, the theory indicates that the curse of high dimensionality in the variables can be counterbalanced by a certain *benefit of high dimensionality* in the samples: to maintain a given phase transition value only a small sample size increase is necessary to accommodate a big increase in biomarker dimension. This benefit diminishes as the sample size approaches the biomarker dimension.
- The theory can be used to determine the right sample size for the experimentalist to be able to reliably detect correlations that are above the phase transition threshold and are therefore both biologically significant and statistically significant.
- This sample sizing can be performed without knowledge of the underlying distribution or true covariance matrix in cases where the distribution is elliptically contoured, the covariance is block sparse, and the collected data are i.i.d.
- If the true data covariance cannot be assumed to be block sparse, this assumption can be replaced by the weaker assumption of row sparse covariance to obtain the same type of asymptotic limit as in Theorem 15.2 (see [9, 10]). However, without block sparsity the Poisson-type rate constant κ_n in the theorem may now be a function of the true covariance and the expression (15.7) for the false positive rate may no longer be accurate. An empirical estimate of the true rate constant may be constructed by estimating a certain "J functional" of the density (see [74]), which may be used to determine proper sample size. This would be practical when sufficient amounts of baseline data exist on which to train, e.g. obtained by pooling publicly available data from prior studies, or when a model for baseline is known.
- If the collected data cannot be assumed to be i.i.d. and have either some temporal or spatial dependency, our correlation mining framework is still flexible enough to deal with such settings. First, correlation screening thresholds from the i.i.d. setup can still be used as lower bounds for the correct threshold when dependency is present. This follows from the fact that screening thresholds are a decreasing function of sample size. Second, if the dependency structure is known such as in an AR(1) model, the effective sample size can be used to adjust the screening threshold appropriately. Third, if the data are a Gaussian temporally stationary multivariate time series then the spectral correlation screening framework of [75] can be applied.

15.4.2 Future challenges in correlation mining biomolecular networks

There remain several challenges and open problems in large-scale correlation mining of molecular networks in biology and biomedicine. Many of these align with currently recognized grand challenges in engineering and the life sciences [76]. We discuss several opportunities and challenges below, with a focus on health and medicine applications.

Correlation mining can play an important role in emerging translational medicine applications such as: personalized medicine [77]; genomic medicine [78]; and network medicine [79]. In personalized medicine, correlation mining might be used to identify personalized molecular signatures and their nominal variations in order to specify a baseline of health for an individual. A principal objective of genomic medicine is early detection of disease by testing for activation of the gene expression pathways in the host immune system that presage inflammatory response and acute symptoms. Network medicine recognizes the dynamic nature of molecular networks regulating pathogenesis and immune response and proposes precisely timed drug treatments that target certain transitions in the network. Owing to the potentially large number of biomarker variables and limited samples, correlation mining can be a key player in determining the complex molecular interactions for these translational areas.

There has been much recent interest in biochronicity and other time-varying phenomena underlying gene regulation occurring in healthy hosts. Biomarkers that manifest as periodically changing over time are plentiful, e.g. clock genes that regulate 24 hour circadian and 12 hour hemicircadian rhythms of monocyte cell-cycles [80]. Quantifying and understanding the role of biochronicity in gene regulation is of independent interest but could also lead to periodic detrending of other biomarkers. For example, this could result in greatly improving sensitivity to subtle pathogen-induced changes and lead to improved early-detection performance. Fourier-based correlation mining extensions, such as developed in [75, 81], could possibly be applied to detecting the biochronicity genes as they manifest their periodic components over time. In this extension, correlation mining is performed independently on each complex-valued DFT coefficient of the time-varying variables. A challenge that would need to be overcome is the non-sinusoidal nature of many biochronicity signals.

There are continuing efforts to establish a biomarker-based baseline of health for preventative medicine. Molecular biomarker signatures are highly variable across time for a single healthy individual and across individuals for individuals in a healthy population. Some of this variability can be explained by fixed factors such as age and gender or deterministic factors, e.g. trait-related gene expression variation [82, 83], or biochronicity, discussed above. Such deterministic factors might be removed by using simple detrending or regression approaches. However, much of the variability of a healthy person's molecular patterns remains unexplained and might be better modeled as random over time but correlated over different biomarkers. Under this model, correlation mining could be used to identify the principal hubs and edges in the correlation network that characterize a person's healthy baseline. If some non-healthy samples were available, these correlations could be incorporated into a logistic regression to detect abnormal deviations from baseline. Alternatively, the predictive correlation screening framework of [84] could be used to tune correlation mining specifically to the task of prediction.

Multiplatform assays probe complementary manifestations of biological system states and behaviors. For example, challenge study protocols may simultaneously assay a tissue sample for gene expression, biased and unbiased protein expression, metabolite levels, and antibodies. Correlation mining and other statistical methods of correlation analysis have been principally developed for a single platform. Development of a theory and practice of multiplatform correlation mining would be worthwhile. Principal hurdles would be the need to account for cross-platform calibration, multiplatform normalization, and different levels of technical noise inherent to each platform. Several approaches to correlation networks for multiplatform assays include: multiattribute networks [85], which create separate nodes for each platform; multivariate normal networks [86] and Kronecker network [66], which accommodate networks with vector valued nodes. While the models used in these approaches are worth considering, the methods are not immediately extendible to correlation mining as they all focus on covariance estimation. Exploring extensions of these models to the testing framework of correlation mining would be worthwhile pursuing further.

Integration of ontological information relevant to biological function is commonly used to place an investigator's empirical findings in the context of the global knowledge base. For example, a gene list consisting of a discovered hub gene and its neighboring genes might be analyzed in the context of previously reported molecular interactions and pathways using PID [5]. More generally, ontologies can reveal transcription, methylation, and phosphorylation correspondences; gene-protein intra-actions and inter-actions; molecular-disease associations; epigenetics; and other types of phenotype-specific data. Such data is often in the form of text, trees, and graphs that can be integrated into the correlation network discovered via correlation mining. However, a major challenge is determining provenance and reliability of the ontological data. Another challenge is the direct use of ontological data as side information that can directly affect the correlation mining outcomes. Possible approaches to integration of such information into correlation mining include: correlation weighting to de-emphasize connections not supported by ontology [87], non-uniform thresholding of the correlation entries, and direct incorporation of ontologically determined interactions and non-interactions as topological contraints [88, 89].

15.5 Conclusion

The biomedical sciences are one of the major generators of big data of our era. An aspect of such data is that with the emergence of next-generation sequencing technology, it is becoming far less costly to collect more biomarker probes (p variables) than it is to collect more biological replicates (n samples). This high dimensional regime will be much better characterized by the "purely high-dimensional" setting (large p and fixed finite n) than the standard "ultra-high-dimensional setting" (large p and large n) common in current high-dimensional statistics. It is in this purely high-dimensional regime that our scalable correlation mining frameworks will play a significant role in extracting information and making sense of the next generation of biological data.

Acknowledgements

The authors gratefully acknowledge several people with whom interactions have been relevant to the work described in this chapter. In particular we thank Rob Brown at UCLA. We also thank Geoff Ginsburg, Tim Veldman, Chris Woods, and Aimee Zaas, at the Duke University School of Medicine, and Euan Ashley, Phil Tsao, Frederick Dewey, at the Stanford University School of Medicine, Division of Cardiovascular Medicine and the Cardiovascular Institute (CVI). The authors acknowledge Michael Tsiang for LaTeX assistance. The authors also thank Joseph Romano for assistance with the naming convention for the different asymptotic regimes. This work was partially supported by the Air Force Office of Scientific Research under grant FA9550-13-1-0043. B.R. was also supported in part by the National Science Foundation under Grant Nos. DMS-1106642, DMS-CMG-1025465 and DMS CAREER-1352656.

References

[1] Gene Ontology Consortium, "The gene ontology (GO) database and informatics resource," *Nucleic Acids Research*, vol. **32**, no. suppl 1, pp. D258–D261, 2004.

[2] G. Dennis Jr, B. T. Sherman, D. A. Hosack, *et al.*, "DAVID: database for annotation, visualization, and integrated discovery," *Genome Biol.*, vol. **4**, no. 5, p. P3, 2003.

[3] M. Ashburner, C. A. Ball, J. A. Blake, *et al.*, "Gene ontology: tool for the unification of biology," *Nature Genetics*, vol. **25**, no. 1, pp. 25–29, 2000.

[4] E. W. Sayers, T. Barrett, D. A. Benson, *et al.*, "Database resources of the national center for biotechnology information," *Nucleic Acids Research*, vol. **39**, no. suppl 1, pp. D38–D51, 2011.

[5] C. F. Schaefer, K. Anthony, S. Krupa, *et al.*, "Pid: the pathway interaction database," *Nucleic Acids Research*, vol. **37**, no. suppl 1, pp. D674–D679, 2009.

[6] E. G. Cerami, B. E. Gross, E. Demir, *et al.*, "Pathway commons, a web resource for biological pathway data," *Nucleic Acids Research*, vol. **39**, no. suppl 1, pp. D685–D690, 2011. [Online]. Available: http://www.pathwaycommons.org/about/

[7] J. D. Allen, Y. Xie, M. Chen, L. Girard, and G. Xiao, "Comparing statistical methods for constructing large scale gene networks," *PLos One*, vol. **7**, no. 1, p. e29348, 2012.

[8] C. Jiang, F. Coenen, and M. Zito, "A survey of frequent subgraph mining algorithms," *The Knowledge Engineering Review*, vol. **28**, no. 01, pp. 75–105, 2013.

[9] A. Hero and B. Rajaratnam, "Large-scale correlation screening," *Journal of the American Statistical Association*, vol. **106**, no. 496, pp. 1540–1552, 2011.

[10] ——, "Hub discovery in partial correlation models," *IEEE Transactions on Information Theory*, vol. **58**, no. 9, pp. 6064–6078, 2012, available as Arxiv preprint arXiv:1109.6846.

[11] P. Bühlmann and S. van de Geer, *Statistics for High-Dimensional Data: Methods, Theory and Applications*, Springer, 2011.

[12] R. Fisher, "On the mathematical foundations of theoretical statistics," *Philosophical Transactions of the Royal Society of London, Series A*, vol. **222**, pp. 309–368, 1922.

[13] ——, "Theory of statistical estimation," *Proceedings of the Cambridge Philosophical Society*, vol. **22**, pp. 700–725, 1925.

[14] C. Rao, "Large sample tests of statistical hypotheses concerning several parameters with applications to problems of estimation," *Mathematical Proceedings of the Cambridge Philosophical Society*, vol. **44**, pp. 50–57, 1947.

[15] ——, "Criteria of estimation in large samples," *Sankhyā: The Indian Journal of Statistics, Series A*, vol. **25**, pp. 189–206, 1963.

[16] J. Neyman and E. Pearson, "On the problem of the most efficient tests of statistical hypotheses," *Philosophical Transactions of the Royal Society of London, Series A*, vol. **231**, pp. 289–337, 1933.

[17] S. Wilks, "The large-sample distribution of the likelihood ratio for testing composite hypotheses," *Annals of Mathematical Statistics*, vol. **9**, pp. 60–62, 1938.
[18] A. Wald, "Asymptotically most powerful tests of statistical hypotheses," *Annals of Mathematical Statistics*, vol. **12**, pp. 1–19, 1941.
[19] ——, "Some examples of asymptotically most powerful tests," *Annals of Mathematical Statistics*, vol. **12**, pp. 396–408, 1941.
[20] ——, "Tests of statistical hypotheses concerning several parameters when the number of observations is large," *Transactions of the American Mathematical Society*, vol. **54**, pp. 426–482, 1943.
[21] ——, "Note on the consistency of the maximum likelihood estimate," *Annals of Mathematical Statistics*, vol. **20**, pp. 595–601, 1949.
[22] H. Cramér, *Mathematical Methods of Statistics*, Princeton, NJ: Princeton University Press, 1946.
[23] ——, "A contribution to the theory of statistical estimation," *Scandinavian Actuarial Journal*, vol. **29**, pp. 85–94, 1946.
[24] L. Le Cam, "On some asymptotic properties of maximum likelihood estimates and related Bayes' estimates," *University of California Publications in Statistics*, vol. **1**, pp. 277–330, 1953.
[25] ——, *Asymptotic Methods in Statistical Decision Theory*, New York: Springer-Verlag, 1986.
[26] H. Chernoff, "Large-sample theory: parametric case," *Annals of Mathematical Statistics*, vol. **27**, pp. 1–22, 1956.
[27] J. Kiefer and J. Wolfowitz, "Consistency of the maximum likelihood esitmator in the presence of infinitely many incidental parameters," *Annals of Mathematical Statistics*, vol. **27**, pp. 887–906, 1956.
[28] R. Bahadur, "Rates of convergence of estimates and test statistics," *Annals of Mathematical Statistics*, vol. **38**, pp. 303–324, 1967.
[29] B. Efron, "Maximum likelihood and decision theory," *Annals of Statistics*, vol. **10**, pp. 340–356, 1982.
[30] D. Donoho, "For most large underdetermined systems of linear equations the minimal ℓ_1-norm solution is also the sparsest solution," *Communications on Pure and Applied Mathematics*, vol. **59**, pp. 797–829, 2006.
[31] P. Zhao and B. Yu, "On model selection consistency of Lasso," *Journal of Machine Learning Research*, vol. **7**, pp. 2541–2563, 2006.
[32] N. Meinshausen and P. Buhlmann, "High-dimensional graphs and variable selection with the lasso," *Annals of Statistics*, vol. **34**, no. 3, pp. 1436–1462, June 2006.
[33] E. Candès and T. Tao, "The Dantzig selector: statistical estimation when p is much larger than n," *Annals of Statistics*, vol. **35**, pp. 2313–2351, 2007.
[34] P. Bickel, Y. Ritov, and A. Tsybakov, "Simultaneous analysis of Lasso and Dantzig selector," *Annals of Statistics*, vol. **37**, pp. 1705–1732, 2009.
[35] J. Peng, P. Wang, N. Zhou, and J. Zhu, "Partial correlation estimation by joint sparse regression models," *Journal of the American Statistical Association*, vol. **104**, no. 486, 2009.
[36] M. Wainwright, "Information-theoretic limitations on sparsity recovery in the high-dimensional and noisy setting," *IEEE Transactions on Information Theory*, vol. **55**, pp. 5728–5741, 2009.
[37] ——, "Sharp thresholds for high-dimensional and noisy sparsity recovery using ℓ_1-constrained quadratic programming (Lasso)," *IEEE Transactions on Information Theory*, vol. **55**, pp. 2183–2202, 2009.

[38] K. Khare, S. Oh, and B. Rajaratnam, "A convex pseudo-likelihood framework for high dimensional partial correlation estimation with convergence guarantees," *Journal of the Royal Statistical Society: Series B (Statistical Methodology)*, to appear, 2014. [Online]. Available: http://arxiv.org/abs/1307.5381
[39] H. Firouzi, A. Hero, and B. Rajaratnam, "Variable selection for ultra high dimensional regression," Technical Report, University of Michigan and Stanford University, 2014.
[40] B. Mole, "The gene sequencing future is here," *Science News*, February 6, 2014. [Online]. Available: https://www.sciencenews.org/article/gene-sequencing-future-here
[41] W. KA, "Dna sequencing costs: data from the nhgri genome sequencing program (gsp)," August 22, 2014. [Online]. Available: https://www.sciencenews.org/article/gene-sequencing-future-here
[42] A. Zaas, M. Chen, J. Varkey, *et al.*, "Gene expression signatures diagnose influenza and other symptomatic respiratory viral infections in humans," *Cell Host & Microbe*, vol. **6**, no. 3, pp. 207–217, 2009.
[43] Y. Huang, A. Zaas, A. Rao, *et al.*, "Temporal dynamics of host molecular responses differentiate symptomatic and asymptomatic influenza a infection," *PLoS Genet*, vol. **7**, no. 8, p. e1002234, 2011.
[44] P. J. Bickel and K. A. Doksum, *Mathematical Statistics: Basic Ideas and Selected Topics*, Holden-Day, San Francisco, 1977.
[45] H. Jeong, S. P. Mason, A.-L. Barabasi, and Z. N. Oltvai, "Lethality and centrality in protein networks," *Nature*, vol. **411**, no. 6833, pp. 41–42, May 2001. [Online]. Available: http://dx.doi.org/10.1038/35075138http://www.nature.com/nature/journal/v411/n6833/abs/411041a0.html
[46] M. C. Oldham, S. Horvath, and D. H. Geschwind, "Conservation and evolution of gene coexpression networks in human and chimpanzee brains," *Proceedings of the National Academy of Sciences*, vol. **103**, no. 47, pp. 17 973–17 978, November 2006. [Online]. Available: http://www.pnas.org/content/103/47/17973.abstract
[47] P. Langfelder and S. Horvath, "WGCNA: an R package for weighted correlation network analysis," *BMC bioinformatics*, vol. **9**, no. 1, p. 559, January 2008. [Online]. Available: http://www.biomedcentral.com/1471-2105/9/559
[48] A. Li and S. Horvath, "Network neighborhood analysis with the multi-node topological overlap measure," *Bioinformatics (Oxford, England)*, vol. **23**, no. 2, pp. 222–31, January 2007. [Online]. Available: http://bioinformatics.oxfordjournals.org/cgi/content/abstract/23/2/222
[49] L. Wu, C. Zhang, and J. Zhang, "Hmbox1 negatively regulates nk cell functions by suppressing the nkg2d/dap10 signaling pathway," *Cellular & Molecular Immunology*, vol. **8**, no. 5, pp. 433–440, 2011.
[50] A. Y. Istomin and A. Godzik, "Understanding diversity of human innate immunity receptors: analysis of surface features of leucine-rich repeat domains in nlrs and tlrs," *BMC Immunology*, vol. **10**, no. 1, p. 48, 2009.
[51] S. L. Lauritzen, *Graphical Models*, Oxford University Press, 1996.
[52] A. Dempster, "Covariance selection," *Biometrics*, vol. **28**, no. 1, pp. 157–175, 1972.
[53] J. Friedman, T. Hastie, and R. Tibshirani, "Sparse inverse covariance estimation with the graphical lasso," *Biostatistics*, vol. **9**, no. 3, pp. 432–441, 2008.
[54] B. Rajaratnam, H. Massam, and C. Carvalho, "Flexible covariance estimation in graphical Gaussian models," *Annals of Statistics*, vol. **36**, pp. 2818–2849, 2008.
[55] K. Khare and B. Rajaratnam, "Wishart distributions for decomposable covariance graph models," *The Annals of Statistics*, vol. **39**, no. 1, pp. 514–555, Mar. 2011. [Online]. Available: http://projecteuclid.org/euclid.aos/1297779855

[56] A. J. Rothman, P. Bickel, E. Levina, and J. Zhu, "Sparse permutation invariant covariance estimation," *Electronic Journal of Statistics*, vol. **2**, pp. 494–515, 2008.
[57] P. Bickel and E. Levina, "Covariance regularization via thresholding," *Annals of Statistics*, vol. **34**, no. 6, pp. 2577–2604, 2008.
[58] O. Banerjee, L. E. Ghaoui, and A. d'Aspremont, "Model selection through sparse maximum likelihood estimation for multivariate Gaussian or binary data," *Journal of Machine Learning Research*, vol. **9**, pp. 485–516, March 2008.
[59] C.-J. Hsieh, M. A. Sustik, I. Dhillon, P. Ravikumar, and R. Poldrack, "Big & quick: sparse inverse covariance estimation for a million variables," in *Advances in Neural Information Processing Systems*, 2013, pp. 3165–3173.
[60] D. Guillot, B. Rajaratnam, B. T. Rolfs, A. Maleki, and I. Wong, "Iterative Thresholding Algorithm for Sparse Inverse Covariance Estimation," in *Advances in Neural Information Processing Systems 25*, 2012. [Online]. Available: http://arxiv.org/abs/1211.2532
[61] O. Dalal and B. Rajaratnam, "G-AMA: sparse Gaussian graphical model estimation via alternating minimization," Technical Report, Department of Statistics, Stanford University (in revision), 2014. [Online]. Available: http://arxiv.org/abs/1405.3034
[62] G. Rocha, P. Zhao, and B. Yu, "A path following algorithm for Sparse Pseudo-Likelihood Inverse Covariance Estimation (SPLICE)," Statistics Department, UC Berkeley, Berkeley, CA, Tech. Rep., 2008. [Online]. Available: http://www.stat.berkeley.edu/~binyu/ps/rocha.pseudo.pdf
[63] S. Oh, O. Dalal, K. Khare, and B. Rajaratnam, "Optimization methods for sparse pseudo-likelihood graphical model selection," in *Advances in Neural Information Processing Systems 27*, 2014.
[64] G. Marjanovic and A. O. Hero III, "On lq estimation of sparse inverse covariance," in *Proceedings of IEEE Conference on Acoustics, Speech and Signal Processing (ICASSP)*, Florence, May 2014.
[65] ——, "l_0 sparse inverse covariance estimation," arXiv preprint arXiv:1408.0850, 2014.
[66] T. Tsiligkaridis, A. Hero, and S. Zhou, "Convergence properties of Kronecker Graphical Lasso algorithms," *IEEE Transactions on Signal Processing* (also available as arXiv:1204.0585), vol. **61**, no. 7, pp. 1743–1755, 2013.
[67] R. Gill, S. Datta, and S. Datta, "A statistical framework for differential network analysis from microarray data," *BMC Bioinformatics*, vol. **11**, no. 1, p. 95, 2010.
[68] N. Kramer, J. Schafer, and A.-L. Boulesteix, "Regularized estimation of large-scale gene association networks using graphical gaussian models," *BMC Bioinformatics*, vol. **10**, no. 384, pp. 1–24, 2009.
[69] V. Pihur, S. Datta, and S. Datta, "Reconstruction of genetic association networks from microarray data: a partial least squares approach," *Bioinformatics*, vol. **24**, no. 4, p. 561, 2008.
[70] D. Mount and S. Arya, "Approximate nearest neighbor code," http://www.cs.umd.edu/~mount/ANN.
[71] J. Schäfer and K. Strimmer, "An empirical Bayes approach to inferring large-scale gene association networks," *Bioinformatics*, vol. **21**, no. 6, pp. 754–764, 2005.
[72] J. Friedman, T. Hastie, and R. Tibshirani, "Applications of the lasso and grouped lasso to the estimation of sparse graphical models," 2010. [Online]. Available: http://www-stat.stanford.edu/~tibs/research.html
[73] J. Lee and T. Hastie, "Learning the structure of mixed graphical models," *Journal of Computational and Graphical Statistics*, vol. **24**, pp. 230–253, 2014.

[74] K. Sricharan, A. Hero, and B. Rajaratnam, "A local dependence measure and its application to screening for high correlations in large data sets," in *Information Fusion (FUSION), 2011 Proceedings of the 14th International Conference on*, IEEE, 2011, pp. 1–8.

[75] H. Firouzi, D. Wei, and A. Hero, "Spectral correlation hub screening of multivariate time series," in *Excursions in Harmonic Analysis: The February Fourier Talks at the Norbert Wiener Center*, R. Balan, M. Begué, J. J. Benedetto, W. Czaja, and K. Okoudjou, Eds., Springer, 2014.

[76] B. He, R. Baird, R. Butera, A. Datta, *et al.*, "Grand challenges in interfacing engineering with life sciences and medicine." *IEEE Transactions on Bio-Medical Engineering (BME)*, vol. **4**, no. 4, 2013.

[77] R. Chen, G. I. Mias, J. Li-Pook-Than, *et al.*, "Personal omics profiling reveals dynamic molecular and medical phenotypes," *Cell*, vol. **148**, no. 6, pp. 1293–1307, 2012.

[78] J. J. McCarthy, H. L. McLeod, and G. S. Ginsburg, "Genomic medicine: a decade of successes, challenges, and opportunities," *Science Translational Medicine*, vol. **5**, no. 189, pp. 189sr4–189sr4, 2013.

[79] J. T. Erler and R. Linding, "Network medicine strikes a blow against breast cancer," *Cell*, vol. **149**, no. 4, pp. 731–733, 2012.

[80] D. B. Boivin, F. O. James, A. Wu, *et al.*, "Circadian clock genes oscillate in human peripheral blood mononuclear cells," *Blood*, vol. **102**, no. 12, pp. 4143–4145, 2003.

[81] H. Firouzi, D. Wei, and A. Hero, "Spatio-temporal analysis of gaussian wss processes via complex correlation and partial correlation screening," in *Proceedings of IEEE GlobalSIP Conference*, also available as arxiv:1303.2378, 2013.

[82] J. J. Eady, G. M. Wortley, Y. M. Wormstone, *et al.*, "Variation in gene expression profiles of peripheral blood mononuclear cells from healthy volunteers," *Physiological Genomics*, vol. **22**, no. 3, pp. 402–411, 2005.

[83] A. R. Whitney, M. Diehn, S. J. Popper, *et al.*, "Individuality and variation in gene expression patterns in human blood," *Proceedings of the National Academy of Sciences*, vol. **100**, no. 4, pp. 1896–1901, 2003.

[84] H. Firouzi, A. Hero, and B. Rajaratnam, "Predictive correlation screening: application to two-stage predictor design in high dimension," in *Proceedings of AISTATS*, also available as arxiv:1303.2378, 2013.

[85] N. Katenka, E. D. Kolaczyk, *et al.*, "Inference and characterization of multi-attribute networks with application to computational biology," *The Annals of Applied Statistics*, vol. **6**, no. 3, pp. 1068–1094, 2012.

[86] S. Zhou, "Gemini: graph estimation with matrix variate normal instances," *The Annals of Statistics*, vol. **42**, no. 2, pp. 532–562, 2014.

[87] P. Langfelder and S. Horvath, "Wgcna: an R package for weighted correlation network analysis," *BMC Bioinformatics*, vol. **9**, no. 1, p. 559, 2008.

[88] D. Zhu, A. Hero, H. Cheng, R. Kanna, and A. Swaroop, "Network constrained clustering for gene microarray data," *Bioinformatics*, vol. **21**, no. 21, pp. 4014–4021, 2005.

[89] A. Rao and A. O. Hero, "Biological pathway inference using manifold embedding," in *Acoustics, Speech and Signal Processing (ICASSP), 2011 IEEE International Conference on*, IEEE, 2011, pp. 5992–5995.

Index

alternating direction method of multipliers (ADMM), 3, 19, 41, 69, 83, 203, 235
Arab Spring, 278
augmented Lagrange multipliers, 224

Bayesian, 285, 342, 365, 392
Bayesian information criterion (BIC), 342
Bayesian networks, 392
Bayesian shrinkage priors, 365
betweenness, 251, 305, 308
big data aware design, 180
block coordinate descent (BCD), 23, 66, 68
block successive upper-bound minimization (BSUM), 66, 70
block-variable optimization, 23

cache, 152, 177, 186, 196
cloud-based radio access network (C-RAN), 181
complex system, 247
consequential, 356

data mining, 37, 187, 201, 303, 409, 418
diffusion network inference, 324
directed acyclic graph (DAG), 137
discriminative factor model, 365, 372, 379
discriminative models, 366, 370
distributed, 37, 87, 101, 138, 148, 161, 181, 198, 203, 217, 232, 328, 368, 418, 423

energy, 181, 205, 217, 260, 400*f*, 400
energy efficiency, 157

Facebook, 301
factor models, 366, 367, 372
false data injection detection, 217
first-order method, 66, 67
full downloading, 161

gene, 8, 337, 365, 391
gene activity profile (GAP), 339
gene networks, 365
gene regulatory networks (GRNs), 337
gene set Gibbs sampling (GSGS), 391, 397
gene set simulated annealing (GSSA), 391, 397
Gibbs sampling, 365, 374, 381, 385, 391, 393, 397
graphic processing unit (GPU), 143
green, 181

Hadoop, 38, 140, 327

independent cascade model (ICM), 320
infectious disease, 365
inference validation, 337
influence propagation and maximization (IP&M), 317
information flow gene set (IFGS), 396

latent Dirichlet allocation (LDA), 281
latent semantic analysis (LSA), 281
least absolute shrinkage selection operator (LASSO), 37, 40
least mean squares (LMS), 37
logical regulatory networks, 340
loss functions, 365

machine learning, 3, 37, 70, 154, 182, 202, 279, 366, 374, 393
MapReduce, 38
Markov chain Monte Carlo (MCMC), 374, 384, 392
Markov chains, 3, 337, 339, 405
matrix factorization, 71, 219, 315, 365, 374, 394
minimum-bandwidth regenerating (MBR), 168
minimum-storage regenerating (MSR), 168
molecular profiling, 391

network analytics, 278
network distance functions, 343
nonconvex, 55, 70, 101, 114, 195, 210
nonsmooth, 101, 105, 120
noncooperative game, 101, 104

objective cost of uncertainty (OCU), 357
opinion leaders, 296, 301
optical, 161, 184

parallel, 6, 47, 66, 80, 101, 138, 150, 182, 217, 327
parallelization, 92, 137
partial downloading, 161
polycentrism, 248
principal component analysis (PCA), 4, 70, 203, 227
probabilistic Boolean networks (PBNs), 337

radio access network (RAN), 66, 72, 208
radio-over-fiber (RoF), 184
rank-one, 7, 15
reliability, 38, 151, 177, 217, 285, 431
rule structure, 337

scalable, 180, 240, 296, 411, 431
scaling out, 137
scheduling, 137, 184, 233
security, 145, 165, 217, 218, 293
security constrained optimal power flow (SCOPF), 217
sequential, 6, 46, 101, 103, 114, 125, 192, 204, 397
smart grid, 217
social influence analysis, 301
social network analysis (SNA), 279
software defined network (SDN), 182
spark, 327
sparsity, 5, 37, 39, 193, 229, 365, 418, 429
spatial-temporal, 248
statistical feature, 181, 200
storage, 38, 46, 102, 137, 148, 161, 205
surrogate function, 101, 105

tensor, 3, 71, 203
tie strength, 307
topic network, 278
transfer entropy, 316
transportation network, 248*f*, 248
Twitter, 278, 301

urban systems, 247
urbanization, 247

variational Bayes, 365, 375, 385

Weibo, 302
weighted cascade model (WCM), 320
wireless, 66, 161, 180